THE PEACOCK AND THE SPARROW

THE PEACOCK AND THE SPARROW

— *A Novel* —

I. S. Berry

ATRIA BOOKS
New York London Toronto Sydney New Delhi

ATRIA
BOOKS

An Imprint of Simon & Schuster, Inc.
1230 Avenue of the Americas
New York, NY 10020

This book is a work of fiction. Any references to historical events, real people, or real places are used fictitiously. Other names, characters, places, and events are products of the author's imagination, and any resemblance to actual events or places or persons, living or dead, is entirely coincidental.

First Atria Books hardcover edition May 2023

ATRIA BOOKS and colophon are trademarks of Simon & Schuster, Inc.

For information about special discounts for bulk purchases, please contact Simon & Schuster Special Sales at 1-866-506-1949 or business@simonandschuster.com.

The Simon & Schuster Speakers Bureau can bring authors to your live event. For more information or to book an event, contact the Simon & Schuster Speakers Bureau at 1-866-248-3049 or visit our website at www.simonspeakers.com.

Interior design by Jill Putorti

Manufactured in the United States of America

1 3 5 7 9 10 8 6 4 2

Library of Congress Cataloging-in-Publication Data has been applied for.

ISBN 978-1-9821-9454-3
ISBN 978-1-9821-9456-7 (ebook)

For Rick and Zev

So I find words I never thought to speak
In streets I never thought I should revisit
When I left my body on a distant shore.

T. S. Eliot, *Little Gidding*

THE PEACOCK AND THE SPARROW

1

I hated the smell of Rashid's cigarettes. He always lit up in my car, a beat-up Mitsubishi Lancer with just enough space to breathe. I hated the smell of his cigarettes, but I always took one when he offered. It was the ability to please that you learned as a spy: smoking a cigarette, offering compliments you didn't mean, falling down drunk from having accepted too many vodkas.

His cigarettes were Canary Kingdom, a cheap Middle East brand that claimed to import its tobacco from Virginia. *Virginia: That's where CIA Headquarters is,* I would inform Rashid casually. Link his source of pleasure to his source of risk, another trick of the manipulation trade. I'd offered to get him real American cigarettes with my ration cards on the naval base, but he'd refused, said he liked his native carcinogens. Anyway, he insisted on green apple, a flavor I'd never find in any of Uncle Sam's packs. It was my misfortune that cigarettes were Rashid's only vice; he was too pious to drink and I was never able to expense alcohol during our meetings.

Green apple had begun to mix with the odor from a nearby dumpster and our stationary car smelled like a rotting orchard. "We will not negotiate until they release Junaid," Rashid was saying, shaking his head and looking out the window. The slums stared back at us, brown and uneven and stunted, as though they'd grown tired over the years, further from notions of a legitimate city. Late afternoon sun turned the car windows, caked with dust, to tarnished copper. I'd convinced myself that the car didn't need a washing, that the dirt helped hide my informants.

Rashid's eyes narrowed, his black pupils reflecting the dying rays of sun like rusty steel blades. He was getting self-righteous and indignant as he always did when talking about Junaid, the dissident poet who'd been rotting in a Bahrain jail since the early days of the uprising.

"Someday the king will answer to Allah for what he has done!" Spittle flew through Rashid's crooked brown teeth. His youngish skin was dark and pockmarked, his curly hair greasy, undoubtedly styled with the cheap gel sold at every corner cold store. He looked leaner than usual—maybe the lingering effects of fasting for Ramadan—the concavity of his chest visible beneath his thin shirt. I never allowed him to wear his preferred white thobe when he met me—too conspicuous.

"If not to Allah, at least Al-Hakim will answer to the international community." I smiled.

Rashid's face turned conciliatory. "I forget you Americans do not believe in Allah. Yes, even the international community has condemned the meritless detention of Junaid." His English was perfect, the product of four years at Oxford—or was it Cambridge? I could never remember.

"And Junaid is not the only unjust detention," he continued. "Four doctors imprisoned last week for treating protesters. Simply providing medical care. Following Hippocrates—"

"Yeah, I heard about it."

"Your country's arms embargo is the only thing that keeps us alive."

"Glad to hear it."

Rashid took a drag, blew a cloud of smoke into my face. "Anyway . . . you understand our position."

I opened the window a crack, threw out my cigarette, returned the pen flashlight to my mouth. "So what about Fourteen February? What's your plan? Continue the war?"

"Yes." Rashid tapped my notebook with his knobby finger. "Write that down. *Inshallah*, we will continue the struggle."

Rashid's silhouette disappeared behind Diraz Cemetery, a dirty ghost among the burial mounds. He was a decent source and an easy one. *Help America learn more about the Opposition*, I'd proffered, and he'd dived in, clothes still on. Pathetically eager to make his case and fund his revolution in the process. A planning officer within Fourteen February, he wasn't perched on the highest echelon, but was good enough to provide the CIA its daily bread.

From the dashboard I removed the safety signal, a pack of cigarettes, took a gulp from my flask, turned my Lancer toward Juffair. The lampless streets were shedding their bulk, becoming paper-thin. I stopped for a soda at a cold store on Avenue 54, one of many Manama streets too nondescript and uninspired to warrant a name. Headquarters liked to remind us to run surveillance detection routes following meetings with informants, but after two decades in the business I could confirm their futility. Fine for younger, fresher case officers who needed the practice. Fine when you were going to a meeting and risked dragging the local intelligence service to your source. But useless after the fact.

When I reached the naval base in Juffair, the white of early evening had darkened to gray. The sky here was never blue. Always hazy and colorless, laden with dust so thick and constant you forgot it existed in the first place. I parked in a dirt lot sandwiched between opulent gated villas. Two-plus months in-country and I still didn't qualify for a parking permit on base; you needed to hold an important position or learn the secret handshake, and I couldn't seem to master either one.

The station was in a small annex at the rear of the base labeled OFFICE OF MIDDLE EAST ANALYSIS. It had the air of hasty and halfhearted officialdom—cheap cubicle walls, clean but shabby furniture, everything unoffensively decorated and slightly dusty, most things in acceptable working order—typical indecisive midpoint between peacetime and wartime operations. A frayed rattan ceiling fan attempted to cool the desert-infested space, buzzing like a dying mosquito.

Rashid's information was thin and it only took a few minutes to type a report. Headquarters had been demanding new dirt, but there wasn't much left. Almost autumn of 2012 and Bahrain had been stuck in a messy attempt at revolution for nearly two years, stalled in an advanced percolation stage: violent but not particularly deadly, a few casualties on either side, recycled rhetoric and sporadic material destruction. International newspapers had stopped reporting on the uprising, relegating it to the bin of petty civil wars. Bahrain's Arab Spring, like its neighbors', was doomed to the annals of inconsequentiality.

The cipher uttered its metallic click and the vault door swung open. Whitney put down his leather satchel, greeted me with a "Hey, Collins." Like nearly everyone else, Whitney discarded my first name, Shane, although coming from his mouth it invariably sounded stilted and unnatural, like a high school boy unsure how to refer to a girl he's dating. Whitney Alden Mitchell had the distinction of being the youngest station chief in CIA history, and if that weren't fodder enough for ridicule, he looked even more juvenile than his twenty-eight years.

"What are you doing here?" Whitney asked as though it were the first time I'd worked past five o'clock.

"Writing up intel from SCROOP." Rashid had the misfortune of receiving one of Langley's uglier code names.

Unbuttoning the top of his shirt, Whitney glanced at my computer screen. He was softly bulging and short and had to lean forward to see what I'd written. "Anything good from our friends on the other side?" *Friends on the other side.* That's what he called the Opposition.

"Not really. Same old shit. Continued protests, demands to release Junaid."

Whitney's blue eyes, framed by girlish lashes, sat eagerly in a round doughy face. Freckles on his neck, cement brown like his hair, quivered behind his starched collar. Every day he wore buttoned white shirts and khakis and penny loafers, and he still had the near-maniacal enthusiasm of a first-tour officer, his smiles uncomfortably close to genuine and his handshakes firm and compulsive. His parents were State Department diplomats, he liked to tell people as though explaining the polish and poise he'd acquired. Then he would make a joke and punch you in the arm, convincing you he had enough grit and humor to be spook material.

"What about Iran?"

I shrugged. "Nothing."

"Collins." Whitney's penny loafers shifted. "We need details on the weapons and money. Where they're coming from. How they're getting here. H.Q. is going to be all over us." He pronounced "H.Q." as two letters, cozily, the way one would say a friend's nickname.

"Understood, Chief." (At my use of the vocative he blinked rapidly, feigned his usual discomfort.) "Nothing so far. But I'll keep my eyes open."

He smiled confidently. "It's there. You just need to find it."

Whitney had arrived a few weeks after me wearing a cheap heavy wool suit in the scorching June heat. Even before attaining station chief status in Bahrain, he'd been dubbed a "rising star," the coveted term bandied about Headquarters, a title he wore with aspirational dignity like a Brooks Brothers jacket that didn't quite fit or that he couldn't quite afford. A lively contrast with me—twenty-five years a case officer, never a station chief. My prior tour in Baghdad had been the latest in a multiyear descent, a descent made worse by the disappearance of a few hundred bucks from my operational revolving funds and an official diagnosis of early-stage liver deterioration. With only a few miles left, I'd been assigned to Manama as the resident Iran Referent, tasked with uncovering the vast Persian conspiracy behind the Shiite uprising against the Sunni monarchy, putting flesh on our fears of Tehran's regional domination. A suitable assignment before I exited the shadows permanently. Manama, as everyone knew, was a halfway house to or from places that mattered: Baghdad, Kabul, Sana'a. Generous living allowances, yearlong tanning, beachside villas—all the facets of decadence rarely seen on a government income—could not disguise the sepulchral reality. Manama was a place where spies came to die. Unless you were twenty-eight and a station chief.

"You do know who Junaid is, right?" I nodded toward my report.

Whitney nodded vigorously, freckles bouncing. "Of course. The poet. Poor bastard's been in jail for . . . what? A year? Two years?"

"About that."

Whitney's hips were wide and curvaceous, I noticed, almost womanly. His belt struggled to circle the rotund waist, a snake clinging to its tree. And a peculiar odor emanated from him—baby powder mixed with sweat.

"You ever read any of his poetry?"

"I've heard his speeches here and there."

Turning to the page Rashid had earmarked, I handed Whitney a leather-bound volume titled *The Manama Verses*. I'd only read a few poems and found them overly lofty and sentimental, but I was a good spy and had mastered enough material to talk intelligently with Rashid, pay homage to his hero. "Read it," I suggested.

Whitney frowned. "It's in Arabic."

I flipped the page to reveal Rashid's translation. Whitney's eyes grew roundly suspicious, as though he were about to view a terrorist bomb plot.

"I can translate if you don't trust SCROOP," I offered.

"Looks interesting, but—"

"Please. Read it out loud." I turned my hands upward: no sinister motive. "I haven't read much of Junaid myself. We can both become enlightened."

Whitney cleared his throat. "Tonight we hear echoes." He glanced up to verify he was reading the correct poem. I nodded.

Tonight we hear echoes.
Tonight we hear echoes across the dunes
From Tunisia, from Egypt, from every maidan where people wear chains
And must stand on footstools
To drink the water from alabaster fountains on high.
Be warned all kings on mountaintops
Whose feet we can see and nothing more:
We are climbing the fountains, flooding your cities with their water!
We are eating your saffron and occupying your checkpoints!
We are shouting to each other, sharing our maps!
You bathe in diamonds but cannot afford freedom.
You are afraid of what will happen when the gems wash down the drain.
We are happening. We are moving.
We hear echoes and we are moving.

"Good stuff, eh? Stuff that would move the masses. SCROOP did a nice job with the translation." I grinned. "Or maybe SCROOP's a better poet than Junaid."

Whitney handed me the book, rapped his knuckles on my desk. "So when can I expect your report?"

"It's done."

Back at his desk, shielded by glass walls, his corpulent fingers began typing. Always busy. The job of a station chief was never done.

I turned off my computer, stared at my image in the monitor's cruel glass. Gray had spread across my black hair over the last few years like taunting cobwebs. My face neither American nor foreign. Nationless. My mother used to tell me I looked like Jack Kennedy but taller and darker, the less classical Irish strain. Women seemed to agree, at least in my thirties. Now I looked like an overtraveled bureaucrat or maybe Kennedy if he'd lived to regret Vietnam. My stomach was expanding and my knuckles had begun to ache in the morning—warnings, undoubtedly, that I was spending too much time at a desk. But I reasoned that I still had a full head of hair for the cobwebs to roost and, at a divorced fifty-two, some residual luck with women.

On the way out I stopped at Whitney's door. A moment of generosity—guilt about the poetry book, maybe. "You wanna grab a drink somewhere? Heat's starting to get tolerable."

Whitney glanced at his watch, thrust his thumb at the monitor. "Thanks. But I really need to finish this cable." His eyes jerked uneasily. "Jimmy left for the night?"

In mock curiosity I surveyed the office for Jimmy, our remaining officer, a counterintelligence guy who always left early on weekends. "It appears he has."

"Appreciate the offer. I'll definitely join you one of these nights."

I made a child's salute. "See you at the Admiral's party."

Usually humming purposefully with sailors and golf carts, the base was emptied for the weekend. It was a small base that people called quaint, picturesque. Palms, cacti, and bougainvillea lent the Fifth Fleet's home a manicured feel; brick patios and grills hinted at a tropical party. INSHALLAH POOL and MOVIE SOUQ were etched onto whimsical arrows pointing in different directions. Names that were too cute, as though someone had used the only Arabic words they knew. Above the exit gate a ticker broadcast a message of good wishes for the weekend, followed by a list of hotels that were off-limits to U.S. military—located in revolution-prone areas, home to a surfeit of prostitutes. BAB AL BAHRAIN HOTEL, SEA SHELL HOTEL, CONCORD HOTEL.

Outside, the taxi stand had filled to capacity. Thursday night, beginning of the Middle East weekend. It was still early and cab drivers were reclining on plastic chairs beneath a crude awning, making persistent inquiries of any passing single male like a fly buzzing incessantly around a ceiling lamp. Piles of trash surrounded the taxi stand in the shape of a fort, too close to the base for Manama's garbage trucks to heed, too far for American jurisdiction, a no-man's-land. From behind the rancid heaps came the sweet contrast of shisha, favorite way to make the night go faster. One driver in particular always sat in the same chair at the same angle, inhaling his sugary mixture with a rapidity and seriousness that betrayed his addiction. Within a few years he'd probably drop dead from the poison, his twisted leathery features permanently at rest, and another driver would take his place, no one the worse for it.

Taxi, sir? You need taxi? Good fares.

American Alley, just beyond the taxi stand, was starting to glow. A mess of massage parlors advertising authentic Asian treatments in brash neon, antiques shops peddling rusty junk, American fast-food joints serving up burgers and fries and Cokes in sprawling quantities. Sailors would soon be staggering down its narrow sidewalks, Saudis in Land Rovers with tinted windows cruising like movie stars.

Thursday night traffic, as always, was a brutish cockfight, and my Lancer crept along Awal Avenue in inches. Tightly packed buildings rose to garish heights beside the street, limbs that had grown too quickly for their body during the oil boom years, an awkward metropolis desperate to prove its bulk and stature. Crawling at street level, like basement vermin, was the third world: tiny photocopy shops, overflowing dumpsters, beggars on bicycles. The city had none of the graceful colonial architecture or curving ancient construction one sometimes sees in the Middle East. Instead, there was a sense of disconnection—noncontiguous streets, noises without origin or purpose, angles that didn't quite fit together—as though everything was an afterthought and nothing worth fighting for.

Outside the Juffair Mall a large screen mounted on a billboard announced the month's events, made possible by King Jassim and the Jasri royal family: a culinary convention, a lottery with staggering rewards

(Sharia-compliant, the video assured doubtful drivers), a garden show. Behind the billboard, plastered to the side of a building, the king's five-story mustached visage stared down, gruesome and misshapen in its elongated form. It was a new poster and the king sported his latest fashionable apparel: sable aviator sunglasses, gleaming gold thumb ring. Directly opposite, a large placard admonished drivers to retain their vigilance amid the advertised pleasures: DOWN IRANIAN CONSPIRACY!

Past the Grand Mosque, domes of Italian marble gleaming in the electric lights. Right on Airport Road. Planes descended toward their runways, armored vehicles ringing the airport in jittery concentric defenses. From minarets across the city the *maghrib* call to prayer somersaulted through the streets, beckoning Muslims to pay homage to their god one last time before the sun went down.

Across Hidd Bridge, north on Dry Dock Highway (south was Dry Dock Prison, Junaid's cozy abode), the Gulf murmuring sweet nothings to the east, calm in the emerging china-white moonlight, the only really beautiful part of Bahrain. Squat date palms rushed by in neatly pruned rows, creating the illusion of a fecund wall, the sole greenery that grew in the arid soil without extraordinary assistance, an unconvincing reminder that this once-feracious island was the storied location of Eden. Along the coast newly erected houses gleamed like sugar cubes, dregs of the decades-long construction boom still riding the last drops of oil. Rashid insisted the monarchy had promised these beachside rent-controlled homes to disgruntled Shia, but when a Saudi royal glimpsed them, he said, they'd turned to luxury properties within weeks. From the top of each home a Bahraini flag fluttered, its zigzag pattern wolf's teeth in the night.

Amwaj Islands drew near, an archipelago of gilded neighborhoods housing an assortment of diplomats, foreign oil executives, and Gulf Air pilots. The safest place to be on the island, at least according to wishful conjecture, a sanctuary if the war ever took a bad turn. RETURNING WATERSIDE LIVING TO THE PEOPLE OF BAHRAIN, a sign proclaimed.

The unarmed guard at the gate waved me through. Creamy villas, windowpanes bright with the refracted light of crystal chandeliers. Alosra, gourmet grocery store stocked with Western favorites like potato chips, frozen pizza, and microwave popcorn. Harsh crooning of a Rod Stewart tribute band from inside the Dragon Hotel. Place was probably crawling with Brits, a rowdy indulgent crowd thrilled to exchange their native rain clouds for sun and palms but who clung to their music and alcohol with Churchillian tenacity.

An Indian loafing on a bicycle jumped at the sound of my car entering Floating City. He emerged from the shadows of a trellis where several other Indians had gathered for the night, their worn jean cuffs rolled up in the unthinking laborer's way, sunburned faces black in the darkness. One of them, I suspected, was mentally disabled; for nearly two weeks he'd knocked on my door looking for odd jobs though I'd told him I had none, and he looked at me through unfocused eyes that never seemed to see or comprehend.

"Car wash?" the brave leader shouted through my open window. He held up dirty rags to advertise his wares, approaching just close enough to make his case, not close enough to make me feel threatened. It was the dance of the impoverished in the Kingdom of Bahrain.

"Another day."

The disabled man stared past me, his starving eyes reflecting my headlights like a nocturnal animal.

Inside my small villa I didn't turn on the lights. Even without looking, I knew exactly what uninteresting furnishings awaited me, lurking like silent conspirators. A worn beige leather couch that had survived all my overseas assignments, more durable than my marriage; a gilded upright globe my son had given me before he'd stopped talking to me; bookshelves filled with a few nonfictions, Arabic textbooks, the occasional Graham Greene or le Carré novel through which I'd futilely tried to make sense of my life choices. My walls were bare; the only art I'd amassed over the years consisted of dated mementos and cheap gifts from informants. At least the place was clean; the maid had come that morning and the cloying lemony scent of her disinfectant hovered in the stillness.

My cell phone rang.

"Collins." Whitney's voice had that officious nasal quality, the kind that makes information seem prematurely unimportant. "Listen, you need to come back in. Just got a call from our friends. Big dust storm predicted tomorrow. Wind coming from the north this weekend."

It was code: There would be a major uprising the following day and a shipment from Iran that weekend—arms, money, some variety of revolution-aiding materiel—information conveniently supplied by our Sunni partners in the Bahrain Intelligence Service, information we received with numbing and comical regularity that had yet to bear fruit.

"Doesn't sound like anything new."

"Collins, I need your help. This is your job."

I hung up. In the Palladian windows that overlooked the lagoon my reflection emerged and disappeared on the black ripples in malicious repetition.

2

Festive prattle hovered like dragonflies. White laughing throats formed half-moons in the large dark windows. Ice cubes made soft dying thuds in their glasses, their half-lives shortened to seconds by the sticky Gulf heat.

When I said I wanted a cocktail last night, I didn't mean a Molotov cocktail!

Damn, we missed all the fun.

Just moved out here from Budaiya—should've stayed there another week.

No one inside was smoking, one of those irritating American customs adopted more stringently overseas than at home. The whole place was a kind of farce, a shinier, cleaner pocket of the Admiral's homeland. Nostalgic paintings blanketed the walls—picket fences, a flag-saturated Main Street parade, aspen-coated mountains—while perfectly creased leather armchairs waited patiently for employment. Neatly trimmed potted palms trembled lightly from the footfalls, their green shadows cowering like villains on the marble tile.

The usual milky parfait of nationalities was in attendance: a thick layer of American, followed by British, topped with a sprinkling of French and German. Loud pastel-collared diplomats, grinning dexterous naval officers, boozy inappropriately dressed civilians, too-relaxed oil commuters to neighboring Saudi, carefully chosen clean-cut enlisted. Indistinguishable from the receptions held nearly every weekend, hosted rotationally by the Ambassador, Admiral, Deputy Chief of Mission, Naval Attaché, and garden-variety lesser officials. Soirees, galas, hails and farewells, shindigs, wingdings.

Whitney Alden Mitchell hadn't shown up. Embarrassed, maybe, by

Thursday night's unnecessary summons. Demonstrations had erupted Friday, but they were tepid and confined to the slums, abating to a manageable simmer within hours. Even the Ministry of Interior's report to station had confirmed the weekend's banality: ROUTINE CLASHES ON BUDAIYA HIGHWAY NEAR THE FISH HOUSE ROUNDABOUT. PARTICIPATION ESTIMATED AT LESS THAN 100 MEMBERS OF THE OPPOSITION. MOLOTOV COCKTAILS, CONCUSSION BOMBS, AND CRUDE HOMEMADE INCENDIARY DEVICES EMPLOYED. No shaped charges, no explosively formed penetrators, not even an assault weapon. No illicit shipments had washed onto Bahrain's shores or shown up on airport X-ray machines. Despite their best attempts, all the king's men couldn't make Humpty Dumpty into a fire-breathing dragon. Our emergency cables to Langley and port-air interdiction alerts might as well have been Christmas cards.

I tapped my beer bottle against the sofa, tiny drops of amber sprinkling the white leather cushions like ugly age spots. The passage of an hour had not prompted me to change my position; small talk had become more repellant as I grew older, drinking and watching easier.

And all the road closures yesterday! Poppy Johnson's voice cut through the din like an electric guitar hitting a sour note. *Had to change my salon appointment three times!* She was swaying on the balcony over the canal, dress hanging sloppily across one shoulder, her husband bracing for a fall in his direction. Burt Johnson, referred to unceasingly and absurdly as "Mr. Johnson," was the senior political advisor to the Admiral, a civilian handpicked for his knowledge of the Middle East. Bottomless, they called his expertise, as though it were a cheap buffet.

What's the point of it, really? All the protesters do is destroy their own country. The voice bobbed across the conversation, ownerless and adrift like the buoys in the canal.

No point at all. It was Jimmy Bakowski, our "broom" and the third spook at our small station. A former jarhead in his early forties, he was paunchy, bald, and perpetually perspiring, a guy who liked to wear weekend shirts with slogans like I'M NOT LAZY I JUST REALLY LIKE DOING NOTHING. *Can't go into sources and methods, of course. But take it on good authority that the protesters are getting a push from the wrong corner. From across the Gulf, if you know what I mean.*

I caught eyes with Jimmy's mousy wife, Gretchen. Jimmy knew better than to advertise his inside access, especially with foreigners around, but he couldn't seem to resist the godlike attention he got from his half-baked insights and cryptic quips.

Road closures or not, I'm switching to the Ritz. Poppy Johnson patted her blond updo like it was a child in need of consolation. *My salon keeps ruining my hair. Bad dyes. And they always make me wait.*

The Admiral's wife shrugged. *Ritz is fine.*

Through the open French doors glided the Chaplain's wife. *The Ritz? Fabulous brunches. Only place you can get beluga caviar. It's banned in the States, you know.* She examined the henna tattoo on her hand, turned to the Admiral's wife. *Do you like my henna? I'm getting it done every week now.*

Poppy Johnson wagged her finger, spilling half her drink. *Be careful. Some of the local girls use black henna. That'll scar you for life. Unless you buy it yourself, you don't know what you're getting.*

Side by side the women's tresses were piano keys: elegant and glossy and begging for touch. Expat wives in Bahrain took mythological pride in their hair, I'd observed, nurturing their manes as though they held the secret to everlasting life. Hues of auburn, gold, and ebony cascaded across chairbacks, swayed in the gusty Manama winds, paraded through cafés like stampedes of vainglorious ponies. Nails apparently exacted equal devotion; I had yet to shake a hand that was not perfectly finished. It was part of the package deal: Here on this island, women inhabited a world far removed from tiny box houses, hands rubbed raw from dishwashing, cardigans, and lawn envy. Here women could sip from pyrite spoons and talk of Michelangelo.

A Filipina maid entered the room carrying a steaming tray of lumpia. She offered her tray to the Admiral, seated by the window with a woman, looking somehow more polished in civvies than in a uniform. A tall man of fifty-three whom people liked to call distinguished, the Admiral had a full head of silver hair and the kind of fundamental muscular build that only gets leaner and more sculpted with time. His face, pleasantly aged like the man on an Old Overholt whiskey label, was tinged coppery red with sunburn. Maybe from a day of training exercises in the Gulf, maybe an excursion on his thirty-foot yacht, *Safe Harbor*. Rumor was that the

Gulf War hero, the man who'd spearheaded a crucial interdiction operation against Iraq, was poised to become the chief of naval operations.

See that house over there? The woman indicated a domed palace across the lagoon. *Owned by the royal family. Did you know Michael Jackson lived there one summer?*

Is that right? The Admiral munched his food, squinted at the mansion.

Can you blame him? No paparazzi. Privacy. Beach right outside your door.

Unable to resist nicotine's claws any longer, I walked onto the balcony. Someone—the Admiral's wife, I suppose—had adorned the deck with citronella candles. Comical: The Kingdom claimed so little vegetation that even insects sought residence elsewhere. But the Admiral's wife, in good delusional diplomat form, liked to construct her own travel brochure, pretend we lived in a tropical paradise. *Land of the golden smiles*, she called Bahrain, smiling to prove her point. Where it never rained and the unending dust created incomparable sunsets.

The conversation outside had grown louder and more raucous, Jimmy and another man apparently engaged in a rhetorical match to describe the Opposition. *Ignorant hangers-on to the Arab freedom fighters. Unwitting—hell, probably witting—lackeys of Tehran. Self-styled, self-defeating, bumbling revolutionaries.* Jimmy was drunk, even drunker than Poppy, his polo-shirted heft moving erratically like a sailboat struggling to enter its harbor. The quips now were mocking, shuttlecocks bandied back and forth on gusts of laughter. *If the shit hits the fan, I'll just jump off my dock and hop on my motorboat. What are the protesters gonna do, chase me across the Gulf? Hell, they don't own boats, do they? (If they do, why are they protesting?)* Laugh, wink.

A pontoon glided by, green lights melting into watery emeralds. Glasses clinking. All the houses in Floating City, I realized for the first time, were variations of the same color. Cream, beige, tan. As though none wanted to depart too radically from the landscape, or from each other.

Salaam! *It's about time!* The Admiral's booming voice. Flock of guests gathering around the door like pigeons to a feeding. *Whitney! He's here! There's our man!*

Busy writing reports, Whitney explained his tardiness to the crowd, apologetically indicating his work attire. He removed his jacket as though coming home to a hot toddy and roaring fire, his fat hands a flurry of handshakes.

And then—lo and behold, the embassy consular officer, a tall brunette, attempting conversation. She wore a tight black dress and cherry-red fuck-me heels. Whitney's face was turning an infantile pink. Poor kid. Jimmy and I used to wonder whether he was a virgin, had gone so far as to place bets one afternoon, then gave up when we realized there wasn't enough legitimate doubt to play. For his sake I hoped the boy would find someone. A spy's job was punishing and lonely, the job of a station chief worse.

Fate was cruel tonight. The Admiral approached and the consular officer beat a dutiful retreat. Beers in hand, the two men retreated to a corner. The vast difference in height between them lent the scene a farcical air. I watched for a few minutes, decided I needed another drink.

"Collins!" Whitney had spotted me, was striding over. The Admiral followed. "Great to see you, Collins. Thanks for the help Thursday night. Solid work."

"Too bad nothing came of it."

"Actually, the Admiral and I were just discussing this very topic." He flashed a loose smile at the Admiral, bowed his head like a prep school boy seeking pardon. "Well, sir, I'll let you use your own words. If I try, *something* will be lost in translation." He punched me in the shoulder.

"Sure thing." The Admiral nodded soberly. "Collins, you're doing a great job—whole station is. Some of the reports you folks put out go straight to SECDEF. Have a real impact on policy." His jaw tensed, eyes squinting in the direction of the Gulf. "Here's the thing. We're getting information through different channels that our neighbors to the north are part of the supply chain."

My eyes were drawn to his gold cuff links. Engraved with Asian symbols, a likely memento from his tour in the Pacific as deputy commander of the Seventh Fleet. A shiny Naval Academy ring kept the cuff links company. Class of '82, he let everyone know.

"What different channels?"

"Well, Collins, you understand the importance of compartmentalization, and unfortunately I'm not at liberty to describe precisely all the sources to which I'm referring. There's information I receive . . . and that Whitney receives as chief. Information that's pretty powerful, that might give you pause when you're out there collecting."

"What's the category? SIGINT?"

The Admiral pressed his whiskey-label lips together. "Listen, just know that we've received additional information. Use what I've told you in your operations. Let it guide your collection, the questions you ask. It's a *target* for you."

"Fourteen February hardly has a full pile of rocks to throw."

The Admiral stepped closer. "Collins, you know what my job is?" His voice sounded like an armored vehicle peeling out on gravel. "My job is to keep the Gulf safe. We have a repressive theocracy a thousand clicks to the north that feeds Hezbollah and comes closer to a nuclear weapon every day. I take that very seriously. When I get word a country like that is fueling revolutions in the neighborhood, I run it to ground. You understand?"

"Yes, sir. Understood."

The Admiral rocked back on his heels. "Then I think we're all on the same page." He put his arm around Whitney, squeezed the soft shoulder. "You're in good hands, Collins. This man here will lead you in the right direction."

Poppy Johnson was already naked and lying on her back when I reached the private beach at the end of the street. Her clothes were balled into a hasty sequined heap, her hair unpinned and spread in a yellow fan. The only thing she'd kept on was her pearl necklace. Despite the growing ravages of middle age—and an artificial tan that only highlighted the discrepancy between youth and senescence—from a distance she looked sufficiently appealing, her small breasts immune to gravity in their skyward position. She'd been one of my easiest targets—even easier than Rashid—a classic neglected wife and stultified mother, minimal manipulation required. All that had been needed was a too-long party at the Ambassador's house, a potent combination of boredom and lychee martinis, and a rooftop deck.

"What did I tell you?" I settled beside her, knocking over the pile of seashells she'd assembled.

She looked around, an indignant child. “What do you mean? I’m exactly where you told me to be. Told Burt I was walking off the drinks.” Her breath stank of too many fruity cocktails.

I nodded toward a villa under construction. “Behind there.”

“But no one can see us here.” Her voice wavered.

“Doesn’t matter to me. You’re the only one with anything to lose.”

Poppy’s eyes wandered mournfully to the scattered shells. The surf lapped at the beach, thin and mumbling. Slowly, deliberately, she fingered her pearl necklace, a signal for me to notice.

“A gift from Mr. Johnson?”

Poppy’s mouth turned up in satisfaction. “No. I bought it myself.”

“At the souq? You went there by yourself? Didn’t even bring the maid?”

“Well, I brought Lata . . . but I drove. She just helped carry bags.” Poppy cast her mascara-caked lashes down. “Figured I’m a diplomat’s wife and it’s about time. . . . And Bahrain has the most renowned pearl industry in the world.”

I started to tell her that Bahrain’s natural pearl industry had died nearly a century ago, crushed by Japan’s cultured pearls, that her necklace was probably a fake, but I simply smiled. “Looks great on you.”

Poppy fingered the bridge of her nose. “Too bad all these pearls can’t fix my nose. There’s this bump. . . . You see it? Right here—” She grabbed my hand, pulled it to her face.

“Jesus, Poppy, your nose is fine.”

“And my boobs . . .” She raised her head to examine her breasts. “Two kids and this is what happens.”

“Your tits are great.”

“All the wives swear by Dr. Pierre. Have you seen the ads? Surgeon to the stars—Doc Hollywood, they call him—trained in the best medical schools in France, here visiting the Middle East for only a short time. Less than half of what I’d pay in the States.” She looked up at me earnestly. “He did the Chaplain’s wife’s boobs.”

“Poppy, you don’t need any of that. You’re fine . . . beautiful.”

“Wish Burt would tell me that. All he talks about is work. Work this, work that. He’s never home, always traveling with the Admiral. They traveled to Singapore last week. Thailand or someplace the week before. Phenom . . . Phenomenal pen?"

"Phnom Penh. It's in Cambodia."

"All I know is he stank of noodles when he came back. Whole suitcase did. I asked him who ordered the takeout." She uttered a tremulous half laugh.

I lit a cigarette. Sex, I could see, would have to wait; I would be her friend and confessor and adoring fan first. Just like with my informants. I felt bitter, resentful of all the hours and nods expended over the years—for this, a world of weak cocktails and fake pearls and faker tits.

Poppy was shaking her head. "*You* never talk about work. And your work is more important than Burt's." Her vacant blue eyes grew lusty. "You're a *spy*."

The stick of tar between my fingers felt suddenly like a nuisance, an unnecessary appendage. I tossed it into the sand. Past Poppy, a few of the Admiral's guests were visible in the canal, their white limbs submerging and surfacing, a late-night dip. A few more weeks and the water would be too cold for their aqueous orgies. In the distance a faint plume of smoke was spiraling up from Muharraq, waiflike, little more than a cloud. I rose to get a better view. Behind the buildings flames had started lapping at the night. Muffled popping: the chirpy soundtrack of Molotov cocktails. With a few frantic splashes the midnight bathers scrambled up the dock ladder and into the villa.

Poppy had crawled over and was watching the scene with alarm. Before she could begin her panicky questions, I cupped my hand over her mouth. Guests had started climbing onto the Admiral's roof deck, staking out prime positions, silhouettes poised against the distant flames like gods gazing down at a dying civilization. A few had lugged chairs and cocktails through the hatch door, and they set up a picnic-like viewing party, hanging over the railing as though at a swanky Manhattan affair, cigarette embers and glass stems glowing gold. Heads tilted back to enjoy the breeze.

I grabbed Poppy's clothes, pushed her toward a cinder block wall. "Get behind there before someone sees us."

The crowd had massed together, leaning toward the attraction in a sloppy tilted pyramid, looking like a parody of the Marines at Iwo Jima. Watching the blaze like morning would never come.

3

Trader Vic's, squeezed into a corner of the Ritz, had only been around for a decade, but its kitschy Polynesian wood statues, pineapple-patterned upholstery, and dimly flickering sconces imparted the outmoded grandeur of an older place. Despite unremarkable food and service, Vic's had earned a devoted following among Westerners, achieving that inexplicable halo, something approaching veneration, that hangs over certain expat establishments in foreign countries, the haphazard product of luck, whim, location, and perhaps an original cocktail. The same unspoken force that once drew British resident ministers to Bombay's Elphinstone cricket club or Saigon journalists to the Majestic: the black fact of exclusivity, of similar elbows rubbing together, everyone sharing the same brand of pants and jokes and adolescent memories. In the last few years shinier watering holes had opened their shutters across Manama—Copper Chimney and The Meat Company, where you could smoke cigars with names like Punch Punch Punch or Bolivar Royal Coronas or, for fifteen dinars, the legendary Cohiba Maduro 5—but expats still patronized Vic's filially and doggedly the way one feels compelled to visit an aging grandparent.

I smelled Jimmy's cigar smoke before I saw him. Habano. He was sipping a mimosa on the veranda, watching the palms gyrate in the breeze around the Venetian pools like belly dancers. Already perspiring, he'd rolled up his sleeves, short forearms bulging, dark stains spreading into flattened spiders under his arms.

The click-clack of hanging wooden beads was deafening to my brain, still tenderized from the Admiral's party, as I ducked through the doorway, grabbed two copies of the monarchy's free *Gulf Daily News*. *Kitchen Fire Destroys Restaurant in Shiite Village*, the Arabic edition declared. *Shiite Thug Vandals Destroy Own Restaurant*, the English edition insisted. I sat down next to Jimmy, lit a cigarette, watched the smoke blur the headlines into gray smudges.

Jimmy reached for the ashtray. The tattoos on his arm, a circle of thorns and *Semper Fi*, peeked out from his sleeve, the words looking forlorn in their partially revealed state like an unfinished crossword puzzle. His shirt, as usual, was comically snug; despite having a naturally fit physique, he'd become softer in Bahrain—too many drinks on the job, he joked—and he unsuccessfully tried to downplay the effects by wearing even tighter clothes.

Across the pool two men in pristine white thobes were tilting back whiskeys, morning sunlight turning their Rolex faces blinding gold, each fingering a loop of agate prayer beads. I nodded in their direction. "Good thing Allah doesn't see across the causeway."

Jimmy raised his mimosa. "Here's to Bahrain never going dry—for our neighborly Saudis *and* for us."

"To the birthplace of original sin."

"Coffee, sir?" The waitress appeared tableside, Jimmy's scrambled eggs in hand.

"Sure." I tossed aside the newspapers. "Black."

The waitress, a petite Filipina wearing Vic's tight tropical uniform, retreated, pausing at a table with three Bahraini women fully clad in abayas and hijabs. The women were sharing a cake, each struggling to ensure that her mastication was covert, that her fork reached her veiled mouth unseen.

"Modesty makes for starvation," I observed.

Jimmy grinned, bits of egg hanging from his teeth. "You ever think about what's under those abayas?"

"Nah. It's like fantasizing about a brick wall. I like to see what I'm getting."

"Still . . . kinda mysterious. Could be interesting. 'Course foreign pussy's just asking for trouble." He puffed on his cigar. "Anyway, you want a quick fix, you got your maid. Hell of a lot easier. Payin' her already, got her passport locked up—and she'll clean up after."

The waitress came with my coffee and tried to adjust our umbrella, a losing battle. I poured a measure of whiskey from my flask into the mug.

"Feel your pain. Rough night." Jimmy leaned toward me in confidence. "Tell you what: Boss man and that cute embassy gal really hit it off." He clicked his tongue as though sorry he hadn't been the one to nab her.

"The consular officer? Something happen?"

"Boss man was a big hit. Who'd a thought he had it in him? Probably buyin' her breakfast right now."

"Hope she likes the base donut stand."

"Eh . . ." Jimmy waved his hand. "He's not such a bad guy."

"Never trust a man who wears penny loafers. And puts pennies in them."

Jimmy snorted, scooped up the last of his eggs. The sound prompted one of the abaya-clad women to glance in our direction. Her eyes, the only feature visible to the world, were trained on us like searchlights. Judging and probing, the eyes of a disapproving schoolgirl, a wrathful god. It was the last thing I remember before the explosion shook our table, rattling our coffee cups, shattering the morning air like a neat champagne cork.

Cars had been abandoned in the middle of Osama bin Zaid Avenue as though their drivers had spotted a looming tornado. Ambulances blared in the near distance. Shopkeepers stood in front of their dingy businesses, arms crossed, watching with the dark removed curiosity that surrounds catastrophes in the Middle East, their Iraqi art, roasting spits, and dusty Moroccan antiques quickly deserted for the sake of front row seats.

The explosion had come from Adliya, a pretentious (some called it fashionable) neighborhood of sidewalk art, faddish galleries, and candlelit restaurants, a lesser man's Latin Quarter, a favorite expat precinct. *Collins, throw on a light disguise and go with the Threat Mitigation Unit.* Whitney's shirt had hung tiredly from his frame, the same shirt he'd worn at the party last night. *You know Arabic. Let us know what's going on.* He'd looked admonishingly at me and Jimmy as though he knew we'd been at Vic's that morning.

Trails of residual smoke, carried by the strong westerly breeze, were drifting toward Exhibition Road. Rashid, that goddamned self-important revolutionary. A major attack on an expat quarter, and he'd given me no warning, hadn't informed me of the Opposition's shift in tactics. Hell, he hadn't even called to see if I was okay—for all he knew, I'd been downing a coffee in an Adliya café when the bomb hit.

A wall of murals signaled the entrance to Adliya. The tech officer driving the TMU van, a lanky pimply kid called Smitty, had succeeded in pushing through traffic with sheer brute force, and now he pulled onto a patch of dirt. The Admiral's aide, a close-cropped unctuous-looking lieutenant, hastily swallowed the last of his Coca-Cola and corralled his bag, moving toward the door as though about to air-drop onto Juno Beach.

We were in the ambitiously termed Arts District; in front of us stood Al Riwaq, the area's flagship gallery. Usually a young, socially determined crowd sat on the terrace sipping complicated variations of coffee and feeling very much immersed in the imagined worldliness of the Middle East. Now the place was empty, minimalist canvases and tangled metal sculptures staring through the windows.

The fallout began at Café Lilou. The café's wrought iron railing and burgundy awnings were intact, but two of its windows were splintered, calligraphic gold letters stumbling across the panes like a fallen dancer. A lone fractured coffee cup lay on the patio, its solo status suggesting a disgruntled patron rather than a violent event, and half-eaten chocolate cake waited to be finished, insisting on civility with its bright mint garnish. Poppy Johnson, I recalled, always raved about Lilou's caramel flan. Behind the café police had cordoned off the alley with bilingual yellow tape.

The alley had the bitter stench of death and wreckage, something like rotted leather, mingled with the lingering fumes of commerce—cooking oil, frankincense, tobacco. Dust from the explosion had not yet settled, combining with the indigenous sand to turn the air sickly brown. Intense heat had burned and dimpled portions of the alley walls, melting awnings like cheap wax creations, and chunks of metal and debris had traveled down the block, piercing dumpsters and shattering windows. A few defiantly erect palms sagged slightly, their trunks black and flaky. It was odd to view the familiar alley through the lens of ruin, to see the backsides of

Lilou and Coco's and Minos transformed into dripping messes. An expat playground suddenly turned to smoke.

On the other side of the cordon an officer wearing the distinctive blue of the Ministry of Interior was issuing tepid commands to a handful of police, pausing every few seconds to regale higher-ups on his radio. He stood apart, with an aura of detached importance. The man in charge.

I ducked under the tape, introduced myself in Arabic. Name only. The officer would lump me with the plainclothes Navy investigators.

His name was Walid Al Zain and he had small muddy eyes and a thick uneven mustache that made him look like he was snickering. I asked him what had happened.

"You are American?" He spoke in heavily accented English.

"Yes."

He looked at me appraisingly as though gauging the truth of my response. "First bomb"—he motioned behind me—"was in the dustbin behind Lilou. Second bomb—"

"Second? How many bombs were there?"

"Five." He smiled coldly. "You did not hear them?"

"Just one."

"Ah, the first. It was the loudest."

"Where were the others?"

"Who knows? The thugs put them all over the alley." Walid gestured impotently at the carnage. He was pushing sixty, probably nearing the point where he was too old to police a young man's revolution.

"Casualties?"

"Yes. One is there."

For the first time I saw the wilted stump on the ground. Still fresh, striped with vermillion and fringed with shreds of flesh, its center stubbornly tan, almost unscathed. About a third of a torso, the chest or abdomen. An expatriate from Asia, Walid informed me, precise origin undetermined. The unlucky man had triggered the first bomb while depositing trash in the dumpster. Two other Asians had been killed—likely laborers from India or the Philippines working in the kitchens and backrooms of Adliya's pricey restaurants (just illegals, Walid assured me)—and two of the bombs had detonated without any known casualties. Information was thin, collected mostly from accounts of eyewitnesses, all of whom had since left the vicinity or gone to the ministry to provide statements. No specific leads. "At least no Americans were killed," Walid concluded, his eyes seeking mine, confident of agreement.

I nodded toward the officers massed awkwardly by the side of the alley. "Your men find any remnants of the bombs? Batteries? Explosives powder?"

"Nothing. It was all destroyed in the blasts." He lit a cigarette, apparently unconcerned about smoking in an ocean of flammable materiel. Following Walid's lead, two of his men lit up, breaking into loud craggy chatter. Urdu. The Ministry of Interior was stuffed with foreign Sunnis, naturalized to fill the ranks, Rashid had told me. Better to have Pakistanis, Jordanians, and Syrians policing the streets than Shia.

"When did your men get here?"

"Ten, fifteen minutes after the last bomb." Walid checked a small notepad with one hand, unwilling to relinquish his cigarette. "Last bomb 9:45, so . . . maybe 9:55."

"Anyone else here when you arrived?"

With sudden impatience Walid shut his notepad. "No. There is no more information. I tell you everything." He glanced sharply at Smitty, who was trying unsuccessfully to communicate with a scrawny Indian in a chef's apron. "We have our own experts, Mr. Collins. Forensics, terrorism. We do not need help. We are well trained. Bahrain is not a third world country."

Local hacks were the same in every country—dogged, territorial, middle-aged, usually overweight. I offered a watery smile. "We're on your side, *habibi*. Just trying to find out the whole story."

The dumpster behind Lilou seemed to have borne the worst of the damage. An explosion had ripped through two of its sides, creating a gaping, jagged-toothed mouth, and the interior was empty, the contents apparently obliterated, blown elsewhere, or cleaned out by Walid's men. I circled the thing then crossed the alley, nearly stumbling over the half corpse. Lying in the gutter so close to Lilou's napoleons and chocolate truffles, the stump looked like a kind of gruesome culinary protest, a poor man's pot roast challenging the sugary delicacies next door. Up close the mottled flesh was a kaleidoscope of ruby and purple and deep brown, all the rich saturated colors of human ruination. I took my camera from my bag, snapped a photo. Something to show Whitney.

Midday heat had begun to infiltrate the concrete blocks, filtering laboriously through the smog, randomly anointing portions of the alley. In another minute this place would be intolerable—the stench, the swelter. Beneath the awning of Let Them Eat Cake bakery, I wiped my face and neck. A few more photos would do the trick: a mess of broken pottery; remnants of a sidewalk art installation; an unlit neon sign, CUCINA, hanging crookedly; trash and rubble tinged an odd shade of purple.

Back at the van, safely away from the incendiary zone, I lit a cigarette. Shop owners were returning to their businesses—to survey the damage, resume commerce, shutter their windows for afternoon *qailula*—streets refilling with sounds and smells. My back was aching. Smitty checked the van for bombs—tailpipe, underbelly, tires—the last officer climbed in, and I reluctantly stubbed out my cigarette. My pant cuffs, I saw, had become soiled with dirt, muck, and traces of John Doe's blood, cakey and brown.

On the way back we heard the final explosion. Faint, a distant drumbeat. Then something odd happened, or at least something I would remember.

"Sixth bomb's a charm," I remarked as Smitty pulled to the side of the road.

The officer next to me, a puny nebbish I hadn't really noticed until that moment, said, "Five. You're losing count, man."

I'd smiled. "That M.O.I. schmuck told me there were five this morning."

"Base watch officer recorded each detonation," the Admiral's aide piped up. "Sent the data directly to our office. Four explosions this morning."

Smitty met my eyes in the rearview mirror. "It's been a long day. Anyone could lose count."

I'd shrugged as if to say, *Oh well, five bombs, six bombs, who gives a rat's ass*. I knew Walid had said five as sure as that disgusting fragment of a man would never walk again. But perhaps Walid, the aging burned-out bureaucrat, was as muddled as his third world investigation. After all, when you hit middle age, you started to lose your edge, drawing perilously close to obsolescence, wading into its muddy shoals.

Whatever the number, the last bomb, we would learn over the next four hours, was what the police termed "insignificant," a victimless detonation triggered by the local hazardous materials team clearing a dumpster two blocks from ground zero. Upon our return to Adliya, Walid informed us the incident had been investigated, no new information found, and the area swept clean for the safety of the good people of Bahrain.

All in all, little worth reporting when I returned to station. Despite the hoopla and fireworks, the incident had not been particularly momentous, the morning's investigation yielding no new insight into the Opposition. Fourteen February had set off bigger bombs in a new location. Still amateurs, still alley cats pawing at trash cans. No Americans had been killed, nothing destroyed that couldn't be rebuilt (hell, with expat money waiting, Adliya's restaurants would be back on their feet in no time), not even a lasting traffic jam. The loops and whorls of Tehran's fingerprints were nowhere to be found in the clunky mess. When the smoke cleared, Manama would be the same plodding place.

And as I headed home that evening down Awal Avenue, the dense haze over Adliya had indeed dissipated into the blanket of smog across the city; routine vehicular squawking had resumed; the white horizon was turning peach and gold in another exquisite dust-filtered sunset; and the *maghrib* call to prayer tumbled unfazed through the hot streets.

4

The day after the Adliya bombings, as they came to be known, the usual predictable security measures were put in place: armored vehicles along Dry Dock Highway and Hidd Bridge, hyperactive helicopters, hastily erected checkpoints. We'd seen it all before, every time the war heated up, and it never seemed to do anything except make expat wives jumpy and worsen the already intolerable traffic.

The threat condition on base had moved to Delta. Outside the taxi stand I tossed a few dinars to the beggar mumbling nonsensical devotions in a seeming attempt to persuade God of his misfortune. Recorded trumpets commenced reveille, inaugurating the American morning inside the fortified walls.

Jimmy was already reading at his desk. Probably paying penance for yesterday morning's diversion. Jimmy and I shared the uncouthness and selfishness of a rough childhood, but somehow he'd learned to please, to act nicely to those who mattered, while I had not.

I poured myself a coffee and went over to Jimmy's cubicle. In the fluorescent light his bald head looked especially greasy, ringed by harsh halos. He was completing his periodic counterintelligence scrub, I saw, a torturous review of all our informants to ensure they weren't betraying us. Jimmy dealt in one of two limitless departments within the CIA—money and suspicion—so he never lacked work or the appearance of importance.

"Carrying on God's work, I see."

Jimmy looked up, muttered something about a deadline. I started to regale him with stories of Walid—his Groucho mustache, B-movie braggadocio—when Jimmy interrupted me. "Your boy passed his poly, right?"

He leafed through a folder labeled SCROOP, holding it close, shielding the contents from my view.

"Jimmy, you were in the next room. You know he passed." I smirked. "Hell, passed quicker than I ever did. An hour tops."

Jimmy scanned one of the documents. "His test results are missing." He closed the folder. "No biggie. I'm sure it's in cable traffic."

"Listen, Jimmy, I know what this is about. I'm meeting SCROOP soon. Maybe today. I'll take care of it." I nodded toward Whitney's office. "Chief in? Didn't he want us here bright and early?"

"Briefin' the Admiral. Phone's been ringin' off the hook all morning." Jimmy winked in shared jest. "Unless he's payin' a visit to the embassy gal. Thought I saw her at the donut stand."

The cipher clicked. Whitney nodded briskly at us. Sweat clung to his face, his cleanly shaven skin glistening like a teenager's.

"Hey, boss." Jimmy quickly straightened. "Packet you wanted is on your desk."

Whitney gestured gravely toward his office. "Come on in, gentlemen."

Various items were arrayed on Whitney's desk like soldiers at attention—a ship in a bottle, a cup of rotund fountain pens, a framed photo of his diplomat parents. Along the wall a bookshelf was filled with standard neophyte stuff on the Middle East, pristine glossy hardcovers, bestsellers that rich white people could read and duly consider themselves experts, a conspicuous compendium of Maurras and Hobbes.

"Gentlemen, I just came from the Admiral's office. Discussed the events in Adliya." Whitney rested his elbows on his desk. "What happened yesterday is very serious. As you know, this is the first time Westerners have been targeted. It's also the first time the Opposition has employed multiple improvised explosive devices, fairly sophisticated weapons."

"It was the work of amateurs."

Whitney looked hard at me. "We think there's something more at stake here. This wasn't just a provocative bomb or two." His eyes shifted from me to Jimmy. "And over the next few months our operations are going to intensify." *Intensify*: a favorite word in Langley conference rooms, one of those punch-packing verbs from the CIA's dictionary, adored by the seventh floor.

"Iran specializes in using proxies to launch asymmetric terrorist attacks. Hezbollah in Beirut. Kata'ib Hezbollah in Iraq. Houthis in Yemen. This latest incident has Tehran's fingerprints all over it." Whitney clasped his hands, an altar boy. "Let's face reality. Iran is an expert player at this game. Iran knows how to manipulate groups for its own ends. A cunning puppet master. In fact, as you gentlemen probably recall, before 9/11, Iran's proxy Hezbollah had killed more Americans than any other terrorist organization."

I picked up the ship in a bottle on Whitney's desk. It was a comical trinket that Jimmy and I privately referred to as the chief's royal schooner. Up close I could see its fine craftsmanship, right down to the gold filigreed sails and etched wooden siding. Probably a childhood souvenir from some noteworthy coastal nation, a fond reminder of a particularly memorable diplomatic posting, one of many Mitchell family overseas adventures.

"9/11, you said?" I flashed a sidelong grin at Jimmy. "I think Jimmy and I remember that day very well. Don't forget, we're old. Hell, I remember the fall of the Berlin Wall. Something your parents can relate to." I squinted in calculation. "So on 9/11 you were . . . what? In college? No, you would've been seventeen, so . . . high school."

"As I was saying—"

"Well, which was it? High school? Or college?"

"College. Princeton accepted me early."

"Ah, Princeton. For some reason I thought it was Yale." I put the ship in a bottle back.

"Collins, let me be frank. We have some concerns about SCROOP. Why didn't he warn us about the Adliya bombings?"

I'd suspected that Whitney's high-minded soliloquy was leading to this, that Rashid was at least half the reason for wasting our morning, and I started to offer my planned explanation, the obvious one, that perhaps Rashid didn't know about the bombings. Then abruptly I realized the answer's inherent weakness: Rashid's ignorance of his own group's plans would be as damning as his deliberate concealment.

"How much do we really know about SCROOP, about his bona fides?" Whitney was going on. "You know, it's possible we're not the only ones paying his salary. Where does he get his money?"

"He's an engineer."

"Where does he work?"

I shrugged. "Here and there. Fixes machines."

Whitney turned to Jimmy. "Does he have any ties to Iran? Relatives?"

Jimmy glanced at me uneasily. "Well, boss, I don't think SCROOP's file is entirely clear in that regard."

"Half the Shia on the island have relatives in Iran," I said. "Doesn't mean they're working for M.O.I.S."

"Collins, you can hardly discount the possibility. And another thing—what if SCROOP was involved in the bombings? Have we ruled out that possibility? We need to do our homework. If we find out SCROOP was involved, we'll have to terminate him immediately. You know the rules: We can't employ an informant with American blood on his hands."

"There was no American blood."

"Pretty damn close—"

"Want to see pictures? Just a few Asian workers killed. Illegals."

"Collins—"

Then it came to me suddenly—a plausible and rather clever explanation. "Maybe SCROOP was afraid we'd share his group's plans with our Sunni friends. Probably felt he couldn't trust us."

Whitney shook his head, dismissing the idea. "Of course we would have shared the information. The bombs were in Adliya and Americans could have been killed. It's our duty to save American lives through any means possible, including cooperation with the local service. And your job is to get your source to trust us even when it's not in his interest—"

"I think I know my job after twenty-five years."

Jimmy stepped in. "SCROOP's file is clean. Don't know about relatives in Iran, but there's no derogatory info. And Collins already has a meeting set up. Fireside chat should do the trick."

Whitney kept his eyes on me. "Either SCROOP needs to start proving himself or we'll have to discuss termination."

"Roger that, Chief. We dismissed?"

"One more thing: Bahrain Defense Forces commander-in-chief has declared a state of emergency. That means a twenty-two-hundred curfew. They've granted me a pass, but I couldn't get one for you two. So be careful. Try to keep off the streets at night. I don't need to tell you how dangerous it would be to get caught with a source." He looked at me sharply. "And please submit your official Adliya report. H.Q. is going to be all over us."

5

Rather than contact Rashid, I drove downtown and checked into the Sea Shell Hotel, my ironclad escape; its status as a forbidden location ensured I wouldn't run into anyone I knew, and the proprietor, a fat Arab who called himself "Charlie," did a nice job of keeping quiet.

In the midst of my three-day vacation, I took an afternoon to meet Rahat, an up-and-comer in Al-Wefaq, the Opposition's nonviolent political "society"—the king had officially banned parties years ago—whom I'd intersected a few weeks earlier at an embassy forum. Something to show for the books, muddy my absence, appease Whitney. In the courtyard of Monsoon, a gaudy Chinese restaurant in Adliya that had escaped the bombings, we engaged in mind-numbing conversation over salty noodles—not enough progress on parliamentary representation, possible boycott on royal businesses—nothing too heavy or obvious, just enough to paint the veneer of a genuine bond and prove my interest, maybe whet his appetite for a relationship. It would all look good in cable traffic. As we cracked open our fortune cookies, I threw out a lie about wanting to visit distant relatives in Tehran (lucky for Collins the Spy, dark hair and a dark-enough complexion made me ancestrally ambiguous); did he know how I could acquire a visa in Manama? There was, after all, no Iranian embassy and Tehran's chargé d'affaires had been banished months ago. Not a clue, Rahat said. He had no connections. I'd smiled and read my fortune: *Your shoes will make you happy today.* After lunch I drove back to the Sea Shell Hotel.

It wasn't until Friday that, half-inebriated, I returned to the office and found the Adliya bombings had brought Bahrain back from the dead. A smattering of worldly left-leaning papers—*The Observer, Le Monde, Süddeutsche Zeitung*—had covered the incident, and Langley had poked its rodent's head above ground, sniffing around with an avalanche of cables.

At five o'clock Sunday evening I signaled Rashid with a coded text message from a burner cell phone. Three hours from time of transmission until our meeting at the designated location.

Few small businesses were open past seven, so I resorted to Seef Mall for a cover stop then a vantage point off the highway for touristy photos of the illuminated skyline. A trickle of worry had begun wending its way down my spine. I couldn't lose Rashid as a source. He was the only informant I'd recruited, my operational raison d'être. And the unfortunate nature of spying is that sources never really fail, just their handlers; if one goes bad, so does the other. (Espionage, I've concluded, is unique in this regard, the only profession where two people, usually strangers, are wholly and synchronically linked, as closely bound as lovers, where individual destiny dictates joint destiny.)

Just before eight o'clock I parked on the north side of Qal'at Al Bahrain, the side farther from the slums. In the darkness the ancient sprawling fortress was barely discernable. When I'd first arrived on the island, the place had been illuminated at night, but weeks later the monarchy began instituting blackouts on the theory that what was visible was also a target for destruction. The fort was a prized tourist attraction, at least before the uprising, and despite the government's botched attempts at excavation and preservation, had earned designation as a world heritage site. It was a status the Ministry of Culture clung to with disproportionate pride as though the thing were the Great Wall of China, its construction by Portuguese colonists a mere trifling detail.

I'd instructed Rashid to meet me inside a chamber in the southwest corner; it sat beneath a parapet bearing the Bahraini flag and was easy to find. But the night was black, Allah insisting on keeping his celestial hints shrouded, every passage end nearly invisible. All I could see were walls.

Sturdy walls, built by determined hands. I felt a momentary dull pity for this country that had belonged to so many outsiders and never quite mastered the art of being a homeland—and equal amusement at the outsiders who'd never quite mastered the art of control.

A noise. A shuffle or a cough. Rashid, maybe, as lost as I was. The wind blowing a piece of litter. An insomniac out for a stroll. Maybe nothing. A police foot patrol. But maybe I was just hearing monsters under the bed. I ducked into a chamber, flattened myself against the wall.

Until that moment, I'd never seriously entertained the idea that Rashid worked for Tehran, that he was anything other than what he claimed to be. Just a planning officer in Fourteen February, a minion, a foot soldier. But now the thought entered my mind like a laughing jack-in-the-box, Whitney's superior grin plastered on its face. Rashid, an M.O.I.S. agent working against the Gulf elite, maybe against its American friends. I smiled in the darkness at my cockeyed fantasy. Fuck if I wasn't turning into Jimmy.

I emerged again into the night. Up ahead the clouds had thinned, revealing the uneven outline of low buildings and a trace of black smoke against the blacker sky. It was the slums, the end of a protest, an arrow pointing south.

Seconds later the Bahraini flag was baring its wolf's teeth, flapping from the parapet like a feral animal desperate to flee its tether. Through the onion arch, into the small chamber. No sign of Rashid.

Wind whistled through the fortress like the laughing dead. I pulled out a cigarette. The room was dank, smelled faintly of defecation. I debated turning on my pen flashlight but figured it would provide negligible light and only draw attention. And what did I need to see anyway? Some bum's pile of shit, a dead rat. Glancing around in the flare of my lighter, I noticed that the bloodstains from Adliya were still on my cuff, tough enough to withstand a washing, emblazoned on the fabric in permanent rusty streaks.

"What is this place?" Rashid's voice echoed as he stepped into the chamber, removed the cigarette from his mouth. He looked different. Dark stubble covered his face, hiding his pockmarks and adding a decade to his age, his clothes entirely black. If I passed him on the street, I wasn't sure I'd recognize him.

"This is the place we agreed on." I checked my watch. "You're fifteen minutes late. Lucky I didn't leave."

"I know Qal'at Al Bahrain. But not so well that I can find one room. At night. It is . . . what is your expression? A needle in a haystack."

"That's why it's a good place to meet."

Rashid's lips curled into a bitter smile. "Next time I choose the place."

I nodded at his cigarette. "Hope you didn't smoke that thing all the way here. It's a glowing beacon."

"Of course not."

"You see anyone on your way in?"

"No one. Why?"

"Anyone know you're out tonight?"

"Only Huda. And she does not ask questions. She knows I do important work for the struggle at all hours." Rashid took a drag, reached his free hand toward his back pocket.

Before I knew what I was doing, I'd grabbed his arm, squeezed his bird-like bones in a vise. He looked at me, surprised. I kept my grip, unmoving, feeling his flesh succumb to my fist like weak dough, wondering vaguely if I'd bruise his fragile limb. Cautiously, like a penitent child, Rashid brought the captive hand around for my inspection. It held a cigarette pack. Canary Kingdom, green apple. "Just wanted to offer you a smoke, *habibi*." His black eyes looked wounded.

I dropped his arm. "*Asif*. Sorry. *Shukran*."

"*Habibi*, something is wrong?"

"Yeah, something's wrong. We've got a problem. Why didn't you tell us about Adliya?" I drew close, stared down at him. "And Tehran."

Rashid blinked, confused. "Adliya? Tehran? Who told—"

I shrugged. "I'm sure you can understand why I can't pay you this month's salary. Come to think of it, you might never see American dollars again. Thing is, my boss is pretty sure you've lost your access."

Rashid raised his chin defiantly. "I have not."

"Doesn't matter how or why. Bottom line: We need someone with the goods. Tell us when Tehran's getting its hands dirty. Don't care whose side you're on as long as you deliver. But you're not our man."

The chamber was silent, our faces disappearing as Rashid's cigarette dimmed. Suddenly, Rashid laughed, shook his head. "Ah, my American friend. Someone has been telling you lies."

"Is that right? Then please, my *Bahraini* friend, enlighten me with the truth."

"I am glad to tell you. I thought you knew." His smile faded. "You want to know who is behind Adliya? Ask the king." Bitterness was creeping into his voice. "You Americans. You think too much of your royal friends."

Slowly, thornily, his words rooted in my brain. I studied his face, the stillness of his mouth. It was absurd, impossible.

"Ask the king and the good Ministry of Interior," Rashid repeated. "Many times it happens. Every few weeks in fact."

"What the hell are you—"

"Usually the bombs are small, not like Adliya. The bombs in Adliya were a surprise. But the ministry has . . . what do you say? In . . . inciters. Instigators. People who pretend to be revolutionaries. Explosions in the middle of our crowds, restaurants and homes in our neighborhoods that burn to the ground. And we are to blame, they say." A pitying smile spread across Rashid's face. "*Habibi*, surely you know this. We did not bomb Adliya. It was your friends."

Despite the bracing wind, the chamber had grown stifling. My fingers itched to grab Rashid again, shake him. Sonofabitch took me for a fool.

I steadied myself, offered my most patronizing smile. "Okay, so the king's men burn your slum restaurants, throw Molotov cocktails into your crowds. Fine. It's possible, an interesting theory. I'll grant you a few minutes of my time for that. But not Adliya. You can't expect me to believe the king would bomb Adliya. He'd never target a Western area. An American area. The king's not stupid. Al-Hakim doesn't want to piss us off." I smirked. "Propaganda lesson one: It has to be believable."

"Ah, but you have just explained it!" Rashid lurched forward, his face electrified with sudden fervor, oily curls falling emphatically across his forehead. "Bombs make you angry! Do you not see? This is the point. You Americans see your coffee shops destroyed, your fancy art galleries burned, and you are determined to *help* Jassim, to give him everything he needs to rid his island of the thugs!"

I scoffed loudly. He watched me. "You do not believe me? Listen to my prediction. Soon there will be a crackdown; they will say Adliya is the reason. One of ours will be arrested. Maybe more than one. False evidence. Blamed for Adliya. Who knows? Maybe they will also say Junaid ordered the bombs and so he must stay locked up another year. And America . . . America will stand by and say nothing. Because now we are a threat." A perverse grin twisted his face. "Or maybe your country will lift the embargo. Give the king weapons. It is all part of his plan."

I shook my head, paced the room. Rebuttals tried to sprout in my mind.

Rashid's voice grew soft. "My friend, you think Fourteen February would throw bombs at your door? Why would we do this? We need you. Your embargos keep weapons from the hands of our oppressors; your newspapers and politicians tell the world when we suffer. Why would we turn against our ally? Our *only* ally! Who else can assist us in our struggle? *Tehran*? Tehran abandoned us long ago! Tehran is afraid; they have revolutionaries on their own streets. *Ya salaam*, our Arab brothers have also abandoned us. You—you are what's left. Americans."

But it couldn't be. Jassim bombing Adliya. I shook down Rashid's words, searching for the places where the wood had rotted. They had to be there.

Honesty for honesty. No anger, a reasonable approach. "My friend, I know you need us. But we don't have to pretend we live in a world of neat fences. Fourteen February bombed Adliya because we cooperate with Jassim. You know it. I know it. We have a naval base here, we like to drive big cars that run on oil, we're nice to Saudi and its little island neighbor. I know how it is. We scratch your backs one day, their backs the next. Our policy is fucked up. Let's talk openly. You can tell me the truth."

Then, when Rashid said nothing, I lost my patience. "For Christ's sake, I don't care if you were out fucking whores and that's why you didn't tell me. Just come clean! Tell the truth and I can work with you. I'm a forgiving man. We'll figure a way out. But keep up these goddamned lies and you leave me no choice."

Rashid moved closer. His face was so near I could see every greasy follicle. "Shane, listen to me. If we had wanted to hurt you, we would have bombed your base. Or your embassy. We would not have bombed Adliya. But the king . . . Think about it, *min fadlak*." His breath was hot, insistent. "For the king, Adliya is perfect. He would not attack your embassy, your naval base. Your people would be evacuated and the buildings closed, yes? No, Jassim would not want that. As you say, he needs the Americans. But Adliya . . . Close enough to make Americans angry. But not so angry that they leave."

And watching Rashid in that fetid chamber, for the moment at least, I could not find any rotten wood. The soft spots had hardened. Even now as I look back, I marvel at the perfect pyramid of his logic, nuances that effortlessly filled the cracks, the elegantly simple foundation. His answer made a fool of Occam, had the natural symmetry of algebra. I'd never seen it until Rashid showed me.

A police siren wailed in the distance and I remembered curfew. I lit a last cigarette, the crack of my lighter traveling across the room. "Why didn't you call to see if I was okay?"

For an instant Rashid looked puzzled. Then he smiled. "My brother, I knew you were safe. You are always safe. If you had been on the *Titanic*, you would have been the first on a lifeboat. I am right? Also, calling you would have been bad tradecraft, yes? You taught me that."

Grudgingly I nodded, removed a wad of hundreds from my pocket. "All right, you've earned this month's salary. Unless I find out you're lying—then I give Jassim your address."

Rashid bent over in a crude bow. "*Shukran, habibi.*" He flattened his hand against his heart. "You will see I tell you the truth. We are friends. I do not lie."

6

The National Theatre and opera house was hosting its opening gala, and Whitney had ordered us to attend. Good diplomacy to show our faces, an opportunity to cultivate foreign embassy representatives, meet the movers and shakers of Manama, show solidarity with our hosts in the wake of Adliya.

The gala marked my three-month anniversary in Bahrain. Three months. A turning point, as any expat knows. The entrance to a long dark tunnel. The point where any extant novelty or exoticism has worn off. Where you sink deeper into foreign soil but it repulses and rejects you, shuns your alien roots. Where you become trapped in the amber of the transplanted elite. Imprisoned in a no-man's-land.

Fifteen minutes late was the preferred timeline for well-heeled expats, a perfect compromise between Western punctuality and Middle Eastern leisure. Much of the crowd was still massed outside the opera house, circling the forest of sandstone sculptures. Multilingual babble flitted like anxious birds. The gala was apparently a post-Adliya proving ground—gowns more opulent, chatter more effervescent. A collective act of defiance against the thugs.

The newly erected opera house, surrounded on three sides by water, looked as though it were floating. A prodigious flat roof extended over the courtyard, supported by insubstantial metal columns like spindly slaves carrying their master. Early evening had turned the building's aqueous reflection into a distorted Greek temple, a wobbly palace appearing twice its actual size. Thin slices of projected white light taunted each other in aerial celebration.

Inside, guests were talking in sparkling beverage circles about Placido Domingo. *"Amapola." "Di quella pira." "O mio rimorso!"* Dueling each other with knowledge of his repertoire, predicting which arias he'd perform come winter. Whispers of the king paying a ransom to bring the opera singer to Bahrain's shores for the holidays. *Worth every dinar. Qatar and Kuwait never get acts like this. We might as well be in Madrid or Paris.*

For the grand opening the Ministry of Culture had transformed the opera house into a showcase for local artwork, all painstakingly hand-selected by the king. *Fall of Culture* posters around the city had advertised the art exhibition and Domingo's upcoming performance as though an entire season of entertainment awaited. It was a humorous slogan—*Fall of Culture*—conjuring, at least in my mind, an image of culture toppling from the heights of Manama's skyscrapers, shattering into thousands of pieces.

Versailles-aspiring chandeliers ate up nearly a third of the breathing space. *Glass from Venice! Czech crystal!* Ornamental white columns half-heartedly attempted to evoke Greece or Rome or other pioneering civilizations, while elaborate arabesque designs assured guests the place had not lost its moorings, that it indeed belonged here. Tuxedoed waiters carried trays of caviar and crackers. A string quartet enthusiastically pumped out Vivaldi's *Four Seasons*.

Spies are nothing if not divining rods tuned to instantaneously detect anything shiny and valuable, and I absorbed the lay of the land, sensing how many potential contacts I could check off the list and later cram into a cable before I could go home. Milling on one side of the lobby was the predictable beau monde: a few members of parliament; the prime minister; the gold-ringed Yousif brothers, friends of the royal family and owners of every major business on the island. Not far away, unsurprisingly, was Whitney, hair slicked back, shoes polished to a blinding shine. The Egyptian ambassador was standing alone and I considered chatting him up, but decided his features looked too serious for chitchat, weightier conversation tolerable only with alcohol. There was the Admiral in full uniform, speaking with the Minister of Defense. As usual the king was absent, attendance too great a risk to his safety.

Jimmy and Gretchen were in the main gallery, idling beside a painting of a mosque, Jimmy's mouth moving slightly as though preparing for any conversation that might be thrown his way. When I clapped him on the shoulder, his face showed relief.

"Well, well." He surveyed my clothes. "You clean up nice."

"Wish I could say the same for you. But the only attractive thing about you is your wife." I winked at Gretchen. She blushed and looked down. I had a soft pitying spot for Jimmy's wife, the skinny colorless woman whose outdated plastic glasses were the most interesting thing about her.

I scanned the pieces around us. "Hoping there'd be a few nudes. Tits and ass. You'd think old Jassim would get tired of his four wives once in a while—"

Jimmy cut me off by asking Gretchen the time. I'd forgotten how paranoid he was, at least when sober. He was like most brooms I'd known—fearful, quick to believe and slow to comprehend, convinced there was a wire hiding in every pocket, eyes behind every curtain. Deportation was always a possibility, and beneath his sporting ten-foot-radius confidence, Jimmy was deathly afraid of losing his cushioned life here, his Cuban cigars, the brilliant blue swimming pool in his backyard.

"What have we here? A confederacy of heroes?" Burt Johnson's raucous laugh assaulted us. He stood next to Poppy like a tall dumb athlete, red-faced yet sober, his rough skin and forced jocularity reminding me of every gym teacher I'd hated. Words always seemed too big for him, didn't fall into place quite right in his mouth.

"Nice to see the Adliya bombings haven't dampened the Bahrainis' spirit." Burt's eyes swept the room. "I tell you, these people have a real joie de vivre."

Poppy edged toward me. "What do you think of Adliya? Pretty scary if you ask me. I was just at Lilou—"

"Yeah, it's too bad. An unfortunate situation." I fumbled in my pocket for a cigarette, excused myself.

The string quartet had stopped for a break and auditory fizz filled the space. Among the tepid art I meandered, among the white thobes and black abayas that speckled the crowd in perfect disharmony, the Gulf's unique checkerboard of domination and submission. I stopped in front of a large mosaic. It was a landscape, a tree on a gentle slope at daybreak. Hundreds of glazed brown tiles formed the trunk, branches extending gracefully like eager hands, alabaster pieces as bright as gold filling the sun's rays, pebbled leaves in blinding shades of green. Hues too rich to label—for when I now recount the moment, I try emerald, lime, jade, but reality mocks the feeble namesakes.

I glanced at the label. TREE OF LIFE.

"Have you been?"

The voice startled me—and then unsettled me. It was a deep voice, not what I expected from a woman. Avalanche-like, the barest hint of an unidentifiable accent roiling its flow—maybe not an accent at all. I turned and instantly experienced that singular combination of alarm, curiosity, and discomfort one feels when confronted unexpectedly with someone disfigured. A gash scarred her left cheek, a fact that should not have arrested me as much as it did, except that it was inordinately long, chin to eye, covering nearly half her face. Or at least it seemed to cover half her face—her features so delicate, so finely angled, they had the perverse effect of amplifying any proximate bluntness. Her clothes, I recall, also struck me. Traditional abaya and hijab, but they were bright paisley instead of black—as though she were trying to reach a suitable midpoint between modesty and glamour. About twenty years younger than me, I guessed.

From my present perch of reflection maybe our first meeting was not nearly so momentous. Maybe it was far more ordinary: American diplomat meets Bahraini mosaic artist at a cultural event; mosaic artist has a facial scar and a contralto voice. But in that crowded opera house I experienced something weightier, something I still struggle to describe.

She was not exactly beautiful, at least not what most men I know would term beautiful. Her face was feminine but sharp, her nose narrow and conspicuous, her chin equally pronounced, jutting out as though trying to flee. Her eyes were almond-shaped and translucent green, the color of the mosaic leaves, ferns when late afternoon sun shines through them. From beneath her hijab a vagrant auburn curl peeked out, just enough to begin a mental picture, and her brows were thick and dark. She had a slender birdlike neck that seemed incapable of supporting her head. Her skin—her skin was the only feature that came close to indisputable beauty. It was golden, the color of sunbaked olives. Skin that eluded nationality, that made her origins indecipherable.

"The Tree of Life. Have you been?" she repeated, her eyes fixed firmly on my face. "You're American, yes? The tree is a popular tourist attraction."

I turned back to the mosaic, mostly to avoid her scarred gaze, quickly skimmed the biographic card beneath her painting. STUDIED IN EUROPE—I was putting my money on the Balkans, maybe Albania—EXHIBITED AT LA BIENNALE DI VENEZIA.

"No, I guess I haven't. I hear it's a long drive from Manama."

"An hour south of the city, in the desert." With no face before my eyes, her voice again struck me. Just a vestige of an accent, almost Western.

"Worth seeing?"

"Yes. It grows in sand dunes without any water. There is nothing living for miles. It is . . . a defiance of nature." She clipped the word "defiance," the way linguistic foreigners unconsciously try to convey meaning through pronunciation. "More than four hundred years it has stayed alive."

I turned to look at her. "Impressive."

"A pea tree. *Prosopis cineraria*." She pronounced the Latin perfectly, as though it had rolled off her tongue a thousand times. For the first time she smiled and I saw that her mouth too was beautiful. Naturally full lips, teeth white and unblemished, free of the unsightly spots caused by too much fluoride in the island's natural springs. A hint of humor was in her eyes. "I only know this because I just finished the mosaic. Otherwise I do not know every plant's genus and species."

The string quartet abruptly resumed playing—this time a harsh piece by Wagner—and our balloon of conversation seemed suddenly to deflate. The woman looked up at me gravely. "So are you interested in buying?"

I glanced at the price: a steep 1,500 dinars. "Well, it's a little out of my budget," I admitted. Then, to cushion the rejection, "But you've convinced me to see the tree. Maybe I'll head out there tomorrow."

"I will give you directions." Her face softened. "If you see the tree, that will be better than buying my art."

I laughed in good-natured doubt, watched her. "Tell you what. I'll see the tree *and* buy the piece."

Her lips pursed in a half smile. Only cash, she told me, but I could pay when I picked up the mosaic at her studio.

She gave me directions to the Tree of Life, her name, the address of her studio. Only the address stuck. Even now I have to pause before I can remember her name in its entirety. It's forgettable, unsuited to her, prosaic.

Months after I left Bahrain I came across a Modigliani painting called *Woman of Algiers* or *Almaisa*. It was the same slightly Semitic face, the same diaphanous intensity. From that moment at the museum she was then, retrospectively, and ever after simply Almaisa in my mind. It's an apt moniker, I've come to believe, a name borrowed from a painting. Because she was, in many ways, the brushstrokes of my imagination: facets of her person, her self, that did not bridge the gap between reverie and reality, that were formed solely from my expectations and interpretations and reflections, my inner artist's eye.

Curfew was looming and the crowd was thinning. I took a last look at my mosaic, bade farewell to Jimmy and Gretchen, walked distinctly by the Admiral so he could give me due credit for showing up. On the way out I passed Whitney. He was chatting up the consular officer, appearing larger than life next to a diorama of the ancient Dilmun civilization. Standing beside each other in their finest, the pair seemed poster-made for the glitzy *Fall of Culture* banner stretched behind them. Good for Whitney—finally putting sexual gratification ahead of his professional duties. I raised my hand and Whitney nodded, smiled his trustworthy diplomat's smile.

A honey-colored moon cut a hole in the sky as I drove back to Amwaj Islands. Skyscrapers lined the coast, onyx creatures with yellow eyes, and the air was mild. Maybe it was just the trick all cities play on you, concrete and glass and sharp angles softening in the darkness, but at night, this night, the city looked strangely palpable, even inviting.

7

About the women in my life.

My photo collection is sparse, but I still have pictures of the women worth remembering. I look at them sometimes, intrigued by what the images show and what they do not.

There are a few of my mother, glossy black-and-white, corners curling with the passage of years. Smiling resolutely in her church-ready finest. Holding me, her dark curls spilling over my bald infant head with near-blissful blindness to the years ahead. Bleary-eyed and blotchy-skinned in a later photo when she could no longer disguise the effects of drinking—like my father that way, but not as violent.

She was beautiful, I now see, something I suppose children can't recognize until they've grown. She'd pushed me out the door in the only direction she could, away from my father and his perpetual stink of whiskey, his endless blows, away from her own purple eyes, away from our rough Jersey neighborhood. The nearest Army recruiting station was the sole future my mother was able to conceive for me, and to her disappointment even that proved unattainable, financial delinquency and a "not insignificant" criminal record keeping the gates closed. But the recruiter gave me a phone number—a place better suited for someone like me, he said, no college degree necessary. So in the end she'd gotten me out.

I've kept a few photos of my ex-wife, Marlene—the two of us grinning next to the reclining Buddha in Bangkok, Marlene's red hair bright and tousled; bundled in heavy scarves next to a colorful remnant of the Berlin Wall, arms around each other, hands in triumphant V-formation. The photos make me uncomfortable in the way one remembers once being comfortable. Maybe for this reason I rarely look at them. Also, Marlene never seems youthful anymore, floating within the frames like a ghostly admonishment, her face merging with my own reflected image, my skin appearing grayer and older, my transgressions more grotesque.

Marlene was never cut out to be the wife of a spook. Weak and nervous, the kind of woman whose naivete and trust only eroded my respect for her, made me wring her even more. Spies are trained to needle, cajole, and persuade, and slowly, before you realize what's happening, it's seeped into every crevice of your life like spilled poison. Taking it home, Marlene called it. The hours, the drinking, the women. The daily dose of manipulation. Manipulation. Nothing less than an addictive substance, the kind of drug easily procured in the dark isolated tunnels of espionage. No one watching, God on your side. A nation has entrusted you with its secrets and its salvation and you *are* God for Christ's sake. And God, I now know, makes a lousy husband.

Then there's Almaisa. Number three. The last—or maybe the first. I have the most photos of her. They are colorful and serious; she wears her signature bold abayas but is rarely smiling, seems distracted at times. Perhaps because of her scar, her face is often turned, not fully visible. Still, her green eyes always manage to dominate the photo, as though insisting on her relevance and existence. Green—the only eye color that changes, I read somewhere.

The photos of Almaisa form only a fraction of her memory—her limbs, perhaps, but not her body. Her mosaics tell more. I was able to keep a few smaller ones: *Linking Rings, Red Abstract, Masked Butterflies.* I store them in my closet, carefully wrapped in layers of tissue. When I take them out, I see her quintessentially: her roots, her lens on the world, the most carefully tended expression of who she was. They bring me instantly to her, to the heat and dust of Manama.

8

Her studio was in the slums. In the fleeting light of the opera house her address had been unrecognizable, a string of numbers without a name. But a map revealed it was at the western end of Budaiya, near the Abdulatif Al Dosari mosque. So she lived among the bottom half, almost certainly Shiite. I'd assumed the only artists deemed worthy of participating in the king's art exhibition were Sunni, and I couldn't decide whether this knowledge cast a different light on Almaisa or the king. Probably Almaisa.

The slums unfurled like a giant moth-eaten quilt as I exited the highway at Al-Maqsha. *Qaryah*, the villages—a country unto itself, whose backroads had become grossly familiar to me, an unwanted second home. Piles of garbage rose from the streets like dying trees. Layers of houses were crammed into alleys; dingy fast-food restaurants and cold stores pockmarked every corner, the emaciated commerce of poverty. No skyscraping hotels here, only stubby broken-shutter lodgings with misleading wistful names—Riviera Palace, Oriental Retreat. The most noticeable feature, a sprawling tattoo stamped across an already-mutilated body, was the graffiti. Bright protests neutralized by government blackouts then covered with more rallying cries, a laughably unavailing cycle. In the end the blackouts mostly prevailed. Above the graffiti, above the jagged rooftops, black Shiite flags tried to best the breeze, vastly outnumbering Bahraini flags, an isolated victory to which the dissidents clung.

When I reached Road No. 5215, I slowed to a crawl. Even for the slums, Almaisa's street was ugly, graceless, stuffed with buildings of radically different heights and materials and shapes. A group of men on the corner was smoking, the same idlers one sees in every slum, the dispossessed, the disenchanted, the unemployed. They watched my car with simmering eyes. A weed-choked lot at the street's end was filled with damaged tires, perhaps a kindling reserve for anyone looking to start a protest fire. Dirty pigeons pecked at the ground.

Few addresses were visible and the only apparent business was an electronics store selling outdated and revamped televisions and computers. I debated walking into the shop and asking for directions—Almaisa's scar would be easy enough to describe—but my better instincts rejected the idea; an American man wandering around the slums was strange enough without raising more questions. And I suppose I felt a sense of obligation toward Almaisa, the slender plucky woman who'd invited me to her studio, a duty to keep her unmarred by a clumsy American's fingerprints.

It was the most decrepit building on the block. I found it by process of elimination—peering into every window, determining whether residential, commercial, or abandoned. Her building was sizable, one of those ancient structures that had been heaped on and chopped up over the years, a graveyard of afterthoughts. The lower floor was built of sandstone blocks, the kind the British used to build up Manama, while the top was framed with sloppily piled cinder blocks, the shoddy construction of later years, slum years. The windows too were clearly postscripts—factory windows, multipaned and grimy to the point of being entirely opaque, a battered air-conditioning unit sagging from one of the sills like a thief trying to climb inside. A few windows were boarded up entirely. The front door was also covered in boards, but as I approached I could see it had once contained stained glass, the jagged remnant of a border visible beneath the plywood, its translucent colors a dirge for better years.

When Almaisa answered my knock, I was somehow surprised; it didn't seem like a place where people answered knocks. But there she was, staring back at me with her green eyes, wearing another bright abaya. She smiled, invited me inside.

Her studio was on the second floor. It was unfinished, easels, paints, and brushes scattered everywhere. Mosaic tiles were arranged on a table in the shape of a fountain, presumably her latest piece, her tools of the trade splayed in mid-use: half-congealed grout, open pliers, large knife, camera and rolls of film. Beneath the odor of cement and turpentine, I could detect the spicy tones of her perfume: jasmine, amber, maybe musk.

"So this is where it all happens." I smiled.

She was combing through a stack of artwork against the wall. Beneath shifting folds of canvas, I glimpsed her works, uniformly arresting even in small doses, windows into absurdly vibrant worlds.

"Yes, this is my studio. Did you find it easily?" Her voice was as deep and velvety as I remembered.

"Sure. No problem." I strolled across the floor. It was covered with fine white dust, patches worn smooth in places, a silent record of her movements. Outside, the voices of two men were shouting, arguing about something. A dog growled in response. As Almaisa glanced through the window, sunlight clarified her face, turned her skin to butter, the texture of newly sculpted pottery, her scar fading to a scratch. She was not quite Arab, I thought for the second time—something else.

"You studied art in Europe?"

"Yes. Florence Academy of Art."

"Italy? Impressive."

She looked up at me, challenging. "You are surprised?"

Mumbling an explanation, I suspected suddenly that her invitation to this studio had been a kind of test, a trial to determine whether an American would brave the slums to pick up a piece of art. In her eyes, I saw plainly, I was like all Westerners: fearful of dirt and uncertainty and streets without names, incredulous that culture could exist in these destitute parts.

"I only studied painting in Italy," she was saying. "When I returned to Bahrain—that is when I began mosaic."

I hastened to regain my footing. "And why mosaic?"

"Because it belongs to my people. The first mosaic was in Mesopotamia. Of course you must know this. Iraq, Syria. There is famous mosaic throughout the East—Dome of the Rock in Jerusalem, the Umayyad palace in Palestine."

"Sure, I saw a mosaic in Jordan—"

"And I love mosaic for other reasons. More difficult to explain. Mosaic is one thing made of many. It is . . . complicated. *Mo'aqqad.* The beauty comes when every tile is in place."

"Well, you've clearly succeeded as an artist," I maneuvered. "To showcase your work at the king's show."

Silence, Almaisa studying me as though deciding whether there was some basis for offense. Finally, she shrugged, smiled candidly. "My mosaics have been shown in Venice, Brussels. The king is impressed. He thinks others will be impressed."

From the Abdulatif Al Dosari mosque the midday call to prayer began. Almaisa showed no signs of wanting to pray; instead, as though suddenly remembering she was host to a client, she offered me something to drink.

I peered through the window. One of the men was standing alone in the alley. "Is this where you live?" I called to her above the running tap.

She returned, handed me a glass of water. Her nails were unpainted, chipped almost to stubs, her hands rough and dry, a different specimen from the polished cuticles I saw every day. Her face too was bare, not a trace of lipstick or blush. "No," she answered tersely. "This is just my studio."

I took a sip. "You're from Manama?"

"I suppose it depends who you ask."

The water tasted dirty, that dull leaden flavor in every third world tap; most Americans sickened when they drank it. Another test, perhaps.

Almaisa had returned to the stack of pieces and, as she leaned over, her hair peeked out from her hijab just as it had at the opera house, a fringe taunting the world. Clumsily she dragged my mosaic from the bunch. I went to help. Then she was staring at me and I realized I hadn't paid her.

"*Asif.*" I handed her the cash. She smiled.

It was at the top of the stairwell, as I prepared to tackle the descent, that I saw the photograph pinned to the wall. Partially hidden by an easel, it nevertheless bore the unmistakable black-and-white image of a child—a girl, it appeared, with one dimple and short dark hair, dressed in a plain white shirt and pants, smiling shyly up at the camera. The photo was odd in its singularity; there were no others in the studio. I ventured the question. "Is she yours?"

Almaisa's face reduced to an assemblage of guarded angles, her features strangely more beautiful because of the shift. "No," she said simply.

At the bottom of the stairs I stopped, lowered the mosaic, fixed my eyes on hers. *"Alf shukr. Shajarat al-haya hatkoon jamila fi beiti."*

It was a last trick, my secret, and it worked. Her face transformed, softened just enough to turn the angles into a masterpiece. *"Afwan, ya sadiqi. Inshallah hashoofak tani." You are welcome, my friend. I hope I will see you again.*

Funny how immersion and reflection are two different planes. It wasn't until later that evening, when I'd stared at the mosaic for some time, that I realized I still knew next to nothing about the woman. She'd talked much of the time—I could remember every word as though I'd just transcribed our conversation—and yet I'd walked away with little more than scattered tiles.

Hundreds of green leaves now stared down at my living room. Hundreds of Almaisa's eyes taking in my sparse bachelor's abode. A red scribble in the corner appeared to be her signature or some stamp of her identity. It was indecipherable. I wondered if I was the first American to buy Almaisa's art.

I'd hung the piece above my bookshelves and was impressed to see how it instantly and sweepingly transformed the room. The bland functional space had changed to something enticing, everything a shade brighter, each inconsequential hue coaxed forth as though the place had been drenched in water. My villa no longer belonged to an aging threadbare bureaucrat; it belonged to a cultivated traveler, someone worldly and discerning, someone able to slip below a country's surface.

9

The Sunday after I went to Almaisa's studio, Said Al Hussein, a scruffy leader in the Opposition societies, was arrested in his Budaiya bedroom on charges of sedition, disobedience, and terrorist acts in Adliya, ministry troops banging down his door and wresting him from his sheets, his wife and children covering their eyes in shame and humiliation, the *Gulf Daily News* reported.

Sunday mornings were the usual time for arrests, coinciding with the start of the poor man's workweek, no time for retaliatory uprisings. (Timing, I well knew, could be as potent a form of manipulation as any classic psychological weapon—duress, blackmail, guilt, flattery.) Still, the king was taking no chances, and precautionary measures were added to existing checkpoints: heightened security at government installations, airports and seaports, Dry Dock prison. Junaid received an admonishing 250 floggings for indirectly aiding and abetting terrorist acts in absentia, and all demonstrations were officially and categorically prohibited in the interest of safety. The American ambassador raised no objections, instead issuing a statement of gratitude to Bahrain for its longstanding support of our naval presence and enduring commitment to regional security. Kuwait offered Bahrain one hundred Mine Resistant Ambush Protected vehicles. It was all playing out substantially as Rashid had predicted.

Initially I felt a certain paternal satisfaction, the pride of a handler in his asset. But as the hours passed, my triumph faded. At the end of the wire I had nothing more to prove the truth of Rashid's charges than his word and a half-fulfilled prophecy. Whitney Alden Mitchell, rising star known for his unyielding standards, would hardly be convinced. Maybe Said Al Hussein was more than a scapegoat, an actual perpetrator. Rashid had duped me, Whitney would say with patient condescension; why believe a lone self-serving informant over a cadre of trusted professionals? One would expect nothing less from a lackey of Tehran, he'd say. Time to discuss termination.

The more I thought about it, the more foolhardy it seemed to reveal Rashid's information at all. Why bother with my Adliya report? Hell, even if I concocted a bland account of the whole affair, left out Rashid's damning accusations, it would amount to nothing in the end; arthritic desk officers at Langley would pore over my information, awed perhaps by the first-hand details, file it in a folder labeled OPPOSITION TACTICS, and return to slurping their morning coffee. No, nothing good would come from writing the report. I would hold the information until I had use for it. It might be inconsequential or it might be vital, but if nothing else it was unknown, at least to Whitney, and in this business unknowns were everything, all the more valuable the longer they stayed that way. The Admiral and Whitney could tuck their information into tidy locked drawers and so could I. It was the architecture of espionage, the house we all shared.

In a last attempt to validate Rashid, I flipped through the watch officer's log. It was entirely unilluminating, its information exactly as the Admiral's aide had said it would be: five bombs in total, each duly recorded with precise times. The Threat Mitigation Unit report was equally useless: three casualties, perpetrators undetermined, on-scene evidence not dispositive but method and motive suggested Opposition involvement, possibility of a future attack against Westerners high. Detailed investigation to be conducted by host country. Signatures of approval from the holy trinity: Whitney Alden Mitchell, Burt Johnson, the Admiral.

"Shisha, sir?"

"No, *shukran*. Coffee, black. And shawarma, *min fadlak*." I dropped a handful of dinars into the palm of the ragged Arab behind the counter. His name was Farouq, I'd learned over the months, but I never used it.

Tornados of stale smoke spiraled toward the ceiling. The walls looked as though they'd been through a war and never a washing, splattered with indeterminate brown stains. Airport Café had little to recommend it—a faceless hole-in-the-wall whose only distinguishing feature was the view of planes arriving at Manama International Airport—but I liked it for precisely this reason, the absence of expats or anything worth talking about.

Photos of the royal family covered the walls, sepia from too much smoke and the wilting Bahrain air, staring down at the vacant chairs and tables as though disappointed by so paltry an audience. The mustached faces all looked a bit too similar, the result of strict matrimonial rules and the shrinking gene pool they created. Rashid loved to feed me graphic rumors about all manner of hereditary diseases and mutations—sickle cell disorders, disfigured body parts—that the family tried to keep secret.

I walked up to King Jassim. Was this a man who would throw a bomb into a crowd of his own people—innocents? In the benign reverential haze of the café, the prospect seemed to thin, another cockeyed Arab conspiracy theory. But his eyes, something about them. He had the shrewd hard eyes of intention. A survivor, a man whose oil wells had dried up long ago but who'd held fast to his throne through a cunning dance of patronage and exclusion, who expelled citizens at a mere word of dissent, who hadn't blinked before selling his soul to the Saudis, calling in their guns and tanks when the fires of protest got too close to his mansions. A man who had four wives and more palaces, who went falcon hunting with the Saudi prince while legions of his people went hungry, the only remaining Sunni ruler to sit atop a land of Shia.

I headed outside to the patio. Someone had left a copy of the *Gulf Daily News*, which reassured readers that the *Search for Remaining Adliya Terrorists Continues* and also *Iran's Economy Just Steps from Collapse.* Outside the airport, I saw, fresh pansy and grass beds had been planted. Landscaping cleverly situated to give newly arrived visitors the illusion of an entire island lush and green. King Jassim's magic show. Like the exotic-sounding names that littered the city—Coral Bay for a park that contained neither coral nor a bay; Exhibition Road for a crowded street teeming with cheap electronics stores; Zinj Garden, a dusty patch that was home to a few scrawny caged emus.

The ragged Arab arrived, set the steaming dishes on the table, careful to place them away from my newspaper. Shadow-blackened palms along Airport Road swayed in the breeze. Twilight crept skyward. It was almost six o'clock, the hour Manama began splitting in two—one half turning dark, avenues becoming alleys, privacy-piercing illumination fading unceremoniously away; the other half coming alive, skyscrapers and luxury hotels humming with neon vibrations, doors of dirty backrooms opening to dust, heat, and oil money. The hour when workers came home and began their nightly revolution. I sipped my coffee, watched the glittering city in the distance, the rose stripes of evening multiplying across the sky like fresh wounds.

Five bombs. Walid had said five, but the watch log confirmed there'd only been four prior to my arrival. Walid's certainty replayed in my mind. A simple explanation: What if Walid had not made a mistake? What if—and the logic began forming effortlessly—he'd slipped up, botched his lines? Walid knew in advance how many bombs were scheduled to detonate and he'd failed amid the chaos to realize the last one had not gone as planned—a malfunction, as it turned out, its explosion delayed, the odd final bomb we'd heard on our return to base. It all made instant sense: the ministry officers' lackadaisical investigation, the absence of eyewitnesses, Walid's efforts to keep us at arm's length, his insistence that we ignore the final bomb. Walid was in on it all, the grandiose play from beginning to end. *At least no Americans were killed.*

I should have seen it before—Walid not a ministry official at all but one of my own kind, an intelligence officer sent to oversee the operation. My bag was at my feet and I pulled out my camera, scanned the photos I'd taken for a clue to Walid's identity: unofficial shoes, uniform worn incorrectly, missing insignia, all the telltale signs of disguise. There was the alley with a few shots of police, the bombed-out dumpster, partial torso, rubble and debris behind a row of planters. No shots of Walid. I went back to the photo of the rubble. Through the camera's lens the streaks of purple I'd noticed looked even more vivid.

Night had arrived when I reached my car. A handful of Indian construction workers waited by the highway for a ride home, their cloth shreds that passed for dust masks pushed tiredly onto their heads, threadbare rags clinging loosely to their limbs. *Illegals.* All the Adliya victims had been foreign laborers. The invisible ants silently carrying the Gulf's heavy loads, easy to step on, their absence hardly noticed.

I swung my car around, rolled down the window. "You boys heading downtown?"

They looked quizzically at each other, then one stepped haltingly forward. "No, sir. Tubli."

"Hell, get in. I'll take you there." They filed meekly into the backseat. "And I'll pay for a car wash too."

10

Jimmy's call woke me. Something about his car breaking down and could I give him a ride. Sunlight streamed through the windows of my bedroom. I drained the dregs of brandy from the glass on my nightstand.

Jimmy lived clear across the city in Riffa, in a gated community more exclusive than Amwaj Islands. A place overflowing with lush purple flowers and sweeping eucalyptus trees. Looking down on a rambling golf course were row after row of newly constructed houses, white stucco with fancy balconies and generous lawns, an uneasy compromise between Eastern grandeur and Western suburbia, inhabited by a mix of senior government officials, Sunni tribal members attached to the royal family, and foreign elite. Layers of security—iron gates, stone walls, guards—surrounded the place.

The Riffa Views guard must have gotten my name from Jimmy, because he let me in without question. Inside the gates, the sweet scent of carefully pruned bougainvillea saturated the air, pungent to the point of being cloying, Riffa's deliberate perfume to mask the sour rot of the nearby Tubli sewage plant. Nestled in a pocket of slums, the sewage plant mostly spewed onto the surrounding Shiite inhabitants, but occasionally the muscular northern winds carried the stink over Riffa View's high walls.

From its perch on a hill the Bakowski villa had a prime view of the golf course. In all seasons the course was kept green as the Emerald City, an exorbitant feat in this parched country, and imported orchids ringed its perimeter, fighting equally hard for existence in hostile territory.

My Lancer sputtered up the driveway. Jimmy stood waiting, briefcase in hand. Grinning conspiratorially, he climbed in, pausing to instruct the Filipina housekeeper to water the flowers before noon. "Yes, master," she answered with a curtsy. Gretchen poked her head out from the curtains and waved.

"Thanks, Collins. You're a pal." Jimmy clapped me on the shoulder.

"What happened to your car?" I glanced at his well-pressed Hawaiian shirt, too tight as usual, clean of any oil or grease.

"Oh, you know . . . wouldn't start. Probably the battery. Couldn't find the goddamned jumper cables. Gretchen probably buried them in the garage with the kids' toys."

We passed Hibiscus Lane, a winding road that appeared to have no flowers. "Doesn't Chief live there? Hibiscus Lane?"

A lone twenty-eight-year-old taking up an entire villa, the same sized villa occupied by his more important neighbors. How could the kid possibly have enough possessions to fill the place? To fill half the place? What did he do in the empty rooms of evening—wander around naked, lusting after the embassy officer or the donut stand lady or any other female he'd brushed up against that day, conjuring far-flung, overly polite, just-tame-enough-for-conscience fantasies?

Jimmy shrugged. "Don't see him too much. Now and then there's a barbecue at the pool house—"

"He's single. How can he afford to live here?"

Jimmy rubbed his fingertips together. "He's chief, that's how. Earns a pretty penny. Gives you some motivation, eh? Earn that G-S Fifteen."

"I am a Fifteen."

Somewhere around Bilad Al Qadeem I noticed Jimmy looking under the seats. He opened the glovebox.

"What the hell you doing?"

"Just checking." Jimmy flashed a knowing grin. "Can never be too sure. This is the car you meet SCROOP in, right?"

"You're checking for a *wire*?"

"Or a beacon. Never know. Just doin' my job." He finished groping the seat and, apparently satisfied, leaned back. Leisurely he stroked the leather briefcase on his lap. "So what was the story with SCROOP, anyway?"

"No story."

"What'd he have to say about Adliya?"

I kept my eyes on the road. "Turns out the sonofabitch was holed up in a hotel for a few days. Whoring. How d'ya like that? I had a suspicion. No biggie. He's still young. You know how it is. Shit like this happens."

Jimmy frowned. "Could've at least given you a heads-up. Signaled you a few days before. Didn't the guy have inside dirt?"

I shrugged. "Bits and pieces. All compartmented, of course. He was waiting to let me know until he found out more."

Near downtown Jimmy abruptly suggested we stop for breakfast at Vic's. I was surprised. In the wake of the Adliya bombings, Jimmy had exhibited a new level of obsequiousness and obedience, double knotting his shoes, buttoning every button, and I hadn't expected him to play hooky so soon.

He grinned. "Why the hell not? My car's broke. Might as well use the excuse while we got it."

The waitress brought us eggs mimosa, mint lemonades, and a Singapore Sling. Somehow Vic's dim interior blurred any distinction between day and night, a vacuum, the normal roots of time and place severed. Two British men were sitting at the bar, agitating their frosty drinks, loudly discussing the price of oil and complaining about the commute to Saudi.

"Not drinking, Jimmy?"

"Nah, not today."

The ice in my lemonade clinked, drowning the mint in soft green eddies. "Jimmy," I ventured. "You're a Marine."

"That I am." He straightened rapidly. During his two years of service Jimmy had never seen combat, but that didn't stop him from considering himself an expert on all things war-related.

"You ever work with explosives?"

"Well, sure, 'course I did. You name a weapon, I used it—"

"Listen, Jimmy, you ever come across something that turned shit purple?"

He chuckled—uncomfortably, it seemed—and pulled out a Habano, sniffing it deliberately, fingering the gold-and-green wrapper as though examining a rare specimen. "Turn things purple? Not sure I read you."

"Explosives. That leave a purple residue."

"Ah." He lit his cigar, squinted at the bar. The bartender was polishing the teak countertop assiduously, his rag repeating the same arc like a car wash.

"Purple, you said?" Jimmy leaned forward, the sweet spice of his cigar wafting toward me. His voice dropped. "Listen, Collins, whatever this is, leave it alone. All this talk—"

The hanging wooden beads clicked and the front door opened, dousing the amber darkness with light and shattering our vacuum. A man walked in, sat at a table diagonal from us. He was swarthy and young, wore tight-fitting slacks and a candy-pink silk tie—European, French Arab perhaps. Ordered a coffee and nothing else. Jimmy eyed the man, puffed contentedly on his Habano as though we hadn't been talking of explosives a moment ago, as though he hadn't suddenly cut me off with an odd warning.

"Now if only I could dress like that. . . ." Jimmy smirked, fingered his wedding ring. "Probably would've had more luck with women."

I sipped my Singapore Sling, forced a laugh. "The Marines would've disowned you."

"Well, you're probably right. . . ." He trailed off, still staring at the man. "And Gretchen would've run like hell."

With a final puff he shoved his cigar into the ashtray, winked. "Listen, I'll get a ride home with boss man. Lives closer'n you. Don't want to put you out twice in a day."

11

It was probably inevitable that my first date with Almaisa was at the Tree of Life. Despite some ambiguity, I feel entitled to call it a date—particularly since I'd gone to the effort of driving to her studio simply to extend the invitation. It was easier than I'd expected; she said she'd be delighted to show me her country's treasure. Maybe she felt a sense of obligation—I had, after all, probably supplied her income for the month—but even now the date seems more than a debt repaid. Not just because my ego resists the idea, but because Almaisa, then and after, never seemed the type to feel beholden to anyone, singularly clean of most humans' need to please.

She'd insisted on meeting me, so I made the long trip alone. The fabled tree was buried deep in southern Bahrain. Past Riffa Views, tenaciously exuding its floral perfume, past the opposing Tubli sewage plant. Past the Royal Golf Club, where antipasti was served on silver trays. Down through the old Harken Oil compound erected when Bahrain was still producing the precious liquid in abundance, now claiming little more than a sprinkling of residents, halfway to a ghost town.

South of the Harken compound was where the emptiness began. I'd never ventured this far. Dunes as far as the eye could see, a forlorn evaporated ocean, endless white sky, heat-shimmering roads. Flatness that swallowed everything.

I approached Jabal Al Dukhan, Mountain of Smoke, a pile of sand that marked the highest point on the island. It was crude country, untouched by the king or revolution. Raw, a Bahrain I'd never seen before.

My first thought was that the Tree of Life was a disappointment. Far from the miraculous aberration of nature promised, an unkempt unholy scene stared back. Low-hanging and indistinctly green, the tree presented a pitiful sight, its trunk scarred with graffiti, the left side sagging so completely it almost touched the ground. Trash littered the area as though discarded by angry fans after a team's defeat. Any romantic mystique was further shattered by the proximity of a parking lot and a thinly clad Indian selling bottles of water for one exorbitant dinar. Perhaps most deflating, the main attraction was surrounded by other shrubs, some equally tall. The Tree of Life was not unique, its likenesses everywhere, mocking minions. The pristine exceptionalism was an illusion.

Almaisa arrived in an old Kia more battered than my Lancer. She emerged wrestling a large manual camera and a bag of food. Her abaya did not disappoint: purple with scalloped trim, a bolt of color amid the dreary monochrome. The first thing I said was that her Tree of Life was better. She smiled, handed the Indian money for two bottles of water.

She wasted no time setting up a tripod at the bottom of the slope beneath the tree, adjusting the angle, focusing the lens. After a few shots she cocked her head, squinted. "I do not know. . . . I should make another mosaic of this tree?"

"No. I like having the only one." I climbed to the top of the hill, blocking her shot. She looked up from her camera, the wind blowing loose a lock of hair, the same dark curl that always seemed to spurn the rules. It moved lightly across her forehead, teasing me with a hint of its sisters.

"So I noticed there are other trees growing here."

"I did not say there were no other trees."

"Well . . . actually you did. You said, 'There is nothing living for miles.' Your exact words."

Her lips pursed in a teasing smile. She nodded toward the tree, blocking the dust and glare with her hand. "You see the way it leans down on one side?"

"Yeah, a few more years and that thing's gonna topple over."

"I think that is the most beautiful part. How it almost touches the ground. It is. . . ." She looked at me probingly. "*Motawade'a.* What is the translation, please?"

"Humble."

"Yes, I think humble. Like it has been carrying the weight of living alone in the desert for many years." She held her palms up, adjusting the level of each like a scale. "And it is uneven. You see how the other side reaches to the sky? One side up, one side down. I like that very much. This is why the tree makes a good mosaic. In art imperfections are better than perfection."

The wind had been steadily gaining momentum, stirring up larger and larger eddies, blowing Almaisa's abaya in purple waves around her, and the Indian pulled a surgeon's mask over his face, watching us as though awaiting permission to seek shelter. The tree's branches began bending violently to the ground. Still Almaisa stared raptly, absorbed by and into the landscape, as much a fixture of the southern desert as the dunes or the tree itself, and from the elevated angle where I stood, as perfectly imperfect. She was beautiful, I remember thinking, and for the first time I saw it wholly, unreservedly.

Almaisa nodded toward her car. "Should we go? There is something else I want to show you."

Somehow she'd taken backroads to reach the first oil well. There wasn't much to see, another disappointment. Except for a sign, the attraction was indistinguishable from the abandoned equipment scattered across the southern dunes, clearly a destination for only the most committed tourists. The well itself was an unceremonious homage to the single greatest stroke of luck in Middle East history, Bahrain's only true claim to relevance. A copper plaque informed us that oil was struck on June 1, 1932, with an initial flow rate of 400 barrels an hour, and invited us to recreate the excitement with a defunct pump. Pipelines were visible in the distance, empty but still intact, crisscrossing for miles like a checkerboard for giants. A colossal coffin for what once was: Manama the boomtown, arteries bursting with lifeblood for the world.

Almaisa seemed to want to linger, so I spread a tattered blanket from her car and we sat on the hood. She gazed at the expanse as though watching a sunset on the beach. I couldn't understand what she saw. It was a land that held no depth for me, little more than a barren wasteland. Somehow this place caused people to see mirages, ghosts. Standard Oil vets I'd met in odd corners loved to regale me with stories of bygone Bahrain: pioneering extraction and refining technology in the seventies, crude and wealth spewing forth from the desert in barrelsful, a mission no less vital than the contemporaneous battle against the Reds. To hear these men talk was to hear them describe a vanished world—a halcyon time, a lively racetrack where aspiring superpowers elbowed and jockeyed their way around bends, where skyscrapers sprouted daily from the sand. No less than the birth of a nation.

"I came to Bahrain because my mother and father died," Almaisa said abruptly. "My uncle was here, so he took care of me."

It took me by surprise and I fumbled with condolences.

"Fifteen years ago. I was thirteen." Incongruously Almaisa reached into her bag of food. "Try this. Golden apple. Grown in Bahrain. You foreigners think nothing grows here."

"Both of my parents are gone too," I offered. "What happened to yours?" Give to get, a manipulation basic.

She bit into her apple, glanced down. "They were killed in Iraq by Saddam Hussein's men. Political prisoners. Enemies of the state."

A ferocious gale swept across the dunes, stinging our eyes, filling our mouths with grit. The point of uninhabitability had come upon us. "*Khalina najlis bi siyartik.*" I held out my hand. She took it.

We sheltered inside her car. And then, while the punishing winds beat the Kia with angry fists, Almaisa told me her story. It came out in an unbroken stream, her low voice turning the years into a smooth elixir. As reserved as she'd been within the confines of her studio, she now poured forth a torrent, as though she (or maybe I) had flipped a hidden switch to raise the fortified gates.

Her father was Sunni Iraqi, her mother British (a fact that explained her ambiguous complexion and accent); they'd met in London while her father was attending medical school and her mother teaching poetry. Al Bayaa City in Baghdad was her childhood home, a comfortable neighborhood of clean maidans and modern buildings, and she grew up speaking English and Arabic, fluent in both but lacking a native's comfort in either.

She'd found the filial conflict between East and West unsettling. Her mother had the enlightened disdain of the expatriate (and here she smiled at me)—outwardly appreciative and inwardly superior—while her classmates and locals looked at her with the same two-faced condescension. Her walnut-hued hair and light skin made her an odd curio. Despite her mother's best attempts—lessons in dance and French literature and art (the only class that stuck)—Almaisa's interests gradually followed her father. She'd worshipped him, had wanted to be a doctor like him. North of the Gulf, she noted indignantly, plenty of women became doctors.

One firm common ground existed between her parents, to which Almaisa was mostly blind. She was aware only that her parents talked late at night after she'd gone to bed, and were often gone at strange times for strange reasons. The night before her twelfth birthday (and as she spoke, she looked as I thought she must have looked then, childlike and disbelieving), two *Mokabarat Mudiriat Arba'a* officers knocked on their door, handcuffed her mother and father, led them to a waiting car. One blond head, one black, disappearing into the night like streetlights expiring. The expression on the Directorate Four officers' faces—this was the worst part, she said. *Bedon ihsas*. Indifference. Indifference to the girl who remained behind, alone and parentless. Cruelly hardened as only intelligence officers can be.

Then, her face changed, Almaisa talked about her uncle. After her parents disappeared, she'd gone to live with her *aa'm* in Bahrain (her mother's relatives in England, she said bitterly, hadn't wanted her), where years earlier he'd fled Saddam and sought a living in the booming oil industry. Her uncle had found work as an engineer for the larger-than-life American company in town, majority-owner of the state refinery. He'd saved money and bought a modest house in Budaiya, where many of the workers lived. Manama was a small town then, her uncle liked to recount. No skyscrapers. Greenery, plants as far as you could see. Not as much dust.

But by the time Almaisa arrived, the country had changed. And suddenly, raising her voice against the wind's cacophonous whistle, Almaisa was telling the story of modern Bahrain. How villages gave way to a brutal expanding metropolis. How the urgent Manama streets overflowed with shopping malls and mansions, more mansions than people. Lush grasslands disappeared, replaced by concrete. How Bahrain decayed and separated like an aging man stripped of his youth and purpose.

Then, strangely, Almaisa's words became Rashid's. They were telling the same story with the same harrowing ending. An inverse fairy tale, a story of riches to rags, of promise turned to betrayal, heartbreak at the end instead of the beginning. The two voices—one present, one spectral—filled the car synchronously.

Poverty began in earnest, the voices said. Cold stores flailed while monarchical conglomerates multiplied. Oil didn't belong to the people; it belonged to the few. Hasty infrastructure spread like a moth's wings; highways conveniently transported travelers from one glistening point to another, avoiding the realms of discontent along the way. Backroads, the villages, became cesspools.

Almaisa pushed back her sleeves. Sweat had glued the stray curl to her forehead. Eventually, she said, the wells dried up and rigs stopped drilling. Pumps slowed to a halt. Her uncle was lucky; he was one of the last to go.

Only the privileged, her uncle witnessed, had survived. Those related to the royal family. Close to the royal family. Knew someone in the royal family. On the right side of religion. Her uncle was Sunni but in the bottom half, outside the sealed circle of allied tribes. Nearly starving in the slums, relegated to odd jobs, he saved every penny he scraped together for his charge, Almaisa. A kind man, Almaisa said with downcast eyes, the sort of man who would feed stray cats when he didn't have enough for himself.

Despite the hardship, her uncle had grown to love Bahrain. Palms bending in the brawny wind, dust settling across the dunes like crushed diamonds, salty rich scent of oil emanating from the land long after the wells had run dry, even the bustling layers of the slums.

Was her uncle still alive? No. He'd died of a debilitating illness—probably cancer, but it was impossible to tell as he'd never been to a proper medical facility. How had she managed without him? A rigid smile. *The villages; we help each other. And my art. You are not the only one who buys my pieces.*

And then she'd shrugged, a signal her story was done.

But it was more than a story; it was a confession, an inside turning out. Raw, unapologetic. In the twenty-five years I'd been a spook in nearly a dozen countries, I'd rarely listened to someone so absorbedly. Almaisa's words had an intimate and unalloyed quality that didn't exist in my world of cocktail parties and sideways glances and cheap riddles, that drew me in.

Before leaving, I asked for her phone number and she gave it to me. I allowed myself a moment of congratulations. Something about me, my approach, seemed to have inspired Almaisa's confidence. I was relieved that my own story hadn't come up, that, for now at least, I could keep my distasteful world a secret. This, and her unexpected outpouring, even diverted me from my instincts that there was more behind the gates.

12

Surrounded by slums and prey to stray rubber bullets, Three Palms had even higher walls than Riffa Views. A gated community in the middle of Budaiya appeared odd only if you failed to realize expats could get obscene gobs of house here for the money: gargantuan villas, real marble floors, an Olympic-sized swimming pool if you were lucky. No greenery, no sidewalks for dogs and kids, daily unrest in the streets, but it was all worthwhile, they swore—the Johnsons, the Chaplain and his wife, other expat-resident faithful. Something exhilarating, I suppose, in the contrast of geographic proximity with metaphoric distance, the slums just close enough to remind Three Palms dwellers they lived in a different world.

I hated coming to Three Palms. It was a security risk, I warned Poppy Johnson time and again. Eventually the guard was bound to figure things out, spill the beans for the right sum or favor. (Poppy naively insisted he would do no such thing, as though he possessed a literary character's unimpeachable loyalty, as though some people didn't have a price.) Then there were her neighbors, rubbing their backs against the windowpanes, noticing when a man's hair grew thinner, a woman fatter. Once discovered, I warned her, our affair would travel through expat circles faster than a refugee at night. Poppy had waved her hand dismissively. *My neighbors wouldn't know if a terrorist pulled into my driveway.*

The compound was divided into neat neighborhoods with French names like La Vie en Rose and Vue d'en Haut and Tranquillité. I parked my car a few blocks from Poppy's house, walked over when the streets were clear. Her mansion was stucco with a tile roof and small windows, a nod to the Mediterranean that came out more like a mockery. Balconies of white balustrade (just like Monaco, Poppy cooed) erupted awkwardly from the sides, offering no real view. Poppy had become indignant when I pointed out the error in design, her fantasy as carefully tended as the bougainvillea in her yard. *Living the dream*, she and her bunco partners liked to reassure each other.

Something was different, I sensed instantly when Poppy opened the door. A tight shawl cocooned her arms, and her face looked puffy and red. One arm dropped stiffly to the side of her negligee. Her breasts. Even draped in the shawl they were impossible to ignore—lopsided and gigantic, like kicked-around beach balls.

"So you did it." I grinned and patted her shoulder, a reassurance that nothing she did affected me.

It would be a tiresome night. Lying on her satin sheets, Poppy prattled on about her hundred tribulations: too many people in her yoga class; Burt was traveling for work again; her daughter, away with her sister at a sleepover, had been having nightmares, something about being stranded in the desert. And then there was Lata, her live-in maid, who'd disappeared in the dark of night. Like something out of a movie, Poppy said. The Ministry of Interior had conducted its semiannual roundup, deporting all hired help who didn't have passports and work visas. Of course Lata had the right papers, Poppy was quick to assure me—*we wouldn't hire anyone illegal*—but they were stuck in the hands of her Bahraini sponsor, a classic Gulfie extortionist. Had she considered giving Lata money to buy back her documents, I asked—and Poppy looked at me as though I'd suggested she assassinate the king. *More money? She was lucky just to work for us—her last owner raped her.*

By the time we took our clothes off, my sexual interest was nearly extinguished, reduced to a trickle of morbid curiosity. Her tits were even more gruesome in the flesh, disfigured and dimpled and bruised, sadly fascinating like a mangled animal. Poppy's tearful confessions and tortured afflictions rushed out: The surgeon had assured her it was just swelling, that it would go down, then had hastily returned to his native France. Or wherever he came from. Doc Hollywood had such impressive credentials, his ads displayed in all major hospitals. It simply wasn't fair; she knew dozens of expat wives who'd paid visiting surgeons for small adjustments—

additions, subtractions, divisions—and they'd all turned out fine. Why had *her* job gone south?—and she was, I think, asking not me but God, who'd apparently singled her out for punishment.

Nine o'clock drew near and I began to plan my exit strategy. I could still escape if I left immediately, maybe even squeeze in a stop at the Sea Shell Hotel if I headed home via downtown. But as I pulled on my pants, Poppy stopped me. She went to the bathroom, returned carrying two packages—purchases from the souq—presenting them to me with a timid smile. Then, as if in afterthought, she let her hair down. It fell to her shoulders in slow sad motion.

One of the products was a colorful cardboard box labeled VIRGINITY SOAP, which promised to turn back time and attract even the most reluctant man; the other, packaged in an equally optimistic bottle, was VAGINAL RESTORATION POWDER. The soap packaging was faded as though it had been sitting in a window display for years, and the powder was a sinister green hue. I laughed, thinking suddenly she'd bought them as a joke. But Poppy's face remained serious, slightly hopeful.

"Come on, Poppy. You don't believe in this stuff, do you?"

Red splotches erupted on her neck. She shrugged.

"Look, this is for local women." I pushed the creepy toiletries away, casually attempting to brush off any residue. "Subservient women in abayas whose lives are pleasing their husbands, who believe in sorcery and witchcraft and all that third world shit. This isn't you."

Poppy gathered up the products. "Is it the girl I saw? The one at the opera house?"

I hesitated, feeling strangely pleased, as though the encounter that night and whatever passed between me and Almaisa had been validated by the world outside rather than born solely of my own reflection. In a surge of magnanimity, maybe pity, I touched Poppy's hair, assured her it wasn't; I'd simply bought a mosaic.

Poppy eyed me suspiciously, muttered, *Let's just go to bed*, swallowed her sleeping pill, and slipped into the sheets. A litany of profanity ran through my mind: Curfew had passed.

An unoccupied room in the dead of night is like a woman stripped naked—vulnerable, forced to bare its secrets. The Johnsons' living room was crammed with junk: gilded candlesticks on worn suburban shelves, gaudy vases, paintings of graceless farm scenes. The carpet was a fiery red shag that Poppy had once defended as the highest form of Arabian luxury. *Like purple used to be in Europe.* I'd told her it belonged in a bordello. Framed photos sat on a cabinet, mostly family shots with the kids: sandcastles on the beach, waving from the bucket of a Ferris wheel, quintessentially American screwball faces. Poppy in a flowing wedding dress, a mess of orchids in her hand.

Next to the living room was the shisha station. Cushions ringing an ornate hookah. Predictable tapestries on the walls, benign scenes of the Middle East—palm trees and camels—the kind made expressly to seduce tourists, sold in the first and largest shop of the souq, where expats could complete all their transactions without penetrating too deeply, without venturing into the belly of the beast.

Burt's office was also on the first floor. The place was dripping with pretense, shelves filled with more books than any reasonable person could read in a lifetime, walls covered with preserved game heads—WILDEBEEST, IMPALA, KUDU, the labels said. So Burt was a game hunter, liked his African safaris, liked to shoot things from the safety of a vehicle.

His computer was on his desk. I turned it on, amused to see how eagerly the machine spilled its contents for my viewing pleasure, no passwords, no locked folders. The esteemed Mr. Johnson would've made a lousy spy. Dozens of photos, mostly Burt and his travels, only a few with Poppy and the kids. A photo with the Admiral—of course there was a photo with the Admiral—the Fifth Fleet commander looking crisp and dignified in his white uniform, Burt trying to keep up in a pinstripe suit, blond hair slicked back. They made a handsome couple. Another pic in casual civvies, background blurry and hard to make out—peeling railing, enough frothy greenery to suggest a hot climate, the asymmetry of poverty, bits and pieces of staccato Oriental letters, a sign featuring something resembling a bowl of slop—somewhere in Asia. There was a photo of a large jovial Chinese-looking man with his arm around the Admiral, and another, slightly blurry, of the Admiral and Burt at a holiday party with scantily clad Santa's helpers. The next picture was of a woman—young, skinny, Asian—whose red tattoos covered more flesh than her clothes, standing in a busy street and looking at the camera with the canny mirthful calculus of whores the world over. I felt a grudging admiration for Burt, then a vague pity for Poppy. It reminded me she was upstairs in her pill-induced coma, the reality and irreversibility of her botched parts buried blissfully in the recesses of her brain.

A few hours later it came to me. Another dream about Adliya, the third I'd had since the bombings. Poppy eating her caramel flan at Lilou with gusto, the fractional corpse mingling giddily with café patrons, then, coating the background like ugly stained wallpaper, faded and blunt scenes of the Baghdad I remembered—tanks rolling, soldiers shouting and gesticulating with death-cusp urgency. I woke up sweating. Rodriguez, an Explosive Ordnance Disposal sergeant I'd met in Baghdad. For a bottle of hooch Rodriguez had handed me one of the military's deadweight informants, someone I could clean up and add to my thin dossier. Rodriguez would know what the hell turned things purple.

Shivering in my pool of perspiration, I considered how to reach Rodriguez. Couldn't contact him from home because the Bahrainis were listening; couldn't contact him from work because Whitney Alden Mitchell was listening. Light streaming through the blinds announced early morning, and Poppy stirred. Before her somnolent haze broke, I mumbled an excuse about meeting an informant, slipped out of bed.

At a small souq in nearby Maqabah, I downed a quick breakfast of labneh and cream, bought a burner cell phone. After a hasty surveillance detection route to a dirt lot, I called Rodriguez's satellite phone, the number stamped on my mind after all these months. No answer. He might not even be deployed anymore, I realized, and I had no idea how to reach him in the real world. The potent white sun rose behind cement flats, clouds of dust stirring and drifting like lazy animals stretching from a long sleep.

My cell rang. It was the number I'd called, and the voice on the other end had Rodriguez's uniquely unfounded hostility. "Who the fuck is this?"

"Collins. From Baghdad."

"Shit. Collins? Where you calling from?"

"Some other dustbowl."

He made a noise between a snort and a chuckle.

"You still in Baghdad?"

"Worse. Jalalabad. Baghdad was the gateway to hell. Now I'm inside the gate."

"Listen, I need your brain."

"Whatcha got?"

"Have any idea what kind of explosives would turn something purple?"

A long pause and then his answer, confident as an innocent man. "Potassium permanganate. Has a violet oxidizing effect. It's your standard disinfectant, fairly easy to get. Combine it with glycerin or sulfuric acid or alcohol and you've got yourself a fire-starting chemical."

"You sure about that? Couldn't be anything else?"

"Sure as I'm in J-bad about to kill another fuckin' haji. Ain't too many explosives that turn shit purple."

Potassium permanganate.

13

The Ministry of Interior was located in the old Manama Fort across from the Manama Cemetery. The fort had thick stalwart walls and windows little bigger than peepholes; satellites and antennae sprouting from the roof were the only signs of modernity in an otherwise medieval structure. At the tall iron gates I showed the guard my Navy badge. He stared at it dumbly, motioned me inside. Confidence was more than half the trick: Act like you belonged and you would; pretend you knew something the other did not and you did.

A deskbound officer kept his eyes on me as I strolled the lobby. Brisk footsteps across the floor. It was the Minister of Interior walking toward the door, followed by a skinny minion. The minister glanced in my direction, nodded vaguely as though he recognized me from somewhere.

"*Salaam 'aleikum*, my American friend." Walid's heavy accent was unmistakable. Turning, I faced a wide smile, all traces of Adliya's tension dissolved and forgotten.

"*Salaam.*" I extended my hand. "*Kifak?*"

Walid's office was spacious, with a leather couch, armchairs arranged neatly around a desk, an exotic leafy plant in the corner. Plusher than I would've expected for a mid-level ministry official. Royal family photos covered the walls, including one man who bore a striking resemblance to Walid.

"The king . . . he is your . . ."

"Cousin."

A poor Arab who smelled even worse than he looked poured tea into cups, offered them haltingly to us. I managed a quick *shukran* before he scurried off.

"So how is everyone at the base?" Walid sipped his tea and lit a cigarette, sat back in his chair. "Good, I hope? Not scared by the violence?"

"Not yet." I smiled. "The bombs will have to get a little closer first. Americans are still safe within their fortresses."

"I am glad to hear it. Let us hope it will not reach those fortresses." He leaned across the desk with a pack of Sobranie Golds. "*Itfaddal*, my American friend." His mustache looked trim, neater than before. I took the cigarette, thanked him, lit it before he could offer.

Walid waved his hand invitingly. "You wanted to talk to me?"

"Yes. Just wrapping up my report on the bombings. Wanted to see if the ministry had any new information."

"Tell me again—what office are you from? When my secretary gave me your name, I could not find it on our roster of American officials." He squinted through the smoke.

"Office of Middle East Analysis. Just arrived a few months ago. Usually takes a while for our names to filter through the system. We're a bureaucracy just like you." I crossed my legs, leaned back. "Heard you arrested Said Al Hussein."

"*Na'am*. Our informants led us to him, praise be to Allah. His *twalet* was full of explosives." He picked up a file sitting conveniently atop his desk. "*Min fadlak*, take a look at our report. You will find all the information in there." He dropped his ashes into a crystal tray. A gold ring circled his thumb, just like the king. "Take your time."

I flipped through the folder, paused at a photo of some dozen bedraggled men in a lineup. "So you found explosives in the shitter. What kind?"

Walid shrugged, pointed to the file. "It is in there, yes?"

I scanned the pages. Captioned photos of a five-kilo canvas sack stored in an outhouse. "Ammonium nitrate?"

"Yes. That is what our tests showed."

"Can I see the evidence?"

Extinguishing his cigarette, Walid got up, stood between me and the wall of photos. "No, you may not examine the evidence."

"Why not?"

"Because our terrorism investigations are our business." Walid looked down at me. "Your only concern should be the welfare of your own citizens. They are safe. Leave the rest to us."

"Thank you for the advice." I took out my notebook, prepared just for the occasion, groped my pockets conspicuously. "Well, would you look at that? Come all the way here and leave my pen at the office. *Habibi*, could I trouble you for a pen?"

Walid stared at me, nodded. "Of course." He opened the upper-right desk drawer, examined its contents briefly, hesitated as if debating whether to search another drawer. Then, like any good intelligence officer caught in an office not his own, he improvised. "You can use mine." He plucked a pen from his pocket. It was embossed in silver with the ministry's seal.

I nodded in gratitude. "Tell me about Said Al Hussein."

"An Opposition leader. Friend of Junaid. You have heard of Junaid? A poet who writes slander about the king. One of our worst *shayatin*—he incites much violence. For a poet, he has not a single peaceful bone in his old man's body."

"I know who he is. Even read a few of his poems."

Walid smiled tightly, indicated a clunky television set in the corner. "We have video surveillance from that morning. Café Lilou's security camera. This also helped us identify the thug. You would like to see?"

"Certainly."

He turned on the monitor, the video perfectly cued up, and we watched the events unfold in blurry black-and-white: 7:24, the digital clock in the lower right corner read. The resolution was poor, the camera range limited, and I could barely make out the rear view of a man—dark hair, thin, average height—as he scrambled up and into the dumpster, plunging into unintelligible heaps of trash. No movement for several minutes, then the man reemerged and I saw that what I'd mistaken for dark hair was actually a mask, his head completely shrouded. Lithely he scaled the side of the dumpster, scurried out of the camera's view. Abruptly the video cut to 9:18 (nothing of interest in the interim, Walid assured me) and an Asian man flinging a bag of trash into the receptacle. Despite his slight stature, the man looked bigger than his remains had seemed to indicate. A blinding white flash, the dumpster instantly erupting in flames, the picture briefly jarred before turning black.

"The camera was damaged by the blast," Walid explained. "That is all the video we have."

"You can identify the guy from this footage? Looked to me like any Mohammad on the street."

"As I said, we also have informants. Of course we use many pieces of evidence, not just one."

Walid studied me. "Mr. Collins, we are anxious to restore peace and stability to this country. For everyone—*all* Bahrainis." He walked over to the window. "Surely you understand the importance of resisting people who are trying to destroy a country. You are American. Terrorists tried to knock down your buildings." He turned to face me and I saw that, despite his best efforts, anger was seeping into his features, igniting his muddy eyes. A man who couldn't stomach or understand foreigners daring to ask questions, foreigners who should have been unwavering allies. A haughty, desperate man who had trouble playing his role.

"You know what the problem with this ministry is?" Walid had turned back to the window. "Our building is not high enough. They moved us into an old fort that is only two stories high. They could have built another story or used another building, yes? But they did not. Who knows why? You see, my friend, the trouble with this place is that there is no view. We cannot see the ocean and we cannot see the rest of the city. Water Garden and Andalus Park, but nothing else. Now from the skyscrapers . . . from the skyscrapers, a person can see everything that happens. Do you see what I am saying? That is our problem." White bars of sunshine crossed his dark face.

Returning the pen and folder to Walid's desk, I joined him at the window. It looked down on the parking lot. "I see what you mean. Not a very good view." I offered my hand. "*Shukran 'ala waktak. Ma'a salama.*"

He nodded at his desk. "You would like a copy of our report?"

"No, *shukran*."

"Thank you for coming. Our American partnership is very valuable. Let us hope it continues, *inshallah*."

Inshallah. The expression—"God willing"—had always irritated me, and now I knew why: It assumed God was on your side.

I picked up my bag. "How'd you know there'd be five explosions?"

He looked surprised. "I do not understand."

"You told me in the alley there were five explosions. But there'd only been four. How did you know there'd be another?"

Walid laughed humorlessly. "What is this? I do not even remember what was said that morning." A twinge of acrimony crept into his voice. "It is not easy to keep things straight on so busy and horrible a day."

Outside, the wind wailed and flags flapped against their posts.

"You are right, *habibi*. Thank you again for your time."

He peered at me as though mulling something over. "You know the Arabic expression *hatha maseroka, lan yufariqka*?"

"Can't say I'm familiar with it."

"If it is yours by destiny, from you it cannot flee."

I smiled. "Who determines destiny?"

He said nothing and I took my leave.

14

I knew I'd have to tell Whitney about my meeting with Walid when, days later, an embassy bulletin landed in our office. The king had deported a British schoolteacher after a black Shiite flag was found in her bedroom; evidence of support for terrorists was the legal justification. It occurred to me suddenly that I too might be declared persona non grata, officially branded an unwelcome person and forcibly ejected from the island. The teacher had been the latest in a long line of recent exiles: Western journalists, activists, aid workers, various and sundry unsavories, anyone who stirred the sand. Informants were probably everywhere. Hell, I didn't need an informant to snitch on me; I'd spit right in a royal's face.

A handful of my colleagues had been declared persona non grata over the years: one for diddling a Russian Svetlana who looked more like a prostitute than the wife of an Army general (a potentially survivable offense in any country except Russia), another for bad tradecraft that proved a diplomatic embarrassment—a "flap," as they called it—openly meeting an informant on the daylight streets of Berlin. Each had lived to tell the tale—forced to do time at Headquarters for a year or so, stripped of his remaining overseas living allowances and any deep cover he might have been using and a reasonable dose of pride, but in the end able to move on with a few irksome scratches. I'd even looked on exiled officers with a bit of envy; they'd officially and formally earned the enmity of an entire country, dug in their heels deep enough to make footprints.

But this was my last tour and I was a fifty-plus man coming in from the cold. Headquarters would have no use for a sick-livered officer expelled from an easy cushioned tour; I'd be walking dead, maybe even fired, barred from collecting retirement, the one remaining thing of value in my life. I had no marketable skills, no human attachments worth their name, no life alternatives. All I had to do was gut out the remainder of my tour the way I'd managed to gut out all my tours—quietly, neither glaringly successful nor unsuccessful, a near-invisible piece of flotsam carried by the stinking sea of sludge. That wasn't so fucking hard.

I had to tell Whitney. He was the only bastard with enough *wasta* and self-righteousness to defend me if the ministry came knocking, who would shake the right hands, get the right people involved, make my sins disappear. Fate-determining, white-collared Whitney. Who never let his people down. Who wore flowing robes and penny loafers and carried the torch of virtue in a debauched garden. Who might as well have been God.

"What's on your mind, Collins?" Whitney crumpled up the paper from his breakfast, tossed it into the trash can. His face had slimmed a bit, I noticed, a few pounds gone from his paunchy constitution, and he was tanner, the probable result of having a woman in his life, someone to see him naked.

"Drove through a demonstration this morning." I sat down at his desk. "Thought the king had banned all demonstrations."

"Was it on Awal Avenue? Between the palace and Grand Mosque?"

"That was the one."

"Pro-government demonstration."

"Ah, that explains *Down with Iran* and pictures of Jassim looking like Clark Gable."

A small muted television in the corner was playing American Forces Network; the ticking headline announced Tehran was protesting an American carrier entering Iran's sovereign waters.

"So there's a real asshole over at the ministry. Walid Al Zain. Think I ruffled his feathers a bit. 'Course, you know how it is—you don't ruffle feathers, you're not doing your job."

Whitney studied me as though devising the most delicate way to elicit more information. "What happened?"

"Oh, you know. . . . I found some inconsistencies in the ministry's report on the Adliya bombings. Didn't match our investigation."

"*Our* investigation?"

"My investigation."

"Your own investigation?"

"Yes."

"What is this, Collins? You never even turned in a report."

"Still working on it."

"What were these inconsistencies?"

"The kind of explosives used."

"You know what kind of explosives were used? You're an expert on explosives?"

"I know a thing or two. Spent a year in Baghdad—"

"I'm aware of your tour in Baghdad."

Whitney's mouth was open, skeptical. His teeth were white—gleaming white—the product of years of unfailingly proper maintenance. Privileged teeth. I had two gold ones in back.

"Chief, you're an educated guy. You got your Ivy League diploma up there on the wall. Has it ever occurred to you the Opposition didn't plant the bombs?"

Whitney put his forehead in his hands, his pudgy fingers nestled in the mop of hair. When he raised his head, he'd turned angry. "Is this more bullshit from SCROOP?"

"There's RUMINT on the street from all different sources. I've got my ear to the ground every day. Out in the slums, the cafés. Stink of it is that dirty old Jassim's behind Adliya." I smiled knowingly. "Gotta admit, Chief, sure makes for a convenient coincidence. Opposition blamed for Adliya, suddenly Said Al Hussein's in jail and protests are banned."

"Collins, do you know how ridiculous that—"

"I know, I know, probably just talk. Everyone's got an axe to grind, huh? Anyway, it didn't come from SCROOP. Guy didn't know a damn thing about Adliya—he was out of the loop for a few days." I got up, pushed in my chair. "Know where he was?"

"No."

"Sea Shell Hotel. Bastard has a weakness for whores. Who'd have guessed? I tell you, these revolutionaries are hypocrites, all of them. At least now we have something to use against him."

Whitney's eyes followed me to the door. "Collins, I trust you saw H.Q.'s latest guidance. SCROOP is officially on probation. He needs to produce solid intel within two weeks or he's gone."

With a reluctant sigh, he picked up the phone on his desk. "I'll let the attaché know the ministry might be calling. I'll take care of it. You were just doing your job."

15

Two weeks. I had two weeks to squeeze good dirt from Rashid. No more moaning about the monarchy's latest injustices, lack of economic opportunities for Shia, the constitution by decree. I needed hard intel. On a burner cell phone I sent a message to Rashid that he should pick up my item at the warehouse at four o'clock that afternoon. In Arabic I scrawled a note on a scrap of paper: *Boss wants you gone. Need info immediately. Make it good.* I shoved the note inside a fire-engine red Mecca-Cola can.

Galali was awash in full morning light as I drove down the coast, the first white flamingos of the season gathering in the Gulf shoals. I checked my rearview mirror. As usual I saw only rusty labor trucks this early, their undernourished human cargo hanging over bed railings. Cover stops at the Ritz and a few other luxury hotels, places willing to serve alcohol before noon, then Gudaibiya to shop for rugs. Between stops I thought about Almaisa. I hadn't been able to reach her since our date. When I'd finally driven to her studio and knocked on the door, there was no sign of life, the only response neurotic barking from the mutt in the alley.

Downtown I conducted a brief foot route, finishing on Ma'n bin Za'idah Street, a deserted passage of tall yellow flats, a half step up from the slums. Two blocks away teenaged boys were doing amateur motorcycle stunts, but they were too distant for concern—if I couldn't see their faces and fine motor movements, they couldn't see mine. Down the alley I walked, avoiding piles of trash, feigning consumption of the Mecca Cola. One, two, three windowsills. There it was—the one with the bent iron bar and dirty lace curtains. Always drawn. I left the can on the sill.

At the end of the alley I turned right, trekked nearly half a mile to the inaptly named Café Rouge, which was neither red nor particularly French. The sun had reached its noon peak. I sat outside and ordered mussels, which were too chewy and salty to eat, so I ordered a beer to take their place. Where the fuck had Almaisa gone?

I knew Rashid had picked up the dead drop when I got a message a few days later from an anonymous number saying my item was ready. We were scheduled to meet at a godforsaken patch of land Rashid had chosen at our last meeting, a soupy mangrove swamp between Tubli Bay and Sitra he assured me was easy to locate and far from inquisitive eyes. I'd never before allowed him to choose the meeting location, but he'd kicked up such a fuss over the last site that I'd decided to throw him a bone, make him feel coddled and appreciated.

Rashid was right: No one would find us. The swamp stretched for miles, all the way to the Gulf, a scraggly half-vegetated wasteland. A fetid sour stench permeated the air, some combination of swamp gas and runoff from the Tubli sewage plant. Ten years from now the place would all be sewage. I parked my car and headed fifty paces south, easily spotting the small island that was home to a flock of orange-beaked Caspian terns, shielded on three sides by tall mangroves.

By the time Rashid arrived, my shoes were caked in mud and the terns had scattered restlessly across the island. "Is this payback?" I dropped my cigarette into the muck.

"*Habibi*, I am following the rules you taught me. Tradecraft." He'd shaved once again, I noticed, the youthful fervor returned to his face.

"Well, this is bullshit. No place to sit. Too hard to take notes."

Rashid shrugged. "You do not like it, we go somewhere else."

"Fine. Let's go."

He followed me to my car. His car was nowhere in sight.

"I know the perfect location." Rashid slipped into the rear seat. "Not far. There is something I want to show you."

"I thought you had intel for me."

"You trust me?"

"No. No, I do not."

Alley upon alley through Isa Town and Jurdab and then, before I recognized the terrain, we were on Avenue 49, south of Riffa Views. "What the fuck is this?" I slowed to a snail's pace. "People I know live here. Expats. Ministry officials."

Rashid shook his head. "You are mistaken. A few unimportant ministry officials, yes, but most of the foreigners live in Amwaj Islands. This I know."

"You going to tell me where my people live?"

"Who lives here?"

"My boss, for one." I nodded toward Hibiscus Lane, its entrance barely visible behind the iron gate. "Right down that street."

Rashid craned his neck. "Very nice houses. Your boss is important."

"Not as much as he thinks."

"I am guessing . . . the big house at the end."

I couldn't help chuckling. "You got it. The biggest one." The villas glided by in a pristine stucco parade. Glancing in the mirror, I caught Rashid gazing out with a faraway look on his face, a combination of longing and disbelief.

We were approaching the eastern end of Riffa Views, near Jimmy's house, and I grew impatient. "What the hell do you want to show me? This place is too hot for comfort. We need to leave."

"New construction," Rashid replied matter-of-factly. "The king is building a new palace."

I pulled over sharply. "Fuck. I'm not gonna risk us getting caught just for a glimpse of a fucking castle."

Rashid was looking abstractedly toward a vacant lot at the end of a cul-de-sac. "There it is. Nothing to see anyway. I do not think he has started yet."

I turned the car around angrily.

Back at the mangroves Rashid offered a weak smile. "I am sorry, *habibi*. I just wanted you to see . . . Al-Hakim . . . he builds hundreds of palaces so he can hide. He sleeps in a different one every night. Afraid of his own people."

I turned to face him. "I don't give a shit how many palaces the king has. You have intel for me? If not, I'm gonna have to fire you."

"Yes, *habibi*. *Asif.*" He paused, lit a Canary Kingdom. "There will be a car bomb next week. Near the Naim police station. Not to hurt, just to scare. We do not kill innocents. We do not lower ourselves to the level of the king."

I looked up from my notepad, surprised at so valuable a piece of information. "What street?"

"900 block of Suwaifiyah Avenue."

"Vehicle?"

"Kia Picanto. Silver."

"Plates?"

"No plates."

"How'd you get this?"

He grinned unexpectedly. "I have been raised up. Promoted, as you say. Now I see details of operations. Not everything, but more than before."

"It's about time. *Mabrook*." My spine tingled. Even better than I'd hoped.

I pushed further. "Who's in charge of this car bomb?"

He offered a few monikers; real names unknown, he claimed.

I raised my eyebrows in amusement. "And you? Now that you've been promoted, what's your moniker?"

He looked away. "I do not have one yet."

It was a lie, but I let it go. Everyone was entitled to keep a few secrets.

I leaned back, studied him. "You're not afraid to tell me this? What if the police find the bomb before it goes off? Your men will know there's a leak."

Rashid waved his hand. "It is a stupid location. I tell them this. Next to a police station is no good. It would have been discovered. They want to show the police how powerful they are, but they are taking a risk. I say to myself that I will tell my American friend since he needs this information."

His countenance grew bitter. "We deserve to have our bomb discovered if we are so stupid as this."

"Well, you're doing the right thing," I assured him blandly. "We're on the same side—America and the Opposition, you and I. The more intel you give us, the more it helps your cause." They were canned lines, the obligatory feel-good script every spy keeps in his pocket, words I knew better than the Lord's Prayer.

I put away my notes. "Keep this up, maybe you'll get a bonus."

Tuesday of the following week local EOD teams swooped down on a silver Kia, no plates, parked quietly on a street near the Naim police station—not Suwaifiyah Avenue nor the 900 block but close enough. With a bit of American assistance they successfully conducted a controlled detonation. Everyone emerged safe and sound, home that evening to a hot meal.

Whitney was thrilled, Jimmy impressed, the Admiral dutifully pleased. *Impactful intel, Collins. Nice work.* I minimized the approbation. *It's all SCROOP.* Rashid and I—informant and handler—were one, the same blood running through our veins, our successes and failures synchronous. Rashid was now safe inside the CIA stable, more than safe—prized—and so, for the moment at least, was I.

Headquarters sent a congratulatory cable lauding me for the lifesaving information and nominating me for an Exceptional Performance Award. It wasn't much—the lowest official accolade, a certificate accompanied by a small bonus—but I'd never gotten one before.

The incident was put to good political use, Jassim's well-oiled media machine cranking out an embarrassing number of headlines on the Opposition's increasingly dangerous tactics, the failed bomb, the superior prowess of local officials. Not long after the news broke, my developmental contact in Al-Wefaq, Rahat, called to let me know his society officially condemned the attempted vehicular attack and that it planned to explore an alliance with Al-Asalah Islamic Society in order to widen its base in Muharraq and Riffa. Whitney gave me a fat thumbs-up when I submitted the information, said it was great "soft" intel, just what policymakers needed to predict the course of government in Manama and even the Arab

Spring. In the blink of a dirty cop's eye I'd gone from producing pebbles to rubies, from a drifting insubordinate to a valued soldier. Maybe, if I wasn't so old, even a rising star.

And then Almaisa called. She'd been traveling, she said, seizing a last-minute cheap fare to Italy to view a new exhibit at the Museo d'Arte Contemporanea di Roma, a name she rattled off with alluring ease. She'd visited old friends, connected with new patrons, collected payment for a piece that had finally sold in a local gallery. Her contralto voice was smooth and canorous over the phone, each word tumbling out like fine spiced wine.

16

Almaisa and I began spending every weekend together.

She showed me what she called the "hidden" parts of Bahrain—parts, she said, tourists never bothered to see. Fishing villages and their wooden dhows, the ancient traders' pearl trail, crumbling blue-tiled houses near the old handicrafts market. Despite the fact that most of the attractions were lackluster and uninteresting, hardly worthy of inclusion in any real country's tour, I played the appreciative audience. The idea that she wanted to share it all, and equally that she had such understanding of and affection for her land, was intriguing. I couldn't muster much more than a modicum of sentiment for my own homeland, which seemed a hell of a lot better than hers.

Our hours together multiplied, and increasingly we spoke in Arabic, yet we seemed to make minimal progress. We still drove separate cars—she was skittish about being seen together in close quarters, I knew—and at times she was distant, like an acquaintance or even a stranger. Other times her natural forthrightness took over and she assailed me with questions that bordered on interrogation. What was my job that I spoke such good Arabic, and why did I hide my linguistic talents? Was I more important than I let on? How had I become a diplomat? Someone with my experience must have an opinion about who would win in the Arab Spring.

Through it all, I found myself whitewashing, erasing, diminishing. For the first time I wasn't tempted to use my profession to enhance my appeal, as an enticement to sex or another date or virtually any object in which the mystique and allure of espionage would aid my cause. I was, in fact, consciously avoiding the truth, suspecting it would drive her away. Maybe because of her childhood, the knowledge that intelligence officers, those in my chosen profession, had stolen her parents in the middle of the night.

Almaisa, truth be told, wasn't easy to understand, to handle. She had none of the triviality or false femininity of American women; neither did she have the humorless affectation of European women. She was moody, often argumentative. I gathered she didn't adhere to any wholesale religion or belief, instead picking and choosing in a manner that only revealed her ambivalence. Stubbornly she was wedded to certain principles and not others, an uneven life view that stretched even my bounds of tolerance (ordinarily and selfishly generous): No alcohol, no pork, her body perpetually covered, and then she would skip daily prayers and wear the most flamboyant attention-attracting abaya from here to Malaysia. A feminist, some might call her (though one, I learned, who recoiled from the label), who drove and worked, cavorted around Bahrain with a single American man and was herself not married, who refused to subject herself to the traditional Gulf woman's life of subservience and confinement. A Muslim man, she admitted more than once, would tie her down. She was the living product of East and West, a combination that often seemed as fraught with conflict as the two hemispheres.

Despite her uneven strictures—or perhaps because of them—I was determined to get somewhere with her. I could do it. Seducing a woman was not unlike wooing an informant: You divine her wants and needs, decode her vulnerabilities, align yourself (sometimes work against her if friction's her thing), make yourself indispensable, fill the gaps and crevices. You do this until you know her better than yourself—better than she knows herself—until the only answer is yes.

I introduced her to Sinatra. She said she'd heard the name from her mother. I told her about my childhood in Jersey, how Sinatra was more revered than Junaid in Bahrain. We'd sat in her studio next to the ancient record player I'd retrieved from a decades-closed box and played my old records. "I've Got You Under My Skin." "Strangers in the Night." Her favorite was "Moon River." I told her it was a good if predictable choice.

When I mentioned I'd taught myself to play guitar, she'd stared at me incredulously, insisted I prove it. I resisted a bit before giving in, realizing that my musical ability was meaningless and the mere attempt would please her, maybe even endear me to her, certainly carry me further toward my goal. So I unearthed my old guitar, played the Stones and Springsteen, sounding better than I'd expected. Playing music for her had thrill in its novelty, like I was hearing it myself for the first time, its euphony fresh, a first sip of whiskey before it descends from pleasure into routine into necessity.

"Sinatra." Almaisa had repeated the name as though trying to brand it on her mind. "You wanted to be Sinatra." She'd looked at me. "So you became a diplomat instead."

One night I cooked dinner at my villa.

Temperatures had moved from tolerable to this side of pleasurable, and we'd eaten on the deck. I'd managed to cobble together my mother's recipe for spaghetti and meatballs adapted from our First Ward neighbors, had even lit a few candles. More effort than I'd gone to in a long time.

Almaisa, it became clear, was in one of her stormy moods. I'd been treading cautiously, careful not to disturb the mines she sometimes buried, our conversation comprised solely of cocktail-party morsels. I was determined not to endanger my progress by a misstep.

Toward the end of the meal Almaisa fell silent, squinting at the horizon. Her face grew troubled. "Look. Those lights in the sky. Do you see them?" An insect-like whir had become audible.

I shrugged. "Just helicopters. Probably an escaped prisoner. We're close to Dry Dock."

"*Sojana*? Prisoners? Is that what you call them?" Her eyes fairly glowed. "*Khara!* The word 'prisoner' is wrong. That means someone who is punished for committing a crime."

I gulped my wine, annoyed by her sudden irritability, the platonic direction of our conversation. The shape of her breasts was barely visible beneath her garment, round and inviting.

"Do you know what most of the *prisoners* inside Dry Dock have done?"

My impatience grew. "Well, let's see, there's the standard stuff: killing and stealing and drugs. And then I guess there's Sharia: no defamation of the prophet, no criticizing Islam. That about cover it?"

"No, it does not." Her eyes were trained on me. "Dry Dock prison is filled with people who have done nothing. Men, women, children—*children*, do you hear me?—who have not broken any laws. Do you know what they are guilty of? They have called for freedom, for justice, for food. They have marched in the streets and spoken loudly against the monarchy. And the king calls it *hartaqa*. What is the translation?"

"Heresy. Apostasy."

"Yes. *Apostasy.*" She said the word as if it dripped with poison.

I put down my fork, nodded gravely, recalibrated my temperament as I did when reining in and appeasing an overly impassioned informant. "No question about it, Jassim's heavy-handed. He's had doctors arrested simply for treating protesters."

She looked at me, surprised. "Yes, exactly. You . . . understand."

"They were just following the Hippocratic oath, doing their duty."

And now she was gazing at me with appreciation and something else—affection. I smiled, got up from the table. "How about espresso? It's from Italy."

"It looks very nice in here." She stood in front of her mosaic. Beside her I could feel the heat from her fully draped body, smell the sharp weight of her perfume.

Deliberately she strolled around the room, taking in the cathedral ceilings, the five-tiered crystal chandelier that came standard with every villa in Floating City. I felt faintly embarrassed by the décor, a décor that was all show—marble floor that was, in fact, just painted tile, wall paint so cheap it chipped from a touch. Somehow this made it worse, the pretense more damning than raw pageantry. Even my worn leather sofa, succumbing to her running fingertips, seemed more supple and luxurious than it really was.

"You are lucky. You can read whatever you want." She'd stopped in front of my books. She spun my globe lightly, traced the gold equator. "This is beautiful. I do not think I have ever seen a globe so fancy."

"Thank you." I hesitated. "My son gave it to me."

She turned around. "You have a son?"

"Yes."

"How old?"

"Twenty-five. No, twenty-four. Just graduated college."

"And what will he do now? Travel the world like his father?"

"Who knows? Jobs are hard to find these days."

She studied me. "You do not talk to him?"

"Listen, you want more espresso?"

She raised her eyebrows as if a thought had suddenly dawned on her. "You are married?"

"No. Divorced."

The globe stopped spinning. "Of course. You Americans are all divorced."

I moved close to her. The wine from dinner was running through my veins, making me pleasantly warm, uncaring. Standing in front of her, I felt a painful surge of desire, the huge disparity in size between us creating a jagged primitive yen, a gulf to be crossed, all the more enticing because of its distance. She was natural and pristine and had the aroma of untamedness. I reached out and grabbed her thin arms—my hands, monstrously large by comparison, struggling to find bone beneath the endless folds of fabric. I pulled her to me and kissed her, a bit savagely, maybe; my patience had long since evaporated. Her mouth was soft, reluctantly compliant. She took a step back. I moved forward again, closing the gap, raised my hand toward her head. Then I stopped, lowered my arm. "Take off your scarf."

Arms hanging limply by her side, she looked at me. Slowly she tugged the back of her scarf, layers of green falling away like a curtain. A titillating cascade of deep brown waves tumbled out, slightly messy from being covered all day, the ends spilling across her narrow shoulders as smoothly as liquid, so thick it seemed to dwarf her frame. The full length of her scar was visible. Instantly the simple act made her mine—a view, an angle, that belonged only to me.

She was an entirely different person. Without the protection of her hijab she looked even more undomesticated, ravishing. Yet she had a vulnerability, almost docility, that I'd never seen before.

I picture her like that now: standing in my living room, hands at her sides, green abaya slicing the drab background while the moon's weak residue illuminated half her body, staring at me as though waiting for direction, on the cusp of surrender.

17

October came, nights grew cooler, Saudi Jet-Skiers disappeared from beaches. The Opposition and monarchy continued to clash like small animals, shedding bits of fur here and there. Rashid complained that Shia were being excluded not only from Intelligence, Interior, and Foreign Affairs, but every government ministry. Despite the ban on protests, Shia took to the streets. Manama's smog grew heavy and ash-filled, hovering like a rain cloud above the city. Conflicting reports issued from the battlefield, each side racing to claim the last dead body as its own. Expat galas rolled on their merry way. I saw Poppy Johnson only in passing, her blond disfigurement surfacing at official events, her name on the tongues of bunco enthusiasts.

Almaisa and I had slipped into a relationship. I'd largely deciphered her code, the asymmetrical pattern of thoughts and behaviors that comprised her. I'd learned to skillfully walk the line between the respect she craved and the direction she needed, to accept the tethers that straddled half of her and the worldliness that unfastened them. Not so different, after all, from the delicate give-and-take dance with an informant, an unending alternation between obeisance and control.

And my progress was visible: a staircase of concessions.

I'd convinced Almaisa to go out on occasion without her hijab. Show her lovely locks to the world. I'd seen enough Arab women on planes swap their abayas for Gucci pantsuits midair to know that even the most conservative secretly longed for some degree of glamour and pageantry. (And hadn't Almaisa asked more than once where American women shopped?) Cunningly I'd drawn an admission that Almaisa's mother had never worn a hijab, that it was nothing more than custom, the Quran silent on the subject, that she mostly wore the garment to blend in rather than out of religious conviction. Despite Almaisa's disdain for Western mores, her aversion to becoming like my female compatriots (whom she accused of hedonism and exhibitionism—and was she in truth so far off?), she eventually gave way.

It wasn't long before I'd persuaded her to ride in the car with me. Stay later at my villa. Even try a sip of wine. She'd made a face, said it would be her last.

The ultimate hurdle remained and I hadn't figured out how to jump it. Sex would require a more nuanced approach. It remained a maddening mirage, appearing within reach at times then receding into nothing.

Still, I was enjoying the game. A spy was a spy, and at fifty-two I could still lure a fish into my net. I hadn't lost it—that ability to please, to find the soft and hard spots and get inside.

18

We were in Almaisa's studio one afternoon when it occurred to me that the young girl in the photo was her. The girl had the same wavy dark hair, the same intense features, and her face, despite the roundness of youth, was sharper and more probing than one would expect in a child, a face of disproportionate awareness.

Almaisa saw me staring, walked to my side. "It is not me." Her denial seemed only half-insistent, as though she herself didn't quite believe it.

"Her name is Sanaa." She touched the photo lightly with her fingertips. "She is my favorite."

"Your favorite?"

"Yes. In the orphanage." She walked back to her worktable. "I teach art in an orphanage, you know."

"No, I didn't know. You never told me." I realized now that the child had distinctly round eyes quite unlike Almaisa's almond-shaped ones.

"In Baghdad I spent two years in an orphanage before my uncle could get me out. There were many papers—problems for me because of what my parents had done." She started to move an easel. I went to assist.

"I wanted to help boys and girls like me." She looked up defiantly. "The Quran says, 'It is an uncaring person who drives away the orphan.'"

"So where is this orphanage?"

"Not far from here. On Diraz Avenue."

"I'd like to see it."

"Then I will take you there." Standing on tiptoe, she kissed me squarely on the mouth. "I knew you would want to see it. *Anta rajol tayib.*" *You are a kind man.*

It was a nondescript building whose only indication of purpose was a small sign labeling it BEDAYAT JADIDAH. New Beginnings. Corrugated metal sheets covered the one-story structure, rusty and graffiti-spackled in spots, breaking here and there to reveal cracked stucco. No windows in front. Laundry lines from an opposite house straddled the rutted street, swooping down to the orphanage roof.

The interior was dark and musty and smelled like antiseptic. A faded Persian rug covered the foyer's chipped tile. Better than the streets, I supposed. The recesses of the building were brighter, rear glass doors allowing a measure of sunshine, opening onto a backyard with a playground and scattered toys.

Here was the dining room—she turned on the lights of a sparse room packed with long tables, cheap folding chairs, and a handful of high chairs, walls painted a suggestively cheerful yellow. Here was the kitchen—no more than a cubbyhole, every inch filled with supplies, edible and inedible, perishable and nonperishable, toilet paper and canned food and diapers. Bedrooms were at the end of the hall: girls on one side, boys on the other. A carpeted playroom (for when the air is too dusty), maps and drawings plastering the walls, homemade planets hanging from the ceiling. And here was the classroom. She motioned me to a door with a window. Rows of students sat obediently at desks listening to a large woman in a heavy abaya drone on about numbers, some kind of math lesson. *Twenty-two children in the orphanage*, Almaisa whispered proudly. *Aged two to fourteen.* One of the students, a girl no more than seven, had lost an arm, her biceps ending in a trickle of flesh.

The large woman gave the apparent signal for dismissal and there was a squeaky commotion as the kids hurriedly gathered their books, pushed in their chairs. Almaisa glanced at me with alarm. "Do not stand so close to me. It would not be good for the children to see."

I grinned, feeling rather like a naughty child myself. "So who am I?"

"You will say . . ." Almaisa nodded to herself, clasped her hands. "You will say the truth. That you are an American diplomat who wants to see an orphanage in Bahrain."

It was surprisingly brilliant. First rule of tradecraft: Keep it simple.

"You got it." I winked, tucked her loose locks beneath her scarf.

A second later kids were pouring out of the classroom like ants in search of food, the hall filled with cries of delight, Almaisa's abaya rumpled by dozens of hands and embraces. The meek of the world adored her. It was as if the kids hadn't seen her in months, though she'd taught a class the day before. *Ommi*, they called her. She greeted the kids in syrupy coos, her Arabic too fast and colloquial to fully decipher. I understood only that she'd introduced me in passing, but the inquisitive faces were easily satisfied. A few shook my hand, their palms warm and gritty, grins full of jagged brown teeth. *Amriki!* they shouted elatedly as though it were a rarefied occasion and I someone famous, the President or an actor. One scrawny boy presented his notebook to me for my autograph.

Then, from behind the crowd, the girl with one arm emerged, edging her way shyly toward Almaisa, and I saw it was Sanaa, the child in the picture. Skinnier now that she'd lost her baby fat, but the same dark curls, same large round eyes. The photo, I now realized, had been taken from an angle to hide the missing limb. Throwing herself around Almaisa, the girl held her prolongedly with one arm, closing her eyes as though trying to shut out the world. Almaisa stroked Sanaa's curls unthinkingly, still talking to the others, asking about their studies, when their next excursion to the wildlife park would be.

We ate lunch with the children, tepid chicken and rice and overly boiled vegetables, less a meal than a cacophonous question-and-answer session in which I regaled the eager faces with American pop culture trivia, pretended not to know or understand differences between my country and theirs, an exhausting exercise in performance and patience. Halfway through the meal the power went out and we were left to eat in darkness, an apparently common occurrence, and the kids, ordinarily impatient with this limitation on their freedom, instead enjoyed the novelty of a stranger experiencing the idiosyncrasies of their world.

The visit lasted through the afternoon, and when we drove back, dusk was dropping its veil. "I am glad you came." Almaisa put her hand on mine. "You see how special the orphanage is."

"The kids love the place. They love you."

"That is not what I meant." She turned and I could feel her penetrating gaze. "The children are different from children in other *dur al-iytam*."

"Different how?"

"Surely you see. Many are orphans of the revolution. They have lost their parents in the struggle—death or prison."

"I didn't realize."

"Also, there are others. Children of *ajanib*. Mothers from Nepal, Philippines, Sri Lanka. Women who work in your villas. You know them? They stay in Bahrain for many years and sometimes they have children. Of course they cannot keep them. And sometimes their owners make them pregnant and force them to hide the baby."

"No other place for the kids to go, I guess."

She glanced at me sharply. "Of course not."

Lights were coming on, alleys melting into darkness. I turned onto Almaisa's street. "How'd Sanaa lose her arm?"

"Infection. *Al-ghangarina*. She was hit by *shazaya* during a protest. Her parents were killed. She did not go to a doctor when she should have. When you do not have parents, no one tells you to go to a doctor."

"Ever thought about adopting her?"

Almaisa frowned, turned away. "I cannot."

"Sure you could—"

"I tell you no. I am young. Unmarried."

We'd reached her studio. She had yet to reveal where she actually lived, answering my questions only with vague directions and general descriptions of a second-floor flat near Al Sater market. She was either embarrassed—the life of an artist hardly lucrative, her residence probably a few boards away from disintegration—or the chains of propriety still shackled her, the slums less hospitable to unmarried trysts than the world of expats.

Almaisa was gripping her purse nervously. "Shane, you must listen. The children who live in Bedayat Jadidah are a secret. On the outside the orphanage is just like the others, but on the inside— The king would not allow this kind of place. He does not admit the war has left children without parents. He would order it closed. Maybe imprison the orphans. You cannot tell anyone."

I offered my most faithful smile. "I can keep a secret."

19

Rashid had momentarily lost his scowl. Even as he sucked his Canary Kingdom, a waxy smile was spreading across his face. I'd just told him he'd earned a raise, an extra five hundred a month. His intel on the car bomb and Fourteen February's plans to covertly approach the king's reformist brother had proved beautiful music to Langley's ears. That, plus Rashid's promotion from planning officer to operations officer, perched now on the highest tier of Fourteen February, made for a banner month.

For the first time we were meeting in Rashid's car, an ancient beater in worse shape than my own. He'd offered a reverse car pickup, something I didn't usually allow, but I could see it gave Rashid a cheap thrill, the perception of advancement and expertise. He was growing nicely into his role of informant and, according to the steadfast laws of espionage, I was reaping the harvest, growing concurrently into my role as successful handler. *A win-win situation*, as Whitney liked to say.

Low music was playing on the radio, a wordless Middle Eastern tune, and the safety signal, a rolled-up *Gulf Daily News*, rocked gently on the dashboard as though in time. Rashid fingered the wad of dollars, a kid with a new toy.

"Don't spend it all on one revolution."

He laughed, held up the money tauntingly. Suddenly, he looked at me with amusement. "*Habibi*, there is something different in your face."

"No difference. If I ever change, I'll let you know."

"I think . . ." Rashid leaned back in the driver's seat, cocked his head. "You have met someone, yes? A woman?"

I muttered something noncommittal, how they come and go. His powers of observation impressed me. Every now and then he could poke his head above the world of revolution.

"A woman is good for you," Rashid was saying. "Every man needs a *mohtarama* woman. Not loose and filled with sin like the American women I see going in and out of the Dilmun Club. Clothes that are tight, always with a different man—" He broke off his lecture, stared into the distance. Lights inside houses were going dark—a signal, sometimes a warning, Rashid had once explained, unsuspicious because electricity went out nearly every day in the villages. Smoke was darkening the sky.

"*Ghabiyin*. The protest was not supposed to start until nine o'clock."

"What difference does it make?"

"Because, *habibi*"—he said the word with a sneer—"I wanted to be there."

I checked my watch. "Well, my friend, you're in luck. Time to head back. Maybe you can still make your riot."

Rashid started the engine then gestured toward the highway. "I cannot drive you."

Eastbound Budaiya Highway, I now saw, was entirely blocked, cars frozen in the rising protest.

Parked north of the highway, we had a direct view of Bean Coffee, the Opposition's favored rally point following the king's demolition of the Pearl Monument last summer (Rashid joked it was only a matter of time before the king, in his desperation, destroyed his own palace, the army barracks, the entire downtown). Dozens of times I'd passed Bean Coffee, even pausing for a drink once or twice as a cover stop. A luckless location, prey to the pollution and noise from the Fish House roundabout, encircled by piles of trash in neighboring alleys that, when combined with the coffee, created a uniquely noxious aroma. Its terrace walls were almost entirely black with government censorship paint.

Rashid pointed out members of the sprawling array of dissident groups that had gathered on the café terrace, articulating their names in proportion to how highly he thought of them—men from Fourteen February,

Al-Wefaq, the Opposition societies—recently coalesced into a single movement they planned to call Tamarod, or Rebellion, a name inspired by Egyptian revolutionaries seeking to remove their autocratic president. Rashid nodded toward a car parked a block away, a plain brown beater with no sign of life, waiting on heavy tiptoes. "Surveillance."

"How do you know?"

"We are good at watching, *habibi*. We have eyes here, across the city. Even someone on the inside." He grinned. "*Ghabiyin*. Fools. In the slums they use the same car every time."

With a loud crash a dumpster was overturned. Almost instantaneously the protest had swelled to more than twice its original size. Hundreds had apparently heeded the silent call, appearing out of nowhere, massing before the coffee shop in a large fan-shaped formation, small groups revolving around the periphery like angry satellites.

Traffic on the highway was steadily backing up, blocked by a pile of tires, a few trash bins, and the shell of a Kia, all set aflame. The rubbish burned easily, leaving a menagerie of frames and skeletons. Curls of smoke crept upward like goblin's fingers, devouring the papery fronds of palm trees in ravenous mouthfuls. Ash was falling for blocks, turning the dingy villages even dingier, and the stink of char seeped through Rashid's leaky car.

Wordless twisting music continued on the radio. I had a vague feeling I should be concerned for my safety—the distance between our car and the flames wasn't much—but somehow I couldn't muster more than a passing thought; the clash still seemed miles away, on the other side of a fine pair of binoculars. Reluctantly I took the cigarette Rashid offered, paltry compensation for the dismal show.

Armored personnel carriers began to roll in from the periphery with their ready shipments of riot police. I waited for action, some spark to liven the evening. Masked men began scurrying to and from the conflagration, adding tires and kindling (looking rather like graceful javelin throwers), careful not to slip on the road they'd doused heavily with oil to spread the flames and stop the police. Shabby clothes, smudged and sweat soaked, hung from thin angular frames. Riot police in smart blue uniforms with armored vests and helmets began exiting the vehicles. A tall protester was piling cinder blocks atop a barricade, while another faced the oncoming forces with hands spread open like a long-awaited messiah. A third man was scrambling rather poetically to hoist the Shiite flag.

Cries began piercing the air like tracer fire. *Oil belongs to everyone! Equality! Representation! Jobs!* Voices throaty with smoke and persistence, somehow managing to triumph over the cacophony. A determined swell was moving toward the APCs. *Makhnouk!* they yelled, a unique Arabic word that means "suffocated." Then, *Bread, freedom, and dignity!*, the battle cry borrowed from Egyptian dissidents in Tahrir Square.

The disgruntled flames were growing now, darting across the road in undisciplined waves, rising slowly like a thrashing multiheaded beast, and the APCs began enclosing the protesters. A few lithe men on the outside were throwing Molotov cocktails, and every few minutes one would land along the perimeter with an amateur-sounding explosion and a brief teepee of flames. The circle of police vehicles grew thicker and tighter, obscuring our view of the violence as though shielding us from Dante's inner ring of hell.

Then we heard birdshot, its whimsical cascade of crackles and pops. More cries, shrill and tortured. Through cracks between the metal dinosaurs I could see batons and arms moving strenuously like threshers in a cornfield. The second salvo: beatings.

"Police hit the healthiest and strongest men first." Rashid blew smoke languidly across the dashboard. "They are the greatest threat, of course. And you see those policemen in back? They are taking pictures. So when they show them to the world, it looks like we have many big strong men. Then . . . you see? The weak men come next. Women and children—" He paused. "You will see what happens to them."

Police had begun removing large black canisters from the vehicles. The canisters had an unmistakable heft. The third salvo. It was this weapon, the reliable last resort of every Gulf regime on its urban battlefields, that I, like any sane expat, sought to avoid. It crept out of nowhere and didn't hit you until minutes later when thousands of needles began slicing your eyes. "Tear gas," I muttered.

Rashid shook his head. "No, *habibi*. Tear gas you cannot smell. Wait. This gas . . . you will smell."

Within seconds an odor of rancid eggs had penetrated our car—sharp, bitter. I blinked, coughed without effect. "What is this shit?"

Rashid shrugged, still smoking. "We do not know. It is new."

Weeks ago we'd received a memo via the Naval Attaché. It came from the Ministry of Interior and was titled *COMPLIANCE WITH INTERNATIONAL LAWS AND NORMS*, describing the lawful and defensive (always defensive) measures employed by the government against the Opposition and enemies of the state, using language like "least violent means" and "fully compliant." Buried in a section called *RIOT CONTROL AGENTS*, I recalled, was a paragraph preparing us for this new substance: an experiment of sorts, a boutique concoction developed over months of battling unrest. *MORE HUMANE THAN TEAR GAS*, the report had assured.

My eyes had started to liquefy, the inside of my nose to smart as though I'd plunged into a pool without holding my breath. A touch of riot control agent here and there wouldn't do any long-term harm, expats liked to reassure each other. Living in Budaiya, Poppy complained about tear gas as though it were bad weather—ruining kids' sports matches, terrace dinners.

"Now come the arrests," Rashid said matter-of-factly.

The police had rounded up some twenty protesters, including a few of the men Rashid had identified and three children who had been energetically rolling tires. Cuffed and stooped, the prisoners were crammed into vans. An early moon had emerged, shrouded by the poisonous gas like a fine gray mist.

"They go to Dry Dock. No bail, no trial. Martial law."

"Where do the kids go?"

"Same as everyone else."

I recall the next moment clearly, like a discrete scene from a dream, separated from the surrounding blurriness by its relevance and quintessential quality. The boutique gas was clogging every pore of my skin, filling my throat, pushing rivulets of sweat down my neck and back, even my forearms. Face flushed, heart beating like a delirious drum.

A small group of protesters was standing its ground like wolves defending precious territory, stubbornly adding to the kindling, throwing calls for rights and equality into the night like inert grenades. Rashid was muttering to himself, searching under his seat for something. Suddenly, he had two thickly knit black face masks in his hand. "Here," he said simply. "Put this on."

Floating in a strange chemical-induced stupor, I pulled the mask over my face. Rashid had crawled to the back of the car, rummaging around the trunk. When he returned, he was holding a glass bottle with a dirty rag stuffed inside, one end hanging limply over the top.

The haze seemed to paralyze me. "You put that thing the hell back," was all that came out. Inaudible, the mask muffling my voice. "You . . ." My throat burned. "You can't carry that shit around. Checkpoints . . . you know what would happen if you got caught at a checkpoint?"

Rashid lifted the bottle toward me in a toast. "This is the best part, my friend." Climbing across the seat, he opened the passenger door, gave my shoulder a nudge. "Get out. You will see. A story for your friends back home."

Inching toward a low cement wall, he motioned me forward. I crawled behind him. Even through the mask the acrid fumes were intolerable, constricting my lungs like a tourniquet. I felt dizzy. Rashid handed me the bottle, pulled a lighter from his pocket. Before I could relieve myself of the bomb, he'd lit it, the golden flames leaping into the air like escaped prisoners. "Throw it!" he shouted.

I wonder now why I didn't hesitate longer, why I was so quick to lob the thing. An infantile reflex, perhaps, the irresistible urge to throw whatever is in your hand—something illicit at that—with impunity. The same impetus that propels all of us, at one time or another, to perform spontaneous bad acts for the sheer visceral thrill of exercising free will and ability, a reminder of the red blood that courses through our veins. The pleasure of wrong, the tingle of good and bad hopelessly inverted, tangled until unrecognizable. No one watching, God on your side.

With a modest crash the bomb landed outside the vehicular perimeter, erupting in a line of flames that died almost instantly. No one was hurt—of that much I'm certain from replaying the moment dozens, maybe hundreds, of times in my mind. I'd deliberately thrown the thing outside the epicenter. Hell, even if it had landed closer, it probably wouldn't have hit anything. Most Molotov cocktails were just for show. If Rashid had thrown it, someone might have gotten hurt. He had better aim, purposeful aim. Purpose that I did not. The police hadn't even noticed my bomb.

More chemicals were apparently needed. Officers were lugging additional canisters from their vehicles, and Rashid and I hastened back into the car. Methodically, efficiently, police sprayed every building around the traffic circle. Shop after shop filled with thick clouds, its occupants streaming outside like hundreds of cockroaches. Stumbling, coughing and sputtering, blind to the damage they would find when they returned. A Filipina clerk wearing a red bandana ran out of her cold store, screaming, clawing at her face.

The extermination worked. Eventually the last protester fled the scene; the burning tires melted to the ground. Bodies lay motionless, dead or comatose. Those remaining and injured would make their way to a nearby makeshift clinic, Rashid said.

The surveillance car peeled out of the alley, disappeared into the brown squalor.

20

That evening I was forced to spend an uncomfortable night at Rashid's. Curfew was close and we couldn't risk getting caught on the ride back. I should've known the revolutionary had a plan for all emergencies; Rashid had built an underground shelter of sorts on his property (most homes in the slums have an escape route or alternate entrance, he told me). Really no more than a concrete basement, hardly hidden from the police, but accessible through a separate back door that I could enter and exit without his family seeing.

The room was small, with a sharp odor of dirt and rat droppings. A prayer mat was rolled up in the corner alongside a tattered Quran and a handful of books and articles about revolution—Marx and Asef Bayat and Rached Ghannouchi, plus an amateurish tract entitled *Al-Ibadah* written in Arabic but published in Tehran (clearly contraband, fair game for imprisonment if found in his possession). Rashid had an irritating penchant for philosophizing about revolution, quoting obscure intellectual tracts, and I now saw the well from which he drank.

Despite exhaustion, I barely slept, passing the hours on the thin worm-eaten mattress, images of the protests intruding into my thoughts—and Almaisa, her soft body asleep in her flat, the flat I still hadn't seen. Then Almaisa as a child came to my mind. I saw the intel officers entering her house, throwing her parents against the wall, handcuffing them, maybe pushing the little girl down in the process. Was that how she got the scar?

Inflicted by an overzealous spook, one poisoned by years of deception and cruelty, too long in the shadows.

When the world was muzzled by the folds of night, I crept outside for a piss and breath of fresh air. Rashid's house, I could see, was a small affair, a blockish cement structure, typical village abode. The backyard contained a rusty swing set for his two young sons. Rashid was an educated man, I mused, who could undoubtedly earn more than the paltry salary he pulled from his family's machine repair business—certainly abroad and probably even in Manama. A degree from Oxford or Cambridge (which fucking ivy-covered institution was it?) carried its weight. But some form of pride or stubbornness or loyalty kept him in his own country, living in this shack, scraping the dust with his fellow downtrodden.

I crawled back inside my hole. When I finally dozed, I had a disturbing dream: Night in the stumpy craterscape of the slums, and I was buried underground to my waist, unable to move my legs. I awoke sweating in the basement's dank humidity. Abruptly I remembered the Molotov cocktail I'd thrown, an indisputable violation of our operational rules, knew I was obligated to report the incident to Whitney and, with equal certainty, that I would not. The memory cut me with a primal sort of pleasure. I emptied my flask and fell back asleep.

Before dawn, just after curfew was lifted, I used a trickling rusty spigot to wash away my sins and sweat and the stench of alcohol. Rashid drove me back to my car downtown. On the way I asked whether he'd gone to Oxford or Cambridge. Oxford, he'd said. On an engineering scholarship.

I could write up what I'd witnessed in an intelligence report, I realized. The waves of my recent success hadn't abated and virtually every worthless nugget about the Opposition I now bothered to include in reports met at least mild praise. Rashid had reached sterling status, considered by Headquarters to be a top producer in the region. I just had to keep the ball rolling.

GOVERNMENT OF BAHRAIN TACTICS, I titled the report, and included details of the weapons used by the riot police (typical characteristics of tear gas absent, I noted with technical-sounding insight), names of persons arrested, a list of surveillance and intelligence service vehicles provided by Rashid. I added Rashid's admission that the Opposition ran both active and static countersurveillance against the king's men, even had someone on the inside.

When I'd finished, I hesitated, wondering if I was giving something away by including information that the Opposition had eyes inside the government. Whitney would almost certainly share the report with his liaison partners. Which meant that eventually, after a thorough internal cleaning, the local service would probably succeed in routing out any agents of the Opposition within its ranks, ship the bastards to Dry Dock. Which also meant that Rashid, in all likelihood, would be blown at some point. But it was an abstract concern, I told myself. A precise domino pattern that might never happen. What Headquarters glibly called a "mere hypothetical." Even if the hypothetical became actual, that day was a long way off, certainly long after I'd left this fucking country. By the time any flap occurred, I'd be back in America, retired, sipping Singapore Slings on the beach. In any case, it was Rashid's problem, not mine; he knew the stakes of the game.

But Whitney rejected the report. *Doesn't meet our threshold.* Our end goal, he reminded me, the real payoff for Headquarters—and our nation, of course—was Iran. My report, while certainly useful at a tactical level, while an excellent description of *atmospherics*, was short of our priority requirement. He'd leaned toward me confidentially, with the generosity that comes from possessing superior information, and informed me that outside channels were still fingering Tehran. Was I aware that the recent assassination attempt on the Saudi ambassador in Washington showed Iranian fingerprints? Details and sourcing, unfortunately, were too compartmented to share with me. *But keep up the good work. We'll get there. Keep your eyes on the prize.*

Then he'd riffled through my report once more, frowned. "This mole inside the government sounds interesting. Too bad you didn't get his name—that's something we could've used."

21

November. In nearly every country I'd served, the month was synonymous with one occasion: the Marine Corps Ball. The undisputed exhibitionist event of the year. A singular opportunity for Americans to shed the weight of any irksome host country shackles and constraints, spread their peacock feathers in full plumage, rise to the pinnacle of expat glory.

The last formal reception I'd attended was a multinational embassy affair, a blur of unpleasant memories. Sitting in Poppy Johnson's bedroom while she got ready, drinking until I was sufficiently numb, half listening to her complain about Burt's absence and making par with other wives. Demonstrations that day blocking roads had pushed Poppy over the edge. *I swear I'm going to kill those protesters myself if I can't get my makeup done!* Furiously she'd tried on one dress after another, testing different angles and poses before her mirror (in the vain hope, I think, that her flaws were the result of position rather than composition), at last taking solace in a pair of gold earrings she claimed were made by the favored jeweler of the Admiral's wife. *I'm sure I got a better price; I bargained him down for hours.* Then the Chaplain's wife had called, offering the services of her makeup artist, and Poppy's mood had lifted. *Problem solved! Burning tires won't get in the way of our fun!*

Almaisa and I had been dating nearly two months, so bringing her to the Marine Corps Ball seemed logical, my right. She'd accepted my invitation with a strange look of resolution on her face, like she was determined to conquer whatever foreign beasts might be lying in wait behind the gilded walls of my community. It was the first time I'd brought anyone other than Marlene. I felt an unexpected surge of pleasure; I'd kept my relationship with Almaisa hidden, instinctively feeling it too precarious for early exposure, and I enjoyed the idea of finally giving it some air. Letting the world know she belonged to me.

The night was balmy and guests were arriving without coats or furs, revealing their finest frocks, purples and jades and ultramarines entering the Diplomat Hotel like dripping too-sweet popsicles on parade. Men wore tuxes, women sleek and high-slit creations—ordinarily too tight and high for aging bureaucrats' wives—reveling in the extra inches of flesh they could display beneath the shelter of the ballroom ceiling, out of the harsh Islamic sun. Tailors in Bahrain were cheap, Filipina seamstresses on every corner, so it was fashionable, I'd learned from Poppy, to order custom dresses in imitation of photos from *Vogue* or the latest red-carpet event in Hollywood. The result was something of a dizzying circus—too many colors, too much misplaced flesh, a sloppy clash of well-fed coarseness with aspirations of grandeur.

Almaisa was wearing a gown that would never be found in the pages of *Vogue* but was nonetheless the most enticing frock I'd ever laid eyes on: pink lace down to her ankles, delectably body-clinging, made more modest by long sleeves and layers of chiffon cascading from the back like a waist-down cape, outline of her breasts slightly visible, wisps and swirls in the lace making them all the more arousing. No hijab, her thick hair styled into shiny waves, and she was wearing makeup—not much, but apparent on her usually bare face. Her scar was somewhat more conspicuous because of it, but the effect seemed only to amplify her exoticism, her feral beauty. To any man's eye she was the most desirable woman in the place.

Hotel security officers were checking cars for bombs. No body searches or metal detectors, almost certainly considered unduly invasive, an improper intrusion into the privacy of privileged lives; even great danger would not justify disturbing perfectly arranged garments and the pleasure of walking into a ballroom uninterrupted. I took Almaisa's hand as we approached the glass doors. I felt the achievement of a hard-won feat.

The ballroom was bedecked in the calculated style I'd come to expect:

just the right balance of pride and humility, glitz and sobriety, a time to celebrate the U.S. Marine Corps and American might but also to reflect on sacrifice. Red lilies sat in graceful ceramic pots on clean white cloths. Crystal chandeliers shed their respectful glow. Marines in dress blues stood on the perimeter, escorting guests to their tables with courteous efficiency.

Almaisa and I were first to arrive at our table, and I glanced at the other name cards. The Office of Middle East Analysis was united: Jimmy and Gretchen, Whitney and his date. Ms. Jordan Crenshaw. So that was her name—the consular officer. A respectable name, well suited to the diplomatic cadre, that sophisticated and erudite circle in which Whitney felt so much at home.

Glancing around the room, Almaisa asked me in Arabic, "Who pays for such splendor?" I shrugged, told her American taxpayers.

In came the Bakowskis. Ingratiating handshakes, forearm grabs. Jimmy's tuxedo was even less forgiving than his regular attire, his barrel chest appearing grotesquely large. Gretchen trailed aft like a tired servant. They greeted the Admiral and his wife, seated at a prominent center table, nestled like royalty amid the revelers. The table's floral centerpiece was the largest in the ballroom, its audacious blooms pointing dumbly at the Admiral and his special guests: the Minister of Interior and his wife.

When the Bakowskis reached our table, Jimmy stopped and stared unabashedly at Almaisa. "Well, lookee what we have here. Collins, who's this secret you been keepin' from us?"

I introduced Almaisa, who'd risen and now extended her hand. Jimmy swooped down and kissed it, grinning exaggeratedly as though the gesture were a farce.

Gretchen's face turned pink. She squeezed my elbow. "She's beautiful, Collins. Really beautiful."

The young Marine behind the bar had botched my Singapore Sling. "Just give me a martini," I said.

"I'll take a rum and Coke." Jimmy appeared beside me. He pushed money across the counter, eyed the bartender's uniform. "How many years you got in?"

"Just three, sir." The Marine handed me my drink.

Jimmy thumbed his chest. "Two years right here. Best time of my life." The boy smiled uneasily.

When I started back, Jimmy grabbed my arm. "Not so fast, Lawrence of Arabia. What's the hurry? I got a few questions."

I'd known this was coming, expected nothing less from Jimmy the Broom.

He squinted at me. "You got us all stumped. A haji! You're seein' a haji! Never woulda pegged you— She a local?"

"Yes."

"Hey, no need to be embarrassed. No need to get upset. What happens to a broad stays abroad, right?" He waited for something more, then patted me uncertainly on the back. "Well, hats off to ya. Thought all the ladies here were covered up and locked at the knees."

"Are we done, Jimmy?"

Suddenly, he grabbed me like an insistent child. "Devil dog! She's from one of the hotels! Thought she looked familiar! I shoulda known. Leave it to you to get a lay and a laugh and rib the rest of us in one clean shot."

I shook him off. "Jesus Christ, Jimmy, she's not a whore. Does she look like a whore? For fuck's sake!"

Jimmy fell silent, grappling with my denial, his face betraying every half-baked juvenile emotion touching his brain: perplexity, discomfort, the smugness of capturing some perceived secret.

I do not fully understand what happened next. How, when I got back, Whitney and his girlfriend had eased undetectably into their seats, slipped seamlessly into the grooves of dinner banter. Jordan was on Whitney's right, Almaisa on his left (had I failed to notice only minutes ago that their name cards were next to each other?), and he was already conversing with my date, clearly past the point of introduction and pleasantries. Leaning in, listening absorbedly, and she was pouring forth with her usual animation, slender hands gesturing earnestly. I sat next to her, plunked my drink down heavily between us.

"Collins! Great to see you!" Whitney reached across Almaisa. His forearm brushed her sleeve. His tuxedo looked off—overly formal, with an absurd shawl lapel.

Jordan was rummaging through her purse, bright-eyed and entirely un-jealous, wearing a draped black gown that revealed muscular tan shoulders. Up close she was even more impressive. Not far from what I always imagined was Whitney's type: all-American complexion, pearls and cardigans by day, lipstick and tight jeans by night, that New England-ish mixture of purity and sin, blue blood yet nouveau, the kind of girl you could bring home but who would hold her own in an argument or the bedroom. Not a bad catch for a twenty-eight-year-old virgin—and the kid had only been on the ground four months. It was because he was station chief, of course. That, and the pixie dust of living in an exotic locale, the strange elixir that perverted one's senses, warped one's judgment, made people choose strange things. A woman like Jordan would never be interested in Whitney if he were back at Headquarters, toiling away in a cubicle under harsh fluorescent lights.

Whitney made a joke about his tux—the first time he'd worn one since high school prom. A chorus of undeserved laughter; when the chief laughed, everyone laughed. Even Almaisa smiled, though surely she had no idea what a prom was. Then she asked whether Whitney and I worked together. "Oh yes, lucky to have Collins on my team," Whitney said.

Indian and Filipino servers began bringing dinner. Pistachio terrine followed by choice of cedar-plank salmon or roasted chicken with fingerling potatoes. Champagne with a raspberry in each glass. Bland tunes poured unobtrusively from speakers. Almaisa was in her element, talking about the virtues of her country—where to buy the freshest coriander, how to get a bargain at the gold souq. Jordan and Gretchen politely engaged her, Jordan with the pretentious interest endemic to government cookie pushers and Gretchen with the incredulity of the sheltered housewife, while Whitney offered sentient quips about the region and incisively compared Bahrain to his prior countries of residence. As the main course was cleared, Whitney raised his glass in a toast to our host nation. Everyone followed. Lime sherbet with mint, the palate cleanser. I was getting steadily drunker.

"So this lovely lady tells me you met at the art show." Whitney set down his sherbet spoon. "An artist! Terrific!"

"Shane bought one of my pieces." Her voice, low seductive ripples.

"Good thing I told Collins to go that night." Whitney sipped his champagne, turned to Almaisa. His face was flushed. "Are you working on anything now?"

"I am honored to be considered for a permanent mosaic inside the new opera house."

Jordan slammed her glass down. "Wow, you're kidding! That's a big deal!"

"I didn't know that," I mumbled.

"Picture of the king?" Jimmy called from across the table.

"No, Jimmy." Whitney shook his head, smiled. "The Quran frowns on artistic visual depiction of the human form."

"It will be a tale from *1001 Arabian Nights*."

"Oh, that sounds wonderful!" Jordan clapped her hands in delight. "Which tale will it be? Love? Revenge? That one with the genie?"

"I have not decided yet."

"Tell you what." Whitney leaned forward, his face inches from hers. "I'm looking to purchase something here during my tour. A memento. Bit of a tradition, actually—kind of thing you do growing up in a family of diplomats. Anyway, I'd welcome your advice. Or better yet, I'd love to see your art. Maybe Collins can take me by your studio sometime."

Almaisa looked pleased as a silly girl. What a coup—the whole expat world suddenly open to her. Fawning, doting, sunburned foreigners exclaiming over the local gem of an artist they'd found. *Authentic*—the well-heeled traveler's favorite label. Partaking of the native wares. She would no longer belong to me.

Dessert arrived, raspberry pavlova with white chocolate ganache, and everyone cooed and shoveled in the food with token protests of satiation. Dishes were cleared; servers retreated politely to the shadows.

The Marines carry forth a long and proud tradition. We are gathered here tonight to honor them. The Ambassador's faint Texas twang was homey and disarming (a political appointee, not one of the rank and file, he had to work extra hard to prove his accessibility); this was a man they knew, his voice assured guests, the neighbor willing to mow their grass when they went out of town.

America would not be what it is without the bravery and sacrifice of these heroic men and women. Behind the Ambassador, slightly obstructed by a glittering chandelier, an enormous screen rolled out images of Marines crawling through trenches, climbing mountains, hoisting flags. *So we say to our nation's finest, both solemnly and sincerely: thank you.* Bursts of applause like firecrackers.

We must also thank our Bahraini hosts, without whom we would not have the opportunity to be here defending, protecting, and advancing American interests. An appreciative gesture toward the Minister of Interior. *And tonight we are honored by a special guest.*

Many of you might not know this, but Saif and I hail from the same alma mater. Small school in Texas by the name of St. Edmund's. In any case, Saif, I thank you for your invaluable service in making Bahrain—this beautiful country—safe for both your people and mine. The minister smiled deferentially, bowed his head. More applause.

A DJ began playing the latest American chart-topper. Enlisted couples filled the floor. Whitney and Jordan joined the throng. I was amused to see Whitney dance just as I expected: movements artless and forced, too-wide smile plastered on his face. Jordan moved smoothly, her sultry quality translating well. The kid barely noticed her.

Almaisa looked at me. She wanted to dance, I knew, and ordinarily I would have obliged. I was decent enough and I enjoyed the idea of holding a woman without having to talk, with the promise of something more gratifying later. But I felt a sudden revulsion to the whole affair. I told Almaisa I was taking a smoke break.

Outside, I lit a cigarette, watched the lights of downtown punch through the haze. It was close to midnight. We lucky ball attendees had been granted an exemption from curfew, and a small sticker on our windshield proved it to any challenging policeman. I had no justification for Almaisa to spend the night at my place. Her lace gown—thin and delicate. Easy to tear off if only I had the opportunity. I'd never waited so long for a woman.

I was on my second smoke when Poppy Johnson sidled up to me, doused in heavier perfume than the whores at Sea Shell Hotel. Hastily she produced a cigarette, asked for a light. Her tight dress rode up her overly tanned legs.

"So I was right." Poppy blew smoke in my direction. "About the girl. Is she . . . local?"

Turning, I saw with some alarm that Poppy's face had changed since I'd last seen her. The ridge of her nose wasn't right. It was swollen and bumpy, as though tiny larvae had sprouted along the slope, and her nostrils had morphed into mismatched mushrooms. Her natural flush had turned dark and ghastly, red capillaries cast across her face like a spiderweb. I grinned. "Doc Hollywood back in town?"

Her eyes grew large. "What do you—"

"Listen, Poppy." I put my hand on her back. "We had a good run. Don't ruin it with expectations."

There are a few moments a man remembers in his life. The way a punch feels on his face, losing his virginity, his first drink and its aftermath, maybe a hard-won athletic victory. Watching your woman dance with another man.

A slow song was playing, Percy Sledge or maybe Elvis, and the crowd had largely cleared the floor and he was holding her. Politely and at a distance, but holding her nonetheless, his hands on her pink lace, her arm, her back. He was saying something and she was smiling. The two glided across the floor, too smoothly, in unthinking turns. They were the same age, I realized. And he was a better dancer with Almaisa than Jordan.

I got another drink, returned to the table, continued to watch Whitney and Almaisa dance. Jordan and Gretchen were chatting as though at a Sunday picnic. The lights dimmed. I stirred my martini in the semidark and began to sweat. Almaisa moved with the same adroitness she seemed to do so many things, her ease and lack of inhibition betraying her Western roots. Somehow I'd known she'd be a good dancer.

"You left her hangin'," Jimmy's chuckling rebuke sounded in my ear. "Boss man had to take her for a spin before the poor girl started cryin'."

Her motion, I now saw, had changed, become more sensual. Uncannily attuned to the music. Her body exquisitely calibrated; if you watched long enough, you forgot she was moving at all. Was I imagining it? Onlookers were staring. She didn't seem to mind, probably liked the attention.

"You hear 'bout boss's promotion?" Jimmy buzzed in my ear again.

"No."

"Just tapped to be the next Chief of Iran Division." He shook his head. "Tell you what: That kid has it made."

The song faded and a roving yellow light caught Almaisa, brought her back to the dance floor. She and Whitney returned to the table.

"You're back!" Whitney exclaimed. He'd worked up a sweat. Almaisa was beaming, her makeup dewy and smudged. Jordan leaned over and straightened Whitney's tie. "Lookin' good out there."

I took Almaisa's hand, raised her from the table. Her feet moved quickly to keep up. "Are we dancing?"

"No. We're leaving."

In another place, another time, through another lens, maybe the events following the ball would look different. So I've told myself over the years. *Perception is reality*, as we used to say in spy school. Still, that night pursues me. It was made of the stuff that keeps a man awake, the darkness in the corner of the attic, so black he's more fearful of what's not there than what is.

We arrived at my villa and Almaisa seemed to know she would not be returning to her studio and also that her protests would be unavailing. She attempted conversation along the way, even after we'd entered the living room—her favorite dish of the night, the lovely floral arrangements—but I mostly kept silent. She was doing her best to pretend everything was normal, that the evening had gone fine if not magnificently. I left the lights off and drew the curtains, long stiff things I never touched that expelled clouds of dust. Almaisa was standing in the center of the room, her pink dress drained of color in the darkness. It was the second time I'd seen her in this position, but this time she looked different, more diminutive and weaker, a child who senses punishment but cannot divine her crime.

I don't remember precisely what happened next, the minute order of things: which hand moved first, what ran through my mind, even a conscious coherent decision lurking in any recess of my gray matter. I remember the sound of her lace dress tearing. It was an intricate weave and I could not rip it easily, but my hands were exceptionally strong and capable that evening, blessed with the power of determination. Satisfaction at the sight of her naked golden body. Untouched, paler than I expected—the color, I suppose, of someone who spends a lifetime covered up, whose skin never sees sunlight. It was a youthful body without the tracks and apologies of old age. Her long dark hair spilled onto her breasts, smooth and slightly limp, nipples large and deep cowering red. It was everything I wanted, there for the taking.

I could have taken her to the bedroom, and as I cannot remember my precise thoughts or actions, I cannot recall why I kept her in the living room. Nor, for that matter, do I understand why I had her lie on the floor rather than the couch or any other more forgiving surface. The cold marble floor—not even real marble but tile painted to look like marble.

The little I recall of the experience was pleasurable. She was silent through much of it. We didn't look at each other; I remember noticing a mole on her neck I'd never seen and that her body was softer and easier even than it looked. I gripped her arms, held her down, and she arched her neck, closed her eyes. I grabbed a fistful of her hair just to feel its silkiness and warmth. Her face grew pink and her scar darker.

When it was over, I rolled off her. She lay still on the floor. Sweat plastered strands of hair to her face. She'd bled onto the fake-marble tiles. I went back and held her.

22

Weeks passed and Almaisa and I didn't talk. My villa became repulsive and I spent as much time away as possible. I couldn't manage to open the curtains and the place smelled bitter. My evening Irish whiskey lost its somnolent effect and the air conditioner failed to stem my profuse nightly sweating. Everything had been moving easily, fluidly. It would have happened eventually if only I'd waited.

The weather turned chilly and still we didn't talk. Days were short, less sullied by dust; flamingos roosted in every lagoon. Headlines announced that an unlicensed clinic in the slums had burned down due to faulty electrical wiring and improper storage of medicine. Two children died in the fire.

I reached a state of perpetual half-drunkenness and spent my weeknights at the Sea Shell Hotel and my weekends at the Dragon Hotel having brunch alongside the Johnsons and Bakowskis. Salmon tartare, carved steak au jus, fruit tarts with cream. Jazz singers tastefully crooning "Some Enchanted Evening." Whitney and Jordan continued to arrive at events together, while the Johnsons put up a united front, Poppy defiantly parading her new physiognomy, Burt clapping shoulders and making jokes no one understood. I was back on the other side of the telescope, in the dismal land of excess, where expats grew richer and fatter and inched toward their long, slow death.

Holiday festivities began. Blur of lights, overly spiked eggnog, drunken carols at Trader Vic's. *Christmas Coming to Bahrain!* A putrid camel on special loan from the king's farm took up residence on the naval base, wearing a bright wrap and reluctantly greeting visitors, part of the elaborate nativity scene arranged by the chapel. An atheist had launched a protest against the display, Poppy Johnson informed everyone, and the wives had united to eject him from base. The Admiral and his wife jetted off to Dubai to ski indoors on fluffy manufactured snow, and droves of foreigners fled the island, the allure of places like Bali and Ibiza too hard to resist. Jimmy declared he would stay—*more holiday pay, saving up for a Porsche.* One cold morning an earthquake shook the island—enough to rattle my villa and wake me from my stupor—and officials drove through the streets at breakneck speeds announcing through megaphones that the epicenter was near Tehran.

And then she came back to me. She showed up at my door one evening with a small mosaic pot of blue and gold tiles that she said reminded her of the guitar music I played, and we went to dinner as though we'd planned it all along. She'd been traveling again, she said, soliciting supplies and funds for the orphanage. When we got home, she spent the night and didn't resist my overtures.

We didn't speak of what happened the night after the ball. It was as though nothing had happened. Maybe it hadn't, at least in the way I remembered. Easily we reverted to our old routine, and the act that had so rebukingly triggered my Western fears and sensibilities began softening in my mind, morphing with the weeks into something more benign, something desired and natural. Instead of battling my touch, she had a new openness about her, a conscious receptiveness. When I caressed her hair, she closed her eyes, smiled. She offered her body to me deliberately.

Based, perhaps, on some vestige of repentance, I went to the souq and bought an expensive string of pearls from Poppy Johnson's jeweler. All real, I assured Almaisa, not cultured; Bahraini divers had plunged into the Gulf to retrieve every one. "Thank you," she'd said. "But there is something else I want."

It was late evening and we were standing in my living room.

She looked up at me. "I want you to stop drinking."

I laughed, surprised. "Sweetheart, I'm not Muslim. If you think—"

"No, Shane. It is not because of Islam."

She'd never said a word about my drinking before. I lit a cigarette. I was perturbed. Not because I felt bad—hell, everyone in my profession drank unless he was too hung up on Jesus or his liver was a complete loss—but because she'd discovered the depth of it, because I, a spook of twenty-five years, had failed to hide my rotten parts.

My customary instincts of dissuasion kicked in—she was entirely wrong, I didn't drink nearly as much as she thought, she didn't understand Western culture (surely her mother drank), alcohol came with being a diplomat. A gentle, or if necessary blunt, reminder that no one was free of vice, including her, and that to strive otherwise was to create the seeds of lifelong unhappiness and disappointment. Avenues of manipulation formed in my mind. It was what I'd done most of my life and it worked.

I looked at Almaisa standing before me. Her resolute frame. She was young and incomplete, roughly the same age as my son, and suddenly I was thinking of him. Had he started drinking? Last I'd seen him he'd sworn off alcohol just as he'd sworn off me (perhaps we were one and the same to him), but that was years ago. Since then someone had offered him a drink, maybe—just one, no big deal—and then he'd gone to a bar, a baseball game, a date, and had another. He'd wind up just like me.

Almaisa wasn't angry—not even a little. In fact, she looked relieved to have emptied her request onto the table. And then I had the realization: Alcohol was to blame for that night. She thought it was the alcohol. A justification. An explanation, an excuse. There was a culprit and the culprit wasn't me.

I saw it so clearly. Why hadn't I seen it before? She loved me and was willing to absolve. All she needed was a story.

23

I didn't stop, but I did taper off, reducing my consumption by a few drinks a day, careful to remain sober around Almaisa. I hid my alcohol in the unused rathole of an attic that passed for a live-in maid's quarters, a place Almaisa never ventured. I could see Almaisa felt a certain satisfaction at my transformation, the elation of playing God, of changing the course of someone's life—and, as though reciprocating, she became more pliable and agreeable. Her unpredictable moods subsided and she regularly cooked dinner at my house, staying as late as I wanted, dispensing with her hijab every time we went out.

To sweeten the deal, I began helping out at the orphanage when I could—lifting heavy boxes, doing odd jobs and repair work. We took Sanaa on a few excursions—Al Areen wildlife reserve, where she fed the ostriches; Prince Khalil Al Jasri Park, where we boated and looked down on Manama from the stumpy observation tower. I made sure not to get too close, ignoring what seemed like occasional hints from Almaisa—*we do not mind foreigners adopting our children; the orphanage makes paperwork easy.*

Meanwhile, the revolution had plateaued, even lost a little steam, local headlines reverting to Zionist conspiracy theories and Israeli plots to undermine the Arab world, stalwart feel-good topics of agreement between Sunni and Shia that signaled relative peace and harmony. The king announced grandiose plans for a cultural center in the heart of Budaiya. Nothing on the scale of Adliya had happened for months, the threat condition on base reverted to Bravo, my contact in Al-Wefaq hinted that negotiations might commence, and I predicted it was only a matter of time before the Arab Spring collapsed once and for all.

When I returned to station one evening after a meeting with Rashid, Whitney knocked on my cubicle and congratulated me. Headquarters had included my latest report, the Opposition's long-term political strategy, in the Presidential Daily Brief—so well received, he said, that Langley was considering keeping me in Manama another year. I wrapped my hands around the armrests of my modular government-issued chair and felt my insides stir. If my last drink depended on it, I never thought I'd want to stay in this goddamned country a day longer than my orders required, but there I was, absurdly titillated by the prospect of another year with Almaisa. The revolution was dying, the remainder of the tour sure to be easy and lucrative.

Then, smiling, Whitney asked if Almaisa and I wanted to join him for drinks with Jordan on Thursday.

"She doesn't drink."

Pink blotches erupted on Whitney's neck. "Of course not. Silly of me." He shifted his leather satchel from one hand to the other. "Well, then, we'd love for the two of you to join us at the opera next week. Placido Domingo is coming for a holiday performance. Saw him once in Spain when was I was a kid—the man can sing."

I started to concoct an excuse, the flimsier the better, then noticed Whitney's face had changed. His eyes were skirting the floor and he looked tired, as though he hadn't slept in days. His tan had faded. He looked up at me pleadingly. "I think Jordan would enjoy the company."

"How is Jordan? Haven't seen her around lately."

"She—she's good." Awkwardly he put down his satchel. "Maybe you could give me some advice, Collins. I don't have a lot of experience . . . and you . . . you've been doing this for a while."

"What's the problem?"

"I'm sure I'm making too big a deal of everything. It's just that she's not always happy—"

"Nobody's happy all the time."

"I know. But it's more than that. This tour . . . she's not looking—" The cipher clicked and Jimmy's voice boomed. *Honey, I'm home!* Whitney stole a desperate look at me, headed over to Jimmy. They exchanged a few words about traffic, what the commute home looked like.

When Whitney reappeared, he was gripping his satchel determinedly and his voice had regained its customary authority, the luster of a shared confidence fading as quickly as it had come. "Anyway, we'd love for you two to join us. Should be a fun night."

I eyed the twenty-eight-year-old station chief. "I'll see what I can do."

All along Awal Avenue strings of red and white lights strangled palms, their fronds dripping with gold like a meteor shower. For Bahrain National Day, December 16, the king had arranged a major celebration, the holiday commemorating not the anniversary of Bahrain wresting itself from British hands but the anniversary of the Jasri family assuming the throne. The fanfare had even traveled across the Atlantic, where the king leveraged the sizable stake he'd purchased in the Empire State Building to ensure it glowed nocturnally with Bahrain's national colors. For the occasion the king had also generously allowed Junaid, reportedly in declining health, to be seen at the Royal Hospital. To complete the festivities an ostentatiously loud parade traveled down Awal Avenue with heart-shaped floats and banners declaring TOGETHER TOWARDS TOMORROW.

Almaisa placed her hand on my knee. "I saw Pavarotti once in Italy. The opera was beautiful. Thank you for taking me tonight."

The parking lot was overflowing with Land Cruisers and BMWs, skinny Indian valets rushing back and forth like firemen putting out flames. I started to open the car door, then nodded at the balled-up hijab in Almaisa's open purse. "Put that on."

She stared at me, surprised. "Why, Shane? I thought you liked my hair."

"Of course I like your hair. Just put it on."

"I do not understand—"

"What's the problem? Put the damned thing on. The world doesn't need to see you. Jesus, aren't you ever afraid of what people think?"

Silently she wrapped the scarf around her head. It was scarlet, louder than a siren. As we walked toward the building, the cold breeze blew the hijab lightly around her face and she looked tantalizing.

There was something satisfying about returning to the place we'd met, this time hand in hand, with a different silhouette as we walked past the floodlights. The opera house was once again overly decorated, the excess of adornment having migrated from *Fall of Culture* to *Winter Holidays*. Plastic green wreaths with red velvet bows hung from the windows; paper snowflakes dangled from the ceiling. In a corner of the entrance hall a puny Christmas tree withered in solitude, its authenticity confirmed by a layer of needles on the carpet, undoubtedly the best that Bahrain's desert soil had to offer. The king was nothing if not eager to please his European and American guests. Last week's *Gulf Daily News* had announced an honorary degree bestowed upon the king from St. Edmund's University in Texas—for enriching the cultural life of his country, it said.

Industrial cloth covering portions of the walls reminded me that Almaisa's art was somewhere underneath. "You never told me what story you picked for your mosaic," I said.

"It is a secret. I am waiting until it is finished."

Whitney and Jordan were already seated when we entered the theater. Somehow the Americans had all managed to find seats together—Burt and Poppy Johnson, the Chaplain and his wife, Jimmy and Gretchen. An usher gave us programs and we took our seats, Whitney and Burt and the Chaplain standing too ceremoniously as Almaisa sat down, Poppy's eyes following Almaisa's every move. Whitney complimented Almaisa on her scarf. *What a great shade of red!*

One of the elite suspended box seats held the Admiral and his wife, chatting with an Arab in an immaculate white thobe. I'd seen him before. A member of the inner circle. Caterpillar-like mustache, leisurely posture, supple leather sandals visible through the railing. Prince Jamil. I'd glimpsed his picture in the *Guardian* last month after an arrest at Heathrow airport for drunk and disorderly conduct: rushing the cockpit to demand better first-class service. The Admiral took a cigar out of his pocket, held it to the light, and the two men examined it.

"Isn't this great?" Burt gestured around the hall. "Opera in Bahrain.

You know, I was here in the seventies—conference on oil—and if someone had told me I'd be watching opera in the middle of this dustbowl, well . . . This country's making great strides, that's for sure."

"I heard the king arranged free concerts out in . . . Buda . . ." Jordan looked to Whitney. "Budaiya," he assisted.

"Right, that's it. The slum area. But I hear no one shows up. Too bad the place is so dangerous—I'd like to see one of the concerts."

The Chaplain's wife put her hand on Almaisa's shoulder. "Are you from Bahrain? We love it here. You have a *beautiful* country."

"Lot of good will here, people trying to do the right thing, help one another," the Chaplain agreed. He was a tall, fit man, one of those God-loving, call-heeding patriots from *Looo-zi-ana* as he called it, his wife a former beauty queen. "Our housekeeper . . . what's her name, honey? She makes all the kids' favorite meals without us even asking. Works weekends sometimes. Told her she should go home for the holidays, but she wanted to stay here. Go figure!"

Poppy leaned across the seats, her oversized breasts nearly falling out of her frock. "Well, I'll tell you what I love about this country. You can wear anything you want!" Her boozy breath tainted the air.

Burt's eyes slid, embarrassed, to see if Almaisa had processed the comment, whether she could gracefully absorb good old American sarcasm. He chuckled uneasily. "Well, now, honey, every country has its own customs—cuisine, architecture, fashion. No country better than any other. Just different."

Poppy smiled as though she'd intended for everyone to misinterpret her remark, as though she regularly spoke in enigmas no one was clever enough to decipher. "That's not what I meant at all."

She thrust her hand toward the Chaplain's wife, light from the chandelier catching facets of a sizable pink sapphire ring. "I mean, look at this. I could never get away with wearing this at China Lake. People back home are so judgmental when it comes to the finer things in life. But here—have you *seen* what women wear here? Chanel purses! Prada shoes! At the restaurants and Dragon Hotel, the Ritz . . . Being able to wear what you want. That's my definition of true freedom."

"Honey—"

"And another thing I love about this country—*religious* freedom! I mean, we can't even celebrate Christmas in America anymore. Because back home we have Jews, atheists, agno . . . agnothisists protesting. Tyranny of the minority, isn't that what you call it, Burt? Chaplain? *Put away your Christmas tree, don't talk about Jesus, separation of church and state.*" She held up her opera program, its cover decorated with cartoonish holly berries and glittering green sprigs. "See this? You're not even allowed to print things like this in America anymore! That's why . . . Oh, where's my petition?!" She reached into her bag, pulled out a crumpled paper. "Collins, sign this. Everyone else has. We're demanding that Santa Claus and baby Jesus be allowed on the naval base where they belong. That goddamned atheist isn't going to take Christmas away from us!"

She thrust the petition in front of me. Then, realizing I had no intention of taking it, she dropped it into the lap of the Chaplain's wife. Sitting back sullenly, Poppy fanned her face with the program. "Santa greeted me at Seef Mall the other day. A Muslim country! Seems pretty damn tolerant to me."

The lights dimmed and the crowd burst into thunderous applause for Mr. Domingo. A rarefied display, an empyrean performer, a flawlessly polished specimen from Europe that so perfectly sustained the illusion. The stage ignited in dripping colors, the walls turned black, and Poppy's ring sparkled brilliantly through the darkness.

When Whitney took me aside after the performance, I knew he had a weighty conversation up his sleeve.

"Something I've been meaning to talk to you about," he said as we walked to the parking lot. Fireworks erupted for the conclusion of National Day, red, white, and gold trails projecting sharply upward from Halab Island.

Whitney nodded in the direction of Almaisa. "Talked to Jimmy. You're going to need to report her to H.Q."

"Why would I have to do that?"

"You know the rules, Collins. Any close and continuing foreign contact has to be approved. I assume she's close and continuing." He looked at me.

A few feet away Burt was peppering Almaisa with questions, Poppy irately lighting a cigarette. I'd never had to report a woman before—for the simple reason that no foreign tryst had been anything more than a one-night, maybe two-night, stand. Close and continuing. Almaisa was close and continuing.

"I'll send in the paperwork Monday."

Whitney punched my shoulder, walked away.

Before I could rescue Almaisa, Gretchen was at my side, her hand awkwardly on my arm. She looked anxious—more anxious than I'd ever seen her—clutching her purse for dear life. "Collins, I'm sorry. . . ." Her voice trembled. "Do you have a second?" Jimmy glanced in her direction as though keeping tabs on a small child.

"Sure, Gretchen."

"It's Jimmy. I just— Have you guys been working late?"

All on her own Almaisa had managed to free herself of Burt and was returning to me. "Sorry, Gretchen." I mustered regret in my voice. "Catch me another time."

That evening I pulled my treasure chest from the maid's quarters and poured myself a drink. Almaisa and I were *close and continuing*. Continuing past the Marine Corps Ball, the gulf between us, maybe even my tour and Bahrain.

A fantasy crept into my head. Langley would approve Almaisa, I would declare myself to her—admit my true profession—clear the smoke and untwist the kinks, start a life with her. Or dispense with the truth and let my past fade away like an uninteresting rumor. I wasn't far from handing in my badge for good and details at this point were inconsequential. Retire with my pension and take her back to Jersey or, better yet, somewhere abroad, where she'd be comfortable, where we wouldn't be shackled by expectations—Italy or France or Spain, some place with art and espresso, better than anything she'd known. She'd have her work and I'd have her. I'd built hundreds of connections across Europe—on the street, in the government, the services—and with a nice retirement package, it wouldn't be too hard. Hell, we could even bring Sanaa. It would be something new, something we could do together, a chance for Almaisa to escape the revolution and the inevitable end of the Bahrain she loved. It wouldn't last forever—anyone could see that—and the Arab Spring would die an unpleasant death, halfway gone already. She knew it was coming. Surely she knew it was coming.

I'd never been one to think about the future—plans were luxuries for prep-school boys and politicians—but sitting in my living room, with her golden skin in my head and her voice in my ears, I allowed myself, just for a moment, to think about the years ahead.

Then, finishing my drink, I remembered what she'd said in the car after the opera, her words like an insistent branch knocking on the window. She'd wanted me to drop her off at her studio—teaching an early class at the orphanage, she said—and a few blocks from her place she'd turned to me. "Did you know that Whitney has lived in more than ten countries? And yet he is so young—not older than me, I think." Unthinkingly she'd unwrapped her hijab, laid it on her lap. "He knows who Farida is. My favorite artist. The queen of Egypt for eleven years. He learned about her in an art class at Princeton."

She'd stared out the window as though searching for something. "I have never known another American who has heard of her."

24

In January Rashid coughed up the name of the Opposition's mole in the government. *Hamid Salabi*, an intelligence officer. Closely guarded information—only a few souls privy—to which Rashid's recent promotion entitled him. He'd withheld it from me, he said, until he knew he could trust me. Now, with both sides once again whispering of negotiations, he wanted America to know whom it could trust among its friends, who had the Opposition's interests in mind and would be a revealing interlocutor. It was a gold nugget—nay, a diamond—the name of a traitor or, depending on your perspective, an ally. Occasionally, if you were lucky, you got a physical description, something uniquely identifiable like a scar or a tattoo, maybe a date or place of birth or moderately narrow description of duties, the information at least two levels removed. But you never got a name.

I drove back to station feeling like a goddamned kid on Christmas morning. Knocked on Whitney's glass wall, opened the door before he'd invited me in, tossed my notes on his desk. He looked at the paper quizzically.

"Backward. Just the consonants. Vowels on the other side. I wrote in English for you."

He turned the paper back and forth, silently piecing together the name, squinted up at me. "I don't understand."

"That's your mole right there. Intel officer."

He stared at it dubiously. "In the service? Working for the Opposition?"

"That's the one, Chief."

Whitney sipped his coffee, nodded. "Nice work, Collins. Type it up and I'll pass it along to our friends." He turned to a stack of reports.

"Our *friends*? This information is for our eyes only."

"It will be safe. I'll let our friends know it's sensitive. Of course we won't reveal the source."

I stepped closer to his desk. "No, we won't have to. Because the mole will be arrested. Jassim's men will leak his name. Hell, maybe they'll even post it in the *Gulf Daily News*. You think the Opposition won't figure out there's a rat in their house? How long you think it will take, Chief? An hour? Two hours? They'll throw SCROOP in a cell and get it from him with brass knuckles faster than you can turn the lights on. Jesus, you want to burn our own goddamned informant? Or maybe you just want to burn the mole."

Whitney blinked, unruffled, put down his mug. "SCROOP's information is still uncorroborated. There's no way to establish its veracity unless we share it." He spread his palms. "Think about it this way: It's an opportunity. Bahrainis confirm the name, we send a cable to H.Q., SCROOP's bona fides are proven once and for all."

I stared at him. "For Christ's sake, SCROOP's proven himself a dozen times over. The best informant we've got in the region. He's the real deal and you know it."

"Even good sources can improve. Maybe H.Q. will agree to a pay raise. SCROOP could be a star."

"Or he could be killed."

Whitney smiled placidly, drummed his white slug fingers on his mug. "Collins, I understand your concern—it's admirable, really—but our friends have a right to know when there's an enemy among them."

"And what about our 'friends on the other side'? We owe them anything—or they just around to provide fireworks for our rooftop parties?"

Whitney shook my notes emphatically. "Collins, we're a neutral party in this conflict. America hasn't taken sides in the Arab Spring. You know that. But the simple fact is that we need the Jasris right now more than we need the Opposition. It's basic strategy, dictated from Washington. H.Q. has made clear that Iran is our priority target. The Jasris—"

"This has nothing to do with Tehran." I placed my hands on his desk, felt the hard surface press back. "It's about hanging our man out to dry. So *you*, Chief, can get credit."

Whitney's face grew dark, a color I'd never seen before, and he rose from his chair. His finger stabbed the air, my notes degenerating into a crumpled mass. "You remember what we learned at the Farm, Collins? Never fall in love with your source. You don't know everything about SCROOP. Just a few months ago—" He steadied himself, smoothed the paper. "Look, Collins, you're a decent officer. And you've done a decent job with SCROOP. But your loyalty is misplaced. Agents come and go. They're dispensable."

I straightened, removed my hands from his desk. Then I did the only thing I could: snatched the paper away.

He didn't resist, simply looked at me. His voice was quiet. "Situations like this, I try to imagine the facts are reversed. And if we had a traitor in our ranks, we'd want to know."

That was the night I asked Almaisa to move in with me.

She said no—too quickly, I think, looking back. She told me it was risky, that an unmarried local woman could only move into the house of a single foreign male on penalty of complete ostracism or, worse, punishment. Cryptically she spoke of having been questioned, but refused to elaborate, leaving me wondering whether she was referring to nosy neighbors, local intelligence, or some anonymous meddling third party.

Rejection was a gift, I told myself after she'd headed home. How else would I have explained the treasure chest in the attic, my trips upstairs during sleepless nights? And Sanaa—the two probably came as a package deal. Eventually I'd have to propose marriage and that would mean bringing the kid along for the ride. I'd been swallowed up in the sinkhole of family life before and forced to claw my way out; what would be different this time around?

Yes, it was a gift. And that evening, to celebrate, I poured myself a tall Irish whiskey, feeling instantly the silken liquid wash through me, that familiar sense of repose and removal, sweeter and more mellifluous than Almaisa's voice.

25

The next time I saw Rashid he had a black eye. He evaded my questions at first, said his kid had accidentally dropped something on him. Kept looking around the car as though someone were hiding in the backseat—this, despite the complete absence of life for miles; we were parked atop a rocky outcrop on the fringes of Saar.

"It was not my son," he finally admitted, eyes sliding to the floor. "Men . . . I do not know their names . . . they find me in the alley. At night, Diraz. I was coming home from a meeting."

"A Fourteen February meeting?"

"*Na'am.*"

"You think they followed you?"

He shrugged, smiled weakly. "I do not know. I am good at seeing behind my back, but not as good as you, *habibi.*"

"They get anything from you?"

"No. I carry nothing. I know better. This is all they got." He touched his eye gingerly. For a second I had a distinct feeling of familiarity, both pleasurable and alarming, the sense I'd seen a face like this before. It was Almaisa, I realized. Her scar, the same eye injured as Rashid.

"You get a look at them?"

He shook his head. "They were wearing masks."

Rashid's face had new wrinkles, I saw. Not the kind from too little sleep—darker, more intractable, etched into his skin like the early stages of a carving. Cheeks hanging like loose clothing. Eyes dropping to the ground as though too enervated to stay aloft. How did they know? I'd taken the mole's name back from Whitney, burned it with my lighter before I'd even reached home. It had been encoded, not a name Whitney could remember from one brief exposure. Or maybe he could.

The wind whined through the outcrops, gently rocking the car, testing the Lancer's cracks and holes. Beyond the edge of the badlands, white stucco corners peeked like timid ghosts, hinting at the meringue-topped villas beyond.

"So who do you think they were?"

"Criminals. From the streets. They want money or a watch or something expensive." He tapped his bare wrist, laughed thinly. As he moved, his collar opened and for the first time I noticed a cut on his neck, small and neat, a knife wound. Rashid wiped his face with the sleeve of his flimsy jacket. Poor protection against Manama's winter winds. Resignation, I realized—it was resignation on his face. The death of the Arab Spring.

"The king's men," I offered.

"Yes. It was the king's men."

And despite his admission, I was glad he had not suspected his own men—for to suspect your own, to realize all is inverted and the sky is the ground and the ground the sky, all jumbled and disarranged and shifting before your eyes, is a far worse thing, a singularly dizzying form of torment, one that keeps you awake at night and glancing in every rearview mirror. Rashid, after all, was still alive, and that meant his men didn't know everything. It was even possible they still knew nothing.

All the same, I suggested he lay low for a while, change his travel patterns, maybe spend a few nights at friends' homes. It was good tradecraft, I advised, even under the best of circumstances.

A few days later Rashid left a dead drop for me—a note in a Muharraq alley beside the Tunisian snake charmer, a wizened fossil of a man who crouched all day playing his peeling pungi and charming his Saharan horned viper into melodic obedience. *Out of the country for two weeks*, the note said, *visiting an uncle on his deathbed.*

In the weeks that followed I said nothing to Whitney about Rashid's black eye and he said nothing to me about the name I'd snatched from his hand. I typed up Rashid's latest intelligence—internal machinations of the Opposition, negotiation strategies, underground lists of demands—while Whitney accepted my offerings and, in the professedly friendly spirit of advice, cautioned against publicly voicing the political opinions I'd expressed inside our safe walls.

26

February 14th marked the two-year anniversary of the revolution. While skirts trailed the floor and Westerners stuffed each other with chocolate hearts and perfumed diamonds, the Opposition rose from its torpor and took to the streets in numbers not seen for months. Police vehicles rolled out; checkpoints and arrests multiplied; passenger flights were relegated to darkness. An inert pipe bomb was found in Juffair, blocks from the naval base. A Night at the Oscars—gaudy parade of Westerners modeling re-creations of their favorite Academy Awards frocks—left its mess at the Four Seasons. The king breathed new life into his Together Towards Tomorrow campaign, festooning squares with streamers and impromptu quartets, turning the downtown into a gold jumble of Disney castles.

In late February Whitney informed me I'd been approved for a meeting. A window, he said, into our most highly compartmented operations. At long last I'd earned access to one of the hidden chambers—*a different channel*, as the Admiral had so weightily termed it. Now, I supposed, I would see what our fears of Iran were all about.

The third floor of the thirteen-story Gulf Hotel was a spacious chandelier-lit lounge ringed by glass cases stuffed with pastries. Napoleons, Bavarian cream puffs, iced ginger cake. An elderly tuxedoed man played soft unidentifiable tunes on a grand piano. What began more than forty years ago as a symbol of wealth and pride for the small island—glistening water views from every room, guest access to a private beach—had faded, degenerated into a landlocked eyesore. The Gulf Hotel no longer looked down on paradise, just city blocks. Yet Bahrain's elite venerated the place as though nothing had changed, as though every window still had a view.

Men in shiny suits and white thobes were speaking in low tones. Burgundy velvet couches, circular like mushrooms, dotted the marble floor alongside coffee tables, creating dozens of fat dancing partners. An obelisk-like structure rose from the center of each couch, which had the convenient effect of providing near-complete cover for conversation, talk on one couch inaudible from any other. It occurred to me that Rashid and his friends would give their eyes to know about this meeting, to have so many targets within one sight.

The Bahraini Chief of Intelligence and Deputy Chief of the Saudi Intelligence Service had already ordered coffee. Rising in his dark polished suit, the Saudi deputy extended his hand to me. "Karim Nayef. I have heard much about you."

Nayef's reputation preceded him: a penchant for bluntness and, in cases of dissidents, brutality. He was the smooth assured face of the new generation of hard-liners put into place by King Rashid's thirty-year-old favorite son, Mohammad bin Rashid. It was thanks to men like Nayef that Bahrain had not succumbed to its revolutionaries, that Saudi troops had rolled across that crucial causeway when the Opposition threw the first stones, supplying the muscle that its atrophied neighbor lacked. A silent bodyguard who waits in the shadows then disappears after throwing the conclusive punch.

Jimmy had already signaled a waitress. "Can't go to the Gulf Hotel without a little sumpin' for my sweet tooth."

Abd Al Matin, the Bahraini chief, was shaking his head, gushing regret about the pipe bomb in Juffair. "Horrible, horrible"—sipping his coffee—"the Opposition is an army of thugs."

When the waitress had brought fresh coffee, Abd Al Matin looked hard at Whitney. He crossed his wrists. "Handcuffs. This is your arms embargo. We do not have the tools we need."

"Certainly understand your position." Whitney nodded. "And your request has been conveyed. As you know, we try to stay neutral in your internal affairs. If we did lift the embargo, the weapons, equipment, and training would only be to professionalize your forces and combat terrorism—and to prepare Bahrain in the event of a regional conflict. They must not be used for political purposes."

"Of course. You will find complete cooperation in this regard. Our only desire is to ensure the safety of our people and yours." Abd Al Matin smiled. Tapping his pack of cigarettes impatiently against his knee, he turned to Nayef. "It is time. We continue our conversation outside."

The duo rose and we followed. No one, as far as I could tell, paid the bill.

Behind the hotel a tan armored SUV was waiting for us, late afternoon sun bouncing off its tinted windows. It was precisely as Rashid had described in his roster of intelligence service vehicles—no plates, small dent in the rear bumper. The young driver quickly extinguished his cigarette as we approached. Nayef climbed into the front seat, Abd Al Matin into the rear.

"Wein rayhin, ya sayyed?"

"Mawqa'a raqim arba'a."

We took backroads to Location Number Four, the enormous vehicle squeezing through alleys across Bilad Al Qadeem and Jeblat Hebshi, new but impoverished neighborhoods packed with bland cement blocks. The trip was slow, the backroads poorly suited for traveling from one end of the island to another. As we approached Saar, the environs began to change, high-rise apartments giving way to crumbling boxes. Shiite flags proliferated, arising like black sunspots after a day at the beach. Abd Al Matin and Nayef kept their heads down. On went the slums, growing steadily poorer until they broke politely for the oddly situated British Dilmun Club, a members-only stable and riding center for expats who needed an equestrian respite amid the wearying conflict.

South of the Dilmun Club, past the Saar burial mounds. Four thousand years ago the Dilmun civilization had existed on this land—a city with a temple, central plaza, houses, cemetery—now simple hills of dirt. North, past the Saar mosque, toward Diraz. We were deep in enemy territory, ground I knew well. The driver's eyes flicked mechanically in the rearview mirror at regular intervals.

"Mr. Mitchell," Nayef called back. "Prince Mohammad bin Rashid wanted to give you warning that Saudi Arabia will be slashing oil prices next week to maintain our position in the market."

Jimmy grinned. "Ain't nothin' wrong with that. Cheaper prices at the pump back home."

In a shaded alley we pulled up to a one-story building marked by a towering heap of discarded car parts. Emaciated stray cats pawed at the trash. A thin nervous man with thick eyebrows and an angular face made almost invisible by a ball cap was waiting next to the pyramid of junk like a Sphinx guarding his temple. The SUV slowed and he climbed rapidly into the back, letting in a blast of cool evening air, pulling off his cap as the car lurched forward. He stank of cigarettes and nervous sweat.

"Qasim, these are our American friends," Nayef said brusquely, his eyes on the road.

Qasim nodded a greeting. "*Salaam 'aleikum, ya sayed.*"

We pulled into a sparse field behind ramshackle cement buildings, parked in the shelter of an overhang. A handful of sizable dumpsters shielded the car from the adjacent neighborhood. Evening shadows were filling the lot, turning us into less than nothing.

"This is Qasim," Nayef finally spoke, turning and removing his sunglasses. "One of our best sources, a double agent within Fourteen February. He has been providing sensitive information on the Opposition for almost one year. No one suspects. He plays his role well—throws the thugs a few scraps about us to make them happy." Nayef nodded summarily to Qasim. "Tell our American friends what you have told us."

"The Shia are planning a very big offensive." Qasim's eyes were on the floor. "Coordinated attacks across the country. Molotov cocktails, bombs in vehicles, EFPs. Also many assault weapons—enough to overpower police patrols."

The man was dripping with grease. I held a special scorn for double agents—despite, or perhaps because of, the fact that on the other end of the barrel, I probably would've been one. Mercenaries with allegiance to none, no belief, no purpose, just deception.

I squinted at Qasim. "A big offensive? Last I heard, the Opposition was considering negotiations."

Qasim's jaw tensed. "Plans have changed."

"What are the targets?"

"I do not know. It has not been decided yet."

"When will the attacks take place?"

"Soon. Maybe weeks. Before summer."

"Please." Nayef gestured in our direction. "Tell our American friends the rest of your information."

Qasim's hands rested tensely on his knees. His fingernails were ragged and black with filth. "Iran is paying for the rebellion. All of it. I do not know exactly how much money. But it is in the millions."

I resisted a smirk. Between its plummeting centrally planned economy and international sanctions, that miserable country didn't have millions to throw at anything. Even the *Gulf Daily News*, which regularly churned out reports of Iran's impending economic collapse, didn't have Qasim's audacity.

Jimmy gave a low whistle. "Just money? Or we're talking more here?"

Qasim smiled for the first time. "Mr. Chief . . ." He looked around as though unsure of which master he should be addressing. "You would not believe it. Weapons, training, supplies. I have here a list of materiel that has been smuggled in last month." From his pocket he removed a scrap of paper, haltingly held it out to no one in particular. Whitney took it, attempting a "*shukran*" in his best Arabic accent.

"Where are all these weapons and supplies stored?" I asked.

"I do not know, sir. I am trying to find out."

"And how are they getting into the country? It's not easy to smuggle when the king controls the airports and seaports." I smiled at Abd Al Matin.

"Speedboats from Iraq. They carry weapons and people. Shia from Iran go to Iraq and join the *Al-Hashad Al-Sha'bi*—Popular Mobilization Units. You have heard of Sepah—Iranian Revolutionary Guard Corps? They run these units. Then the units travel to Bahrain. I am trying . . ." His English faltered, the last word lost down his throat. "I am trying to get more information. Here in Manama there is a small group inside Fourteen February that controls the line—"

Qasim looked anxiously at Nayef and Abd Al Matin as though seeking guidance on how much he could reveal. The intelligence officers' countenances remained firm, stony.

"Less than a dozen men," Qasim finally said. "This information—very precious. Even men in Fourteen February do not know about this group and what it does. It is secret, the most secret part of the Opposition."

"Give us some names."

"I do not know."

"At least give us the name of the group."

Again he looked to his Arab masters for direction, again received nothing. "Saraya Al-Mukhtar. Islamic Front for the Liberation of Bahrain."

I thrust my face in front of Qasim, stared into his small gritty eyes. "*Habibi*, tell me: How did *you* get this information?"

Nayef cut in. "That is privileged intelligence. Qasim is our source and we must protect details of his access. He has told you enough for now." Clasping his hands across his knees, Nayef leaned against the leather upholstery. "Now you understand the gravity of the situation. This is very serious. It is getting worse every day. Even every hour."

Suddenly, Nayef turned to me. "*Sayed* Collins,"—it was the first time he'd used my name—"I know you remember the Cold War."

Jimmy chuckled as though the comment had been addressed to him. "Hell, I remember it. My old man damn near died of jungle rot in 'Nam—"

"Exactly. You Americans understand the Cold War. A war of ideas. And ideas—they spread easily. So that people become involved who should not be involved. The wars become indirect. They are fought through . . . What is the word?" He tilted his head in my direction. "Forgive me. My English is not as good as your Arabic."

I watched him, his slits of eyes, his impassive mouth, the impeccably coiffed waves of his hair. "Proxies. That's the word you're looking for."

"Ah, *shukran*. Proxies. The disagreement between Americans and Soviets continued for years through proxies. Very bad, very serious. *Alhamdulillah*, nuclear war never happened, but it was a possibility." Nayef unclasped his hands, smiled. "Mr. Mitchell, Mr. Collins, surely you see we are fighting a second cold war. A war of shadows."

"It is a war between Sunni and Shia, between the southern Gulf and Iran," Abd Al Matin picked up the baton. "Also a battle of ideas. Territory and power. It is a battle that has become bloody and will not end peacefully. Iran has moved to one side, and Bahrain, Saudi Arabia—our merciful protector—have been forced to the other. There is no in-between. Only the Arabian Gulf."

"And do you understand how this new cold war affects you?" Nayef's voice rose sharply as he seized the conversation again. "Our fight against Iran, my friends, is your fight against Iran. You fear Tehran as much as we do. Tehran is ruthless, aggressive; soon it will have a nuclear weapon. Today Iran and the Shia target us—the Sunnis. Tomorrow they go after Americans. This is why you must pick a side. The right side."

Jimmy exhaled heavily. "Wow. That right there is some scary shit." He patted his jacket for a cigar.

Nayef turned from me to Whitney. "So you will let Langley know. And Langley will let the White House know. America's support is necessary. *Mohim*. Without it, this kingdom will lose the fight. America too will lose the fight."

Though it was dark out, Nayef put his sunglasses back on. "You Americans do not learn your lessons in the Middle East. 1979. You had an ally in Iran, a man who supported America. But you could not keep him in power. You did not give him the resources he needed. So you suffered fifty-two hostages and a brutal theocracy ready to make war with you and all its neighbors. We cannot let that happen here." Scorn crept across his features, his mouth curling into a half smile. "Or maybe you believe in the Arab Spring. You think you will get a democracy like America." He leaned forward, bared his smooth hard teeth. "Let me tell you. There is no Arab Spring. It is an illusion."

Nayef signaled the driver and the car backed out of the shadows. Abd Al Matin passed a thick bundle of American dollars to Qasim. "Good work," he said simply. Then, to the driver, "You can let Qasim out here."

27

Immediately after the meeting with Qasim, Whitney fired off multiple cables to Headquarters about the impending offensive with words like "indisputable" and "highly convincing" and "all due haste," while Jimmy crafted a counterintelligence assessment concluding there were no grounds for suspecting our Saudi and Bahraini friends or their double agent of fabrication. NO UNDERLYING MOTIVES OR DECEIT NOTED; TRACES ON QASIM LAST NAME UNKNOWN PRODUCED NO DEROGATORY INFORMATION. INFORMANT FULLY VETTED BY HANDLING SERVICES. Langley bestowed unto Qasim an official code name and asked for biographical details. Experts at Headquarters agreed that Saraya Al-Mukhtar, the Islamic Front for the Liberation of Bahrain, was a resurrection of the Iranian-funded militant group of the same name that attempted to overthrow the Bahraini royal family in 1981.

For several hours each day Whitney was gone from the office, engaged in heavy discussion with the Admiral, and by the end of the week our threat condition was elevated to Delta, all Navy warships temporarily withdrawn from Gulf ports, a Marine exercise in Jordan halted, and nonessential military air traffic in the region grounded. Local and some international media picked up the story, reporting indications of an as-yet-unknown threat that was both "serious and credible," details still emerging. Checkpoints sprinkled the roads like birdseed; blinding lights secured curfew with unmovable fists.

SCROOP needed to drop everything, Whitney said—no more strategic plans or political machinations—and focus all his time and energy on Iran. Perhaps he could insinuate himself into Fourteen February's innermost sanctum, Saraya Al-Mukhtar.

Vic's never seemed touched by the world outside. Stopping for a drink after work, I was amused to see the place unchanged since the holidays, still decorated in a Christmas-in-the-Tropics theme, ornament-strewn palm tree anchored by starfish-dotted mounds of cottony snow, pineapple sconces stocked with red and green bulbs, origami snowflakes and silver tinsel hanging from the teak rafters.

I ordered a Samoan Fog Cutter, pulled Qasim's crinkled paper from my pocket that Whitney had instructed me to translate. The handwriting was hard to decipher, but I could make out a list of weapons with numbers, presumably quantities. If it weren't for me, I mused, Qasim could provide a goddamn grocery list and Whitney wouldn't know the difference. I held the paper close to the light.

Arms (170 Kalashnikovs)
Ammunition (52 boxes)
Magazines, loose (30)
RPGs (12)
Mortar rounds (93)
107mm rockets (34)
Plastic explosives (? Kg)
C-4 (? Tonnes)
Detonators (63)
Ball bearings (14)
Grad missiles, BM-21, 122mm multiple rocket launchers (2)
Anti-aircraft missiles (3)
Long-range Fajr-5 missiles (1)
Passive infrared sensors (41)
Hand grenades (10)
Explosively formed penetrators (?)
Ammonium nitrate (? Kg)

If true, it was a decent cache, certainly not something Rashid had ever revealed. If true. Qasim—his downcast eyes, his discomfort that bordered on shame. What did we know about this man? His information supported the monarchy's conspiracy theory, that was for damn sure. And it didn't add up. Too much money involved. How were all these weapons getting onto the island? A few speedboats couldn't explain it.

Convenient, it was all so convenient. Everything spic and span, perfectly tailored. *Listen to my prediction,* Rashid had said in Qal'at Al Bahrain. *One of ours will be arrested. Blamed for Adliya. And maybe America will lift the embargo.*

I drained my glass, headed outside. The sky was clear, the Fog Cutter electrifying my brain. Suddenly, there it was—a passage, smooth and easy. A way to the truth. Why hadn't I seen it before?

28

My rearview mirror showed little on King Faisal Highway. Too early for the Middle East to wake up. A black sedan was far behind me, but it turned onto Al Jaser Avenue. No turn signal, so probably not surveillance, which invariably obeyed every traffic law, even in countries with no traffic laws, in an absurd attempt to prove its banality. Off Al Fardah Avenue I pulled into my cover stop for a pack of cigarettes. It was an unusually hot day for March, the sun beating down with the ferocity of midsummer.

"Coffee, black, *min fadlak*. Man'ousheh—plain. *A'albat wahidat* Gold Marlboro Smooth," I ordered from the man behind the counter. Stacked along the rear in gold-and-green rows were packs of Canary Kingdom. "*A'albat wahidat* Canary Kingdom," I added. "Green apple."

The cold store was grimy and overflowing with third world products: herbal toothpaste, gray paper goods, ancient candy. A slim Arab stood in the corner examining something. The cashier slid the Canary Kingdoms across the counter.

Inside my Lancer I shoved the doughy man'ousheh into my mouth. It was overly spiced and sprinkled with stale cheese. All along King Faisal Highway police were setting up checkpoints in the thinning dawn, vehicles inching into place like clumsy black dinosaurs, purple-haloed headlights piercing the smog. The Gulf sparkled gray and placid. As I passed the King Faisal corniche, a policeman atop a vehicle followed my Lancer with the barrel of his Kalashnikov.

By the time I arrived, the souq had begun its daily buzz. Late morning heat hung in the air like cobwebs. Throngs had gathered beneath the Bab Al Bahrain arch, shouts growing louder as I approached. *Sir, I give you good deal on diamonds. For your lady, sir. Finest gold at best prices!* Fragments of Tagalog, Hindi, Sinhalese. Clusters of birds had nested in the scraggly leafless trees, and their strident chirping blended with the sales pitches to create a frantic disharmony.

Some combination of forces—teeming masses of vendors, furnaces cooking crisp and sugary snacks, narrow alleys that left little room for cleansing breezes—had created a heat vacuum inside the souq that surpassed even the deserts of southern Bahrain. Spice alley was fragrant and crowded as usual, burlap sacks threatening to spill their vivid contents: crimson chili peppers, gold saffron, mountains of dried cranberries, coppery turmeric. Many of the vendors, I saw, had shuttered since the last time I'd been there. Lost business because of the unrest.

Our meeting location, a dark alley on the fringes of the gold souq, was blocked by dumpsters, inaccessible. Bad fucking luck. I sheltered under an awning outside a trinket shop, wiped my forehead. The sun was reaching its apex and even the shade offered no relief.

Three minutes. From around the corner Rashid appeared. His black eye had healed and, in clear defiance of my orders, he was wearing a freshly laundered white thobe. His eyes slid casually in my direction, registered the adjusted meeting location. Halting beneath the awning a few feet from me, he stared intently into the trinket shop's dark windows.

"Welcome home. How are things?"

Rashid pulled a Canary Kingdom from his pocket. "Fine, everything is fine. No more problems like last time."

"How is your family?"

"Good, *alhamdulillah*."

"And your sick uncle? Recovered, I hope?"

"He is dead." Rashid blew smoke into the heat.

A group of men walked past us, laughing, sandaled feet peeking beneath their thobes. I waited until they'd reached the end of the street. "Fourteen February planning a big offensive?"

"Big? No. We have a few small plans. I will tell you when I know more."

"That's not what I hear."

Rashid's head turned slightly in my direction. "You have another informant? Someone who tells you different information?"

I said nothing.

"I will find out. I missed many meetings while I was away. Maybe there is information I do not know."

"Fine." I lit a Gold Marlboro. "Then let's talk about Tehran. That's why I signaled you. Tell me about Tehran."

Rashid, I could see peripherally, tensed, as though he'd stepped on hot coals. "I told you, there is nothing. Why are you asking me this again?"

"Sometimes things change."

He chuckled bitterly. "Believe me, *habibi*, if Tehran offered support, we would say yes. But we have received no offers."

"Or you just won't tell me."

From a nearby mosque, speakers broadcast the midday call to prayer, the muezzin's beseeching voice carrying through the souq. Rashid tossed away his cigarette, pulled a mat from his bag, and kicked off his shoes, falling to his knees, his body turned briskly to the west. Rather than attract attention with my perpendicularity, I too dropped to the ground, the cement pressing painfully into my aging knees.

Suddenly, Rashid turned to me, his face a smudge of dark angles against the sunlight. "You pay me money and I give you information." His voice was rising. "I do everything you say. I tell you the truth. And eight months we have known each other. You are my brother now. Brothers trust each other." Angrily he turned back to the pavement. I could feel his wrath rising like heat from the sidewalk.

When prayers ended, I stood up. Slowly, as though in deliberate misstep with me, Rashid followed.

"Listen, you want the truth? I'm going to tell you something I shouldn't. I'm gonna level with you." I paused to prepare him for the weight of my words. "My headquarters doesn't take the Opposition seriously. This revolution thing—it's getting old. Two years now and what do you have to show for it? Nothing.

"Now look," I continued. "America's like any other country—practical. We want to bet on a winning team. And right now you're not winning. Which means you're losing. Get it?

"So here's the deal. We keep hearing rumors about support from Iran—money, weapons. . . ."

Rashid scoffed. "You want me to lie? You want me to say we have something we do not?"

"Because, my friend, if you *did* get help, that might prove you had some teeth, that there's strength behind your movement." I shrugged. "Might even convince the President to throw more support your way. Know what I mean?"

Rashid was silent for several seconds. When he spoke, his voice was low, incredulous. "You are telling me America would support Fourteen February if *Iran* is behind us? This is . . . crazy policy. It makes no sense."

"Well, it's more complicated than that. Iran—sure, we don't like Iran. But what do they say? The enemy of your enemy is your friend? Something like that."

"Yes. Ancient Sanskrit proverb." Rashid's tone was growing warm, superior.

"Right, that's it. You see how it is? If Tehran wants to send money—or even, say, weapons—to a worthy cause, we're not going to complain. In fact, it works out pretty damn well for us: We get reformists in power without showing our hand. Without taking sides. You see? America's on the side of the Arab Spring; we just can't let the goddamned world know." The words were pouring out now like gold. I was Ezekiel. "Can't make any promises, but show me we're on a winning team and things could start to change pretty quickly."

The hum of the marketplace had surmounted its midday peak, dwindling to a lazy buzz, *qailula* minutes away. Sweat trickled into my eyes. I'd made a play—a losing play, undoubtedly—but a play nonetheless, and there was a certain satisfaction in the movements. This was the game I'd played all my life, and I knew it better than myself.

Rashid was mulling over my words. I feared I'd lost him, my pitch failed. Finally, he said, "I would like to show you something."

"What is it?"

"I cannot tell you. I must show you. You come back here tomorrow, same time."

Something stirred inside me, a small engine kicked to life. This was it.

I would return tomorrow, be finished with it all, and have done a damn good job of it. Whatever Tehran was throwing at the Opposition—fifty Kalashnikovs or just a box of slingshots—I would be the one to uncover it, dig up the hard truth.

"Okay, here's what we'll do. You leave left, I leave right. I'll pick you up tomorrow, same time plus one hour, behind the Bab Al Bahrain Hotel. If it's too crowded, I won't stop. Walk to the old synagogue and I'll pick you up there."

Rashid nodded.

I was a masterful manipulator, deep red blood coursing through my veins, puppet strings of woven silk at my fingers. Manipulation is like a thief breaking into a house: He studies its weak points—lighting, locks, visibility, times of habitation—decides what he wants to steal, how he's going to do it. I'd executed an elegant theft. Espionage, in the black of night, was about outsmarting your opponent. After all these years, I could still do the job.

29

The alley behind Bab Al Bahrain Hotel was not crowded and Rashid slipped into the passenger seat swiftly. He wore a scarf around his face. "Go to the villages," his muffled voice said.

West on Budaiya Highway to the Markh exit. Knee-deep in the slums. My Lancer sputtered as we slowed in the alleys—piece of tin was sure to break down any day now.

Behind the Abu Subh Mosque, between two sagging houses, was a small carpentry shop. I'd passed it before. It had one of those lost-in-translation names that stuck in your memory: Builders from Heaven. When I started to pull in front of the shop, Rashid shook his head. "Too close. Next block."

Before we got out Rashid pointed to my ballcap. "Pull that over your eyes"—glancing out the window—"and turn off your cell."

It was the first time he'd issued orders to me—a strange feeling, like your child telling you what to do. I pulled my cap down as far as it would go. "This do, Chief? Cell phone's already off."

He nodded. "Fine. Now we go inside."

Surprisingly little furniture was on display—a few desks, large cabinet, mismatched chairs without tables. All built of the cheap composite material necessitated by the shortage of trees, the same material seen in nearly every Bahraini household save those of expats and the royal family.

A thin man with close-cropped hair wearing a faded polo shirt and jeans rushed from the back. "Sirs! How can I help you? Please have a seat."

"We are here to see Mohsin," Rashid said.

The man hesitated, eyes digesting my cap and sunglasses and bag, and I understood that Mohsin didn't exist.

An oversized cheap bookcase covered the rear wall, filled with odds and ends, copious quantities of industrial junk. Pushing aside a large toolbox on a middle shelf, the man leaned over, his lean shoulders covering his movements. Flash of a rusty door lock, a key coaxed inside. The rear of the bookcase swung open, bringing the shelves along with it. There was a small rectangular opening in the newly uncovered floor, and Rashid reached into it, produced a flashlight. He nodded to the man. "*Shukran. Tagdar tsekker al-bab waranna.*" *You can close the door behind us.*

I removed my sunglasses. We were in whole gritty darkness. "*Ana bothaq fik*," Rashid said simply. *I trust you.*

Only a few feet inside, a short flight of stairs emerged from the darkness like a bat awakened. Rashid's weak yellow circle danced erratically as we descended. The stairs ended on a platform, a long passage before us, mine-like and humid, suffused with the odor of metal. Another door at the end. Rashid rummaged in his pocket, pulled out a key. No single person had all the keys—good tradecraft.

A cavernous dirt room stared back at us: dank, near-broiling, the air choked with small particles of debris. Three kerosene hurricane lanterns were the only light source. Rashid ignited them, the weak flames flickering to life, barely penetrating their dirty glass prisons, throwing dim arcs on the soil walls. He grinned and his teeth looked like proud yellow pearls in the darkness. "Welcome to the revolution factory."

Spread across the room were dozens, maybe hundreds, of glass bottles, their uneven heights and varied hues conjuring an eclectic soda factory. Piles of dirty rags and drums of gasoline lined the back wall. In the middle, a toolbox revealed a haphazard assortment of nuts, bolts, and nails. Two ten-gallon plastic jugs sat in the corner.

I walked over to the jugs, crouched to get a better look. "C-4?"

Rashid shook his head. "C-4 is too hard to get. We are not Al-Qaeda. It is our own creation. Homemade. Fertilizer, paint remover, sanitizer the *Amrikiyin* use for their swimming pools."

This was their revolution factory, the formidable enterprise so feared by the king and his men. It was skeletal, pathetically raw, fashioned from remnants of bigger wars, from leftover materials and knowledge. A fraction, a sliver, of what Qasim's intel suggested—and none of it imported, all homegrown. The most efficient assembly lines within these four walls would produce little more than nuisances, thorns in sides, certainly not enough for the widespread full-scale assault foretold by Qasim. A few Molotov cocktails against a sea of gas. A clumsy explosive device against automatic weapons. Might as well be slingshots against nukes. Staggering asymmetry.

Easily the temperature had reached 100 degrees in that room and I removed my cap, wiped the sweat from my face. Almost funny when I thought about it: how wrong Whitney was, how stupefied he'd be standing in my shoes, dumb with disbelief.

A pile of shallow wooden boxes was stacked in a corner. I picked one up, examined it. It was held together with thin wire, constructed of the same flimsy composite material as the furniture upstairs. "Pressure plates? For IEDs?"

"Yes. That is our plan. We are still learning."

"Who are you getting lessons from?"

Rashid opened a drawer of the toolbox, handed me a tattered manual, hand-bound and sloppily assembled. "The internet. And some of our men have traveled to other countries—Iraq, Syria." His face glowed with reverence.

The pressure plates were crude, far below the caliber I'd seen in Iraq or of any insurgent group worth its name. Still, fingering the device, I felt a modicum of respect. There was a certain resolve, determination, in the gadgets' roughness.

"What is so funny?" Rashid's voice was caustic. I hadn't realized I was smiling.

"You think our weapons are a joke? Pathetic?" His voice was rising, his fingers twitching.

"Not at all." I indicated the boxes. "Hungarians used less than this to stop the Soviets in 1956. They put porcelain plates on the Budapest streets and that was enough to spook the Russian tanks."

"Exactly! They used their brains. They had courage." Rashid gestured around the room. "Now you see."

"Not bad. Fourteen February has done a lot with a little."

"I knew you would appreciate our work. I was right to bring you here."

My eyes scanned the chamber again. "This is it, huh? No other secret room where you're storing shiny stuff from Iran?"

Rashid looked at me stonily. "My friend, you tell me we get more support from America if we prove we are . . . powerful. I bring you here to show you we are powerful. But not because of Iran. Because of what is inside this room, this *factory*—courage, *indibat,* commitment. And we are powerful because of *this.*" He gestured toward the wall behind me.

It was then I saw, behind a stack of crates, the wall plastered with photos: hundreds of sepia faces staring at the crude materiel, some glossy, some matte, their expressions grave and piteous as only true believers can be. In the center, the apparent object of the two-dimensional shrine, was Junaid, watching over his fellow revolutionaries with the calm confidence Rashid found so inspiring. The picture looked amateurish and outdated, Junaid in his youth and at his best, but I recognized him immediately, his face ingrained in my mind by the monarchy's dogged propaganda.

I moved to the wall. "Martyrs and . . . ?"

"Prisoners who can no longer help the cause." Rashid walked to my side. His voice dropped to a whisper. "Do you see? The people. Not weapons. People. This is our revolution."

For several minutes we were silent. Finally, Rashid spoke, staring penetratingly at the photos as though trying to bring them to life. "We will keep fighting. And we will keep dying. We can outlast the king's men. If you have the people on your side, you cannot lose."

And in the worshipful hush that followed, I had a flash of inspiration. "My friend,"—I put my hand on his shoulder—"can I get a picture of the factory?" Then, quickly, "I just want Langley to realize your potential, see how much you boys have done. This isn't half-bad for what you're working with. It might help. It's why you took me here, right?"

For a moment Rashid seemed to consider the idea, then he shook his head. "No. This is a secret location and we cannot risk compromise. You have seen it with your own eyes. Now you can tell your people."

"No pictures then. You're the boss." I put my cap on, moved toward the exit. Waited a moment then turned around. "What if I promise it will stay at station? No one except my chief would see it. Guaranteed." The instant the words tumbled out, I knew I'd opened the safe. In the final calculus true believers can never resist.

When I was done snapping photos, Rashid picked up one of the smaller pressure plates, held it delicately as a baby. "Take this too. Show your people we are serious. Show them our determination. *Tasmim.*"

I didn't know what Whitney would make of a few pieces of wood and wire, but I put it in my bag.

30

We hadn't driven far beyond the revolution factory when we saw the smoke. Riots on Budaiya Highway. Confined, modest in size, easily avoided. I reversed direction, headed east on a parallel road. Flashing yellow orbs appeared ahead, sweeping the decrepit neighborhood. Hastily erected barriers rose from the haze. In the periphery one shadow hunched over another, moving up and down like a relentless oil derrick: a cop beating someone. It was a checkpoint.

Rashid spoke my thoughts. "We abandon the car."

I pulled behind an assemblage of construction equipment, grabbed my bag, stowed the memory card from my camera in my pocket. I could discard it if necessary.

"Exit on my side," Rashid instructed as I reached for my door handle. "Farther from police."

I snickered, dubious about his attempts at invisibility (there is no true invisibility, I always reminded him), but hoisted myself nonetheless over the parking brake and out his door.

Crawling on all fours like we were goddamned GIs in the jungle, we ambled across a dirt lot, behind a cement wall. There we crouched, the wall's vexed graffiti a colorful backdrop for our afternoon drama, the audience—lizards, various and sundry vermin—waiting for the story to unfold. I slumped to the ground. Rashid had inched to the edge of the wall. All safe, he signaled with a thumbs-up.

Checkpoints—worst nightmare of handler and informant. Passages that determined destinies, friend from foe, sometimes life from death. A checkpoint in the slums carried its own particular risks. Rashid was now high enough in the insurgency hierarchy that he would undoubtedly be made an example if caught—humiliated, life for his family made unbearable, rock-bottom prison conditions. Or worse. And my own unpleasant future if discovered with an informant—persona non grata at a minimum, a headlines-topping scandal, maybe even imprisonment.

I lit a cigarette, remembered the gift I'd bought Rashid the day before. "Green apple," I proffered, producing the pack of Canary Kingdoms. Rashid looked surprised, pleased.

Chants of protest, quieter now, carried from the highway, punctuated by the familiar pop and hiss of riot control gas. Smoke hung in stagnant loops around us. Across the lot a weeping eucalyptus taunted us with a glimpse of its feathery headdress, gray-green leaves drooping over a wall, looking strangely beautiful in the poisonous mist.

"This country," Rashid smiled. "Not the island vacation you thought it would be."

I shrugged. "My expectations weren't too high."

He sucked thoughtfully on his cigarette. "Maybe, *habibi*, this is your problem. You have no expectations." He swatted at the haze. "You are like this air. Empty."

"This air is pretty potent."

"Yes. I am right," Rashid persisted. "You just exist. You do not believe. You do not take sides."

"Well, my wise friend, you're right on that count. But my job is not to take sides. It's to find people who take sides. And pay them. I don't get involved."

"If you do not have something to believe in, you have nothing."

A siren wailed nearby. While we were talking, night had crept into the slums, turned the splintered houses into angry faces. The whites of Rashid's eyes stood out like moonstone. "You do not care about this revolution, what will happen," he said.

I looked away. The chants were dwindling now to a distant song, and the faintest suggestion of traffic was discernable. Rashid's concrete shack was nearby. His wife and kids would be there, maybe wondering about his whereabouts, his safety.

"How's your wife?" I offered.

Rashid hesitated, looked away. "Not well."

"Something wrong?"

"She lost another baby."

The words hung in the air. I offered a bungled apology. Then I realized he'd said "another."

"She's lost more than one?"

"Five."

"Five . . . Jesus. You take her to the doctor?"

"Yes. It is from the gas." He gestured toward the highway. "We live on the front lines. Huda has gone to protests with me. She is fragile. The king's poison has made her sick, taken away our future children." From his wallet he produced a wrinkled dirty photo of his wife. She was uncovered, wearing a well-used apron that hinted at her culinary predilection, a soft unassuming figure that looked as though it absorbed everything equally—food, embraces, tragedy.

Suddenly, Rashid withdrew the photo. "You have never asked about my brother. Do you know that he was killed on the Day of Rage? The day the revolution began." His lips quivered. "My mother, may Allah keep her, is dead. Tuberculosis. She had no medicine. My brother—he is dead. For my father, for my sons, only I am left." His face, resistant until now, crumpled. "You know why I join the revolution? This is why."

Our cigarettes were glowing like dying fireflies. "I'm sorry," I said.

"The checkpoints are gone, I think." Rashid spoke softly. "I will just walk home."

Minutes passed before I pried myself off the wall. The air had cleared, but somehow it felt heavy. My Lancer waited for me, undisturbed.

I hadn't driven more than a few blocks when a circle of light appeared. Broad diameter, the source far away. Instinctively I slowed, extinguished my cigarette. Police still looking for protesters—someone specific, maybe. Then the searchlight disappeared and the only sound was glass breaking in the distance.

At the edge of Karranah Cemetery the yellow circle materialized again.

This time it was trained directly on my car, the windshield electrified by its inquisitive glare. I turned off my headlights, slowed to a crawl, put on a relaxed genial face. The danger was over—Rashid had gone home—and I had my cover story: visiting a friend in one of the gated expat communities. The Johnsons if pressed, lost on my way back if pressed further. All very plausible—I was in a car, not on foot. If all went as planned, no sane Bahraini officer would think to push an American beyond that or to search my person. My memory card, the revolution factory, was safe. I proceeded toward the beckoning light feeling an overinsistent sense of security, mustering the smugness to which all my years in operations entitled me.

"Halt!" A rough voice carried across the darkness in accented English. I braked, reached for my bag on the passenger seat, rummaged inside for my diplomatic passport, a sure and easy ticket out. The pressure plate. I'd forgotten about the pressure plate. It had slid from under the passenger seat into the anterior foot space. I needed a drink, something to keep my hands from shaking. My cover story was lousy. Americans didn't mosey around the slums during a protest; they avoided the streets, stayed home until the tumult had died a certain death.

My passport was proving elusive. It had to be there. I could feel the supple worn leather of my bag, the sensation heightened as though I were touching expensive cashmere. My vision was winnowing and expanding, the cemetery's stumpy tombstones growing alternately sharp and fuzzy. It was in a side pocket, maybe—

"Get out!"

Willing my hands to a manageable tremor, I exited. My flimsy beater was probably arousing suspicion; the cop had surely pegged me as a local. The light blinded me, hid the operator's figure. Rapidly the circle moved toward my face. He would see I was Western—or at least hear it in my voice. I prepared a smile, started to ask directions. *Good evening,* habibi. *I seem to be lost. Can you help—* The gunshot sliced the darkness.

Silence, the slums releasing their hot wet breath in a long sigh. A muted thud and the yellow circle dropped to the ground, rolling in gentle arcs.

And then Rashid was beside me, his hand still gripping the pistol, offering a simple explanation. "He was armed." He made a motion of shooting, shook his head. "I saw him from the side. Nervous policeman."

Rashid relaxed his hand and I took the pistol from him, stared at it. A gasp like a drowning man coming up for air—the policeman had stirred. A second shot. I hadn't even realized I'd done it. I glanced around: no one in sight. Sounds of the protest had disappeared entirely. The night was hot, too hot for March, and the air smelled of gunpowder. "Come." Rashid snatched the weapon back, shoved it inside his pocket. "We get the body."

The man had fallen almost face-first onto the ground, the dark pool of once-sustaining fluid creeping quickly away from his body as though fleeing the scene. There was indeed a gun, a Caracal pistol, still in his flaccid grip. He was a lefty like Rashid. Using his foot, Rashid prodded the man until he turned over. One of the shots had felled him squarely in his chest, unprotected by armor. It had made a mess of his uniform, blackish blood splattered onto his face, rendering him unrecognizable. Amid it all, his eyes were intact and open, staring straight ahead—at me, of course, still trying to figure out if I was friend or foe. Rashid stomped on the man's left hand until it opened like a compromised safe. He kicked away the gun, swooped down, stowed it in his jacket. Turned the policeman's hand radio off, a precaution that would have escaped me.

Rashid was kicking sand into the seeping blood. In the dry air the dirt soaked up the liquid almost instantly. He spat on the man—"*Khayin!*"—then, raising his face to the sky, muttered an incomprehensible prayer. He knelt down and grabbed the man under his arms, nodded for me to get the legs. The policeman was not big—modest in stature, reedy, little more than a collection of jangling bones—but carrying him was difficult. Dead bodies, I suppose, are always heavier than they look. Sweat glued my clothes to my body, preventing me from moving too much or too rapidly. I remember being surprised at the relative ease with which Rashid carried his end of the deal—limberly, almost effortlessly, with what seemed like preternatural strength.

We stumbled across the street, our package swinging clumsily, not unlike the sensation of carrying a heavy golf bag, its clubs rattling and shifting unpleasantly inside the malleable sack. Rashid was moving in the direction of a dumpster on the corner. I stopped. "We're putting him in the trash?"

"It is okay, my friend. No one will find him. In the slums they never take away the garbage."

I put down the man's feet. "This won't work. They'll search the area."

"You have a better idea?"

The cemetery behind us now seemed like a cruel joke. I turned around, verified that my trusty Lancer was still there, parked conspiratorially between two streetlights. "I'll drop him in the Gulf."

"No, I will do it."

"Your car is too far. You're more likely to be followed. I'll do it."

Without another word we changed course, shuffled back toward my car. Clouds shrouded the moon entirely, turning us into common thieves, lying silhouettes, making our work charmingly invisible. "Watch out for your clothes," Rashid called. "Keep them clean."

We reached the Lancer and I opened the trunk. "Wait"—Rashid grabbed fistfuls of dirt and shoved them into the man's wounds, making him even more grotesque, stemming the residual fluids still attempting to escape. I removed the plastic tarp I kept in back, folded it into a bed. Rashid nodded approvingly. With a few swings for leverage, we slung the body into the car. Rashid stared at me with inky eyes. "There was no choice, *habibi.* He would have shot you."

I sat in my car emptying the flask in my glove box until I was sure Rashid had made it home. My hands were rattling on the steering wheel as I drove through the slums. The yellow circle still seemed to dance before me, imprinted on my retinas, bobbing in the darkness like a drunken fairy. I'd drained the flask too quickly. Vaguely I wondered what time it was, whether it was past curfew, somehow unable to check my watch. In the distance I could hear faint braying. More protests? No, the king's five-acre camel farm, not far from here. I pulled onto the highway.

The city was dark and quiet as I drove south to keep my promise. I would never get there before curfew. My iron grip on the wheel seemed to spur my Lancer forward like a racehorse. The wind cut my face with hundreds of small knives through the open window. Down through Tubli, past Riffa Views where Jimmy and Whitney slumbered peacefully. *There goes Collins with the corpse in his car. That guy was never quite right.* Checking my rearview mirror. South of Awali dozens of twinkling lights emerged near the scrapyards. Faint burning smell—mechanics working overtime or maybe trash fires.

I stopped in Askar, the farthest I could go, I reasoned, without straying too far past curfew. Drove through stumpy neighborhoods until they petered out and I'd reached the sparse rocky coast. A few dhows slept in the water, their weathered wooden bottoms rising and falling like somnolent sea creatures. An old Shiite fisherman who supplied shrimp to the fancy restaurants in Adliya lived in a shanty here; I'd met him early in my tour and used him once or twice as a support asset, a courier of sorts, someone to buy burner cell phones when I could not. It had been months since I'd contacted him and I wondered distractedly whether he was still alive on his subsistence diet of half-rancid leftover catch.

Still I drove until the last dhow was out of sight, until the only life-form was families of cormorants, perched like ghosts in the warm shoals. The aroma of salty shallows signified the tip of the mangrove fields. At night and without the help of Manama's city lights, the mangroves were a dense forest of black thorns, thousands of angry claws wrestling each other, the tops iced gray in the dim moonlight. My tires had become grinding teeth; I'd reached sand.

I'd driven onto a small promontory and now I parked. The body was far heavier coming out than when I'd lifted him with Rashid (how was it possible, I again wondered, that a slight man could have such strength?), and he exited feetfirst, torso and head clumsily following. Quickly I rummaged through his pockets. Empty as a beggar's. How could the man have no identification, no coins, not even a lousy cigarette?

The tarp had worked—caught the blood—but now the rear bumper was covered in the slimy stuff. I pulled the tarp out, trying to mop up some of the mess in the process, and wrapped the body in it from the underarms down. With the near-Herculean strength of necessity I grabbed the policeman under his shoulders, lugged him to the edge of the promontory. Some of his heft, I realized, was due to the uniform he wore—even without armor, there was a holster, truncheon, radio. Better to let it weigh him down than try to dispose of the telling items.

The tide was weak and sluggish. I waded backward into the shoals, the putrid water swirling around and up my pant cuffs like hungry wolves sensing blood. Water had reached my knees by the time the corpse was submerged and I could relax my grip. *Always place a corpse faceup in the water so the lungs can release air and fill with liquid*, I'd learned in training—for what purpose, I don't know; spooks weren't supposed to kill, just defend themselves if necessary. But that's what this was. Self-defense.

I resigned myself to ruining my clothes and waded as far as I could from the shore. Almost immediately the body started to sink. His face was the last to go, eyes still shocked and open, face bloated, the dark eddies of watery dissolution tightening their protective caress. A wide diluted cloud of blood was spreading across the surface, turning the gray water burgundy. As the body floated gently to the bottom like an autumn leaf, I kept him underfoot, vaguely fearful he would suddenly and rashly attempt to rejoin the world of the living. He reached the bottom with a soft squish and I kicked him down the sloping Gulf floor toward deeper waters. At least a few days would pass before the body began to decompose and resurface, maybe more because of his uniform, long enough to drift far from the sands of Askar. Even if someone found him, the dead cop couldn't possibly be traced to Rashid. No way he could be traced to me.

My pants were pasted to me like a mummy's rags, and slogging back to the shore was worse than walking through dense mud. I collapsed on the sand. The effort had emptied my insides. I stared up at the starless sky clothed in layers of darkness, clouds atop clouds, always shades of gray, never blue. The clouds were marching slowly southward. Good. The corpse would drift toward the bottom of the island. No one lived there.

When I finally mustered the energy to rise, my muscles screamed, an ache unlike anything I'd known in the last decade. A half crawl was all I could manage to the car. The Gulf had cleansed my clothes of the worst of the blood, and for a moment I considered keeping them, too exhausted to struggle with the wet rags. Then I saw the stains on the rear bumper and, summoning the vestiges of my strength, peeled off all my garments except my shorts, wiped the bumper clean. Tomorrow I would hire the disabled Indian to wash it thoroughly. Put some money in his pocket. By the time I'd balled up my clothes and shoved them under the seat, it was well past curfew. Find a hunker-down point a safe distance away, sleep in my car.

That junkyard I'd passed in Awali—I would go there. I started the engine and drove instinctively. My thoughts turned backward. I hadn't killed him—Rashid had. No reason to feel anything at all. The man was just a cop, someone I knew nothing about, an armed lackey who would, as Rashid insisted, probably have shot me. Not an innocent. Nobody in espionage could justify sympathy. Right and wrong existed on paper, but in reality all that existed were the amorphous masses of in-between. He couldn't tell I was American, wasn't well trained, operating solo in a dangerous area. A time and place where you killed or were killed. He would have fired the gun. There was no choice, as Rashid said. Certainly, *I* hadn't had a choice; I wasn't the one who first pulled the trigger.

My hands, which had been shaking since Budaiya, relaxed to a faint tremor and my heart slowed, effects usually impossible without several flasks of alcohol. A strange sensation crept through me. It was a feeling like floating in water or reaching a mountain summit after a backbreaking hike. My sweat dried and my legs felt rested. The lights of Manama were sparkling in the north. I watched the moon emerge from the clouds. It was undeniable: the child's excitement growing inside me, the sense that I could outwit the smartest opponent, escape the meanest trap, knock down a champion fighter with a few well-placed blows. The whole of the day, the nearness of it all. The kind of exultation that hadn't touched me for decades. What veteran spooks describe as the moment hairs stand tall on the back of your neck. The meal to satisfy a dying man's craving.

That evening a miracle happened. Rain fell heavily, bathing the dusty land, the morning awakening to marbled gray. Pitter-patter on car roofs in the junkyard roused me just before dawn, swelling loudly like an African drum circle. For several minutes I sat in my half fog, staring at the hollow vehicles and sideways spouts and muddy rivulets, moving my stiff muscles, peeling my bare back off the seat. My stomach was full of acid, like I'd swallowed a vat of batteries after a bender. I watched the drops splatter across my filthy windshield, cleansing my beater, cleansing the whole island, washing away any residue of sin in the junkyard, on the Askar beaches, in the slums.

The rain continued all the way home. Manama had grown soggy, the city's half attempt at a drainage system rapidly succumbing to the overflowing streets and sidewalks. The Muharraq souq was nearly deserted, slanting in the rain. I parked in a muddy lot, paid a nearby Indian sheltering beneath an awning twenty dinars to buy me clean garments and a pack of cigarettes. *Rain ruined my clothes.* When he returned, I got dressed, drove to Busaiteen, found a never-emptied dumpster, and buried my old clothes deep inside it.

Back to Amwaj Islands, the rain petering out to a trickle then nothing at all. The dust had cleared and everything looked startlingly, unrealistically, clear and crisp. Colors long hidden emerged from the desert, painted my windshield. The gritty mix on the streets crunched beneath my tires. When I got home, though my car no longer needed it, I paid the disabled Indian for a washing. He stared at me in gratitude and I gave him an extra fistful of dollars.

Just before sunset the clouds departed and the sky turned glassy blue and clean, cleaner than I'd ever seen in Bahrain.

31

The following morning the only major story in the papers was a tribute to the Gulf Cooperation Council's progress on nuclear energy. No mention of the missing policeman.

When I got to work, Jimmy was sitting at his desk, muttering to himself, Whitney typing away in his glass office. The coffeepot was gurgling and the ceiling fan was whirring. No one the wiser. I busied myself for several minutes, then showed Whitney the pressure plate and photos of the revolution factory. He seemed duly impressed by my access, asked for details, told me to send a report to Headquarters and run it through him. When I told him I wouldn't reveal the location, that the photos were close-hold, for our eyes only, he nodded, said he understood.

"Pretty convincing evidence, don't you think?" I pushed. "Opposition's not working with much—certainly not an arsenal from Tehran. Not what our friends Abd Al Matin and Nayef would have us believe."

Whitney shrugged. "Maybe. Or maybe SCROOP's just not privy to everything."

I put the pressure plate on his desk. "Why don't you keep this as a souvenir? It will look great next to your ship."

Word reached station that a high-ranking member of the Peninsula Shield Force, a senior aide to Major General Khalid Al Hasan, had been reported missing when he didn't show up to work the day after recent protests.

Speculation, given his high position, that he'd been kidnapped, a ransom, political demand, or other scheme planned. A sweep of the protest area, where he'd reportedly been supervising operations undercover as a low-ranking policeman, was underway.

The news was poison in my stomach. Fucking cooler's luck that the bastard had been top brass; any other puke and no one would've noticed or cared. I signaled an emergency meeting with Rashid, ordered him to lay low again—sleep elsewhere, at friends' places, as far from home as possible.

Three days later a body washed ashore in southern Bahrain below Al Dur, several miles from Askar. A lance corporal from nearby Shaikh Isa Air Base discovered it; as an official servant of the king, he'd reported the corpse with all due haste. The face of the dead man was horribly mangled and distorted—unrecognizable, papers said—but the knowledge that an officer was missing and the coinciding facts were enough to identify him. That and his uniform. I hadn't removed his uniform. The sturdy blue garments had remained remarkably intact, reported the *Gulf Daily News*, a testament to the foresight of the Minister of Interior and his meticulous battle considerations. A bullet wound was found—almost certainly a homicide—and an investigation was underway commensurate with the gruesome nature of the crime. *Evidence indicates the cruelest possible methods were used. Only barbarians would defile and dispose of the body in this manner.* Location of the presumed homicide and cadaver disposal point still unknown. And checkpoints spawned manifold, and metal birds swooped and careened in vulturous arcs.

It wasn't long after that Whitney called me at home and told me to come into the office. To discuss a "crucial matter," he said gravely. When I told him I was taking a sick day, he said he would come to me. "Forget it," I gave in. "I'll meet you at Vic's. Twenty minutes."

I grinned boozily at the newest addition to Vic's, a security guard out front, shoulders slumped, hand resting lazily on his weapon. The hanging wooden beads clicked a leering welcome. I was fully aware of what the contents of this conversation would be, braced for the banal revelation that the Opposition had been involved in the police officer's murder. Maybe there was more. Maybe eyewitnesses had placed Rashid near the protests; maybe the Bahrainis had fingered him even without witnesses. Hell, maybe the guy I'd shot had reported a dubious American on his radio before I'd stepped out of the car. I stood in the doorway, blinking as my eyes adjusted. Was it possible, I wondered, that the king's men were savvy enough to put the pieces together? Maybe the Indian at the Muharraq souq where I'd bought dry clothes had snitched on the suspicious man who'd hired him. Maybe the dumpster where I'd deposited my clothes was finally emptied. Had someone been following me, my powers of surveillance detection atrophied to the level of an amateur cop? Maybe a local had spotted me on the beach or sleeping naked in the junkyard. The country was crawling with informants. Rashid had an alibi after he'd left the scene; I did not.

The desperate thoughts of the damned scratched my mind, the pathetically wild hypotheticals that emerge when one is cornered. Maybe, even if everything were lost, I would only get a hand slap. Easily I could come up with a story—after all, it was self-defense. No one had seen what happened. No one could prove anything.

Whitney and Jimmy were at the end of the bar, far from the thin crowd. Whitney's handshake was clammy, not as firm as usual. "Glad you could make it. Hope you're . . . feeling better."

Jimmy ordered a London Sour and a Habano cigar, I got the house special, Whitney a beer. Jimmy rambled verbosely through his drink about the cherry-red Porsche he planned to buy, how the Yousif brothers, whom he'd come to know personally, had promised him a deal. Cigar smoke encircled us with cloying ghost fingers, shielding us from the others.

Whitney's face grew officious. Jimmy avoided my eyes, sipping his cowardly fruit drink. Glowing bottles behind the bar blurred to a distant city, an unreachable Shangri-la.

"Collins, I'll get right to it. Have some bad news."

I glanced around the room, half expecting to see Walid or another blue-clad officer ready to take me away. Almaisa. I couldn't leave her. The cop hadn't been worth it. Why hadn't I let Rashid get rid of the body? I knew better than to get involved.

"Got a communication from H.Q.," Whitney was saying, "and I'm very sorry to tell you that your close and continuing contact was denied."

Jimmy's eyes flicked skittishly from me to Whitney.

"What did you say?"

"I'm sorry, Collins. I know how much you liked her."

My thoughts stumbled to their feet. "My close and continuing . . . ? No. It's a mistake."

"Sorry, brother." Jimmy patted my arm. "C.I. issues."

Someone in the crowd laughed loudly. I placed my glass down hard on the bar. It slithered, snakelike. "Tell me, Jimmy, what issues did you and your C.I. pals at Langley come up with? Gotta have something to show for your tour, don't ya?"

"Now Collins, this is a conversation best had in the office. I can show you the paperwork—"

"Tell me now. You brought me here on my sick day."

Jimmy straightened, cleared his throat. "Well, Langley did traces on your gal. Standard procedure, of course. Turns out"—he flashed a nervous grin—"her uncle was suspected of underground diss. . . . What's the damn word Headquarters used? Dissident activities. Against—well, I guess against the Jasris. Rumors he was working for a third country. In and out of prison. Wound up dying there."

"He was a saboteur in the oil industry," Whitney chimed in. "Part of the Arab nationalist movement. Ever heard of the Popular Front for the Liberation of Bahrain? Anti-British group in the seventies with communist ties."

I gripped the bar to steady myself, my fingers slipping as though the teak were coated in gasoline. "A *saboteur*? That's fucking ridiculous. Where'd you get this shit? The Bahrainis?" The words limped out of my mouth. "Anyway, what the hell does that have to do with—"

"There's unfortunately no debate over this. H.Q. was quite firm in its answer." Whitney was watching me closely. "They view your relationship as a conflict of interest. Too much allegiance to our friends on the other side."

"'Course we can do a bit of due diligence on our end," Jimmy offered with forced reassurance. "Make sure there's nothin' we're missin'."

"Casual contact is permitted," Whitney went on. "You can buy art from her, have a coffee now and then."

My second drink arrived and I downed it.

Whitney eyed my empty glass. "And another thing I wanted to discuss while we're here, Collins. This drinking of yours . . . a bit excessive, don't you think?"

I smiled placidly. "No. I do not think so."

He leaned toward me, lowered his voice. "Collins, we both know you have a problem." The bartender came by, cleaning with his rag. *Swish swish.* Circles that made me nauseous. Whitney cocked his head, eyed me appraisingly. "Not trying to ruffle your feathers. Just want to see you use your full potential. Look, you still have time left in your tour. Why don't you take this opportunity to turn things around, clean up your act? Focus on work. It's not too late."

"Kid . . ." My words had become invalids. "You don't know what the hell . . . what the fuck you're talking about. Who are . . . where'd you . . . ? You learn all this at Princeton?"

Whitney was shaking his head, anger starting to pinch his infantile features. "You know, they warned me about you, but I gave you a chance. Said you were dead weight. Space filler. Just biding your time until retirement. Collins, you have no idea how many times I've defended you. And you thanked me by going out and getting too wasted to work. You only have one informant. You falsify your time cards. Who do you think you are? The CIA isn't what it used to be. You can't get away with that crap anymore."

I pulled myself together and focused on Whitney's face, his round clear eyes, the eyes of someone insulated from slings and arrows, from advancing years' cruel chisel, a father's blows, the desperate impotence of irrelevance. All that came out was, "You can't take her from me."

"Consider this a gift, Collins."

Vic's, its pineapples and tiki torches, disappeared entirely, drained into a sinkhole, and I didn't even notice my fist curling or cutting through the air until it made contact with Whitney's face—a fleshy, slightly wet transaction that I recall feeling somewhat like an altercation with a soft fruit. He was on the ground before anyone knew what had happened, the stool behind him crashing down in the process. There he lay, shaking, a mound of pasty injury, his hand nursing his cheek and mouth, incredulously investigating a small trickle of blood. Jimmy stared at me, wide-mouthed and stupid.

I wouldn't last much longer on my feet. I stumbled out the door, setting off the avalanche of wooden beads.

32

"Why must we always turn the lights off?" Almaisa stared up at the ceiling. It was covered with a sprawling web of cracks and brown stains. Fissures and holes throughout the place let in the Gulf breeze. She shifted impatiently and the dingy mattress creaked beneath her.

"I told you. It's safer this way. People don't like to see us together. Something could happen to you." The tide along the Askar coast went *thrush-thrush* outside the shack. "You're the one who told me that. You couldn't move in with me because it's too dangerous. I'm just protecting you."

Her eyes roved the walls uncomfortably. "And your friend does not mind us in his house? Who is this friend?"

"It's fine. I did him a favor once."

The fisherman had done a good job of hiding his belongings—only a few pieces of furniture visible, no clothes or even food, items he probably didn't possess anyway—and the end result was a decent safe house. A shed in back stocked with shrimp cages was perfect for hiding vodka. Hundred bucks a month—more than his monthly income—and I had unlimited access, door always unlocked, only a half-hour warning necessary. It wasn't so bad really—there was electricity, plumbing, and it was hidden, a place at the end of the world that prying eyes would never find. As long as I didn't get sloppy, Jimmy and Whitney would never know.

I'll forget what you did. I know you took the news hard. We both said things we didn't mean. Just do me a favor—stop seeing her and do your job, and we'll call it even. Water under the bridge. I need you in this fight.

Listen, I'm heading to the Gulf Cooperation Council conference with the Admiral for a few days. Take this time to cool off and break the news to her. You help me, I'll help you. We can turn your career around. We'll make it a win-win situation.

"Are you done with your mosaic for the opera house?"

She nodded.

"Well, what is it? You've kept me waiting for months."

"The Peacock and the Sparrow. You know the story?"

"Can't say I do."

Almaisa sat up, warming to the explanation. "So the peacock is the king and the sparrow is his helper."

"This gonna be all about birds?"

She smiled. "One day the sparrow tells the peacock he is afraid of a net—a trap—near his nest. The king tells him to stay home or go to work but do not go anywhere else. Do not go near the net. You follow the story?"

"More or less."

"So the sparrow follows the king's advice and he goes to work every day and comes home at night. Then one day he is walking to work and sees two sparrows fighting. He is very sad and he speaks to himself, 'How can I, who work for the king, watch sparrows fighting in my own neighborhood? By Allah, I must make peace between them!' "

"Let me guess: The net catches him because he strayed from his path."

"No, Shane, you are wrong. The sparrow is caught in the trap, that is true. But then the story changes. You think the sparrow regrets that he did not listen to the king. But the sparrow is angry at the king! You see why? Because the king made him believe he was safe when he was not. The sparrow learned the lesson *al-ihtiyatat bedoun faida did al-qadar*. How do you translate?"

"Precautions are useless against destiny."

"Yes, exactly. The sparrow was caught, but he had to suffer hope—false hope—first. You see?"

You'll forget about her. The mission—that's the important thing. That's what I've learned as chief.

I shook my head. "Too much for me." I kissed her. "But I'm sure your mosaic explains it perfectly."

"You must come see it when it is unveiled. And your friend Mr. Whitney—I think he would also like to see it, yes? He knows many things about art."

"Sure, maybe. But he's usually busy. Works a lot."

She nodded, accepting.

The low promise of the Gulf outside, more reassuring than an oath of silence, the sand and grit that stuck to our bare feet when we padded about the fisherman's shack—for nearly four weeks we had that and nothing else. Four weeks of calm, contentment. Easy to cling to the illusion that there was still a way out, a satisfactory ending.

33

April 14, 2013. *RAID UNCOVERS MASSIVE WEAPONS FACTORY.* The story made the *New York Times. Deep within the confines of a purported furniture shop in the Budaiya neighborhood, Bahraini authorities discovered several rooms containing a prodigious weapons cache, including 102 Kalashnikovs, 88 kilograms of C-4 and plastic explosives, two anti-aircraft missiles, some $40,000 worth of hydraulic presses, three metal lathes, two long-range Fajr-5 missiles, and more than 50 explosively formed penetrators, Tehran's signature weapon in guerilla-style conflicts.* The list went on, detailing dozens of weapons and providing a few choice accompanying photos, mirroring Qasim's intel almost word for word.

Preliminary evidence confirms connections to Iran, Abd Al Matin's intelligence report to station said. *Electronics found inside select equipment was still in its original Iranian military packaging and instructions were in Farsi. According to our forensic analysis, all C-4 recovered was traced to two known manufacturing sources inside Iran and several unfinished bombs were identified as identical to those supplied by Tehran to Shiite insurgent groups for use against U.S. troops in Iraq. The quantity of C-4 discovered is the largest to date, equivalent to the amount used by Al-Qaeda to bomb the Navy destroyer USS* Cole *in 2000. This cache significantly upgrades Oppositional terrorist capabilities. We believe the Opposition intends to conduct more lethal and effective attacks against the Bahraini government and its allies, as well as shift its focus from local unarmored Bahraini forces to international targets. This level of advancement is highly unlikely to have been reached without outside support, guidance, and training. We view the results of this successful raid as proof that relentless Iranian actions are attempting to undermine security and stability within Bahrain and the wider region.*

Back in Washington the President emerged from hibernation to condemn the Opposition, while the Secretary of State declared that, based on the recommendation of his ambassador on the ground, immediate sanctions would be imposed against the leader of Fourteen February. The Defense Secretary followed by announcing human rights would be lifted as a condition for F-16 fighter jet sales to Bahrain and combat ship sales to Saudi Arabia. In retaliation, Iran publicly announced it would accept Russia's offer to join the Collective Security Treaty Organization, the aspiring counterpunch to NATO, and Russia welcomed Iran with a sales contract for an S-300 missile system, 11 MiG-29s, and 20 T-72 tanks.

Curfew had been moved up an hour and police were out in force, spread across the city in a seething dragnet, the search for the police officer's assassin still in motion. The usual meeting sites were too dangerous, so I'd texted Rashid coordinates for the fisherman's house.

Inside the shack it hit me immediately, something wrong. The odor. Habano cigar smoke. It was unmistakable—sharp, sweet, spicy.

Why the fuck had Jimmy been here? How did he know? I walked through the place, checking behind every door, under the sofa and bed. Slowly, tentatively, my alarm gave way to rationality. Jimmy was not the only bastard to smoke Habanos. Sure they were expensive, popular with expats, but plenty of locals partook. The fisherman, maybe. What did I really know about him? Or one of his friends, family.

Knock at the door.

"Meen hnak?"

"Ji'to li'altaqit quota' al ghayar." I came to pick up the spare parts.

I opened the door. Rashid walked in, looked around, made a show of sniffing the air. "You are smoking cigars, *habibi*?"

"No." I closed the door behind him. "Guy who owns the place does."

Rashid shed his jacket. He was wearing a white thobe again.

"Anyone follow you here?"

"No one, *habibi*. I would not be here if someone had followed me. I am still not sleeping at home. Staying with friends, as you suggested." Settling on the sofa, he lit a cigarette. "You heard that our revolution factory was discovered."

"Yes. That's why I called you here."

He tapped his ashes into the cheap plastic vase on the coffee table. "It is all lies, of course."

"Which part, exactly, is the lie?"

"The weapons they found." He waved his hand in the air dismissively. "Explosively formed . . . what are they called?"

"Explosively formed penetrators."

"Yes, that. Anti-aircraft missiles. Kalashnikovs. *Ninety* kilos of explosives?" He shook his head, chuckled. "What we would give for ninety kilos of explosives." He looked at me. "You saw the factory. Why are we discussing this?"

"*Habibi*, you've been telling me it's all lies for a long time now. I don't think it's all lies. Maybe there was another room in your factory. Maybe you didn't show me everything. My boss thinks you pulled a fast one on us. There were pictures in the paper—chemical drums, bags of white powder." I paused, chose my words carefully. "You want America to protect you, help you, we need to know the truth. You blindsided us. The world expects us to respond."

Calmly Rashid held out his cigarette, examined it as though evaluating a newly discovered artifact. "I know where the information came from."

"Oh, I see. And where would that be? Maybe *I* spilled the location."

"A man named Qasim. Yes? You have met him?" The smoke careened in lazy circles around Rashid's face. "One of ours. A double agent. He gives the king's men information and we get information from him. More from him, of course. You know how it works." He reached into his pocket, handed me a document. A list of weapons—identical to the one Qasim had given me. Elegant and irrefutable proof.

"Qasim. . . . Why didn't you tell me?"

"Only a few people know. We must keep our secrets. What is the word you use? *Compartmentalization?* But when I heard the news, when you signaled a meeting, I knew I must tell you."

It took a minute to digest the information, to recast Qasim in my mind as a true believer, an Opposition sympathizer rather than the greedy unctuous pawn I'd thought him to be, his overgrown brows and emaciated flesh seeming suddenly to reflect poverty and the strain of dancing between worlds rather than the cheapness and servility of lies.

"So all those weapons—the EFPs, the missiles, the Kalashnikovs, they are—"

"Imaginary. Created out of thin air, as your Hollywood says. Nayef and Abd Al Matin gave this list of weapons to Qasim. They tell him to show it to the Americans. Very convincing, yes? And the pictures in the newspapers—these are convincing too, I think. Photos from the ministry files, maybe. Or maybe the Saudis—they have plenty of explosives to photograph."

"But why would Qasim reveal the location of the revolution factory?"

"We had to give the king's men something. If Qasim did not give valuable information, they would have terminated him. Maybe worse." Rashid shrugged resignedly. "It is the way the game is played, yes? Anyway, not so bad. We left a few things for the police to find, but most of the weapons we moved. Different location."

"This is crazy. Impossible. You can't tell me the king is making it all up."

"Have you seen the confiscated weapons with your own eyes?"

"No. But I could. Simple call to my contacts in the ministry."

"*Min fadlak, habibi*. Please look at these weapons. Then you can report back to me. Ask them please if they would allow you Americans to look at them with your forensic tools." He tilted his head. "My friend, why would I tell lies if you can see them with your own eyes? And ask yourself: Why have we not used the weapons? The revolution has been for two years. What are we waiting for? Trust me, we would not let good weapons go to waste."

"So where's the new revolution factory?"

"This information you do not need to know. For your own protection, my friend. You understand."

"Qasim. A double agent. No, a triple agent." The revelation was still wobbly. "He told us— Was everything he told us lies?"

"A little lies, a little truth. Like most things." Rashid grinned. "When you are the weak side, you can afford to give information. You have more to gain than to lose."

"And Iran? Was it true? Millions of dollars and training from Tehran?"

"No." Rashid's face turned grave. "I have told you the truth. You see how poor our revolution is. We get nothing from Tehran. But Qasim tells Abd al Matin and Nayef what they want to hear. This is why they come back for more."

I stared at his face. It was unwavering. "So Qasim leaked the location. That's why you knew it wasn't me."

"Yes, *habibi*, I knew it was not you. I know you would not betray me. Just like you did not betray me about the policeman, *alhamdulillah*. We keep each other's secrets."

We locked eyes in the smoke-filled hovel. The *maghrib* call to prayer rose out of the evening. It was this, maybe—the intangible contours of a moment that even now I cannot precisely identify or describe. The point of pivot, when there was no turning back.

Or maybe it was Rashid's words as he prepared to leave.

"There is something else. I do not know if I should tell you."

"What is it?"

"It will help you decide who to believe." His eyes remained on my face. "Which side you are on."

The sun's last rays crept through the windows and across the walls. Rashid lit another cigarette. "You remember I said we had an inside man."

"I remember."

"He has given us important information. About your Admiral. He is not who you think he is, this man with the golden stripes on his sleeve."

34

Check out the Oasis Rug Shop." Whitney called me into his office on a Sunday morning. "Our friends tell me the Opposition is using it as a cutout for money and supplies. And start looking for new informants—SCROOP is definitely losing access or giving us the runaround."

That same Sunday morning, start of the poor man's workweek, the Minister of Justice and Islamic Affairs announced the investigation into the senior security official's death was complete and all available evidence indisputably pointed to Junaid. The Opposition leader had directed the operation from his prison cell using an intricate network of thugs, the minister said, with the advice and support of Iranian operatives in-country. And the deceased, right-hand man to Major General Al Hasan, had clearly been targeted for assassination, a deliberate attempt to undermine the highest levels of authority and tarnish the reputation of Bahrain's fine fighting forces. Details of the investigation could not be revealed out of consideration for sources and security needs. Junaid was scheduled for an immediate bench trial in a special military court—a "national safety court"—pursuant to the ongoing state of emergency, closed to bystanders and the press, with due process necessarily abbreviated. Three judges would preside—a military officer and two civilians appointed by the Advocate General—proceedings to be held at the Bahrain Defense Forces complex in Riffa.

The trial occurred swiftly. Human rights advocates reported to international papers that Junaid had been robbed of his right to defense counsel and barred from examining the state's witnesses or testifying in his own defense. The state presented a web of circumstantial evidence and deep-pocketed snitches—many, fellow prisoners promised exoneration—eager to attest to their firsthand knowledge of Junaid's activities, Iranian contacts, intricate plans to foment unrest, destabilize, and overthrow royal authorities. Junaid had planned to kill dozens of senior officers that day, the prosecution alleged, and only by the grace of Allah was just one murder successful. The prosecutorial presentation was apparently so rich, descriptive, and emotive that insiders said it was worthy of Junaid himself, on par with his best poetry—the brutality of the crime; how Junaid, with the help of Iranian operatives, had plotted the incident months in advance, including securing the murder weapon, charting escape routes, and teaching surveillance and countersurveillance methods; how Junaid's lackeys had mutilated the victim's body after the fact and ignominiously dumped the corpse in the cruel innominate waters of the Gulf.

The verdict was unanimous: aggravated murder, taking up arms against Bahrain, facilitating enemy entry into Bahrain, cooperating with a foreign power, leading and forming an armed gang in an attempt to overthrow the constitution and Amiri rule, assisting the enemy in weakening the Armed Forces and adversely affecting military operations. Sentencing would be determined at a separate hearing, but the death penalty would necessarily be considered.

Guilt had no place in my mind. To ensure this, when Junaid's conviction was announced, I set forth on a weeklong binge, shuttling between Vic's and the Airport Café. I told Almaisa I was busy with work. I was a spook, for Christ's sake, immune to remorse or liability. Behind me was a lifetime of deflecting notice and attribution, perfectly angling the mirrors so nothing was revealed—not me, not my informants. It was in my blood, my flesh. It was who I was, and there was no reason it should suddenly become distasteful to me. Junaid had his conviction coming with or without the dead policeman.

When Rashid scheduled a meeting, I felt relief. An urgent compulsion to see Rashid had been building—he who shared the muted deed, the only other living witness to our sins. Externalization, the camaraderie of conspiracy, of diffusing and dispersing weight until it becomes negligible, until one can almost pretend it doesn't exist and that the too-large headlines are, in fact, true.

35

The setting sun painted tiger stripes across the stubby yellow turrets of Dry Dock Detention Center. "Shane Collins. U.S. naval base." I held up my diplomatic passport.

The surly guard compared me to my photo. *Taqdam, Sayed Collins.*

In the lobby a serious Pakistani searched me, dirty hands patting me down, while his canine assistant sniffed my shoes. I emptied my pockets and the Pakistani picked up my pack of Marlboros, shook the contents onto the counter. "No cigarettes."

"You're kidding." I grinned. "Want me to smoke one, show you it's real?"

Without a word he swept the pack and its contents into a trash can, nodded toward the metal detector. On the other side a guard waited.

Our footsteps echoed on the concrete floor. Skinny arms, elbows, knees poked out from behind bars—the lower-security wing. Women and children. Doctors too, probably. An unhinged, incomprehensible shout spiked the silence.

Then we were in a different wing where bars gave way to solid doors, and the few small exterior holes that passed for windows disappeared. There was a sooty dankness and a feeling of claustrophobia. Drug users, thieves, traitors, a few Salafis, Shiite dissidents, all the misfits of the Arab world crammed between thick walls like rancid sardines. Cell after cell passed like compact perditions, small slats offering glimpses of inmates pacing in rooms furnished only with sleeping mats and Qurans. None called out or showed any sliver of interest in the well-fed footsteps outside their doors. Not unlike a whorehouse, I mused—the heavy stench of sin and sweat and overflowing toilets, simultaneous isolation and brotherhood, things that can't be spoken too loudly. Coughing, shitting, jacking off, all the disgusting noises of the body one hears among those removed from society and the mindfulness of others.

We'd reached the visitation room and the guard deposited me in a cracked plastic chair. "The prisoner will see you in a minute."

It was a strange sensation to see the man in the flesh, like viewing banned material. He was underwhelming, the physical face of revolution not what I'd expected. Aged and dissipated rather than sprightly and hopeful and years-defying, as I'd always imagined impassioned leaders to be. Junaid's fifty-something shoulders had grown stooped and sickly, and his once-thick snowy hair, full and paternal in younger pictures, was almost gone, bones sculpting his face in sharp and irreversible angles. Talk in Manama was that he had only a few years left, whittled away by cancer, inadequately treated in its early stages by underground dissident hospitals then ignored by prison doctors, but even this (and not merely syphilis, as the *Gulf Daily News* labeled Junaid's affliction) was perhaps propaganda planted by the Opposition to generate sympathy, or by the government to discredit and weaken the ailing leader. It had all seemed inconsequential until he walked toward me, his slight frame hobbling determinedly, face trained on mine. Even through his ragged uniform, I could see he was emaciated, a sliver of a human, the ghastly result of a weeks-long hunger strike. The death penalty, I saw plainly, might never even reach him.

Junaid approached the table blinking rapidly, an instinctive reaction to light after a long period in darkness. Dark splotches of urine stained his prison trousers, and the frown I'd mistaken for an expression of discontent was, I realized, a permanent facial distortion, some kind of downward tic. With a noisy clink he extended his shackled right hand, we shook, and he sat down across from me. I could smell his sickness and ill health like death being kept unsuccessfully at bay. The face of revolution had one foot in the grave.

I introduced myself, then, forgetting the confiscation of moments ago, started to offer him a cigarette. He moved to accept, but the guard standing sentry a few feet behind him rushed to halt the transaction. Junaid grinned, his disfigured mouth trembling with the effort, and muttered, "*Al-tajreed min al-insaniya.*" *Dehumanization.*

I held up my hands to show the guard they were empty, turned to Junaid. "How are things?" I asked in English. "They treating you okay?"

"I am their most important prisoner. What more could I ask for?" He tried to smile again.

I switched to Arabic. "As the authorities probably told you, I came here to make sure you're being treated decently. Verify Dry Dock is observing human rights protocols." I scanned his torso, his lower half and its urine stains hidden beneath the table. "You getting enough to eat? Getting some sleep? You look like you're in okay shape." I couldn't meet his eyes.

"What is your name again, my American friend?" Junaid squinted at me. Handcuffs jangling, he placed his hands on the table; they were covered with the unmistakable marks of bondage: swollen maggot-sized welts, red and white stripes corresponding to pressure points. In Baghdad the Ministry of Interior had used electric shocks to extract confessions and information. Dry Dock probably had its own cozy chamber.

"Collins. Shane Collins."

For a few minutes we talked in bland terms about prison conditions: his cell furnishings, bathroom breaks, how often and how long he was allowed to go outside. The guard was getting restless, his eyes shifting around the room.

This is our chance to turn Junaid. When I'd said the words, Whitney had put down his pen, set aside his work. *That bastard's never been in a more compromised position. We know he won't work for the king, but he might just work for the Americans. With our help he could avoid the death penalty. And if we don't recruit him as a full-time informant, at least we'll get dirt. Find out once and for all what Tehran is up to.* Whitney's eyes had positively glowed. It was nothing less than the best idea he'd heard in his career, he'd said. Glad to see I was back in the game. He would brief H.Q. immediately, making sure I received due credit, and make the necessary arrangements with Dry Dock. We'd need a cover story, I reminded him.

I was an American official paying a routine visit to observe living conditions and compliance with international and military protocols, assure our human rights groups that appropriate procedures were being followed. If someone along the way suspected my true affiliation, as would invariably happen, it only worked in our favor; I was getting intel, and intel from Junaid was all to the authorities' benefit (of course we would share). To sweeten the deal, Whitney would drop a reminder that America was reconsidering its wholesale weapons embargo. When I'd finished, Whitney had looked at me with newfound appreciation. Maybe something more. Envy. He wished he'd come up with the idea.

I raised my voice slightly, just enough for the guard to hear. "*Habibi*, I have another matter to discuss with you."

Junaid raised his brow.

"Allow me to be frank. You've been convicted of several crimes. Serious crimes. You and I both know how this will end."

His dark watery eyes stared at me in their hollows, challenging my assumption that death mattered.

"No one wants to see you dead," I went on. "Not your Opposition, not America, and—believe it or not—not even the king." I leaned forward, lowered my voice. "The king doesn't want to turn you into a martyr. Hell, that'd be worse than letting you rot in prison."

"You think I fear death? Death is a great honor. Allah will take me when he is ready."

"Ah, so I hear. You would not starve yourself if you feared death. But *habibi*, you do not understand me. You must consider more than yourself. Many people depend on you. An entire nation, actually." I allowed my words to ripen. "And you are human—a man, like me. This I know: Every man wants to do something, make his mark. The dead make no mark. This is your chance, my friend. You can do something. Stay alive."

Junaid said nothing. He was listening.

"So here's my proposal. You help us. The Americans." I held his eyes. "Not the king. The Americans. We're neutral. We don't take sides. All we want is information. You can trust us. Our only loyalty is to the truth. In return, we help you."

"You think you—"

"Please. Think about it. I don't want you to decide now. I'll be back in a few weeks. You can tell me your decision then." I nodded to the guard and he hastened forward, grabbing Junaid by the arm and pulling him erect. Whether because of the angle or the roughness, the action nearly pushed the elderly revolutionary over, his knees buckling abruptly, his torso flopping forward like a toy with too-loose hinges, and for an instant I saw the pathetic flesh and bones of my scapegoat staring up at me. This man was not a sacrifice for the Opposition; he belonged entirely to me.

"Just think about it," I repeated. *"Al mawta la yatrikun atharan."*

The dead make no mark.

36

Something had shifted between me and Almaisa. We saw each other less, and our hours together were strained, almost unpleasant. Was it the wear, the humiliation, of sleeping in a filthy safe house? Of a tryst imprisoned behind closed doors with no end state? The war was taking a toll. Airport and seaport restrictions had severely limited charitable donations from abroad. The orphans were subsisting on small quantities of humanitarian aid orchestrated by the Red Crescent, Doctors Without Borders, and Global Society. *Sanaa is unrecognizable from the last time you saw her*, Almaisa told me—*thin, pale*. Children in the streets were suffering from malnutrition; others were mowed down in protests. The worst conditions she'd seen since arriving on the island.

The conflict consumed her. Politics became Almaisa's singular topic of conversation. She voiced louder opposition to the monarchy, repeating stories of her uncle, his slow demise as a result of the Jasri family, the patterns of abuse and oppression that had continued over the years. Her talk was so volatile that I made her promise not to repeat her words outside our doors.

It was during one of these conversations that we had our first fight. Amid a volatile invective against the Gulf tyrants who'd destroyed the lives of her uncle and parents—King Jassim no better than Saddam Hussein—she'd turned on me suddenly. *How can you understand what it is like to lose your family? You do not even talk to your son! Do you know that I still have nightmares? About the intelligence officers who took my parents. I hate intelligence officers. Here in Bahrain, they are everywhere. They watch every house, every movement in the slums. They are cockroaches—who turn people against each other, make them informants. Informants snitched on my parents. Intelligence officers are even worse than soldiers; at least soldiers have the courage to pull the trigger.*

Her eyes had smoldered. I'd walked out of the room.

One afternoon she greeted me with a box of food. *Compliments of Queen Fatima.* A Bangladeshi cook for Fatima, one of the king's four wives, regularly brought leftovers to the orphanage, Almaisa explained, a token of gratitude for sheltering her illegitimate child. Almaisa had spread the rich morsels on the safe house table. *Please help yourself.*

Eighty chefs, she'd informed me, in three kitchens—Arabic, Moroccan, European. Kitchen floors of Italian marble. Solid gold utensils, crystal glasses from Bohemia. The meal I was enjoying consisted of eggplant with pomegranate, scallops with garlic and lemon, ground lamb in citrus tahini, barbecued Cornish hens, walnut and cardamom crepes, pastry with rosewater cream (rosewater ordered specially from Paris). One dish enough to feed ten families. A few leftovers given to servants, the rest loaded onto vans. Did I know where the vans went? To the Jasri family and their friends. To show the king's generosity. Poor families and orphanages received nothing—except what the friendly Bangladeshi cook brought back.

Then Almaisa proceeded to regale me with descriptions of caviar favored by the royals—beluga, osetra, sevruga. Last spring, she said, a servant was not careful and left a palace gate open during a nearby protest. Demonstrators caught a peek inside: fragrant lush courtyards, sunken tropical groves, banana and olive trees. A delivery truck arrived with tins of caviar from Russia. Did I know Junaid? Because he wrote a poem about this caviar. *We have glimpsed the rubies*—Junaid's famous line. She went on, weaving Junaid's poetry into her own recital of monarchical excess.

It's impossible for me to now determine whether Almaisa's radicalization occurred rapidly or slowly or not at all. For all I know, sitting here and turning over pieces of the past like mismatched socks, she'd always had the fire inside her, the revolutionary impulse. Passed through generations, perhaps, a nonnegotiable fate. Maybe I simply began to listen. Whatever the answer—and I have no confidence in any ultimate resolution—things were changing between us, the unanswered questions that had always floated in the gap now developing sharp edges, spreading and lengthening until they filled the space entirely.

37

It was in May that the betrayal began. May was when heat threatened to scald the land, spring mildness lost in the shuffle, when the sea breeze turned the sky to malt, when fishermen set forth early morning on their rickety dhows.

Late dusk, a trickle of moon spilling onto the Gulf like wasted gin. I'd spent the evening with Almaisa at the safe house. She'd arrived before me. When we left, she'd climbed into her car with unusual haste, mumbling something about making it home before protests began. I watched her drive away, the trail of dust from her car a reminder of her perpetual decampment, the feeling she might never return. I headed home.

There's an extra sense you have as a spy. Dictionaries call it intuition, but it's more like the jumbled sum of years of following and being followed, wrapping your life in layers. You detect aberration as easily as a crooked stop sign; deceit and disguise reveal themselves without prompting. It wasn't that I'd forgotten my guitar, as I told myself at the time (who really needs a guitar?); something else was pulling me back. The hour, maybe. Curfew was approaching and I realized Almaisa wouldn't make it home in time. Never had I known her to ignore or forget curfew. She wouldn't risk getting caught by the police. She must have somewhere to stay. Or—she'd returned to the safe house. Crossing Hidd Bridge, my thoughts skirted the previous hours. She'd been jumpy, nervous. Too eager to please. Anxious to verify that she'd used the correct safety signal.

I turned my car around and headed back toward the safe house. My stomach felt turbid. Nothing on the radio, just one of the king's talking heads babbling about an upcoming food festival. *To be held at Bahrain Bay. Live cooking shows and a celebrity chef.*

Somehow I'd covered two-thirds of the island in what seemed like minutes, and now the fisherman's shack emerged from the darkness. Silent, as silent as the dead, and I felt relief. Shutting my headlights off, I turned down the perpendicular alley and parked. I opened the safe house door, didn't bother to turn on the lights. There was my guitar, waiting on the sofa like a patient child. Before taking it I walked into the bedroom, stared at the bed. Almaisa hadn't wanted to have sex that night.

Front door opened. The crack beneath the bedroom door, slightly ajar, filled with yellow light, flickered with the shadow of movement. The owner had returned, maybe. A low female voice. And a male, his voice too faint to identify. Chatter. Even with the edges of the words dulled, the cadence was unmistakably English. Voices swelling, becoming more animated—or maybe they were arguing. The bedroom grew light, my head weightless. Louder and he was laughing. I hear it now, clear and robust, as though right beside me. Laughing in the way only a station chief can laugh. Confident that everyone will laugh with you. Vomit rushed up my esophagus and I swallowed it. Sweat dripped down my back.

Clumsy sounds, a large object being moved. Amateur chords strummed on a guitar—my guitar. He was playing my instrument. He played without talent or attention, a kid who seeks only the thrill of frivolous disharmony. The two laughed at the attempt. With each chord my brain swelled. The sound—the sound—a thousand alarm bells, too loud and discordant to bear. A blow worse than my father's drunken punches.

Footsteps, movement. They'd put the guitar down. The swelling in my brain subsided long enough for me to realize they were probably headed to the bedroom. Why else were they here? A few quiet strides and I reached the bed, climbed onto it. The mattress was squeakier than I remembered, but their inane banter and the virulent tide outside masked my sounds. When I opened the rotted wooden windowsill, a shower of chipped paint fell onto the bed. They probably wouldn't notice it in the heat of the moment. The thought carried a perverse satisfaction. Through the window I hoisted myself, barely able to fit, landing on my shoulder outside with a heavy thud. I nursed my wounds, listened. They were closer to the bedroom, maybe inside it, continuing to talk, the conversation flowing more easily than before. Words still indecipherable. Slowly I rose and, seeing they weren't yet in the bedroom, gently shut the window. Edged over to the shed full of shrimp cages, and took deep breaths of the brackish air, every inhalation a gulp of morphine. In the distance, on the horizon, I saw her body—smooth and golden and unblemished, only it was unblemished no longer, sullied with his greedy youthful touch.

Protected by darkness, staying away from the few small windows, I stumbled around to the front of the house, searching for answers in the scene before me. There was his car, a black sedan. License plate unreadable. I memorized its position, angle, as though collecting surveillance data. If I moved closer, I risked being seen. Light from the living room spilled onto the dirt yard. Her car was nowhere to be found—she must have parked farther down the street, or maybe he'd picked her up somewhere—and her lie of a departure was suddenly apparent. She'd used tradecraft. She'd driven away for the express purpose of deceiving me.

I stood for some time—maybe minutes, maybe more. Then I began padding about the area, cowering behind one piece of junk after another—a collection of old buoys, a pile of fishing equipment—until I determined that the crawl space beneath a dumpster provided the best covert view of the front door. Lying flat on my stomach, I inhaled the hot fetid air and had the sensation again that I would vomit.

It wasn't until after midnight that Whitney finally emerged from the house. I could see only his feet through the gap. He was jangling his keys and whistling. A good whistler, powerful. Beethoven's Ninth Symphony, each note slicing the night. Sound of his car starting. Circling in front of the safe house, headlights swabbing the exterior, painting the concrete yellow. As he swung around, the lights ran across the dumpster, narrowly missing me. I flattened myself against the ground and slithered back, farther into the shadows.

I wormed my way out. My injured shoulder throbbed. Breathing fresh air briefly turned me into a human again. Whitney's car, I realized, had stopped at the end of the street, and I was faintly fearful he'd seen me. I entertained the idea of him exiting his car, dutifully investigating the suspicious form behind the dumpster, then receiving a swift blow to his unsuspecting head. The ground—what was on the ground?—desert, not a stick or large rock to be found. I could use my hands—blessed, I knew, with strength in determination. My hands around his pulpy neck. I had the element of surprise on my side. I'd killed one; what was another? Wasn't that the way things worked?

But he drove away, blind to my lurking presence. I peered around the dumpster, watched his headlights bob up and down as the tidy black sedan jostled along the poorly paved road. How would he pass the checkpoints after curfew? Of course. The ministry had granted him a special pass.

Lights inside the house turned off. She was spending the night. She didn't have a pass like Whitney. I thought of entering the house, confronting her. But numbness had overtaken me, and my will dissipated, seeping through my pores like day-old alcohol. I'd head to the junkyard, hunker down till morning.

He knew a lot about art, she'd said. He'd lived in more than ten countries by the age of twenty-eight.

38

Constant surveillance. Observing patterns of life. Watching others when they think no one is watching. A woman putting on lipstick, puckering her lips exaggeratedly in the mirror; a man picking his nose or fondling his genitals, staring unrestrainedly at an attractive woman. Whitney looking a little less cocksure without an audience; Almaisa more confident, more hurried, alone.

For less than 1,000 dinars I bought a beater, kept it parked on a remote sand dune in Al Dair, arrived each morning to swap cars and begin my daily stakeout. I split my surveillance between the two, usually beginning in the parking lot of a bridal shop near Riffa Views, electrifying my brain with a breakfast of drink and cigarettes, waiting for Whitney's black sedan to come rolling down Shaikh Salman Highway. And come it did every morning, sunglasses hiding Whitney's face behind the dusty windshield, his neatly combed hair shining with fresh gel. Typically, I followed him to the base, worked enough hours to be convincing, then resumed tracking during the evenings. Because of Whitney's training in countersurveillance, I was particularly circumspect, hanging back on the streets, occasionally losing him.

For nearly a week Whitney's activities were disappointingly routine. Other than traveling to and from work, he attended a meeting at the Bahrain National Security Agency, dinner at the Admiral's villa in Amwaj, and an apparently solo visit to Vic's. Jordan had disappeared entirely from the picture.

One night when I trailed him to Riffa after work, I noticed an unusual number of stops—for Chinese takeout in Adliya, a quick dart into a cold store (odd since the kid didn't smoke and could get better and cheaper wares at the commissary)—then watched as he drove past his neighborhood. His journey ended at the Shaikh Salman bin Ahmed Fort, a turreted stone structure sufficiently isolated that I was forced to park some distance away. I sat in my car sweating, waiting to pick up the eye again, wondering what the fuck he was doing in there. It was dark, hardly the time for sightseeing. Forty-five minutes later he emerged, returned to his car. When he was a safe distance ahead, I pulled onto the road behind him. In my rearview mirror I glimpsed another man exiting the fort, his face only visible for a moment. He had a mustache I'd seen before. Walid Al Zain. The intelligence officer who'd overseen the Adliya bombings. Was it Walid? But the connection was too fuzzy, too paranoid. A stranger, maybe.

Contrary to my expectations, Almaisa proved more difficult to shadow than Whitney. During work hours I headed to the slums and staked out her watering holes. Though I often caught her, it was hard to hold on. She was no clumsy tourist, infuriatingly elusive, darting in and out of slum alleys like a jackrabbit, as well versed in the streets as any urchin. She went to the usual places: her studio, the orphanage, one trip to the opera house with a box of art supplies. But I would lose her for long stretches, her car missing from all her haunts.

One time she called me as she exited her studio, where I watched from my perch at the end of the street. She was wearing one of her ridiculously bright abayas, a surveillant's dream, and even from a distance I could see her dark locks taunting me beneath her scarf. Now I carried my camera everywhere and I snapped a few photos (for what purpose I don't know; somewhere in my mind I surely knew the photos wouldn't serve as evidence but only as a reminder of her). Watching the phone ring on the seat next to me, I'd felt bitter longing, more gnawing than a dry day, and a sudden urge to drive by and wrestle her into my car—or maybe run her over.

Through it all, I still had no idea where she lived. Curfew curtailed my surveillance before I could learn the truth.

After little more than a week, my initial suspicions had begun to develop fault lines. Even taking into account Almaisa's erratic routines (and doesn't any person's life appear a bit strange when viewed under a magnifying glass?), the patterns I witnessed were sufficiently uninteresting that I started to entertain the fantasy I'd misinterpreted the whole thing. The affair, after all, didn't make sense. Why would Whitney risk seeing Almaisa after Langley had deemed her a counterintelligence risk? Surely the youngest station chief in history, rising star, wouldn't jeopardize his hard-won reputation.

When I regurgitated that evening at the safe house in my mind (as I still do years later), images and sounds became blurry, less certain. I hadn't seen Almaisa's car. Was it possible Whitney was there with someone else? The low voice I'd heard eliminated Jordan as a possibility, but Almaisa was not the only alto female on the island. But how would Whitney have known about the safe house if not for Almaisa?

Or maybe it wasn't Whitney there that night. His car—was it his car? On my morning stakeouts near Riffa, his black sedan had appeared dingier than I recalled it that evening. Was it simply the light of day that accounted for the change, or was it an entirely different vehicle? If only I'd gotten a plate number. The sole identifier I'd seen was his shoes, and now they turned indistinct in my memory—brown (or was it black?), the kind every Western man wore.

Was it then possible that the woman chatting inside the house was not Almaisa and the man strumming the guitar was not Whitney? They spoke English—of that much I was certain—but so did thousands of others on the island. Maybe the goddamned fisherman had wised up and decided to rent his house to another wayward expat, someone too tightfisted or paranoid to shell out dinars for a fancy hotel—a randy married Brit, a European spook, a senior naval officer constrained by fraternization rules. Any number of permutations existed in the garden of sin, and the fisherman, who peddled his wares daily at the local markets, need not search far to find a willing customer. Unlikely, but possible—increasingly possible, it seemed, as the theory picked up mass in my mind.

I began to take comfort and relief in the possibility I'd imagined the entire thing, or at least so disfigured the events that reality was nothing resembling my musings.

And then I learned where Almaisa lived. It was after a night of work at the orphanage. She'd spent the better part of an hour unloading boxes labeled GLOBAL SOCIETY from her car, the same boxes that filled the pantry shelves, charitable food and medical donations. No one had come to her assistance and I'd longed to jump out and help her, wrest the weight from her insubstantial hands. By the time she'd finished, curfew was still two hours away. I followed her car a short distance behind, too close for comfort, but I was determined not to lose her.

Halfway down Bani Jamrah Avenue, she parked in front of a lopsided building flanked by an alley on one side and three gangly palms on the other. A lone streetlight cast a weak orb on the building's cracks and general state of decay. A smattering of dark holes cascaded down the side, depressions from bullets; Bani Jamrah had been the scene of violent clashes when the uprising began two years earlier. Still, it wasn't a bad building, in better shape than its neighbors. Red balustrade framed a tilted balcony that ran across the width of the second floor, accessible from several shuttered French doors that gave the place an incongruously quaint appearance.

She'd disappeared into the building, and a few seconds later yellow light awakened the second floor, spilling through gaps between the shutters. Her silhouette, disconcertingly large and bulky in her abaya, moved from one side to another, adjusting the shutters, then disappearing. I lit a smoke and enjoyed the smug satisfaction of having obtained a long-elusive piece of information, the treasure at the end of an arduous journey.

Headlights filled my rearview mirror and I slid down in my seat, waited for the car to pass. But it parked some distance behind me and I remained balled on the floor, extinguished my cigarette. Door slamming, footsteps. I raised myself, peered through the window. A man in a hat carrying a briefcase was approaching Almaisa's building. She'd pulled the shutters until the slats were exactly horizontal, I realized, the safety signal I'd taught her.

By accident or design, the streetlight directly illuminated entrants, and the halo formed a malignant cone around the man's back. As he opened the unlocked door, face still shaded by his hat, he turned slightly and I saw his shoes. Instantly the memory punched my mind, a lens coming into focus—beneath the dumpster, his feet visible through the crack. The same shoes, penny loafers.

He was across the threshold, short frame moving hastily and furtively, the singular hustle of a lover en route to his tryst. Briefcase in hand. Then there were two silhouettes instead of one, moving closer, drawing apart.

Only with great effort was I able to pull myself onto the seat. If not in plain sight, my car was in discoverable proximity, and I needed to leave. But I couldn't manage to turn on the engine. He'd thrust his fat white maggot fingers into her loamy warmth, gone where he didn't belong, where only I'd gone. The cozy dripping world of expats, that's where he belonged. A smooth-skinned boy attempting to navigate a forest too dense for him. The silhouettes reappeared and something tumbled inside me. A loud crash—just a mangy dog pawing at a trash can. In rote motions I started the engine, jerking my beater to life, pressing hard on the accelerator as though willing it to defy me.

Driving back with the traitorous puppet show seared on my brain, I thought I was livid, maybe even jealous—all the usual emotions to be expected—but I see now I felt something else. Something darker, perhaps—for the simple reason that it was more concentrated; it demanded swift movement, action. A sense of urgency.

Rashid's words in the safe house: *It will help you decide, maybe. Which side you are on.* Qasim's list of weapons lying on the table in silent reproof. Rashid had offered up his information on the Admiral, the conclusive evidence, a final persuasion. Opportunities were developing quickly, he'd said, and the time had come to choose. The king, the Admiral, my station chief—or him. Now, as I turned onto Road No. 3901, the slums fading to rubble in my rearview mirror, the answer was emerging.

39

In June the declining oil prices set in motion by Mohammad bin Rashid hit a record low. To counteract the loss of revenue, Bahrain and Saudi Arabia had been quietly reducing their mollifying subsidies and raising taxes through their rubber-stamp legislatures. Now local pump and food prices climbed, forcing small employers to lay off laborers. Skirmishes in the villages escalated and cries of protest, hungrier than usual, could be heard on almost every corner. Near-starvation conditions had taken root in parts of the slums, Rashid reported, black markets rampant and multiplying daily. Temperatures rose and the heat of looming summer spurred protest fires higher. Sandstorms swept the land, blurring buildings to indistinction. American Alley grew putrid and fermented, its stench carrying across the high naval walls. Security at Dry Dock was tightened—no visitors allowed—and my follow-up meeting with Junaid was postponed. The Bahrain Defense Forces commander-in-chief prohibited all males under the age of thirty-five from entering or leaving the country. Direct flights to and from Asia were banned—too many waypoints for terrorists, regional airport security inadequate.

The only other passengers on my flight were jittery aging Europeans and Saudi wives. *Your most important meeting yet. Highly compartmented informant handled by Sarajevo has intel on a major weapons shipment from Iran. H.Q. wanted me to go, but the Admiral needs me here. Too much going on. I trust you to take my place. You can make the arrangements—cover story, meeting location.* Clap on the shoulder, no trace of resentment over our months of sparring or the blow I'd inflicted on his soft Ivy League face, no hint of the duplicity that lurked deep in his bowels, the green shame of deceit. A perfect case officer if I ever saw one. Truly a rising star.

So be it. I would use this assignment as an opportunity. Level the playing field. Whitney and the Admiral had their purposes for this trip; I had mine. They had the king in their hand. But I had the ace.

The plane spiraled upward in its evasive corkscrew and the island shrank below like trash dropped down a well. I leaned back on my extra-soft pillow, opened the in-flight snack. Yogurt and honey, Bahrain's signature gourmet treat. The seat-belt sign blinked off.

Change planes in Brussels. Fourteen hours until touchdown.

40

Cambodia, Day One

Phnom Penh's airport had improved considerably since I'd last been. No squawking chickens, beggars mostly relegated to the sidewalks outside, working air-conditioning, more glass and granite and people in suits than I remembered. A country, like Bahrain, determined to put on a good show.

Monsoon season had begun and light drizzle was falling, but the sticky heat outside still hit me like a bucketful of tar. There were the beggars I recalled, stooped and crawling, many of them children, many without the limbs we Westerners swung so cavalierly upon exiting the terminal. This was the Cambodia of my past, the Cambodia that never changed. I was a young spook the last time I'd walked these streets, married and my son still talking to me. He hadn't yet learned who I was.

On the corner an emaciated tuk-tuk driver sat waiting. For a few thousand riels I hired him to take me to the Kabiki Hotel. Wipers on the tuk-tuk moved grudgingly across the small windshield, failing to clear the drops, rain grazing me from the open sides. When we pulled up to the hotel on the outskirts of town, it was just as pictured in the guidebook: a bungalow with teak-framed windows, short palms swaying in happy welcome. A quaint sign in the lobby admonished guests that WE STAND FIRMLY AGAINST SEX TOURISM.

The interior was as picturesque as the exterior, my room offering an ample bed anchored by carved wooden posts and cloaked in mosquito netting, colorful framed photos of quintessential Cambodian scenes: smiling peasants, a glistening fish hanging from a bamboo pole, steaming bowl of noodles. A pristinely immersive experience, *a favorite with expats*, the guidebook had said. View of the courtyard, just as I'd requested. Perfect accommodation for an American diplomat in town to discuss routine matters: port access, coordinating regional naval policy.

Cambodia, Day Two

Rain was still falling the following morning. Uncle Sam was paying, so I took an air-conditioned taxi for the three-hour trip to Sihanoukville. *I'm a diplomat. I talk with your people so we can use your port,* I answered the nosy driver when he asked my business at the Ream Naval Base. I'd held up my briefcase as evidence, the simple but bulletproof cover story I'd suggested to Whitney.

The guard at Ream barely glanced at my diplomatic credentials before waving us through. Past shipping containers and swinging cranes. In front of the Ship Support Office I instructed the driver to meet me outside the gate in an hour.

Uncle Chang, the office director, lived up to his name, or at least what I imagined his name represented: a fat, swarthy Chinese man with pockmarked skin and a Fu Manchu mustache that stretched below his chin, longish black hair slicked back with enough grease to lubricate a small engine. He greeted me with a cocked head and a smile, revealing unusually straight teeth for this part of the world. His clothing was ill suited for the time of day and profession—an expensive-looking shiny black suit that veered dangerously close to a tuxedo, screaming-red satin handkerchief. His unbuttoned white shirt showcased an ostentatious gold necklace with a large cross pendant luxuriating in a nest of hair.

Ushering me into his plush office, Chang thrust offers of cigars and liquor upon me. *Mr. Collins, I have been expecting you! Have one of these—Cohiba Siglo. You know it? Best you can find in Cambodia. How was your trip? Are you enjoying our beautiful country?* The cigar box he held before me like a sacrificial offering contained easily $2,000 worth of smokes. He lit my cigar with something resembling a miniature blowtorch and, when I expressed admiration, presented a card with the name of the shop where he'd purchased it. More than $700 American, he said, but for me the shop would give a discount, maybe even offer one free. Busily he shuffled around the room, his sizable bulk shifting oddly, and I saw that at least a small portion of his heft was accounted for by a sidearm strapped beneath his jacket.

The walls were covered with framed letters from various U.S. admirals, trios and duos of stars heading the pages like rarely seen constellations, thanking Uncle Chang for his unmatched hospitality and generous hosting, peppered with phrases like, *Many of the crew are still talking about the great adventures they experienced, You are as much a member of the U.S. Navy team as any of us, We are all proud to call you Shipmate.* There were photos too—Uncle Chang with this admiral or that admiral, shameless grip-and-grin shots. Hung in a shrine-like center position was a photo of our own Fifth Fleet Admiral from his days commanding the Seventh Fleet, his arm around Chang, looking somewhat diminutive next to the obese man but still monopolizing the picture with his million-ship smile.

"I miss your Admiral," Chang said, watching me. "Last I saw him was months ago. He has been a good friend to Cambodia and to me."

I smiled and took the seat he offered, assured him the Admiral was well, and yes, I was enjoying everything in Cambodia. I puffed on my expensive cigar lit with an even-more-expensive flame, sipped the cognac he'd poured me. Leaning back in the leather chair, I reminded him I was acting as the Admiral's emissary on this trip, representing the Office of Middle East Analysis on issues of wider Indian Ocean policy, in town for a bit of coordination with the embassy, wanted to stop by and introduce myself to America's point man in Cambodia. Several minutes of extolling our East-West partnership, the friendship that had blossomed in recent years, Ream stepping in as the fortuitous replacement of Subic Bay, the invaluable provision of refueling, resupply, joint training for the U.S. Navy—details entirely unimportant, the conversation serving only to validate my trip and cover story, polyester stuffing for the pillow.

Then Chang leaned forward, refilled my cognac, told me he was making progress on assembling a patrol ship of armed guards—British-trained Gurkhas from Nepal—available for hire to fend off pirates; as he'd told the Admiral, the Fifth Fleet would surely find it invaluable when traveling around the Middle East and Africa. Might even come in handy against those Iranian patrol boats clogging up shipping lanes in the Gulf, taunting and threatening U.S. vessels. And had he mentioned that his company was expanding, serving ports in the Admiral's neighborhood? By next year he would be up and running in the Gulf, ready to service our ships in Manama. Amid his pitches, he chomped vigorously on fried crab and curry delivered by a skimpily dressed young woman, offering at least three times to share, apologizing lightly that he hadn't been expecting me until later. *Tiger penis*, he explained a crispy brown phallus on one plate, a Cambodian delicacy, praising its flavor and mythical virtues until I agreed to try it.

When talk had petered out, he asked, "Where are you staying?"

"The Kabiki."

He stroked his mustache. "You are happy there?"

"It's fine."

"You want a nicer hotel?" He reached for the phone. "I have many connections. I get you a room at Raffles."

"Next time."

He nodded haltingly, withdrew his hand from the receiver. "Driver? You need a driver?"

"Tuk-tuks and taxis suit me."

"Well, you must promise you will try a meal at the The Grill Room." He plucked a few American hundreds from his wallet, pushed them in my direction. His fat arm struggled to cross the desk, and a shiny gold cuff link emerged. It was monogrammed with Khmer letters.

"Those are very nice." My eyes arrested his movement. "What do they say?"

Chang jerked his wrist free of the jacket sleeve and glanced down, smiled nostalgically. "Uncle." He grinned. "What else?"

"Made around here?"

"Yes, Mr. Collins, as a matter of fact, they were made by one of our best jewelers. Would you like to pay him a visit? I have here his card—"

"How much would those set me back?"

He frowned. "I will be truthful, these do not come cheap. Solid gold, 24 karat. I'd say . . . at least three grand for the pair."

I muttered something about operating on a bureaucrat's salary, scooped up the bills in front of me, thanked him.

"Enjoy our women, our food. Honeys and bunnies and yummies. You know what I always say? You can't put a value on happiness. Listen, you want me to send a girl to the Kabiki?"

"I think I'll explore the city myself."

"What's your preference?"

I hesitated then said, "Women."

Raucous laughter. "Then I suggest the Walkabout. You will find it at the corner of 51 and 174. Of course you tell them Uncle sent you."

Chang's cell phone rang abruptly, Lee Greenwood's "God Bless the U.S.A." He promised to call back. By the time he hung up, I'd headed to the door. He followed me, extended his squat hand. "You will give the Admiral my best? You will tell him I treated you well? And tell Mr. Johnson his friend has been asking about him. She would like to see him next time he is here."

"Will do." I smiled and, as though the idea had suddenly occurred to me, patted the briefcase I'd brought. "Listen, you have any material on this patrol ship you're leasing? I think the Admiral might be interested. Maybe I can arrange for you to come out and brief him."

He was pleased, gripped my forearm with the familiarity of an old friend. "Certainly, Mr. Collins. Just let me look. . . ." He produced a thick leather portfolio filled with documents.

"Thanks." I slipped the portfolio into my briefcase. "This is perfect."

The tuk-tuk driver was waiting outside the base as I'd instructed. I approached his vehicle obliquely, in a rough circle, glancing at the gas tank, the wheel rims. Just before entering, I feigned an inadvertent drop of my briefcase. Glanced underneath. Clean to the eye. No beacons, no bombs. The driver could've left his vehicle for a stroll, a smoke, and I couldn't take any chances.

The steam from the shower turned my small hotel room into a jungle. I changed my clothes, hailed a tuk-tuk, headed downtown. The drizzle had ceased, leaving a fine gray mist, and Phnom Penh was faded and subdued. This was indeed the city I remembered: blinking neon fighting to be seen, careless vehicular screeching, humidity so thick the windows of air-conditioned bars fogged up instantly, thousands of blurry eyes. Shabbier, grittier, more resilient than its Asian cousins—Singapore, Tokyo, even Saigon. Rough around the edges, its secrets only half-buried. I told the driver to take me to Street 130.

The mist began to lift, and glowing blue, red, and yellow signs emerged, elbowing their way toward the middle of the street, twilight's repose giving way to the hustle of evening and Phnom Penh's nightly wages of sin. Striped awnings and broken shutters framed crumbling colonial balconies, the love-hate flirtation between East and West built into the architecture. Girls in miniskirts that verged on panties swarmed every passing male, unabashed and insistent, groping themselves and uttering outlandish promises in a desperate attempt to beat the odds of mismatched supply and demand. Vietnamese girls dominated the crop, easy to spot with their light skin and delicate features, smug with superiority to their rougher Khmer sisters in the hierarchy of sex workers. Red light cast by hundreds of Tiger Beer signs turned the whores vulnerable, made them softer and more beautiful than they actually were. On every corner alluring pictures of dove-white Asian women smearing lotion on their faces proclaimed Pond's Cold Cream a magic elixir.

The streets stank—I'd forgotten how putrid Cambodia was, with its piles of refuse and fecal matter, rancid noodles and produce scraps heaped in every alley, cooking all day in the sun to an infernally ripe temperature, the process of decay accelerating unnaturally like a child who instantly becomes an adult. Between the piles of trash, street vendors rolled their carts, hawking their late-night wares: fish, lok lak, pork buns, fried crickets. To the east the Tonle Sap River shuddered, the promenade's bawdy colored lights painting its ripples the colors of a kid's crayon box. A beggar holding a cup with gnarled hands stood on the corner, glanced in my direction.

I bought a skewer of sausage and a Tiger beer, and ate as I walked. A tantalizing pair of Vietnamese twins approached but turned elsewhere when I didn't show immediate interest; they were attractive enough that they didn't need to cajole or pander. I tried to avoid the girls who looked like children, though such was the Asian complexion and stature, made worse by juvenile clothing, that it was nearly impossible to tell young from old. A few other men, mostly British from their looks, were strolling the vicinity, and we avoided eyes, the politeness and respect of the mutually damned. Hostess bars, karaoke joints, beauty salons winked at me like a cheap carnival, flashing their benign if not wholesome euphemisms for the filthy pleasures Westerners couldn't bring themselves to spell out. SPICY GIRLS. HELLO SWEETIE. FIGHT NIGHT.

A few more blocks and I found Walkabout, a corner spot with a dingy red awning and a terrace overcrowded with perverts and potted palms. Inside, I ordered a beer, enjoyed a few sips before a swarm of whores descended on me like wasps. I ignored them, closely guarding my wallet, until the crowd thinned out. Then I bought a drink—a "lady drink," they called it—for one of the remaining girls who looked a bit older and more sober than the rest.

When she'd finished her drink, I smiled. "I'm looking for a girl with red tattoos. You know her?" With my hand I drew a figure on my biceps. Many things had changed, probably, in the months since I'd seen the photo on Burt's computer—the woman's weight, hair color—but her tattoos would still be there.

The girl at the bar, who was curvy and short and had grotesquely fake lashes, blinked up at me. "You want to fuck?"

"No, I'm looking for a girl with red—"

"I have red tattoo." She grinned and sidled up to me, her hand crawling along my leg. I removed it and pulled out my wallet, handed her a hundred-dollar bill.

"I just want to know about the girl."

She shrugged. "Many girls have tattoos. Red ones, black ones. Too many. But mister, you have good time with me. I show you." Her hand crept back.

"Not just one tattoo. All over her body." I motioned up and down my arms.

She shook her head. "Mister, your girl not here anymore. We come then we go. We *freelancer*."

A loud hiss, then cheers from the back of the room, a handful of Khmer men celebrating another bout of the bar's spectator sport: continuous Madagascar cockroach fighting. I finished my beer, handed the girl another fifty before leaving. The Angkor Noodle House was just across the street, its green sign proclaiming WE LOVE NOODLES!, enticing passersby with a picture of an overflowing bowl, a pair of chopsticks standing erect in its center. It was unchanged from the photo. She wasn't at the Walkabout, but she had to be somewhere. The answers were here in this simmering city.

The east side of Tonle Sap was apparently the rougher side, a dumping ground for those who couldn't make it on the west or, judging from the unapologetically brazen ads on every hotel front (SEE GIRL WHO PUTS PING-PONG BALLS IN HER PUSSY! YOUNG YOUNG YOUNG!), nirvana for the fringes of the depraved. The Yellow Serpent Hotel was no exception—not a vestige of yellow nor any hint of a serpent, a tenement-style gray cement building cluttered with decrepit awnings, tilted at such an angle that it looked in danger of tipping over. I'd finally struck paydirt at my third hostess bar. *You look for Chanlina,* a girl lying naked on a pool table had informed me. *She have tattoos alllll over. Yellow Serpent Hotel, mister.*

A skinny, nervous Khmer man opened at my knock, ushered me inside. Some dozen girls stood waiting, all outfitted in the skimpiest of garments to ease the process of discernment. I scanned the group. "Girl with the red tattoos," I said to the Khmer.

"No girl with red tattoos here."

One of the whores jerked her head up. "Chanlina."

The Khmer's face grew dark. "Chanlina here. But she busy with customer." He gestured toward the girls. "Here many beautiful ladies. They all satisfy."

"I'll wait." I pulled out my cigarettes, offered the Khmer one, sat down.

"Sorry, mister, but you pick girl here." The Khmer hovered above me like a nagging mother. "Chanlina will be busy all night. Ver-ry popular."

I handed him a wad of bills from my dwindling supply. "The girl with the red tattoos is the one I want."

With a nod from the Khmer, one of the girls scurried upstairs.

Morning came and still Chanlina remained hidden, brown dawn creeping through cracks in the boards across the windows, dispelling the mask of pleasure-blindness and turning the place old and haggard. Sunlight amplified the heat, turned the stuffy room sour. A steady stream of men had paraded down the stairs and past my perch, slinking toward the door with the self-loathing euphoria that comes from escaping one's own transgressions. Mostly to placate the Khmer, I kept myself plied with liquor, a rancid house-made concoction chockfull of debris that tasted like piss mixed with dishwater.

When I was sure the last man had left, I asked again where Chanlina was. The Khmer shrugged. "She is ver-ry popular, I tell you already."

Girls floated before me, draped over chairs, their cigarettes dangling from orifices, dejected and rejected, the youthful energy and temporal beauty summoned by their survivalist instincts now entirely dissipated. They'd become old women. Without thinking I rose from my chair, bounded up the stairs.

Hazily I scanned the architecture of the sex den, figured I knew my way around a brothel better than the next guy. The Khmer's panicked footsteps hurried behind me. A closed door at the end (was it the only one closed, or did I simply guess?) and I headed toward it. Turned the knob—unlocked. Probably none of the rooms had working locks.

A second was all I needed to see that the girl passed out on the mattress was Chanlina. The tattoos covered nearly every inch of her exposed body, conspicuous as the smell of incense and sex and the foul open latrine at the back of the room, deep red calligraphic evidence across her smooth dewy flesh. In person she looked leaner—wasted was probably a better description, the prematurely aged hand of addiction having caressed her whole body—also prettier in a strange, vulnerable sense. She was Vietnamese, the only one I'd seen in the place. I barricaded the door with a chair, yelled to the Khmer that I just needed a few minutes. He began knocking and the noise woke Chanlina.

"Listen, sweetheart." I crawled onto the mattress, stroked the nipples on her implanted breasts, shoved a few bills into her fist. "A friend of mine says you're the best." The Khmer's knocks were growing louder; bastard probably wanted more money. "You know my friend? American like me. About my age. My height. Gray hair—grayer than mine. Very important man. He came with his friend a while ago. Also American."

The girl had probably been in bed with hundreds of graying American men. Smiling dimly, she nodded. "You friend of Admiral? Uncle Chang send you?"

Fucking commander of the Fifth Fleet. It was all true.

"Yeah, sweetheart." I offered up my biggest grin. "Uncle Chang sent me." I lay down next to her. "I only have five minutes. Let's make it quick."

When I left the brothel into the squawking morning, the city was rising. Fastboats were arriving from Saigon, bobbing erratically among the early fishermen, tiny Cambodians emerging from their bright patchwork houseboats for another day of heat and toil. A pack of French sailors emerged from a hotel, strutting languidly into the blazing sunshine, one zipping his pants and laughing loudly. At a street vendor I got bai sach chrouk for breakfast. The vendor nodded toward the French sailors, informed me that a carrier had arrived at Sihanoukville the night before.

I walked along the riverbank while I finished my breakfast. Hailing a tuk-tuk, I saw the beggar with gnarled hands in my periphery, watching me with the penetrating gaze of the impoverished, the exiled of society. When I turned to face him, he scurried into an alley. It was the same beggar I'd spotted the night before. Either the most peripatetic bum I'd ever seen, or a paid surveillant.

41

An essential digression at this point, a brief trip back in time to Manama, ten days before I went to Cambodia. Not long after I'd discovered the affair. The impetus to action coursing through me.

I wanted it to be true—everything Rashid had told me at the safe house. I wanted to know the Admiral was treacherous and filthy and unworthy of the starched uniform he so smugly wore. That sanctimonious Whitney had blindly and selfishly picked the wrong side. The pupil no better than his teacher. Both hollow men, stuffed with straw.

I wanted everything Rashid had told me to be true. But a good spy does his homework. Checks around every corner, kicks open every door, before he enters the room. I would find out for myself.

So I'd gone downtown. A writer once said that one loses perspective after spending time in a brothel. I might counter that whatever one loses is not nearly so great as what one gains.

From outside, the Sea Shell Hotel looked like any other ramshackle building in the poorer areas of downtown Manama. Soot-streaked walls, dingy windows. I took a certain pride in my familiarity with the place, just as I took pride in knowing the seedy netherworld of every country I'd been to, the hookers and drug dens, the muttering retreats of restless nights, the substratum most expats never penetrated or did so once or twice for the sole purpose of journal entries and bar stories back home.

Inside, visitors entered a large room, dimly lit like all whorehouses. Faded floral wallpaper, a few prints on the walls—Asian-looking waterfalls and landscapes more at home in a dentist's office.

Charlie, a fat, mustached Lebanese man, had welcomed me with his usual sickening effusion. *Robert, so good to see you! Long time it's been!* He'd offered me a seat in one of the broken plastic chairs that looked as if it had been recovered from a dumpster. A few other men sat at slight distances from each other, their smoke forming elongated shapes that stretched across the room.

Some dozen women sat in a row—Thai, Filipina, Russian, Chinese, mutts—facing the men like nervous schoolgirls getting their picture taken. Three more customers arrived, no more than fifteen or sixteen years old, crisp white thobes and diamond-encrusted Rolexes betraying their origin. A nighttime getaway across the causeway to Pleasure Island. The boys sat behind me, their hot drunken breath wilting my collar. The women commenced birdlike tittering, crossing and uncrossing legs, running tongues over painted lips.

"I'll take her." I pointed to a petite Chinese girl swinging her legs from the side of her chair. Her skin was pale, features small and sharp. Easily the prettiest in the bunch, with naturally curving lashes and a cherry-shaped mouth like a cruel doll.

"Dai-Tai." Charlie smiled.

Upstairs, Dai-Tai's room was tiny and sparse. She'd thrown red scarves over the lamps, giving the place a cheap funereal air. From down the hallway came a triumphant, overly masculine shout—one of the Saudi boys.

The mattress crinkled, the only sound in the room. She had flat Asian breasts and they barely moved, jiggling ever so slightly like slumbering beetles as I entered her. Her body was rail-thin and brittle, not healthy and smooth like Almaisa's.

When I awoke the next afternoon, Dai-Tai had lowered the curtains and I could see only gray, the already desiccated room drained of its last dregs of color. Dai-Tai struggled into her heels for another night of work.

Downstairs I took Charlie aside, ordered his most expensive drink. *Listen, Charlie. The Admiral—he ever been here? Need to know if I'm ever gonna run into him. Wouldn't want him to see me here—or vice versa.*

Charlie had grinned, shaken his head. *Robert, you know I do not tell names. Just like I do not tell anyone your name.*

I nodded understandingly, ordered another drink, put a few hundred dollars into his sweaty palm.

He comes one time, maybe two times, each month. And only for Dai-Tai. He comes through the back door so no one will see. A man I do not know makes the arrangements.

A back door, eh? Didn't know there was one—would be helpful when I visit.

Ah, it is just for very important customers.

I'd thanked him, told him I'd be returning the following night with a friend.

That evening I sat motionless in my living room until the late hours of night, staring at the lagoon, watching the room grow dark, the shadows lengthen.

Charlie had lied to impress me, perhaps. Given me the answer he thought I wanted to hear. But there were too many pieces to ignore, too many for coincidence. It was all spread before me like a trail leading in only one direction, Rashid's revelations banging insistently around the basement of my mind, urging me onward. *He goes to the Sea Shell Hotel. Our inside man makes arrangements for the Admiral. And of course we have our own eyes on the street. It is part of the deal, my friend: The Admiral gives Bahrain weapons; Bahrain gives him cash and a place for sin. He is very corrupt, your Admiral. They say he has a bribe waiting in every port.*

42

The morning after my conversation with Charlie, I awoke at dawn, washed off the stink of Dai-Tai's perfume. The *Eisenhower* was due at port in two hours.

American ships arrived in Bahrain almost monthly, but this was only the second visit of the USS *Dwight D. Eisenhower*. The initial voyage four years earlier had been dubbed "historic," the first time a nuclear-powered Nimitz-class carrier had pulled into the island's port, Bahrain officially important enough to warrant a place on the Navy's map. Now the *Eisenhower* was returning to visit its Gulf friend—well timed amid the current conflict, the Admiral had publicly noted, to show American support for stability in the Middle East, and to mark the ten-year anniversary of Bahrain becoming a major non-NATO ally. To honor America's special relationship with this tiny yet pivotal country.

The port in Sitra was located on a peninsula off the eastern shore, colored this morning by a brilliant pollution-fueled sunrise. Slowly the carrier was decreasing speed, the engine's whir dulling to a quiet roar. Sailors were spread in an even line along the upper deck like tiny painted figurines, shoulders square, hands at sides, proud emissaries of the most powerful nation on earth. A horn announced the *Eisenhower*'s arrival in a confident baritone.

Shots carrying mooring cables jetted from tiny portholes across the water. Seamen in yellow-and-blue vests coiled lines around horns protruding from bollards on the pier. *One-two-three-heave*. Flawlessly executing the bidding of higher-ups—chain of command, a coterie of admirals, Uncle Sam himself.

Sailors disembarked. Those on the lower deck began unloading large crates. Protests had erupted in Sitra that morning and instead of the usual crowd of lipsticked women and eager children sporting glittery signs and American flags, a single angry plume of smoke greeted the sailors. They'd undoubtedly been warned of the insurgency, the situation explained in neutral understated tones a weatherman might use to describe a mild storm front: Watch for sudden swells, worsening conditions possible but unlikely. Certain areas would be off-limits, and Navy personnel were to respect local customs: no excessive drinking (though Bahrain was by no means a dry country), no vulgar tee shirts. Same guidance they'd received at every port from Manila to Okinawa.

Walking down the gangway, Campbell looked just like my buddy Rodriguez had described him: tall and lanky, tight buzz cut, enough of a loner that even in the crowd he walked several steps from the nearest sailor. Standing to the side, he moved his sea legs jerkily. *Campbell's your man*, Rodriguez had reported back after I'd placed another call to Jalalabad, explained I needed someone with access to the ship's cargo. I knew that well-connected Rodriguez—who hadn't failed me when I needed informants in Baghdad or expertise in purple explosives—would have the goods, know who could be trusted. He was one of those guys who claimed a friend, or at least a coconspirator, in every port. *Campbell's got the keys to the kingdom*, Rodriguez had assured me. *I'll grease him for you.*

I headed over, clapped my hand on the kid's shoulder. "Campbell? I'm Collins. From the Office of Middle East Analysis. Friend of Rodriguez. Here to show you around and welcome you to our island." I thrust my hand in front of him.

The naval base was jammed with the influx of seamen, all scrambling to spend money, eat perishable food. Campbell was no less immune, strolling through the aisles with hungry eyes. In anticipation of the carrier's arrival, the exchange had stocked its shelves with luxury goods: Rolexes, Coach bags, diamond necklaces for girlfriends and wives. *Best prices in retail*, the exchange boasted. No matter that most sailors were enlisted and could hardly afford the necessities of life; they'd been quarantined in a maritime prison for the last several months and the deprivation was enough to distort the minds of even the most frugal. Campbell selected a bottle of cologne and a few CDs, and I graciously took care of the bill.

Evening was falling when we left the base, the nightly taxi stand solicitations rising into the air like swallows. I told Campbell we wouldn't be partaking of the nightlife just yet. Twenty-some years in this business hadn't been for nothing; I knew the way to a lonesome sailor's heart.

Ric's Kountry Kitchen was packed when we arrived. A short walk from base, the place had perfected everything a sailor could want, from greasy comfort food to whiny Filipinas belting out Western pop tunes. Signs on the walls boasted that Rick Abernathy, legendary retired Marine, was the only American restaurant owner in Bahrain.

Somehow the layers of nostalgia compensated for an otherwise subpar establishment. Moth-eaten velvet drapes covered every surface, the once-red carpet black from years of foot traffic. Plastic-covered tables were crammed end-to-end, and American flags littered the walls. Between them one could glimpse garish murals of Western landscapes so incongruous only an Eastern artist could have painted them: cacti sprouting from fertile green fields, wide blue river lined with vibrant red roses. The food was unapologetically inedible and the prices exorbitant, but the menu offered American staples as rare and valuable in Bahrain as diamonds: thick-cut bacon, baked ham with honey glaze, deep-fried turkey, mashed potatoes with gravy.

Campbell shoved a spoonful of baked beans into his mouth. He'd been complaining about ship life for the last twenty minutes. The piles of food on my plate looked sickeningly uniform in color.

He put down his spoon. "So what is there to do in this sandbox anyway?"

I smiled. "Depends what you're into."

Campbell's eyes wandered over to the Filipina singers, crowing and swaying in miniskirts and sequined tops.

"There's more of them, if that's your thing."

"Oh yeah, that's my thing. China girls." He bit into a large hunk of bacon.

Briefly I entertained the idea that his stated preference was just for public consumption, that there lurked within him some more forbidden impulse that he'd learned to disguise or suppress over the years. But I decided that my offerings, if not perfectly tailored, would do the trick.

I raised my glass. "I have just the place for you. A little spot called Sea Shell Hotel. But"—and here I leaned forward with the conspicuous air of confidentiality—"it's off-limits, so it stays between us."

"Hey, what happens on shore stays on shore." Campbell grinned. "I know the drill."

"Of course you do. You boys are good at keeping secrets. As am I."

The singers paused for a break and I lowered my voice. "Did our friend Rodriguez tell you what I do on base?"

The poor kid looked respectfully uncomfortable, reached for his beer. "He mentioned . . . yeah, I guess he told me."

"And did he tell you about our special project?"

Campbell glanced skittishly around the room. "Something about finding marked boxes?"

"Exactly." I shrugged to show how simple his assignment was. "We arranged for certain boxes to be delivered on your carrier. For a covert op. Wish I could tell you more, but of course I can't. Just need to look at the cargo before we hand it over to the locals. Quality control, make sure those assholes at H.Q. got it right, that kind of thing." I winked.

Campbell scooped up the last of his beans and bacon. "Why me? Why don't you go through my chain? Must be someone higher'n me involved."

The question caught me off guard and I took a long swig of beer, buying time. "You want the truth, Campbell? Here's the dirty truth. This op . . . it's so compartmented, we didn't tell a goddamned soul on your ship. My guys at station didn't want to tell you, but I said we could trust you. Shit was supposed to be handed off to the Bahrainis right away, but those fuckers at H.Q. . . . well, let's just say they've screwed up right and left on this op and now we need to check their work, make sure everything's in those boxes before we give the green light or look like a bunch of assholes. Know what I mean?" I stared hard at him until he flinched. "Now Rodriguez said you could be trusted. Hope he's not wrong. This is important stuff here. Our nation depends on you—to help us and keep your mouth shut. This is it, my friend. The big leagues. You ready to be James Fucking Bond?"

The Filipina singers returned.

"Let's get out of here." I nodded toward his empty plate. "Curfew's at twenty-one-hundred and Sea Shell Hotel is waiting."

43

Cambodia, Day Three

The outdoor range was hot and muggy, my hangover from the long night at Yellow Serpent Hotel squeezing my head. The interminable walk from Prek Thnot market, where I'd instructed the tuk-tuk driver to drop me off, had done nothing to help. I'd arrived an hour early to stake out the place, uncover any setup—the peripatetic beggar had spooked me, but something usually did in these kinds of operations—and I hadn't detected anything amiss. The remoteness of the range—a small outpost in the middle of dense jungle—made countersurveillance easier, provided further reassurance.

I was saturated in sweat by the time Sven arrived. I complimented him on his choice of locations. He had rented the place for a song, he chortled; knew the owner and we were guaranteed no visitors. *American dollars hard at work*—and he tossed his thick blond hair from his eyes, grinned boyishly.

Neither Whitney nor any of the cables from Headquarters had prepared me for Sven's appearance. I'd known only that he was a Swede, and had accordingly expected a wizened veteran of espionage, a sophisticate who'd been around the international cocktail circuit a few times or, as sometimes happens among the Aryan lily-white types, developed an overzealous sympathy with the Arab world and its poor huddled masses—not this man before me: no more than thirty, short and overly muscular, resembling a cleaned-up Popeye, his white teeth grinning at every word spoken, jesting or serious. He had a faint Swedish accent, that singsong cadence endemic to northern Europe, and was perversely cheerful, too buoyant for this line of work.

Within minutes I'd learned that, his youth notwithstanding, Sven was indeed a seasoned denizen of the underground, the netherworld of spies, a gunrunner who proudly informed me he auctioned his services to the highest bidder or whichever country happened to be on his vacation list, wanted by Interpol under a different name for supplying the weapon used by a terrorist to spray a Copenhagen café with ninety-six bullets. If business was slow he sometimes dabbled in the trafficking of drugs and little girls. But he drew the line at little boys. Since I was from the U.S.A., he could also confide that he held a special place in his heart for my country out of respect for American mafia types of yore and their methods of human disposal. Was I familiar with the historic mob practice of cementing an offender's feet into his shoes and tossing him into a river? *We call it* ståplats i Nybroviken *in Sweden*. Then I'd offered him a cigarette and he'd refused, genially informing me he never touched cigarettes, drugs, or alcohol.

For the better part of an hour we stood behind a makeshift counter of hastily hewn logs, under the shade of a straw awning, firing rusty Kalashnikovs and Dragunovs into the distant mountains, a blurry green graveyard of old bullets, drawing steadily from the box of greasy rounds the owner had left for us. Each shot rang out like an explosion. Cement Shoes seemed to take a singular pleasure in firing the aging weapons—left over from the country's civil war, he informed me, from the Khmer Rouge. The Killing Fields were not far away; had I been there, felt the dead bodies beneath my feet? *It is marvelous.*

Our ammunition nearly exhausted, my hearing all but eviscerated, I stepped back from my weapon. "Let's talk business."

Sven put down his gun. "Of course, Mr. Collins. But first, money."

I handed him an envelope. "You'll get the remaining fifty grand in Sarajevo."

Sven leafed leisurely through the bills, nodded. He gave me a manila envelope. "You have the key at your station?"

"Yes." I opened the envelope, withdrew a blank piece of paper.

"*Bra.* So you will use the first list of numbers I gave to your man in Sarajevo, add seven to their sum, go to that page. Then you take the second list of numbers . . . yah, you are with me?" His tone grew strident as he tapped the paper in my hand. "This list right here. It is lemon juice so you heat in the oven. Old-fashioned, but it still works, yah? Those will take you to the letters. Start at the top of the page. Finish when the message is done. Do not go backward. Simple. This will give you time, place, and method of the weapons shipment."

"The name of the cutout too?"

"Of course."

"Impressive. How'd you get all this?"

Suspicion crept into his features, turned his boy face into a criminal's. Quickly I added, "My boss is gonna want to know. Headquarters always wants a sourcing chain. Vetting and documentation. You know the deal."

"Vetting is not necessary. Talk to your man in Sarajevo. He will tell you—I am never wrong." He picked up his Kalashnikov, traced its muzzle lovingly. "I work directly for Sepah. You know the IRGC? No middleman. I am the middleman, yah? That is all you need to know."

I nodded, folded the paper, slipped it into my shoe. "Listen, Sven." I smiled as he took a package of pickled herring from his pocket, tore it open with his teeth. "Once I get back to station, our broom is going to grab this paper from me before I can even take a piss. I'll never see it again. Don't keep me in suspense. How the fuck are these guns getting into the country?"

Sven eyed me as though gauging the authenticity of my inquiry. Then he glanced at my shoe containing the hundred-thousand-dollar secret, perhaps determining logically that he'd already passed me the information and, more importantly, I'd passed him the money, and he had nothing to lose. Leaning his elbow on the shooting stand, he munched his herring, grinned. "First, you see, they come by boat. Then . . . well, it is very clever. You would never guess."

44

Guns. They could indeed travel in clever disguises. Eight days before Cambodia, following our visit to the Sea Shell Hotel, I'd discovered this with my new friend Campbell.

Our flashlights didn't illuminate much of the base warehouse, but it was enough to see dozens of rust-colored shipping containers crammed into every square inch. They were constructed of thick steel, bolted shut. Campbell jangled his keys nervously and I offered a reassuring grin. "Doin' good, sailor. Keep it up and maybe you'll get some kind of medal out of this."

His flashlight beam was alighting on random containers like a jittery insect. "So what markings are we looking for?"

Rashid had been maddeningly vague about the markings; his inside man knew only that it involved numbers—misplaced numbers, he said. Standard white digits were printed on every container—maximum cargo weight, maximum gross weight, maximum capacity—how the fuck was I supposed to decipher the code? I strolled between the boxes, shining my light up and down, ignoring Campbell's questions as though they were too elementary.

"These came from Singapore, right?" I asked. Maybe the cargo's origin would provide a clue.

Campbell nodded, fiddled with the manifest in his hand.

The Admiral. A man of impeccable composure, unequaled capability, boundless ego. How would a man like the Admiral choose to mark the designated containers? In a way that reflected him, his prowess, his cachet.

If I occupied the Admiral's brain, how would I wink at the world? Numbers, only numbers on each container. A man who owned a yacht named *Safe Harbor*. A man who wore monogrammed cuff links. And suddenly I saw it: right in front of me, so simple I'd almost missed it, like giant letters sprawled across a poster that you can't see because you're staring at the details. Of course. 1982. The year he'd graduated from the Naval Academy. Painted in white on a container below the rows of actual meaning. 1982—it indicated neither weight nor capacity nor any other quantity. A useless number that would not cause anyone to blink an eye, that hid in plain view by virtue of its meaninglessness, telltale in its very unremarkability. But that carried the weight of the Admiral's importance.

Or maybe I was wrong. I indicated the container before me, which had 1982 printed below the maximum capacity of 1,173 cubic feet. "Here's one."

And then yellow was splitting the darkness. "What the hell . . . Who's in—? Oh, Collins. What are you doing here? I saw lights." It was the Admiral's aide. Nosy unctuous fuck.

I smiled, shone my flashlight in his face. "Getting a few cartons of smokes for one of my men. Our top dog. You know the guy—saved everyone's ass from a fucking car bomb. Admiral loves him. That okay with you, chief?" Act like you belonged and you would.

The Admiral's aide looked uncertainly at me then Campbell, who had turned away and was staring at the ceiling.

I gestured toward a container. "Wanna stick around, get a pack for yourself? Some Coca-Cola, maybe? We won't tell."

The Admiral's aide shook his head. For several seconds he stood motionless, unsure what to do, maybe gauging his desire for free soda. Then he surveyed the warehouse, nodded at me. "Just make sure you lock up." Campbell, I could see in the half darkness, was sweating.

When the Admiral's aide had left, Campbell and I each raised one of the right-side latches on the container, turned the handle until the bolt was disengaged. The door swung open and I shone my light on dozens of stacked crates. Too heavy to lift, so Campbell reluctantly retrieved the warehouse forklift. When we'd placed the first crate on the floor, I pried open the lid with my knife.

Neat rows of blue canisters labeled DIBENZOXAZEPINE stared back at me, naked and impotent in their unexpectedly discovered state. CR gas. Campbell glanced at the manifest then at the container number on the side. “Says here this one’s supposed to have mashed potatoes.” He looked at me, confused. “This ain’t no mashed potatoes.”

I glared at him dismissively. “Well, sure, that’s the point. We’re not gonna advertise our covert op.”

The next crate contained weapons sights and night-vision goggles, the third, silencers and ammunition. By the fifth crate I felt Campbell’s nerves fraying, so I slammed the lid shut and stood back, satisfied, declared this was exactly what we’d wanted. Verification complete. Climbing into the container, I scanned the crates one last time. One looked large enough to house a helicopter, or at least its parts, maybe small artillery.

“Job well done,” I congratulated Campbell. He said nothing. It was a dirty business, no way around it.

We headed outside. The base was nearly empty.

“Listen, kid. Here’s a little extra spending money for your shore leave. Don’t spend it all on China girls.”

Even in the moonless dark I could see clearly the way he looked at me. It was a look of disgust—not, as I would have expected and even preferred, disgust at me, the questionable spook who couldn’t get a handle on his own operation, but disgust at himself, the stooge who’d succumbed to his weaknesses, his vanity and sexual desires, his pesky human desire to be important, he who’d ignored his better instincts, did that which he knew he should not.

I told myself again it was a dirty business. Then I turned away and lit a cigarette. “Get lost,” I said harshly. “You’re done.”

When he’d scurried off into the darkness, I took the handheld GPS from my pocket and marked the location of the warehouse: easternmost point on the base, just behind the Sentinel Hall Temporary Quarters, north of the CIA station.

45

Cambodia, Day Four

Outside my window at the Kabiki I saw only the palm-choked courtyard, heard the bustle of late afternoon. I closed the shutters, pulled the curtains across the French doors. Checked under the bed and nightstands and coffee table, behind the paintings, clock, and lamps, felt the cushions, examined the mirror. When I was satisfied the place was clean, I swapped my tee shirt and jeans for a black linen shirt and tan pants, my sneakers for loafers. I slicked back my hair with heavy gel and shaved until I had less growth on my face than a twelve-year-old.

The power went out—a regular occurrence, judging from the candles and matches in the nightstand. By flickering yellow light I dug into the mess of guidebooks and cigarettes and maps in my suitcase, pulled out fake round eyeglasses, a small roll of duct tape, ten thousand in cash. I surveyed myself in the mirror. Not too different—a radical change was never believable, didn't mesh properly with one's body and demeanor—just enough to trick the eye. Keep it simple.

I duct-taped ten hundred-dollar bills to my thighs, divided evenly between right and left to minimize bulk. Insurance cash, a ticket in or out of somewhere. Smoothed out Sven's paper, placed it inside my diplomatic passport, locked the passport, my wallet, and all identifying material in my briefcase. Put only as much cash as I could fit in my pockets without creating an undue bulge. When I'd tidied the room, I pulled the curtains back from the French doors, headed outside. Through the deserted courtyard behind the hotel, under the arch, into the city beyond. Light rain was falling again.

On Samdach Phuong Street I hailed a tuk-tuk. We headed west, turned onto quiet leafy Pasteur Street, where the driver traveled some ten blocks and let me out. I'd remembered correctly, landed within a few feet of Heart of Darkness, which I was pleased to see was essentially unchanged: an unassuming pink stucco building identified chiefly by the crowd loitering outside.

The interior was similarly untouched, a cavernous dark space punctured by red lights that somehow evinced Conrad's cross-continental world more faithfully than simple blackness, multicolored baubles hanging at dizzying uneven heights, oversized disco ball in the center of it all. Even on a rainy weeknight the place was packed to the gills with its signature mix of high-pitched uniformed Cambodian schoolgirls, skinny blasé Asian men and ladyboys in cropped tank tops, lipsticked drag queens circling like flamboyant vultures, a smattering of sloppily clad expats, recreational users, a few aging European fetishists. In the corner a DJ wearing sunglasses and a skull tee shirt spun sugary music, while on a platform an earnest group of Asian girls did its best imitation of an American band, their incongruous facial expressions and mispronunciations quickly shattering the illusion. Behind them, three women in traditional gold lamé and elaborate pointed hats gyrated as if to ancient music.

I made my way to the bar, ordered an Angkor. Candles stuffed into bottles dripped wax, their flames casting fireflies on the counter. Several women and one or two ladyboys approached me, tried to entice me into buying them a drink, but I just smiled, made enough small talk to justify my permanent presence at the bar, sent them on their way. Close to midnight a wiry, dark-skinned man sidled up to me and ordered a beer.

"You are a tourist?" he shouted over the din in a muddled accent. He wasn't touching his drink.

"That I am. Stopped here on my way to Saigon." I pushed my glasses up my nose.

He inquired about Vietnam. Had I been there before? What was it like? Was I traveling by plane or boat? The necessary shouting was making him uncomfortable and beads of sweat were breaking out on his young face. Feigning a few sips, he pushed away his beer with one hand and a 50,000 riel bill toward me with the other, soaking it in the mess on the bar. "Thank you for your advice. Please have a drink on me."

I swept up the wet riel, pocketed it. "Beer not to your liking?"

He frowned. "No. Not my taste. I do not think I will come back here. Thank you again for the travel advice." *He will be nervous,* Rashid had warned. *He is only twenty-seven. A refugee in Berlin. He has applied for asylum so he is scared to travel outside Germany.*

After the man had left, I finished my beer slowly, enjoying the scenery. The glorious stream of excess, the sweet tangy fragrance that only comes from an undiluted cocktail of pleasure and sin. It was Phnom Penh, Heart of Darkness, and I luxuriated for a moment in the feeling that time stood still, that nothing existed before or after this moment, this place. There'd been no Sven yesterday morning, no Whitney or Almaisa, nothing awaited me tomorrow. I could stay here forever, buried in the windowless pulsating bar where I couldn't see outside and had only the vertiginous lights to guide my way.

Placing a few bills on the counter, I went to the bathroom, read and memorized the words scrawled on the riel, flushed it down the toilet. Outside, the rain had grown heavier and the crowd had thinned to a trickle. Along Pasteur Street drops between leafy acacias fell on my face. Shadows of middle-of-night locomotion flickered across the pavement. It was that gray colorless time that was no time at all, too late for night, too early for dawn.

The rain and the hour must have winnowed down the number of tuk-tuks because there wasn't a driver to be found. I cut across narrow Jayavarman Street to try my luck on Preah Norodom and caught my reflection in an illuminated glass storefront. My altered appearance startled me, though it should not have. I felt a jarringly pleasant sensation, the feeling I'd had in the bar that none of it was real, that I was free from the tangle of mistakes, promises, and defeats that grips a man's heels, taunts him with flashes of its sinewy strength.

Turning away, I walked a few paces and lit a cigarette, or at least got the lighter as far as my mouth. That was when the blow hit me. It was to my head—that much I recall—with a heavy blunt object that knocked me instantly to the ground. All I could see and taste was brown pavement, and my consciousness must have slipped rapidly because I can't remember any sound or words or much at all except hands groping me, searching my clothes. Then it was black, and the Heart of Darkness, still cradling my mind, slipped entirely away.

46

Cambodia, Day Five

When I woke, the rain had stopped and it was almost morning, rays of light tentatively revealing the alley where I lay splayed on the pavement like a discarded playbill after a performance. My unlit cigarette and lighter floated in a fetid puddle next to my glasses, which were smashed almost beyond recognition. One side of my face was plastered to the wet pavement and I peeled it off with great difficulty, bits of gravel and dirt implanted in my skin. My mouth tasted coppery, sooty, and swallowing was dry and painful. I rose slowly as if emerging from a grave, half-dead and battered and bruised enough that I wanted to return. My arthritic joints had been pushed beyond their limits, threatening to abandon me. I stumbled to a shuttered cart, leaned against its damp side with all my weight.

I resisted the urge to vomit. My palms were scraped and bloody and I wiped them on the cart; I still had the sense not to draw more attention to myself with stains on my clothes. Tucking in my shirt, I felt my front pockets—the throwaway cash gone, of course, but my reserve still stuck to my thighs. Everything else locked away at the Kabiki. I took a piss in the alley, removing the emergency cash.

Before returning to the Kabiki, I stopped for a drink at a 24-hour dive bar, the kind of place where my appearance would go unnoticed or at least ignored. Alcohol, as always, repaired my insides, greased my brain. I reviewed the last several hours, weighed the odds of a random hit against a calculated one, determined I would never know for sure, and since my pockets hadn't held anything important, it didn't matter. It must have been the pockets they were after because otherwise I would've been dead.

I tried to convince myself shoddy tradecraft wasn't to blame—the assailant had come out of nowhere; a surveillance detection route before or after Heart of Darkness wouldn't have done any good—but I knew it wasn't true, that I should've seen the guy. I consoled myself with what I'd done right: flushed the refugee's information down the toilet (the most important thing, after all) and carried nothing worth a shit on my body. The real tradecraft was ahead of me and I was equipped to handle it. Fuck, if I couldn't handle this operation, I was worthless as a spy, and I knew I wasn't worthless.

The Blue Lime Hotel welcome sign was in Khmer and English. Tall and blockish, with worn white balconies, the place had none of the charm of the Kabiki and all the anonymity the Kabiki lacked. At the front desk the receptionist informed me, Mr. Robert Thompson, that a room was available with a prizeworthy view of the boulevard and yes of course cash was accepted.

My room was sparse, not nearly as tastefully decorated or expat-pandering as the Kabiki. I put everything away—not too neatly—turned all my zipper tabs in the same direction, placed my briefcase on the table. The afternoon lull was approaching as I walked onto the balcony, tuk-tuks circling idly on Preah Ang Yukanthor Street below, looking for passengers, killing the hours in more or less the same rhythm as the taxi drivers outside the naval base.

Fatigue had hit me with a cruel blow and I allowed myself a half doze, but restlessness and pain woke me an hour later. Just before three o'clock I washed my face, combed my hair until it gleamed, put on my spare pair of glasses (second pair clunkier than the first but serviceable). Outside the hotel I headed north on the first leg of my surveillance detection route.

Brief stops at the Why Not Club and Sharky's Bar and Mango Mango Lounge before hailing a tuk-tuk to the 9 Dragon Restaurant, where I forced a few bites into my protesting stomach. No one on my tail. When I left the restaurant, the sun was dyeing the dirty colonial buildings orange. West on Russian Boulevard, past the used car lot, right onto Street 58P, an alley like so many in the Manama slums, too nondescript to merit a name. The kind of street that in a few letters and numbers delineated the world of a spook: a place without identity, untouched by bright lights or maps or tourists.

On a narrow makeshift bridge that crossed a stream of waste, I leaned over the railing. I was still black. The military market vendors were clustered at the end of the street, their scattered lights and chatter and goods-hawking cries injecting life into an otherwise hidden dead end. I lit a cigarette and headed to the market.

Made-in-Thailand military knock-offs littered tables and hung from awnings—army commendation medals, boots, tin dog tags, camouflage lighters, Copenhagen chewing tobacco, ponchos, holsters, belts, Kiwi shoe polish, all the paraphernalia a combat voyeur could want from a country ravaged by war. From behind one of the stalls a skinny junkie emerged and, catching up to me, motioned his head foggily in the direction of an alley, muttered something in incomprehensible English. I told him to shove off, grinning just enough to let him know I meant it.

Mismatched BDUs were strung in a line like clothing at a dry cleaner, the tabletop crammed with stacks of ballcaps. An old woman stood behind the counter, her skin wrinkled and gray, eyes beady and squinting beneath a wide coolie hat. No sun touched that alley, even during the day, the hat undoubtedly for the benefit of tourists seeking a genuine mamasan. My disguise must have worked because she showed no sign of recognition as I approached.

A few seconds perusing her wares, then I said, "*Kaphleung*." She stared at me, sizing me up, probably wary of police. My skin color worked in my favor; gunrunners in Asia were invariably less skeptical of a plainclothes Westerner than a plainclothes native. She nodded her head toward the shuttered building behind her.

Down a long hallway, last door on the left. Dozens, maybe hundreds, of Kalashnikovs and small arms and grenades, alongside a sizable quantity of TNT and ammunition, leaned against the walls, covered the dank cement floor. I scanned the goods, turned to Mamasan. "Documents?" I folded my hands into a book, made the motion of flipping pages.

She looked at me blankly and I feared she didn't understand—or maybe Cambodia's black market had transformed radically in the years since I'd last been, her goddamned son no longer in the business. Hell, maybe he was in jail or dead. But she was moving toward a rear door.

Munny looked up from his desk when I entered the room, a plastic shaded lamp casting a cruel glow over his face. He was a spare ugly Khmer with scraggly black hair and a twisted mouth, whose face revealed his meanness at all times, even, I suspected, when he slept. The place was barely lit by a few hanging bulbs, the only decoration a framed well-preserved vintage movie poster by Moeun Chhay (with the rich detail of an original—payment, perhaps, from a cash-strapped customer): a creepy and oddly colorful depiction of a mother cradling a half-dead baby.

Like the old woman, Munny didn't recognize me—this much was evident when, upon my entrance, his hand flew instinctively to a drawer. Before he could shoot, I smiled, said, "Dollars," removed the clump of cash from my pocket. "I hear you're the guy in town for a passport."

He studied me closely and for a moment I thought he might have glimpsed the familiar man beneath the glasses, smooth bruised skin, and impeccable hair, the man who'd come to him twice before, seeking services for official U.S. government operations. He was the best in the business when it came to stealing and forging passports, sought out by the finest terrorists, sex traffickers, drug dealers, gunrunners, and spies in the world.

"How you know about me?" he demanded.

"You helped out a friend of mine. Ten, maybe twelve, years ago. American guy." I let him see my impatience. "Look, buddy, if you're gonna ask questions, I'm gonna walk. I didn't come here to talk."

Before I saw it coming, Munny was frisking me roughly while the old woman stood and watched; if I were a betting man, I'd wager she had a sidearm herself. I held tightly to the money in my fist, my only weapon worth a shit in this place. When Munny was done, he took a step back, squinted at me. "You not too hard. You dark hair, middle age. Look like from different countries. I have many for you to choose. How much money you spend?"

"Depends what you got."

"Belgium, right? Everyone want Belgium. Belgium cost you lot of money."

"I don't need Belgium. What else do you have?"

He unlocked a desk drawer and removed a box, artfully shielded from my view, sifted through its contents.

"Africa? Africa cheap. I have Angola—"

"Not Africa. Somewhere in Europe."

"You want stolen or forged?"

"Stolen. I'll never get through Europe on a forged."

He looked appraisingly at me as though weighing my ability to wield an illegal document. "Some countries not bad. Greece, Spain. France getting worse, but still not bad."

"I want stolen. Blank."

"Ah, blank? Blank will cost you." His mean mouth twisted into a grin. It was an answer I was expecting; Munny, the wizard at using black marketeers across Europe to steal small batches of blank passports from embassies, consulates, town halls, any officialdom that carried the paper gold, charged a small fortune for his premium product. But to anyone seeking more than an offbeat souvenir, it was well worth the cost; genuine passports, blank but filled with false data, were the only ones that stood a chance.

He'd found something in his inventory, and now he held up the maroon booklet. "You lucky man. I have one Switzerland left." He waved it in the air, gloating at the holy grail.

"How much?"

"For you? Ten thousand."

"You're crazy. Five. Not a penny more."

He returned the passport to the box, locked it. "You wasting my time."

Bastard was just as greedy as I remembered. I shrugged, made for the exit. The old woman, watching our conversation like a bystander at a Ping-Pong match, shuffled hurriedly to follow. When my hand reached the door, I turned around. "Seven. Final offer. And throw in a *kaphleung*." Good insurance against another assault in an alley.

Munny surveyed me, gauging whether I was his match, or close, in hustling, whether he could haggle me higher, extract an extra thousand.

He broke into a terrible grin. "Business slow. So I give it for seven." He shook his head, opening the box again. "You steal from me. So cheap . . . I make no money."

I counted out the cash and he ran a long yellow fingernail along the collar of each bill. He placed a pen and paper on the desk, indicated an empty chair. "Now we talk about what you want. Name, birthday. Better not to wear glasses."

47

Cambodia, Day Six

A morning of tourist attractions around downtown Phnom Penh and now the tuk-tuk driver let me out at the Choeung Ek pagoda. Rain again, dampening any rational tourist's penchant for sightseeing, and I explained to the driver that an old friend of mine had served in the war and I wanted to pay my respects. When the driver had left with my generous tip in his pocket and I'd perused the pagoda, I walked south to Bird's Nest Cafe.

A classic Asian shanty, the café doubled as a residence, the kind of place that served only noodles and had a mattress and hole-in-the-floor latrine steps from the kitchen. I ordered a drink and a bowl of kuy teav.

The rain let up and I sat beneath the lone umbrella outside, watched cars speed by. A few stopped at the petrol station across the street. Searing heat rose from the ground, producing pockets of steam, and distant Choeung Ek Lake emerged. Checked my watch. Had the refugee at Heart of Darkness written 1:30 or 2:30? The liquid soaking the bill had blurred the numbers. 2:30, no question. My noodles arrived and I slurped them down. Finished my beer, put on my jacket, walked to Choeung Ek, the sun reaching its high point, beating down on me.

The Killing Fields were deserted. The ground was still wet from the morning rain and I wondered whether, as my guidebook cautioned, the precipitation would propel small pieces of skull and bone to the surface. I wandered the area, taking time to read the tourist signs and explanations. When 2:30 approached (I knew it instinctively, didn't need to look at my watch), I walked to a large chankiri tree—the Killing Tree, as it was known—where a sign explained matter-of-factly that this was the tree against which executioners beat children. The growth was large and sprawling, the only one of its kind for some distance, not unlike the Tree of Life, which I reflected with some amusement was as close to the Killing Tree's arboreal nemesis as a plant could come.

Rainbow ribbons covered the chankiri, tribute to the thousands of juvenile victims, and I pinned my green ribbon next to the others. From the north side of the tree I walked one hundred paces. When I reached a spare grove, dark and scraggly, I found the widest trunk in the center, shielded by a small canopy of leaves. Leaned against it, lit a cigarette, watched as the last vestiges of clouds fled the muscular sun. The rising heat was making me sweat and my glasses slid down my nose, an unfortunate consequence of plastic frames. But plastic was essential; wire-rimmed frames made long-lasting indentations on one's nose, a potentially fatal tell in the days to come. I pushed them up slowly, conscious of the movement and its foreignness, its unnaturalness, suddenly concerned that, to an observer trained in detecting fiction, my lie of an identity was all too apparent. The sun beat down.

My contact approached with no apparent pretense of sightseeing (though somehow he seemed to fit seamlessly into the landscape), walking with a slightly uneven gait that suggested a later-life, maybe war, injury. Or perhaps it was simply cover, the result of filling his right shoe with stones. Rashid had refused to reveal anything about the guy, claimed he didn't know—no background, no details, not even his hair color.

The man was Russian, I could tell before he'd opened his mouth—distinctive light olive skin, mixed features of Occident and Orient, heavy lids, tightly knit brows, square Slavic jaw. Unkempt greasy brown hair fell across his eyes, perhaps to diminish their conspicuous blue. His eyes were cruel, discerning, ready to detect the smallest movement within a mile radius. Thin and wiry, he wore a battered army jacket that looked as if it had been plucked from a local market.

The Russian cocked his head, examined my cigarette pack. "Green apple? I have never tried that before." His accent was moderate, probably heavy if not for years of exposure to black market English.

I looked at my pack as though seeing it for the first time. My glasses slid down a bit. "Guess so. Usually I smoke Marlboros, but a friend gave me these." I offered him one. He took it, our oral paroles complete, the mere six syllables of "green apple" and "Marlboro" having set our train in motion.

As we smoked I wondered if the Russian bought the fiction heralding my arrival: a businessman in need of a solution. Probably not. Probably didn't matter. The Russian handed me a small manila packet. I turned it over, slipped it into my pocket. To avoid any confusion I repeated what I'd been told about his payment—half before, half after, doled out by my company.

"Yes. It has been taken care of." He moved toward me and I instinctively reached into my jacket for the Colt .45 I'd purchased from Munny. But a strange conspiratorial glimmer was electrifying the Russian's blue eyes.

"Listen, stranger." His eyes flicked to the envelope in my jacket. "I have a friend who will buy that from you. Three times what your company is paying. We split the difference. Fifty-fifty. You say it was lost. Stolen. No one knows."

"Who's your friend?"

"You are interested?"

"Depends who's asking."

"Does not matter. But I tell you if you agree."

"Why didn't you sell to him yourself?"

"I am sure you know this is not good business for breaking contracts. Future business. I need to think of future business. And I think of my life. Yes? You understand? This is why I do not sell to my friend. But if I deliver the package as agreed and then the package is lost"—he shrugged to indicate surrender, forces beyond our control—"what is to be done?"

I took a drag on my cigarette. What was going on here? A test, maybe. To see if I could be trusted. For a moment I even wondered if the Russian was a plant, if the real middleman was decomposing in a Phnom Penh alley, replaced by this wily-eyed, limping charlatan. "Not this time, comrade."

The Russian looked at me shrewdly, extinguished his cigarette. He reached out his hand. "So good luck."

After he'd left I smoked awhile longer. Then I hiked the remainder of the Killing Fields and caught a tuk-tuk back.

Exhaustion made me tempted to skip a cover stop after meeting the Russian, but I knew better. The driver let me out downtown, and after a short foot route, I stopped at a newsstand. Topping the front page of the *International Herald Tribune* was a headline about Bahrain. I bought a copy, finished my route back to the hotel.

Dissident Bahraini Poet Sentenced to Death. King Jassim and the Saudi grand mufti had made the announcement, declaring the sentence "a mercy to the prisoner" that would save him from committing further acts of evil. Death would be by firing squad.

Following the proclamation, the article reported, protesters had flocked to the streets in the greatest numbers yet seen, spurred by still-increasing food prices, still-falling subsidies, rapidly deteriorating conditions in the slums. Riots had spread to eastern Saudi Arabia, where youth were demanding democratic reforms. Iran-backed Hezbollah had issued a statement calling Junaid's planned execution an "assassination" and an "ugly crime," and declaring that those who carry the "moral and direct responsibility are the United States and its allies who give support to the Bahraini regime." Subsequently, King Jassim had severed the few remaining diplomatic, trade, and air links with Iran, and angry rioters in Tehran had responded by torching the Bahraini and Saudi embassies. Then, buried within the article, was mention of the Admiral. He'd received an early promotion. In a long-winded quote, the U.S. Secretary of Defense praised the Admiral for his leadership in the volatile region and willingness to tackle Iran, America's most formidable adversary since the Soviet Union.

I stood before the hotel room mirror. I ran my fingers through my hair, breaking the greasy mold, watched the strands beat a slow retreat. My face still looked unclassifiable, outside the expected bounds of any group or country—on close inspection, didn't even look like it belonged to me. Hollow cheeks—I must have lost weight in the last weeks—deeper pockets. The whites of my eyes no longer white but the color of old paper. Somewhere between the lines I could see the smoother valleys and hills of earlier years, the time before it was too late, a distant land that had become unrecognizable. Nausea gripped me and I lay down, shut my eyes, clutching the Russian's packet. Rashid's voice was in the room. *The choice is yours.*

Did you have any trouble getting here? He'd asked as though he were the handler and I the informant. He'd dragged me to the *ard wa'ira* of Saar, fabled location of the Garden of Eden, now bearing no resemblance to its former state. Burial mounds, sand, rock, and debris had overrun the patch of heaven.

Paradise, yes? Rashid had nodded outside as though reading my mind. *Adam and Eve. What we believe, my people and yours.*

I'd shrugged. *Some of us.*

Here is the beginning of life.

Or the birth of sin.

His face had grown serious. *My friend, the time has come. We need your help. Just a few things.* He'd folded his hands calmly. *This trip you say you must take. There is something I would like you to do for me while you are gone. For the revolution.*

Which is it: for you or the revolution?

Is there a difference?

I'd said nothing.

Do it for the woman.

I'd lit a cigarette and asked how he knew. He'd stared at me as if I should have known better, as if we were brothers and knew each other's thoughts and actions instinctively.

This is your chance. There will never be another.

Night was descending on the badlands, gilding their supple contours.

I am going to tell you our most important secret. You can either help me or you can betray me. You can be a hero or a traitor. The choice is yours.

A sound woke me. The neon clock numbers said 3:19 a.m. I tried to turn on the light, but the power was out again. Murmur of voices next door. My eyes adjusted and I noticed the rear zipper on my suitcase was turned in the opposite direction. Had I failed to notice that earlier? Surely it meant nothing; no anti-tampering warnings were fail-safe. The French balcony doors were ajar. Had I left them open?

Removing my weapon from under the pillow, I walked outside, gray night filling the gap, inchoate street sounds floating upward. My free hand gripped the balustrade until my knuckles turned white. The potted plant was rolling on its side. It was like that before, maybe. Or did that account for the noise I'd heard? The nausea I'd felt earlier returned and I suddenly remembered the bartender at Heart of Darkness, the too-blank look he'd given me when I'd ordered, the way he'd turned his back completely while he made my drinks. The Angkor beer had tasted strangely bitter. Sickness and paranoia clawed at my esophagus. This was what happened in the business, I reminded myself: You stay in long enough, tiptoe down enough dark alleys, and you become convinced every shadow is following you. Odd circumstances and behavior are usually just odd, nothing more.

But there was, in fact, something below my balcony. The outline of a figure moving beyond the hotel gate on Preah Ang Yukanthor Street. It paused beneath a palm tree, in a dark patch between streetlights. A car drove by and in the headlights I thought I glimpsed the junkie from the military market.

None of this would do any fucking good. It was either something or nothing and there was no way of knowing and no benefit to wondering. A world of unexplained devotions, as a spy writer on my shelf once described it, probably the only world I'd ever inhabit.

Then I remembered the packet. I'd had it in my hand on the bed. Rushing inside, afraid to try the lights, my hands running across the covers like an addict who has lost his stash. There it was, exactly where I'd been lying.

Delicately I held it, examined it for the first time. The seal was fairly basic—easy enough to steam or pry open and look inside, reseal if necessary. I turned it over two more times. Curiosity had killed better men than me. The packet was nondescript, something you'd find in any office. Knowns and unknowns were everything in this business. The architecture of espionage, the house we all shared. I ran my fingers over the seal. Sometimes unknowns were better than knowns. Another rule of espionage. No place for curiosity, only need-to-know. Sometimes knowledge caused unpleasant consequences. Liability, questions. Doubt. The sequence had begun and there could be no doubt. I would not open it.

I locked the French doors, placed my weapon and the packet under my pillow. For distraction I leafed through my guidebook. Small particles of sand between the tattered pages brought Manama into my room. The book slipped from my shaking fingers and I realized I was drenched in sweat, the dense Asian heat invading every pore. I sifted through the drinks I'd bought from a street vendor and downed half a bottle of Sombai. Lying on the bed, I watched the lights of passing cars slither across the ceiling. They turned into the dead policeman's searchlight, trained on my face. The alcohol sped through my veins and my limbs relaxed. I couldn't resist any longer. My thoughts turned to Almaisa. She came to me in her purest form, untarnished with betrayal, cloaked in the unending devotion she showed to all that mattered: her orphanage, her art, her adopted homeland, the slums. And her devotion to me—despite everything, despite my flaws. I couldn't move.

Then her betrayal returned, along with my nausea, and I summoned enough energy to finish the Sombai. It was enough to get me through the night.

48

Cambodia, Day Seven

The tremor in my hands had resumed when I awoke the next morning. Splashing my face with the tepid Cambodian tap water, I could hear the city stirring outside, high-pitched birds igniting treetops. I packed my things. My nerves were kicking in earlier than I'd expected.

I braved a look in the mirror. Shaved, combed my hair, adjusted my glasses. I was Robert Thompson for the last time. Mr. Thompson would be seen leaving the Blue Lime Hotel.

I opened my briefcase. Inside the portfolio containing Uncle Chang's promotional literature, I placed my new Swiss passport, Sven's paper, and the Russian's bulky packet, smoothing down the entire affair until my palms burned. When I'd fastened my briefcase with the cipher lock, it looked entirely normal—except for the Great Seal of the United States. My unassailable insurance. Trusty eagle with olive branches in one claw, arrows in the other. Not worth a second glance, maybe not even a first.

Despite the burning sun, I walked to the Raffles Hotel Le Royal for breakfast (a must-try for expats, my guidebook said), briefcase in one hand, suitcase in the other. Textbook stairstep pattern for my surveillance detection route. Nothing could be taken for granted. Every window a mirror, every street crossing an opportunity to turn my head. No room for shortcuts or truncated corners. By the time I reached the yellow colonnades of the hotel, I was drenched in sweat. At least I was clean. No shadows.

At the Elephant Bar I ordered scrambled eggs and bacon and a Femme Fatale, the bar's signature drink honoring Jackie Kennedy's 1967 visit to

Cambodia. Jackie Kennedy hardly seemed the femme fatale type, but I supposed she, like every woman, had her own singular powers of persuasion. Scooping up my food, willing it to make the journey to my stomach, I surveyed the place. Leather-backed furniture held a few well-dressed expats, gilded frames showcased black-and-white photos of famous visitors, potted palms speckled the mahogany floor. Like the potted palms at the Admiral's house. Trembling lightly from footfalls, green shadows cowering like villains.

A figure in the corner. Dining alone, except he wasn't dining. A plate full of untouched food. An open *International Herald Tribune* covering his face. His shoes. I'd seen them before. The same ones Walid was wearing in his office. Black, well polished, about an American size 11. European size 45. I signaled for the check. Walid Al Zain. Of course he was here. A senior intelligence officer trained in surveillance. Assigned to tail me—by the Bahrainis, maybe by my own side. I'd beat him at his own game. Finishing the last of my eggs, I gathered my things with deliberate casualness, rose from my stool. The man was gone.

The polished marble bathroom near the lobby was empty, too early for regular traffic. Had I imagined it? Did I really remember Walid's shoes with perfect clarity? Even if Walid had followed me to Cambodia, he surely didn't recognize me this morning in disguise. I was Robert Thompson, not Shane Collins. Raffles was merely a coincidence. Paranoia. Just paranoia.

Inside one of the stalls I removed my glasses and linen suit and shoes, set my Colt on the back of the toilet. Farewell to Robert Thompson. Adieu. *Ma'a salama*. Stuffed my clothes and weapon into a plastic bag, donned a diplomat's outfit—casual but respectable buttoned shirt, khaki slacks, fully loaded penny loafers. They'd bring good luck. Hadn't the fortune cookie at Monsoon predicted my shoes would bring me luck? Or maybe it said they'd make me happy. Outside the stall I straightened my clothes, parted my hair, splashed water on my face to rinse away the grime from my glasses. There was no sign of Walid as I exited Raffles.

Twenty minutes of walking along Preah Monivong Boulevard and I knew I was black, no tail, but the feeling of unease wouldn't leave me. Behind the night market bus stop I dropped the plastic bag with my disguise and gun into the Tonle Sap. I emerged onto the quay lighter but somehow more weighted. By the time I walked into a convenience mart for cigarettes, my body was on high alert.

Exiting through a rear door, I found myself in an alley bright with laundry and hanging plants. Mismatched entrance and exit was an aggressive move, sure to trigger suspicion if eyes were on me, but I had to be certain. The only movement came from a woman pushing a bicycle laden with bananas.

Orussey Market, off Charles de Gaulle Boulevard, was a beehive. Sweat dripping into my eyes obscured a clear view of the place, turned it into a sea of overheated activity. Squinting in the sunshine, I tried to pick definition from the scene. It was impossible. Probably needed to drink more water.

And then a flash of warning in my guts. The whole thing was too risky. Too difficult. I had to abort. My instincts were all but unerring; I goddamned knew when I was being followed. Walid, the beggar, the junkie, my assailant . . . Too much for coincidence.

The briefcase started to slip from my perspiring fingers. Precious cargo inside. And then suddenly—the revelation, the tactile sensation. It injected me with potency, banished my cowardly fantasies of abandonment. I carried the treasure. My downfall was also my salvation—and the irony replayed in my mind like a newly discovered tune.

You never had to worry about third world Asian airports, certainly not if you were an American diplomat. Singapore, Korea, Japan, maybe. But not Thailand, Malaysia, Cambodia. Gunrunners and drug traffickers welcome. Escorted in and out with a red carpet. The reason Manama had banned direct flights to and from Asia now worked in my favor.

Would I, Phnom Penh's honored diplomatic guest, be checking any baggage today? No, not today. I placed my suitcase on the conveyor belt and, indicating the Great Seal of the United States, informed the officious-looking Khmer that my briefcase was a diplomatic pouch. The officer glanced at my diplomatic passport, nodded, waved me through with nary a second glance. I smiled and proceeded through the metal detector, gripping my briefcase, retrieving my suitcase on the other side. The sensor made no sound.

On the plane I read the ubiquitous luxury magazines then tried to sleep, achieving only a half state of stupor induced by a Bloody Mary and the cloudless blue sky. I lowered my shade. Everyone has to take a side eventually, I reminded myself. Even if only in secret. Whitney, Rashid, the Admiral, Almaisa—they'd all taken sides. Immolated vendors had ignited the streets, poisonous clouds darkened sunsets, the world had moved to one end of the room or the other—and I? I'd sat in the middle watching the show with a drink in my right hand and a cigarette in my left. Biding my time until I retired on a government pension in some reasonably priced city in Florida, playing bocce every Wednesday with creaking men, stoking my fading lust in cheap strip clubs, spewing choice war stories when even minimal interest reared its head, reclining by the pool as the sun burned my chest, waiting for my liver to finally give out. This would change it all. I was taking a side. It would make a difference. It would bring her back to me. I would become something out of nothing.

Clouds materialized and grew thicker as day turned to twilight. The captain announced thunderstorms up ahead—"weather," he called it—and a bit of turbulence, possibly a delay. The cabin grew dark. I closed my eyes.

I am not supposed to tell you this, habibi, *but I think you need to know. You have heard the Saudis are building two nuclear power plants? Extracting uranium? This is not the only thing they will do. The Russians are helping your Saudi friends build a nuclear weapon. The Russians—they help the king, they help us, they help anyone who can pay them. Of course you know the purpose of this weapon. It is to scare the Shia. To intimidate, force us to retreat. No more Arab Spring. This will be the end of hope in the Middle East.*

Habibi, *this is why we need your help. We have a defense. It is outside the country. We need you to bring it in. A small defense, but* inshallah, *it will be enough.*

Jesus Christ, Rashid, I'm not smuggling a goddamned nuclear weapon into the country.

My friend, it is nothing like that. Do you think I would ask you to do that? It is simply— Do you want to know?

No.

We do not have much, but we have this defense. We have Allah on our side. And we have you.

The seat belt sign blinked on. We were descending into Brussels. The refined city sprouted from the earth, neat boulevards emerging in broken squares beneath the clouds. Everything, as Rashid would say, was in God's hands now.

It was a rough landing amid precipitation and lightning. I disembarked into air-conditioned halls, the cool breeze of civilization imparting a sense of calm, a temporary haven from the sweltering heat of Cambodia and the promise of an equally punishing sun in Bahrain.

Upon landing I learned I'd have to pass through security despite the fact I was merely transiting through Brussels. Heightened tensions in the Middle East was the official explanation, an added layer of security to prevent dangerous sympathizers from traveling to the region. I'd prepared for this possibility. There was more than one transit point in Europe between Cambodia and the Middle East, and I'd chosen Brussels deliberately. A spate of recent bombings had starkly exposed the airport's weak security, fundamental flaws that were unlikely to be rectified for years. No cause for alarm. It would be as routine as every other flight I'd taken. With a few hours to kill, I drank two Black Russians at the bar, watched the lightning tear the sky into jagged blue pieces.

Flight 835 was boarding, the voices in the ceiling announced. Already the line through security had formed. The briefcase in my hand felt heavy. Weighty. Like a spinning game show wheel: capable of any result, benign or awe-inspiring. It could be a suicide bomb, I realized—the Russian's packet could contain a suicide bomb, a smaller, smarter device than authorities had come to expect, the kind born of innovation in the Iraq conflict—and I the unwitting vessel. Timed to detonate with maximum casualties. On the plane, in the airport, perhaps. A page torn from the Palestinian playbook. A bomb in the Bahrain International airport would make quite a splash. *American Diplomat Makes Ultimate Sacrifice Aiding Cause of Arab Spring.*

At a convenience store I bought an international phone card, headed to the nearest pay phone. My fingers were clumsy, too many Black Russians. The number came to me instantly—even through nerves and everything else this nasty business dumped on me, I'd always been able to remember his number.

A long pause and a few clicks and the line rang. I listened to the banal trill summoning the one person I wanted to talk to, the sole voice I wanted to hear. Lightning gave way to heavy rain and the downpour drowned the sound of the ring, sheets of water slamming against the airport windows.

"Hello? Shane?" My voice came out louder than I'd intended. I couldn't tell if he'd picked up or not. For fuck's sake, if I wasn't going to exit the goddamned plane alive, I wanted to hear my son's voice. Crackling at the other end and a muffled sound—a voice. "Shane! Are you there?"

Silence, then a click. I stayed on the line waiting for the voice to come back, for Shane to realize the mistake. He hadn't known it was me. How could he have known? The dial tone returned. I went back to the bar, ordered a last drink.

A young slim blonde carrying two purses stood in front of me. She reeked of perfume—European perfume, earthier and milder than Eastern scents. Behind me an older man noisily folded and unfolded his newspaper. On the wall a glowing advertisement showed a man in an impeccable gray suit confidently brandishing his flawless TAG Heuer watch. The line snaked around the corner—at least one hundred people, the usual backup caused by a change in procedure. One hundred people blasted to pieces would certainly warrant an international headline. Ticking—was the briefcase ticking?—and then I wanted to laugh out loud at the cliché, the inanity of it all, the hackneyed paranoid fantasies filling my mind, almost too entertaining to dispel. My hands felt weak from gripping too long. Something unbearably isolating about points of transit. No-man's-land, in-between stretches, abysses. Where you were sandwiched between worlds, where you didn't belong.

The man behind me cleared his throat. The noise reminded me of my father. He'd smoked rapaciously, had a raspy voice, always a tickle in his throat. Probably half of what killed him three years ago. I hadn't gone to his funeral. It was a cheap affair that old neighborhood friends told me was sparsely attended—a few poker and drinking buddies, nobody had much liked the guy. What had I missed except his waxy made-up face, courtesy of Eternal Rest Funeral Home? What would I have brought to his coffin? A failed career, a failed marriage, a failed attempt at parenthood. A life no better than his. I drummed my fingers against the briefcase. If my father were watching now, he'd feel a grudging admiration. Good, bad, my actions didn't matter—I was on the streets, doing something. Something that would be in the papers. My son would read about it. Marlene, maybe. And Almaisa—she'd read the papers too.

A representative from the airline walked past the waiting passengers, announced a further delay due to weather on both ends: thunderstorms in Brussels, dust storms in the Gulf. The man behind me continued to clear his throat—habit, not necessity. Habits crippled you, turned your life into a foregone conclusion; he needed to do something about that cough.

I tightened my hold on the briefcase. Portfolio containing materials from Uncle Chang—that's all security would see in the unlikely event it violated international protocol and opened my diplomatic pouch. I'd prepared for even this remote eventuality. But of course it wouldn't happen.

Glanced at my watch. Maybe I hadn't given myself enough time. My loafers stared up at me, pennies intact, reassuring reminder of my unperturbed diplomatic façade. The line wasn't moving. What was the holdup? A problem, someone taken into secondary, perhaps. Tremors were beginning, sweat hurrying to my pores as though summoned by a silent call.

Ten people or so in front of me. My heart began banging, its insistent knocking a kid trapped in a closet who realizes the game he's playing is no longer fun, that the darkness is stifling and he can't get out. This was madness. It was all madness. Around me were blissfully normal people—or if not normal (for who among us is truly normal?), people who carried normal luggage. People who would not bring a universe-changing briefcase into a country. An officer, I saw, was working his way down the line with a K9, directing it to each person's luggage.

Five people. Down to five. The dog had stopped at the woman in front of me and was sniffing vigorously. She looked down, annoyed, and incautiously moved backward, bumping into me and causing me to drop my briefcase. It landed with a clatter on the floor. I bent down, picked it up.

The K9 had alerted its handler to the contents of the woman's suitcase. Mumbling fragments of haughty confusion, the woman followed the security officer toward a side room: secondary. Down to four. Four sets of arms and legs until my own set walked through the metal detector, dip pouch in hand. Like every other diplomat around the world. How many government assholes were walking through metal detectors right now with briefcases under their arms? Tens at least, maybe hundreds. Fuck if I hadn't done this a dozen times—carried covert material through an international checkpoint, sometimes odd-shaped or oddly weighted packages in far worse conditions than this. All I faced now was mildly increased security and a slight delay due to weather, small bumps in the road. I knew the drill, looked the part. Tradecraft, manipulation—things I knew better than any profession in the world. Espionage was a game and if a spook wasn't playing, he was either irrelevant or dead. My foot tapped impatiently, a bit giddily. I felt the urge to hum, recite something.

Be warned, all kings on mountaintops
Whose feet we can see and nothing more:
We are eating your saffron and occupying your checkpoints!
We are happening. We are moving.

And then I was at the front and the Belgian security officer, a boy in a blue-and-white cap with an FN hanging across his juvenile frame, was motioning for me to put my luggage on the conveyor belt. I hoisted my suitcase onto the moving strip with my left hand, redoubled my right grip on the briefcase.

"Your other bag," he ordered.

Calmly I pulled out my passport for inspection, its black leather and shiny silver embossed letters protecting me, warning the boy not to lift the lid, shine a light in dark corners where he had no rightful business.

He took the passport, looked at me. "Boarding pass?"

I handed him the document.

"Zhane Colleens?" he asked in a light French accent, frowning.

"That's me."

"You are coming from Cambodia?"

"Yes, sir." The man behind me was clearing his throat again.

"Sir, what business did you have in Cambodia?"

The hair at my temples was wet; I was sweating more than I'd realized. "I'm a diplomat." To emphasize the point I tapped the passport in his hand forcefully. Too forcefully. My vision was narrowing, the world turning black and white, zigzagged. "You know what a diplomat is? *Diplomate.*"

The French—it was probably the French word, I would later muse. Foreigners should know better than to attempt linguistic abductions of the eggshell-fragile language, particularly if the audience is a young officer of the state, a baby-faced bureaucrat anxious to prove his worth.

The boy looked at me sharply—my trembling, I knew, was visible—and signaled to another officer behind him. "Mr. Collins, we need to ask you a few questions. My colleague will show you the way. S'il vous plaît."

The second officer was a stout milkmaid with round hips and long hair forced sternly into a braid, a probable attempt to persuade her male colleagues of her seriousness and parity. She wore sensible shoes, clunky matronly things. You could always tell by the shoes. Like me. I was wearing fucking penny loafers, for Christ's sake. I was an official representative of the U.S. government.

I blinked, trying to see the fuzzy milkmaid. My shoes floated above the floor, weightless. I had a feeling of separating from myself, watching my body as an outsider: a middle-aged courier in diplomat's clothes carrying too much luggage and too many pounds around the waist and sweating drink and not knowing what the next words out of his mouth would be. Delusions of finally making a dent in the world, but in truth nothing more than an inconsequential mule.

By the time I reached the small interrogation room, my hands were so slippery I could barely place my bags on the table. The milkmaid sat on one side, indicated I should sit on the other. Opposite me a black window gaped at the scene, the Belgians' not-so-secret eyes.

"Mr. Collins, you seem nervous." Her accent revealed she was German.

"White-knuckled flyer, I guess. Bad weather does that to me."

She tapped her pen on a notepad. "Something more, maybe? Tell me about your business in Cambodia."

"I'm a diplomat. Did you see my passport? Stationed in Manama. Traveled to Phnom Penh on official business. Can you tell me what this is about?"

"You work at the embassy in Manama?"

"Naval base. Office of Middle East Analysis."

Her pen was scratching away. "Who did you meet in Phnom Penh?"

"I don't see what—"

"Answer the question, please, Mr. Collins."

"Francis Chang. Why are you asking me all these questions?"

"And he is . . . ?"

"Director of the Ship Support Office."

"What was your business with him?"

I sighed heavily, exaggerated my exasperation. Probably blew a gale of Black Russians her way. "Port access. Coordination of regional naval policy. Can't go into more detail—we're talking confidential U.S. policy. What's the issue here?"

The milkmaid looked at me appraisingly beneath her blue-and-white cap. "Did you go anywhere else?"

I squirmed deliberately in my seat. "Restaurants, hotel, the usual—"

"Where did you stay?"

"The Kabiki." She wrote down the name. A search of hotel records would show I'd paid for all six nights of my trip.

"Did you go to Russian Boulevard?"

"Russian Boul—? Look, I know what you're getting at. I went to one hostess bar. Just one. She was legal. Twenty-five at least. Probably closer to thirty."

"Mind if I look through your bags?"

I adopted bewilderment, gestured invitingly toward my suitcase. "Go right ahead with this one. But my briefcase—sorry. Under strict orders. It's a diplomatic pouch and I'm an official courier. You know how that goes."

She offered a noncommittal half smile and, snapping on a pair of rubber gloves, unzipped my suitcase. Briskly she riffled through my clothes, toiletries, guidebooks. Deeper she dug, her hands groping the bottom with tactile severity. No drugs, no weapons, no contraband, nothing that even hinted at the black market. Using a wand, she ran a cloth swab across the contents.

A blue machine the size of a cash register sat on the table and the milk-maid coolly inserted the swab as if validating a parking ticket. The device made its destiny-determining calculations.

"You are testing positive for explosives, Mr. Collins."

My surprise was genuine and I held up my hands. "Well, I'll be— Are you sure? No idea why. Obviously I don't have explosives in there. Your machine must not be accurate." Fucking Russian Boulevard, the military market—traces from vendors or Munny's warehouse must have rubbed off on my belongings. Or maybe something in the Russian's packet.

"There are errors, that is true. Sometimes dirty laundry can trigger the machine. Grease and dirt have an effect similar to nitroglycerin."

I laughed—too loudly. "Guilty of dirty laundry. Lock me up!"

She didn't return the laugh, didn't even smile, instead turning to my diplomatic pouch. "Mr. Collins, in light of the swab results, we will have to look at your other bag. This is our protocol."

"That's not possible. As I told—"

"I understand your wishes, Mr. Collins. We abide by the Vienna Convention of Diplomatic Relations whenever possible. However we are not obligated to honor all articles of the convention when we have reason to believe the passenger presents a threat to our national security—"

"*A threat to national security?* What the hell are you talking about?"

"Or the passenger or his documentation might not be authentic."

"Not authentic?" I was a ridiculous parrot, stalling for time. "Here." I pulled a folded piece of paper from my pocket. "You don't believe me? Here's my official letter."

My safety net, my last resort in black and white. One diplomatic pouch, accompanied by Mr. Shane Collins, signed by the Ambassador himself, please direct any questions to the Regional Security Officer at U.S. Embassy Manama.

She examined the letter, eyed me intently, inserted a new swab into the wand. "Mr. Collins, would you mind if I swabbed your hands?"

Hesitantly I placed them on the table. They shook, left nervous smudges on the unblemished steel. "Listen, am I going to miss my flight?"

She ran the swab over the back of my hands. "Please turn them over."

Palms up, sweat visible in the creases. My hands looked old. Older than fifty-two years. Too old to be shaking this violently, proffering explanations, convincing inquisitors. Hands on the cusp of guilt. Too late to wipe clean. The small rectangle of fabric skirted my damp flesh.

"Mr. Collins, how much have you had to drink today?"

Suddenly, I detected an opening. An excuse. Just as Almaisa had blamed alcohol for the night on the faux-marble tiles. People saw what they wanted to see. Good spies turned vulnerabilities into assets. And I was a good spy.

"Officer, I'll admit I've had a few too many. Had a lot of fun in Cambodia . . . guess I wasn't ready for it to end." I grinned, embarrassed.

She studied me, the wand resting in her hands like a truncheon. The door opened and the boy handed the milkmaid my passport, said something in French. She inserted the swab into the machine.

"Your passport, Mr. Collins." The milkmaid reached across the table. "And your official letter."

It took me a moment to register the transaction.

"Sorry for the delay." The milkmaid was still watching me. "Drugs and weapons from Cambodia are a problem. We have to be careful."

The thrill of victory pulsed through my veins, better than any drug I could smuggle from Russian Boulevard.

Somewhere over Bulgaria the pilot announced that, due to the severe dust storm in the Gulf, a diversion to Istanbul was likely. I gripped the briefcase on my lap and watched the graying sky. Would a further lag throw off the timing? Maybe we would all plummet into the sea after all. Die a cold, or at least lukewarm, death because of a change in schedule. Vaguely I wondered what had happened to the blond woman pulled into secondary. I closed my eyes, exhaled. I was black. After all the ghosts, after the milkmaid's scrutiny, I was still black.

Despite avoiding a diversion, we were delayed more than an hour, our plane circling above Manama at 14,000 feet, waiting for the clouds of sand below to dispel. Fuel was undoubtedly running low. Flames from the last stalwart refineries along the Askar coast were all that was visible through the haze, reaching nervously for the sky.

Close to midnight the pilot began our descent. Impossible to tell whether the air had cleared or we were close to empty. I made my way to the lavatory, briefcase in hand. The plane lurched as it angled downward, the outer tentacles of the storm roiling the air.

Inside the lavatory I promptly vomited. Splashed cold water on my face. Shane Collins in the mirror. Some version of Shane Collins. The embossed seal on my briefcase looked up at me. I could open the Russian's packet now. I'd passed through security; what difference did it make? The seat belt sign deafened me. *Ladies and gentleman, the captain has turned on the seat belt sign for our final descent. Please return to your seats.* With unsteady feet I walked back. Turbulence shifted passengers and overhead luggage. In a rough corkscrew we descended, cutting through the sand eddies, attempting to evade any hostile fire that might erupt from the simmering island below.

The blond woman who'd been taken into secondary reappeared—this time on television screens in Bahrain International Airport. Pippa Smith, Al Jazeera reported, a thirty-one-year-old British national with a history of supporting Arab nationalist movements, intercepted and arrested at Brussels Airport, large sums of cash found on her person. Exploitation of her cell phone and other documents was underway, details being withheld in the interest of the ongoing investigation.

Two? Two mules on the same flight? Tradecraft so sloppy a new puke would've known better. One arrest, increased layers of scrutiny—that's what came from this cockeyed plan. I should never have agreed to it. What else was hidden in the woman's two purses, or strapped to her body, or shoved uncomfortably into her orifices? Did she know about me? Would she crack under interrogation, divulge dirt? Sights could be trained on me even now as I proceeded, unassuming, playing the part of a U.S. official returning to his post.

The airport had changed radically since I left. Troops, mostly Pakistani and Jordanian, lined the walls, gripping their Kalashnikovs with trigger-ready hands, with the detachment of well-paid outsiders. Eager dogs pawed the floor, panted, waiting for the opportunity to pounce, uncover. And there was an X-ray machine for arriving passengers that had not been there before.

"Passport, sir."

Black booklet, shiny protective seal.

"Thank you, sir."

Passport back in my hands. Almost done, a drink waiting at home.

"Please put your bags on this belt, sir."

"This one is a diplomatic pouch. It stays with me." Home turf, easy territory. I'd passed the high hurdle.

The Pakistani looked dumbfounded—too little training, didn't understand. "Excuse me, sir, I need to talk to my supervisor." He turned toward a mirrored room and a man in plainclothes.

"Diplomatic passport," I reminded him, holding it up in exasperation. But the plainclothes official was already heading toward me. It was proceeding with a life of its own, simultaneously rapid and labored, a movie playing on fast-forward that stops every goddamned second. I produced the letter from the embassy, edges badly worn from the long flight. "Here. This should take care of it."

The plainclothes official examined the letter, disappeared into the mirrored room. The Pakistani smiled apologetically, offered me a chair. I shook my head, watched the mirrors. No one failed in the eleventh hour. I would pass the checkpoint or wind up in Dry Dock.

The plainclothes man returned. "Mr. Collins, I called the number on your letter. The Ambassador's office has no record of you or your diplomatic pouch."

I forced a knowing grin. "Nothing like bureaucracies. Call again. Tell them I'm with the Office of Middle East Analysis. Remind them I work for Whitney Alden Mitchell."

He nodded hesitantly, disappeared again behind the mirrors.

Passengers arriving, moving through passport control and security, big dogs eyeing each person, jerking forward on occasion to sniff a suitcase or purse. Businessmen, Arab women once again ensconced in their abayas after days wearing designer frocks, not as many Europeans as usual—one, maybe two. Older. The ban on young men still in effect.

The plainclothes official emerged and handed me the letter. "I am sorry, Mr. Collins. We called the embassy back and they apologize also. They know Mr. Mitchell. They confirmed you are working for him. We apologize for your inconvenience. As you can see, we have several new security procedures in place—"

I snatched the letter from his hand. He bowed awkwardly.

A putrid coppery stench hit me when I opened my trunk. Vestiges of the policeman's corpse. The odor hadn't seemed this bad when I'd left just days earlier, but maybe I'd failed to notice, accustomed to it. I threw my suitcase into the backseat.

It took five attempts to start my Lancer. Old thing was nearly at the end of its life. Puttering and bumping my way out of the parking lot, I navigated the snake of tanks constricting the airport, their turrets poking through the haze, aimed at the trickle of exiting vehicles as though daring each one to run the gauntlet unscathed. Clearly the king had called his friendly neighbor across the causeway; Bahrain had just a few tanks to its name.

The air was dusty enough that I didn't see or hear the swell of protesters until I was nearly in the thick of them, massed just outside the airport and along Road No. 2403, thousands strong, entangling every car trying to enter or leave. Young men, lean and angry, carrying signs with both the usual slogans and some I'd never seen before. *Free Junaid. USA Stop Arming Killers. Get Out Western Infidels.* Pictures of Uncle Sam looking greedy and malignant. Several carried torches they jabbed threateningly at the sky as though daring Allah to stop them. Curfew had apparently been lifted or, more likely, disregarded, another useless barrier knocked down.

Entering the fray, I sensed the mood immediately. I'd seen it before—the dark roiling rage that builds in large numbers at a certain velocity, a certain momentum, stoked by years of fist-banging, suddenly presented with a time and place, an outlet, a stage. A lethal combination. My car inched toward the traffic circle, began to take a few rocks and fists, the old beater reeling disproportionately from the blows like an overused punching bag. I tried not to make eye contact with the tinder sticks around me, but a particularly irate boy no more than fifteen saw me through the windshield. *Ajnabi! Kafir!* he yelled, and a large stone crashed through the front passenger window. It landed with a thud on the floor. Instinctively I looked to the police station across the street, where a single tank stood sentry, outmatched, a pathetic onlooker abandoned by its brethren guarding the more valuable airport. The protesters' cries were becoming audible. *Al-sha'b yurid isqat al-nizam! The people want to bring down the regime!* The cry that had begun in Tunis and spread to Cairo. It was happening. The Arab Spring had finally found its feet on this pile of sand.

Gently I stepped up my acceleration, just enough to carve a path through the throng without provoking anger or alarm. My dying vehicular friend wouldn't last much longer, maybe not beyond the riot. Another rock, mercifully smaller this time, ricocheted off the rear windshield with a pop. Just a few yards to go. The briefcase was still on the passenger seat and I placed my hand on it as though it were a child in need of protection. Agitated murmurs were sweeping the crowd, commotional scuffle, shouts. In my rearview mirror I saw a small detachment of riot police delving into the masses. Birdshot, a hissing explosion of tear gas—or whatever the boutique gas du jour was, sharp and potent through my broken window—cries of anguish, a few people stumbling blindly into my car, then an earsplitting roar, the crowd moving abruptly toward the police in an angry wave. It was my opportunity and I took it, pressing the accelerator steadily, forging a path out of the chaos, until I was past the center of mass and threading my way through the fringes. Jasri Al Kabeer Highway. I was free. Slowly I removed my hand from the briefcase.

Barriers closed off Arad Highway and I was forced to weave my way through backroads. Stopping by the station to drop off Sven's information would be impossible. Whitney and the Admiral would just have to wait. My eyes were tearing from the chemical agent, and the poorly lit streets were blurring. On Road No. 4403 I saw that Arad Fort was burning, or at least something near it was burning, the conflagration enveloping the entire structure in smoke, a volcano-top of flames spouting imprecisely from the haze. The eastern end of Arad Highway was open and I was able to turn back onto the main road. Most of the streetlights were broken and, despite the fires, the area was drowned in darkness. Dry Dock Highway was closed, lines of APCs facing the distant prison, ready for any attempted mass egress.

I headed north, the Gulf to my right, stubbornly peaceful and quiet beside the unstable giant taking its first shaky steps. Dhows bobbed unconcernedly, the white beachfront homes still pristine sugar cubes, their Bahraini flags looking only slightly tired and dingy. For the first time I'd seen, the Amwaj gate was closed, and the rent-a-cop on duty looked at me carefully before nodding me through. He was carrying a weapon, laughable protection against the kind of seething masses I'd seen at the airport.

Archipelago life appeared unstirred, light spilling from open windows, the Tea Club balcony on the lagoon filled to capacity, a Rolling Stones cover band belting out "Satisfaction" from the depths of the Dragon Hotel. The Indians under the trellis jumped at the sound of my car, ready to assist, concerned only about meals and roofs, not the war outside their benefactors' gates.

Before unpacking, even before getting a drink, I carried my briefcase upstairs to the maid's quarters, unfastened the diplomatic seal, took out the portfolio. Unzipped it, removed the Russian's packet. I ran my hand over it. Not now. Maybe later. In the shelves inside the closet, I placed the packet and my Swiss passport behind a stack of books. I kept Sven's paper in the portfolio with Uncle Chang's literature. Slowly I felt the grimy vise of paranoia and uncertainty loosen its grip. I was intact; my packages were intact; I'd made it past the checkpoint. No one had succeeded in stopping me. I was a good spy.

Downstairs I poured myself a gin and tonic and watched the lights tickle the canal outside. On the balcony I surveyed the quiet rows of villas. I was home, I realized with bitterness and amusement, or as close to home as I'd ever be.

On a roof deck across the canal a circle of expats talked loudly and clinked glasses, their chairs facing Arad Fort, the flames surely visible from their towering perch.

49

By morning the dust storm had begun to die down. But the stubby legs of war that had grown overnight were walking now, streets choked with protesters, smoke columns puncturing the landscape like geysers. The city moved about its business in a heightened state.

THREAT CONDITION DELTA the sign outside the naval base declared, and I smiled my usual greeting at the guard, clutched my diplomatic pouch, safe now on friendly territory.

Whitney was gone when I arrived at station. In a damp Hawaiian shirt Jimmy greeted me with Chihuahua-like nervousness, a state that seemed to have become his norm in recent months. "How was your trip?" he asked.

"Good. Got the Swede's package. Chief in?"

"Briefin' the Admiral. Shit's heatin' up outside, ain't it? Trying to get updates on the unclass machine, but net's been shut down across the island. Too many broadcasts in Diraz against the king. Calls to arms." He wiped the sweat from his head. "Thought you were gonna bring the package to station last night. Boss waited for you."

"Roads were closed. But don't worry, birdwatcher, I double-wrapped it, slept with it under my pillow. Even cooked it breakfast."

Jimmy grinned.

A click and Whitney walked in. He motioned me into his office. "Lots going on. Let's talk. Tell me about your trip."

I unfastened the pouch, removed the portfolio with Sven's paper inside it, placed my transliterated copy of *The Manama Verses* on Whitney's desk, relayed details of the transaction. Important to give information accurately, leave no room for doubt, mistrust. "I took the liberty of cooking it this morning," I added. "Seeing as how we don't have an oven at station."

Whitney looked at me questioningly. "Lemon juice?"

I shrugged. "It works."

He stared at the baked brown digits. "Sarajevo sent their numbers." Tapped a cable on his desk. His hands—his hands that had touched her.

I repeated the Swede's instructions: Add seven to the sum of the Sarajevo numbers, go to the corresponding page. Whitney verified our chicken scratch with a calculator. Painstakingly, as though cracking a nuclear code, he counted letters across the text. Nearly thirty minutes passed, more than fifty pages covered. An answer. Cargo ship, Khalil Al Jasri port. July Six, Zero Three Four Three. Two weeks from the day. Whitney smiled, almost chuckled. A puzzle we'd pieced together, a spy game to be won.

The cutout remained. Eagerly Whitney leafed through *The Manama Verses*, matching numbers to letters, decoding the final piece. When he'd finished, he frowned at the words.

"You know it?" I searched his face.

"Never heard of it. I'll get Jimmy to sniff around, see if he can learn anything before the shipment arrives."

He punched my shoulder in satisfied conclusion. "We found the pot of gold. Everything we need. Interdict these weapons—interrogate the crew, find out who's behind it—and this uprising will be D.O.A. Nice work, Collins."

I got up to leave, handed Whitney the literature from Uncle Chang. "Picked this up for the Admiral. For my cover story. Thought it might interest him." Whitney fingered the material, nodded distractedly. His eyes had moved back to the prophecy before him. His rising star had just reached its apex.

50

It would be a simple matter of steaming, I'd concluded. Even evidence of tampering probably didn't matter, so long as the packet ended up unmolested and in the right hands. Surely Rashid had considered the possibility I would open the thing. I turned the packet over in my hands until it was nothing more than an abstract object, something to be examined and investigated with clinical detachment.

Do not call me when you return, my friend. No contact except our meeting. He left me no choice.

I decided the heat of steam might damage the contents, so I resorted to carefully opening the flap with a letter opener. Tearing was minimal; I had surgeon's hands tonight. Inside was a box no more than one inch deep, ten inches long. It opened easily. Five unlabeled dropper bottles rolled out, milky brown glass filled with liquid. Raising one to my nose, I caught a sharp sweet scent not unlike jasmine. The bottles rolled to a gentle standstill on my countertop.

The banality of it all prompted a strange sense of relief. I'd expected hollow-point bullets or explosive powder, thermite maybe, the milkmaid's swabs having suggested nothing less. But these vials, these tinctures, whatever they were, seemed somewhat less instrumental, more benign. A desperate revolutionary's delusion. In a country consumed by tectonic forces, entire legions clashing in the streets, how much harm could five small bottles do?

51

Insistent knocking at my front door woke me. The clock said 8:07 a.m. I threw on yesterday's clothes, peered through the window. Whitney was on the doorstep, his dark blue jacket an oozing stain on the street. He checked his watch, glanced around.

"Collins!" Whitney exclaimed when I opened the door. "Wasn't sure I'd find you here."

"My car's in the driveway."

He turned around. "So it is. Anyway. Can I come in?" He was nervous.

"I was just getting ready to leave—"

"It won't take long. It's important. I wanted to talk to you in person."

"Problem with the Swede's intel?"

"No, nothing like that."

I stepped back, watched his short frame shuffle inside. Despite his better-fitting clothes and weight loss, Whitney had the same awkward diffidence I remembered at our first meeting.

In the living room he stopped, stared at the wall. "Is that . . . hers?"

"Yes."

"Incredible." He moved closer. "The colors—"

"What do you need?"

He motioned haltingly toward my sofa as though he were the host and I the guest. "Please. Let's sit."

I offered him something to drink. He declined, and I went ahead and poured two cognacs.

"Listen, Collins, this isn't an easy conversation." He placed his drink, untouched, on the coffee table. "I got a cable from H.Q. last night. You got an offer of early retirement. Curtail your tour, return to Langley, badge out."

From inside his jacket he produced a thick document, smoothed it down. It had the unmistakable markings of a communication from Langley, blackened in parts where Whitney or maybe Jimmy had redacted names, codes, sensitive information. Whitney read the enumerated grievances aloud: inappropriate relations with a married individual and spouse of a staff officer to the Admiral, falsified time cards, misuse of official funds, excessive alcohol consumption resulting in probable medical disability due to liver malfunction (prior pre-cirrhotic condition noted), and frequent habitation of restricted assignation houses including but not limited to the Sea Shell Hotel.

He folded up the cable. A fleshy god broadcasting my sins, officially damning me, declaring my death as a spy.

Possible moles ran through my mind. Charlie the pimp with his slime-dripping teeth. Poppy, jealous, trying to stir envy in her prick husband. Maybe one of her neighbors or even that goddamned guard at Three Palms, the one I warned about. Everyone has a price. Maybe Burt had gleaned the dirt all on his own.

I sipped my cognac, mulled it over. "This isn't really an offer."

"No, I'm afraid not. H.Q. has the final word." The freckles on his neck giggled.

"How much time do I have?"

"Two weeks."

I smiled. "Two weeks. When our friends will be interdicting the weapons shipment."

I was numb. I was walking dead. The end of the road. I stared at Whitney's shoes. Penny loafers, copper coins now dull but still in place. He'd been good inspiration—a solid muse—and I'd done a decent job crossing the checkpoints, putting on a diplomat's face. The packet. I still had the Russian's packet. The thing Whitney and the Admiral and Burt knew nothing about. Its effects would be here long after I was gone.

"Were you behind this?"

Whitney's neck stiffened in surprise. "Me? Of course not. I—"

"Dead weight. Space filler. That's what you called me."

His mouth twisted into an uneasy smile. "Look, Collins, I'm not proud of that night. You're probably not either. But trust me, I need all the help I can get. You did solid work. You picked up vital intel from Sven. And Junaid—the op you planned with Junaid—didn't pan out, but it was a terrific idea. You did an impressive job with SCROOP. No one thought he was worth a damn." He pursed his lips, looked at me. "Collins, I want you to know I fought for you."

I downed the last of my drink. "I know about Walid."

"What? Walid . . . who?"

"Walid Al Zain. The ministry official who's really an intel officer. I saw the two of you at Salman bin Ahmed Fort."

Alarm crossed Whitney's face. "Collins, I never met Walid at Salman bin—that fort. I met him a few times after the Adliya bombings. Is that what you're talking about? You asked me to—said there was a flap, that you'd pissed him off. I smoothed things over."

"Walid was in Cambodia. Following me."

"*Cambodia?* Collins—"

"You took her."

Whitney's hands rested calmly on his knees. "Look, I know H.Q. rejected your request. But I had nothing to do with that."

"You were with her at the safe house."

"What are you talking about?"

"In Askar."

"Collins, I have no idea what you're referring to."

"And at her flat."

His face relaxed. "Her apartment? I was interested in buying one of her pieces. She's very talented. How did you—?"

"She keeps art at her studio, not her flat."

"Well, I saw a piece at her flat." His answers were quick, easy. He watched me steadily. "You don't think. . . ."

I hadn't realized how hard I was gripping my drink, moisture spreading beneath my fingers, until suddenly the glass was on the floor, shattered. Maybe I threw it.

"Collins." Whitney's voice softened, a patient father trying to rein in an unruly son. "She's in love with you. I wouldn't have stood a chance." His face wore a strange expression—resignation, maybe regret.

"So did you buy it? The piece at her flat."

"Oh." He shifted almost imperceptibly. "No, actually I didn't. We're still negotiating. Her prices . . . a bit steep, don't you think?"

I mustered a smile. "I suppose they are."

52

Two weeks. I still had two weeks.

I waited until early evening to begin my surveillance detection route. The city was breathless, dust from the storm lingering in pockets like stale smoke after everyone has left the bar.

My first cover stop was Airport Café for a coffee. The place was bustling with a scruffy chattering crowd. Protests at the airport had dissipated, I saw, but armored vehicles still ringed the premises. Twenty minutes, down to the second, and I drove to the opera house.

Lights were blinking to life, igniting the terrace and its gleaming sculptures. Briefcase in hand since I'd left my villa. Inside, a few painters were on ladders, oblivious to the war outside, finishing gold trim around the exhibition hall. My shoes echoed on the marble floor.

The moment I entered the hall I could see it, stretched across an entire wall. A sparrow fleeing a nest of gold sticks, the net a billowing checkerboard of black and white, small crimson and silver birds twittering warnings in the background. But as any observer would agree, the real masterpiece was the peacock, its feathers so vivid and intricate that the bird looked ready to fly off its tiled perch. Arresting, demanding attention like a bright light in a dark room. THE PEACOCK AND THE SPARROW, 1001 ARABIAN NIGHTS read the card. And then a thought occurred to me. I'd never realized it. The fabled book wasn't about tales at all. It was about distraction, Scheherazade's ability to tell stories so scintillating they would stave off execution. It was about manipulation, bending another's brain. The revelation left me with a strange feeling as I exited the opera house.

Dusk was alighting. South to my third and final cover stop.

The Gulf Hotel was strangely deserted, only a few men sipping steaming mugs in the café. On a crushed velvet seat I ordered a pastry and cappuccino. Piano music in the next room amplified the emptiness, the odd spectral feeling of a cavernous space with too few people, a soundtrack to some forgotten movie. When I tried to order a Whiskey Sour, the waitress informed me the hotel no longer served alcohol. I took a last sip, checked my watch. Time for the final leg.

Through Bu Ashira, onto Oman Avenue. The tail end of a brown beater turned the corner—the surveillance vehicle I'd seen outside the protests. No. Brown beaters were a dime a dozen. Gremlins tickling my brain again. I'd conducted a solid surveillance detection route. I was black. All the same, I parked on the border of Umm Al Hassam, walked nearly half a mile to Adliya, a stairstep route with mid-street crossings, for good measure. Road No. 3610. Near the end of the block, just as Rashid had said. Oasis Persian Carpets, sandwiched between a frame shop and a florist. Cover story in my back pocket: checking the place out per Whitney's instructions. A bell tinkled when I opened the door.

Hundreds of rugs were piled haphazardly inside the small shop, crammed so tightly one had to weave between them as through a maze. Small cameras hung from two ceiling corners. Behind the counter a man in a white thobe who could have been Rashid's cousin—for all I knew, he was—measured a rug. He looked over the glasses perched low on his nose. "Good evening, sir. Welcome to Oasis."

I gripped my briefcase. "Do you have any Tabriz rugs? I'm looking for something for my living room."

"I am sorry, sir. Right now we do not have any Tabriz. But we have a lovely Isfahan rug that just came in. Perfect for a living room. Would you like to see it?"

"Please." I followed him behind the counter. He pushed aside a hanging maroon rug, unlocked a door in the back wall. Inside, the man pulled the cord of a dangling bulb. It was an unfurnished room with a cement floor and bare walls, a stack of dusty rugs in the corner. The bulb swung gently, stretching and contracting our shadows. He locked the door. "You have the package?"

"Where is Rashid? He told me he'd be here."

"He could not make it." The man pushed his glasses up his nose.

"I only deal with Rashid. This isn't what we agreed on."

"He is busy. It is impossible that he comes. It is safer this way."

"Who made this decision?"

"Our leadership."

"Not Rashid?"

"Of course. Rashid as well."

His gaze was hard and patient, the bulb illuminating his face in uneven swaths like car headlights. Slowly I set the briefcase down on the stack of carpets, unlocked it, handed him the packet. He turned it over immediately as though looking for code or signs of tampering, and for an uneasy moment I thought he could tell I'd opened it. But he simply looked up, smiled. "*Shukran, habibi.* You are a true friend of the revolution." He shook my hand.

"Tomorrow please go to the Tree of Life at two o'clock in the afternoon. Someone will meet you. You know where the tree—"

"I know it."

I'd handed over the package. It was done. As we emerged from the back room, the maroon rug blocking our view, the bell on the door tinkled. "I am sorry you did not find what you were looking for," the proprietor spoke loudly to me. "We will get a Tabriz next month, maybe."

Though the piles obscured the doorway, a mirror high on the wall revealed the back of a man's head. As I moved closer, I saw the customer was an Arab man wearing Western clothes. He turned his head slightly in our direction, nodded. I muttered a thank-you, headed toward the door. Behind me the shopkeeper's cheerful voice rose. "Welcome to Oasis, sir. How can I help you?"

Happy hour at Vic's was good cover, a convenient stop on my way home. Show my face, prove I wasn't taking early retirement too hard, enjoy a last visit before leaving this world forever. Undoubtedly Whitney had invited me to flaunt his good will, take the edge off the blow, drown any lingering hard feelings in expensive tropical drinks. Assure me I was still part of the crowd. I lit a cigarette. It was done.

New developments greeted me in the lobby of the Ritz—bag check, metal detector, guestbook—fragile self-assuring measures against the tidal waves outside. I signed Robert Thompson. How meaningless and malleable were names, how easily they transformed one's identity—pedigree, background, purpose, degree of danger posed. Mere stroke of a pen.

The hanging beads clicked. Everyone was there, rubbing their backs on the windowpanes. Same jewels and drinks as the Admiral's party nearly a year ago, but the words faster and more staccato, laughter more desperate. The celebration was in danger of ending. Whitney was at the bar, shirt unbuttoned and sleeves rolled up, talking to Jimmy. Gretchen stood behind the two, stirring her iced drink, looking dejectedly around the room.

"Just sayin' how things won't be the same without you." Jimmy grinned as I approached. "Tough break, eh? Good thing you were close to retirement anyway."

I shrugged, ordered a Moscow Mule.

"Matter of fact, now is probably the perfect time to leave," Jimmy kept on. Whitney nodded.

My drink arrived and I excused myself, joined Poppy and Burt Johnson talking with the Chaplain and his wife. *How bad can it get? Anyway, we have an airtight evacuation plan in place. Durrat Al Bahrain! You know the place? Manmade islands, southern tip of Bahrain. Horseshoe-shaped! Looks like horseshoes from an airplane! The king's renting them out—of course there's a waitlist, but we've already put in our names—only a few thousand bucks a week. Just hop on a boat and you're there in minutes. Full facilities—fresh food, satin sheets, housekeeping. Great place to wait it out. Jassim's got to give Americans priority—look how much we've done for him. What have the Brits done?*

I placed my hand on Poppy's arm, unabashedly ogled her patchwork body stuffed into a tight red dress. "You look good enough to eat, Poppy." I raised my glass. "Lose weight?"

She turned in surprise, started to utter something, then stared into her drink. The Chaplain and his wife looked away.

"Sorry to hear about your curtailment," Burt said acidly. "Always sad when that happens. On to greener pastures."

"Any place is greener than Bahrain. Am I right?" I clapped my hand roughly on his shoulder. Then, finishing my drink, I kissed Poppy hard on the lips. Burt opened his mouth in protest, but I was already heading toward the door.

Whitney had moved to the Admiral's side when I left the bar. It was one of the last times I would see Whitney and I knew it. Maybe for that reason the moment stays fresh in my mind—the damp wrinkles of his white shirt, ironed pleats of his khakis, gleam of his watch. He had an earnest, slightly troubled expression on his face. He was just a kid, I saw for a moment—not, as I'd seen him before, a gold-plated diplomat's son who lived in conference rooms and quoted textbooks and counted accolades on his fingers before bedtime. Just a kid plunked down in the middle of a grown-up poker game.

I switched on the lights in my villa. My couch was upside down. Books had been thrown from their shelves, chairs strewn across the place as though dispersed by a violent gale; the globe my son had given me was broken, still rolling on the floor in crescent arcs.

I righted the couch and surveyed the damage. It could've been worse. Could've happened before I'd delivered the package that morning. Or when I was here, the same damage inflicted to my body, broken skull instead of broken furniture. And then I realized I'd left something damning in the house, something worth taking. In the maid's quarters I opened the closet, felt the passport hidden behind the books. It was still there.

The other rooms of my house were in similar upheaval. It was a thorough job. Either side could have done it—just depended on who'd suspected or gotten wind of the Russian's package, who wanted to take it off my hands before I lost or found my nerve. There were scads of soft spots in the operation, leaky holes, perforated security.

I doused my face with cold water. When I looked up, I realized the mirror too had been broken, my reflection splintered into dozens of shards. Dripping silver claws turned my physiognomy into a monster, something less than human. It was all a mess, too big to clean up at this point.

And suddenly—my Lancer. Why hadn't I thought of it before?

Tailpipe, wheel wells, fenders. Underbelly. A small black oval-shaped device. Sending signals, broadcasting my transgressions, a trail of breadcrumbs to all the places I'd gone.

I ripped off the beacon, raced inside, threw it against the living room floor, then took a hammer to it until only a pile of wires and plastic remained. Who knew how long the goddamned thing had been on? Maybe months. A significant tradecraft failure—I'd checked for beacons in Cambodia but had never thought to check back home. The device had likely tracked my visits to the Sea Shell Hotel; maybe that's how they knew. A counterintelligence investigation authorized by Headquarters. Or maybe Burt was to blame, craftier than I thought, orchestrating an entire operation to entrap me and Poppy. And then I remembered the day Jimmy's car had broken down, when I'd given him a ride. Oddly he'd wanted to stop at Vic's for breakfast. An opportunity for someone to install the beacon while we were inside—fixed window of time, eyes on the target. He'd checked for conflicting beacons inside the car. He hadn't needed a ride home. It had all been staged, a well-maneuvered plan by someone to keep tabs on me.

The revolution factory—was that how the king's men knew its location? Maybe not Qasim after all. But I couldn't go down that path.

The contraption was splayed across my floor like a mutilated insect. At least the delivery of the Russian's packet was safe from the telltale beacon; I'd conducted the final leg of my route on foot. The most they would know was that I'd parked in Umm Al Hassam, and that could mean dinner in Adliya, repairing electronics, even shopping for a rug. I was still black.

53

The landscape of southern Bahrain was the only thing unchanged since I'd last been to the Tree of Life. On the route down there'd been a strange mix of too many checkpoints and not enough cars. The Gulf Hotel had been a bustling exception, a line of trucks parked outside as I drove by, the front steps a hive of activity, a complete reversal from a day earlier. In Isa Town I'd encountered a crowd of manual laborers trailing bearded men in thobes. As I'd slowed to let them pass, a pamphlet had fluttered through the hole in my passenger window—an amateur drawing of a clamshell with a stop sign in its mouth. AL-'IFFA was written across the top. *Chastity.*

Few cover stops existed in the deserts of southern Bahrain, but I spotted a scrapyard in Awali, sifted through the junk, found a board to cover my broken window. Stopped for a drink at the Awali Golf Club but was flatly refused; the place had stopped serving alcohol. By the time I reached the Tree of Life, I was parched and irritated.

Rashid was crouched under the tree as though warming himself at a campfire. If I hadn't been expecting him, I might not have identified him; he'd grown a long beard and was wearing a white kaffiyeh, starkly conspicuous against the drab landscape, nothing like the tattered aspiring revolutionary I'd once known. He embraced me tightly, kept my arm in his grasp. His face was harder, leaner, more purposeful, the tentative ravines and valleys turned permanent as though having irrevocably crossed the border from youth to middle age. Faintly he smelled of defecation, sour and unwashed, as though he'd been living in the open. "*Habibi*, I am so glad to see you. I heard the trip was a great success."

I cupped my hand and lit a cigarette, nearly impossible in the gale-force winds. "This place going dry or what? Couldn't get a goddamn drink all day."

Rashid looked at me sternly. "Strong drink is Satan's handiwork."

"Satan and I are old pals."

"We are cleaning up the streets. Last night we walked through Sitra. We saw men drinking and we told them, 'We know you are unhappy. The situation here is not good. But this is not the solution.' And you know what they did? They cried. They begged forgiveness. We told them if they join the revolution, Allah will forgive."

A gust extinguished my light. I said, "I saw Qasim at the rug shop."

"He was early then. Or maybe you were late?"

"I wasn't late."

Rashid shook his head. "Too many people at the same place and time. Bad tradecraft. I will tell him when I see him."

"Why weren't you there?"

"I had important things to do. I have been busy down here. Also I do not like to show my face these days."

"You staying safe?"

"As safe as I can."

I eyed the surroundings. There was no other car around. "How'd you get here, anyway?"

"With my feet, *habibi*."

"You *walked*? From where? Only thing here is dust."

Rashid smiled. "Our training camps. I can tell you about them now. They are no longer secret." He nodded to the east. Through the haze, west of Jaww and Al Dur, I could barely make out the beige silhouettes of what appeared to be large tents and a few vehicles.

"Training . . . for what?"

He looked at me pityingly, as though the answer were obvious. Then he turned toward the camps. "There is something we say in our culture. *Hatha maseroka, lan yufariqka*. You know what this means?"

"If it is yours by destiny, from you it cannot flee."

"I am impressed."

"I've heard it before."

"My friend, you understand. When the Kingdom of Bahrain falls, we will reach our destiny. *Al-qadar.*"

I watched him, the granite certainty in his eyes. "Listen, Rashid, there's something you should know."

"What is it, *habibi*?"

"Jassim's men know about the weapons shipment from Tehran. Time, place. I had to give the intel to my boss. I couldn't lie. Too much risk tampering. It wouldn't have worked." I avoided his eyes.

Rashid smiled. "Ah, Sven. You do not think Sepah would feed him the correct information, do you?" He put his hand on my shoulder. "But thank you for telling me this. You are a great help to the revolution. Our leadership is very pleased. It is because of you that—"

"Tell me something." I nodded in the direction of the tents. "Sepah. Tehran. Helping you all along. I was just too blind to see it, wasn't I?"

Rashid studied me. "No, you were not blind. Our luck changed. *Al-qadar.*"

His face softened. "*Habibi*, you have plans to get out?"

I grinned. "Leaving in two weeks. Just in time, eh?"

He shook his head. "Listen to me. You must leave before then."

"I'll be fine."

Rashid stomped out his cigarette. "Leave. You understand? I cannot guarantee your safety." The wind blew his kaffiyeh in white waves behind him, an imperious prince.

"Just stick to worrying about yourself. Where will you go if it turns bad?"

"Me?" He motioned toward the Tree of Life. "I am like this tree. Four hundred years without water. A survivor."

"Guess we're two of a kind. I'll outlast the cockroaches."

Rashid laughed.

"All right, my friend, I'm going to leave before the wind picks up my car and dumps it in the Gulf."

He gripped my arms, planted a kiss on each cheek. "*Shukran.*"

At my car I turned around. "Question for you. What was my name?"

"Your name?"

"My code name. Nom de guerre. Pseudonym. You were SCROOP."

Rashid looked amused. "Our American friend."

"That's it? Our American friend?"

He shrugged.

Our American friend. The most mundane moniker. Nothing to distinguish me from the rest. Just a name for the rosters.

54

Layers of smoke were darkening the sky from dozens, maybe hundreds, of fires across the island, so thick that even the hallowed air of Amwaj had become tainted. *Gulf Daily News* headlines boasted a slew of arrests, but a cold-store owner confided he'd witnessed masses of idle soldiers in the streets. Garbage was accumulating on corners. Outside the naval base the taxi stand was empty for the first time I'd seen. The ticker on base informed personnel that the causeway to Saudi Arabia was closed to all traffic except incoming military vehicles. I couldn't wait any longer.

Navigating the slums would be difficult, maybe even impossible, but it would only get worse. Better daytime than night. I headed west on Budaiya Highway. I would retrieve her, shelter her. What was past was past. I had a way out. Females could still board planes. I would tell her what I'd done and she wouldn't be able to refuse. I was a spy, a master of persuasion. We'd escape. It wasn't too late.

I knew the slums better than my own son, the intimate knowledge gained from months of meeting Rashid and Almaisa enabling me to take backroads and avoid the main arteries clogged with protesters. My beater provided dependable camouflage. Even the alleys were barely navigable, overflowing with smatterings of hurried ragged men carrying sticks and Kalashnikovs. Black flags flew from every rooftop and balcony. Jumbled brown buildings, clotheslines, graffiti at long last shining through the government blackouts. I knew the land, the streets, the polluted air. I'd burrowed deep into this foreign soil, gotten dirt under my nails. It had ceased to be alien and I was part of it—the texture, the vibrations.

My knocks on her studio door went unanswered. The street was deserted and it took me only a few minutes to pick the lock. The place looked the same as last time I'd been there, which now seemed eons away—four weeks? five?—the only difference the artwork scattered across her table. A new mosaic was almost complete, a paradise of sorts: feathery palms, unearthly blue water. Almaisa's Bahrain, the land she'd pieced together from her uncle's memory. A delusion, a wish long ago extinguished. She still didn't know it was a delusion, that the Arab Spring would never bring it back. She'd done her part. Now it was time to go.

I turned to leave and saw the photo of Sanaa. It had a different effect on me this time—a feeling between regret and indifference, a sense of futility, the assurance we give ourselves there was nothing more to be done.

The orphanage was empty, but it had recently been occupied—door unlocked, lights still on. Antiseptic wafted through the empty halls. The silence was cloying. Just to be sure, I checked each room.

Boxes were piled nearly to the kitchen ceiling, some empty and scattered across the floor as though opened in a hurry. I scanned the labels. Diapers, toilet paper, first-aid kits, penicillin, canned lentils, figs. All stamped with the clasping hands that constituted Global Society's charitable logo, the cargo's suit of armor, insistence to the world that its contents were intended to help and not to harm.

A box of packaged food teetered at the top of a pile and I pulled it down, opened it with my pocketknife, rummaged through the mess of packaging. Contents materialized like figures from mist—smooth, shiny. I felt them, turned them over. Checked my watch; my hunt for Almaisa had taken hours. It was nearly night.

Darkness was dropping when I reached her flat on Bani Jamrah Avenue. The streetlight cast its weak eye on her door.

No answer to my knocks and I tried to pick the lock. It didn't work, deadbolted, surprising for this neighborhood. Her studio had been easy; why would her flat be so difficult? I surveyed the place. A back entrance. Nearly every place in the slums has an alternate entrance, Rashid had said.

And there they were—rear stairs to a second-story landing. The door at the top was easier to pick than the front. I was in.

It was a modest space, messier than I'd expected, a pile of unwashed abayas in a corner, art supplies strewn across the floor. Dark, illuminated only by the streetlight outside, but I didn't dare turn on the lights. No television, her walls nearly bare and bereft of her own work. Then I noticed her bed in a corner—nothing but a mattress with a colorful woven quilt, exquisite in detail, probably something she'd made. I was not the first man to see it.

The kitchen was off to the side, little more than a pantry, cooktop, and icebox sandwiched together. Another small room in back, shielded by a heavy curtain—a makeshift darkroom, pungent with chemicals. A desk covered with books. Hardbound European books about art, *Orientalism* by Edward Said, *Les Misérables*, a worn but richly decorated Quran—passed down from her uncle, I recalled. A biography of Sinatra. I picked up a framed black-and-white photo of a young couple. Her mother and father. Even sporting a thick black mustache, the man in the trim white suit bore a striking resemblance to Almaisa: same slender build, same penetrating eyes and angular face. The woman was beautiful, a real stunner—creamy skin and delicate features, wavy blond hair fashioned into a bob, blouse revealing a hint of cleavage, slacks effortlessly hugging her hips. No, Almaisa wasn't the imprint of her father; she was the perfect blend of the two.

Taking a seat in her chair, I opened the top drawer, turned on my pen flashlight. An unlabeled manila folder was on top. Inside was the Tree of Life, the photo she'd taken on our first date. Me in front of the tree. Then a photo she'd taken of me in Hidd. We'd brought Sanaa to the park that day, climbed the observation tower, and fractions of Manama were visible in aerial view behind my shoulder. I looked tanned, relaxed, Sanaa in my arms so she could get a better view. Next were photos of Sanaa when she was younger, variations on the picture hanging in Almaisa's studio. Me playing guitar at the safe house. I remembered the moment—looking up as she'd called my name mid-song. A photo of my villa, taken from the far side of the canal. I stopped, stared at the black-and-white image. Next photo was also at a distance—me having coffee on a terrace with a suited man. The man's face was too far away, too blurry, to make out, but I recognized the café. Le Chocolat in Karbabad. I'd only been there once—to meet Rahat, the up-and-comer in Al-Wefaq, the nonviolent Opposition movement. Almaisa had been following me.

A fuzzy shot of the Trader Vic's terrace at night, its tentlike roof aglow in black and gold, colorful expats clustered like bunches of wildflowers. Taken from outside the hotel grounds, probably with a telephoto lens from a backroad in Seef, almost artistic in its composition. Next, the tail of my Lancer, poised to enter a compound. The gate was high; a guard leaned into my window. Three Palms. I took a long drink from my flask. She knew everything.

Next was Whitney's villa, taken from outside Riffa Views, the same route I'd driven with Rashid months ago when he'd insisted on showing me the king's latest palace—nothing more than a fallow lot, construction not yet begun. I'd reached the last picture and I found myself staring into Whitney's face, a close-up in black and white, the photo's subject clearly witting this time. My light danced across his face, illuminating first one feature, then another. He wore a look I can only describe as disarming—eyes direct and candid, mouth relaxed, hair combed but not too neat. He looked young and guileless, sure of himself for the simplest reason—that he didn't yet know what to doubt or fear. The photos were fluttering in my hands as though someone had opened a window. They slipped from my fingers and dropped to the floor. Something else was in the drawer.

The Practice of Recruiting Americans in the USA and Third Countries. An ancient KGB manual, no less than fifty years old, its pages tattered and yellow and smudged. Instructors at the Farm had shown us a copy once, presented as a relic that belonged in a museum: anachronistic but weighty and relevant enough to preserve for posterity. I flipped through its well-worn guidance, underlined and starred passages winking here and there. On a dog-eared page I paused, read the highlighted text.

> *In studying Americans, our Residency in Italy identified a number of places visited by Americans working in target installations of in-*

terest to our Intelligence Service. It was possible to determine that Americans in Rome systematically frequent the same bars, restaurants, and places of recreation. Americans feel almost at home in these places: They drink a great deal, are very free in their conduct, and frequently sing. American women, especially the wives of Americans who are away on temporary assignments, drink and have relations with other men. Similarly, married men have relations with other women, and single men have relations with married women. Many are divorced.

I reread the passage several times, amused by its near-satirical accuracy, its timelessness. Fucking Russians knew a thing or two. A trickle of moonlight spilled onto the desk. It reminded me of the time. I put the recruitment manual back in the desk, slipped the photos under my jacket.

On the way home I realized suddenly where she was. It hit me with a sharp blow, forced me to pull over. In the distance a sizable fire was rising from the rooftops of eastern Budaiya, steel skyscrapers downtown faintly reflecting gold, cowering before the conflagration.

The shutters were closed, no safety signal, and the place looked deserted, not a car within sight. All signs pointed to a dry hole. Still, I knocked on the door. There was a muffled commotion inside, a male voice, insistent. Something crashed. She was there. I'd been right. No one came to the door, and after waiting long enough to create the appearance I'd left, I quietly opened it.

Jimmy looked up with a mournful apologetic look, a disobedient dog seeking forgiveness from his master. Clumsily he was zipping his pants, rivers of sweat streaming down his face. His mouth opened but nothing came out. Behind him a skinny Arab boy, naked from the waist up, quailed, looking from Jimmy to me as though seeking guidance. The air was stifling hot and the buzz of a fly was audible.

"Collins . . ." Jimmy trailed off.

I said nothing, retreating backward through the door, somehow sensing that if I turned around I might not make it out. The boy's smooth brown arms were carved into my memory with a sharp serrated knife. This. This was what the bastard had gotten from the beacon. My safe house.

I drove even faster than when I'd dumped the corpse. As though I'd come face-to-face with a dead man. But when I look back now, I realize something else was pulsing beneath the repulsion and disbelief and horror. Relief. The only thing that could eclipse the magnitude of Jimmy's offense. Relief that it wasn't Almaisa in the safe house. She was still out there.

55

First thing I remember on June 25, 2013, is being awakened early morning by the call to prayer. Even in my half-conscious state I could tell the entreaty was louder than usual, more orchestrated, an apparently perfect harmony of all mosques on the island that seemed to fill every square inch of air, my villa nearly vibrating with the synchronicity.

Driving to station was impossible. Fires and protests choked every artery. Hidd Bridge was blocked and I was forced to abandon my vehicle in a dirt lot and hoof it to the naval base through Al Fatih. The sky had grown mottled like an industrial city with too many noxious factories. The call to prayer continued.

When I reached Juffair, I was soaked and exhausted. Protesters had begun swallowing up the neighboring streets like a slow-motion tsunami, a dark sea peppered with torches. Women, usually absent from protests, speckled the crowds, faces painted in jagged red-and-white stripes. Clinging to the top of a fountain outside a gated mansion, a man shouted through a megaphone. Riot police, their ranks considerably thinner than last I'd seen, had moved in to stem the tide but were quickly engulfed, overwhelmed. *Ela al-amam!* protesters chanted. *We are moving!* I kept to the shadows. Just a few blocks from base.

American Alley was alive and well, whores still hawking their wares, infinitely more resilient and undaunted than the taxi drivers next door. A splinter group of protesters had spilled into the alley, and some began turning against the prostitutes, shouting them off the street and back into their dens of iniquity, even inflicting a few blows.

Head down, careful not to draw the blunt and indiscriminate ire of either side (united only in their inability and unwillingness to differentiate too carefully between friend and foe—and which, after all, was I?). Even as I neared the shelter of the U.S. military's strong arms, sounds of battle carried through Juffair, knocking against America's fortified walls. *We are moving.*

The shredder was working at a feverish pace. Whitney didn't greet me, barely registered my presence. Paper upon paper disappeared into the machine's teeth like prey offered to an angry beast. Jimmy was nowhere to be seen. I began stuffing a stack of documents into a waiting burn bag.

"Prison break this morning," Whitney said. "Led by Junaid."

"How many prisoners?"

"All of them. Dry Dock is empty." With a quiet whir the shredder died and the room turned black, ceiling fan slowing to a stop. "Power outage. Second one this morning." Whitney's voice sliced the dark.

For a moment we were silent, our eyes searching the leaden heat. We found each other. Then the base generator kicked on; lights flickered and the shredder belched. Whitney looked at his watch. "Lockdown will be declared any minute now. Might as well go home while you can."

In the fog of war I was only fleetingly aware of sights, sounds, smells—the instantaneous. I understood neither the full scope of forces sweeping the island nor the significance of my sensory input. Over the ensuing weeks and months I read papers and reports with the intensity of a fanatic, piecing together every fact uncovered, every theory postulated, every detail added, consumed with understanding how it had happened, what role each of us played.

Walking to my car—which I now marvel was not stolen or destroyed—I watched riot police flee with uncanny speed into corners and alleys, places they'd once sent others. Jassim's five-story paper visage on Awal Avenue was succumbing to flame, first his face and sable sunglasses, then his shoulders and hands, until only the scorched side of a building remained. Pockets of flame flared in every direction (Manama's buildings, the world learned, were coated with a cheap flammable mixture of aluminum and plastic that burned to the ground with lightning speed). Heat and aridity accelerated the destruction, the rapidity with which a modern Gulf city—its showy downtown, its shimmering hotels—could come crashing to the ground. Some would later describe the widespread conflagration as *fitna*, a Quranic word that means discord or inextinguishable burning.

Billboards were painted over with victory slogans. *Laqad dhahaba al-tugha! Mabrook 'ala thawratakum al-tarikhiya! Tyrants are gone! Congratulations on your historic revolution!* On the steps of the national library an enormous pile of books burned, everything that was not an Islamic tract. Piles of defecation fouled the streets and a general stench filled the air, as though the uprising were so important that its executors had no patience or time for a toilet. Through a megaphone Junaid's voice traveled around the blocks, urging the masses forward. Amid the chaos an entire lot of Egyptian horses escaped from the Al Rashediah stud farm in Budaiya and ran loose across the city.

Laqad mata al-malak! The king is dead! The jubilant chorus rose from the streets as I made my way back to Amwaj. A chilling cry, different in tenor and conviction from the dozens of chants that had flooded the island over the last twelve months. The symbol of royal oppression had fallen. Later I would learn it was only a matter of hours, in some cases minutes, before the remaining riot police and low-level ministry officials, mostly foreigners bereft of any national allegiance, abandoned their posts. Al Wefaq and Al-Asalah Islamic Society united in their call for the Sunni government to step down; most obeyed and those who did not were locked up in cells at now-empty Dry Dock. Saraya Al-Mukhtar, the Islamic Front for the Liberation of Bahrain, had taken over the royal palaces by force. The vanguard established its headquarters at the Gulf Hotel, the highest building on the island, higher than the Ministry of Interior, from whose windows one could see everything.

All media outlets were overtaken, jammed with continuous propaganda and reports from the field. Every radio station as I drove home was playing the same program: broadcasts from Tehran in Farsi and Arabic, senior

Revolutionary Guard commander Alborz Mirzadeh declaring to the world that Manama was the new capital of Greater Iran in the Gulf. Bahrain was simply an Iranian province separated by colonialism, he announced, and would hereafter become a bulwark against further aggression by infidel countries, including the damnable Saudi Arabia and United States. Street signs were knocked down or painted over, their names changed. IMAM HOSSEIN STREET, AZADI AVENUE. Roads that had long been too ordinary to identify were finally worthy of a name.

Overnight Sharia had become the law of the land. In the course of hours mosques were overtaken, liquor store owners assaulted, strict Muslim dress codes enforced upon penalty of summary death, Sunni tombs desecrated, politicians and human rights activists locked up. Prostitutes and lawyers—representing the two equally anathema forces of sin and rule of law—were among the first to go. Not quite Shakespearean, but close. Brothels, along with their inhabitants, were burned, homosexuals arrested or shot. Manama Cemetery near the Ministry of Interior became a dumping ground for bodies.

The revolutionaries were irrational, unrestrained, brutal. They had the savagery of forces with nothing to lose, for whom the land is not home but territory. They'd joined a cause and the cause had become everything, supplanting logic, compassion, decency, all the calculations of daily living. The excesses of belief.

Amwaj Islands was deserted. Villas empty, windows wide-eyed and injured. Kayaks and paddleboards bobbed mournfully in the blue canals. Alosra grocery was dark and shuttered. Long white linens blew across the vacant patio of Dragon Hotel, sun-tattered edges tickling the surface of the swimming pool. From the penthouse of a newly constructed high-rise came the sound of reckless shots, a victorious erstwhile sniper firing at nothing. The Indian laborers had abandoned their roost beneath the trellis. At my villa I stuffed the belongings I'd set aside into my suitcase.

Explosions, I remember countless explosions—dull concussive thuds, ground shaking—as I pulled up to the embassy. Polluted sky made worse by a prodigious and unending stream of smoke fleeing the embassy chimney, ghosts of classified documents. Evacuation helos, I mused, might not be able to fly in the haze. My belongings still in the trunk, I joined the dense crowd of evacuee-hopefuls on the terrace waiting to enter the stately alabaster building. Americans, Europeans, diplomats, oil men, bankers, doctors, everyone who hadn't believed it could ever happen, who'd stayed on their rooftops with a cocktail until the views turned unsightly. Their dream had been knocked abruptly off its feet, their live-in maids and fancy villas and Olympic pools incinerated into nothing more than nostalgia. A row of vans waited to whisk the lucky lottery winners to safety. Helo ride to a waiting cruiser in the Gulf, a seaborne journey home.

The Chaplain and his wife spotted me instantly, moving closer as though together we had a greater chance of getting out. *The Johnsons didn't make it. Stayed home too long. Trapped when Three Palms was overrun. A real bloodbath. Really, they should've known better—living out there in Budaiya, middle of it all. At least we got out in time.* The crowd was anxious, sweaty, sun beating through the haze on skin unused to so much exposure, hands desperately gripping well-made luggage. Cartons of American craft beer were tucked under arms; a plastic Christmas tree poked its tip above the masses. Laughter. *Did you hear they blew the causeway up? That big boom an hour ago. No more help from our Saudi friends, I guess. Just as long as the thugs don't mess with my yacht—couldn't bring that beauty with me.* Flasks were opened. Shouting, whistling, singing. A few sat down, wilting in the dripping heat.

Women, children, embassy personnel! A Marine bellowed from his hallowed perch in front of the gates, his uniform dark with perspiration. I elbowed my way through the throngs, the Chaplain and his wife watching me with wide eyes and salivating mouths, ravenous animals. I gave my name to a second Marine with a clipboard. He scanned the pages, shook his head.

"Office of Middle East Analysis," I prodded. He glanced at me, opened the gate.

The din of shredders inside the political affairs office drowned out even the high-volume televisions broadcasting the revolution in real time. But no soundtrack was needed, the photogenic carnage its own storyteller, a smugly gory reminder of the world's underestimation. The room, like the hallways, reeked of char, too much paper for an unprepared incinerator. An absurd number of personnel was crammed into the space—from consular, economic, public relations—frantic and harried, a human factory destroying documents. Explosions were multiplied: Bombs outside the windows reverberated on TV seconds later, death and destruction delayed like profanity on a radio station.

The Bakowskis were at the far end of the room, the only ones seated. Jimmy looked over then quickly away. Jordan was in a corner, sifting through a stack of papers.

"You made it." She handed me a pile to shred.

"When's the first helo out?"

"Don't know. Everything's been delayed by the attack on base. You heard, right? Storage facility hit by indirect fire—couple of rockets or mortars. Massive explosion. Destroyed nearly half the base. We could hear it all the way over here."

"What was in the facility?"

"Not sure, but it made a hell of a bang."

Scenes on the televisions cut to the opera house, which was no longer a theater but a pile of rubble. A scruffy man in a bad suit was reporting in awkward, serious-sounding English. *As you can see, the National Theatre has been destroyed. It was erected by the king, built of pure gold using money stolen from Shia. In this theater*—he made a grand gesture behind him—*was shown many infidel performances that were antirevolutionary, blasphemes against Allah, and Western propaganda. Under our new revolutionary leadership, we do not need a theater. We will have only performances that glorify Islam. Now Allah smiles down on us. . . .* His words faded away in footage of the debris. Almaisa's mosaic was buried somewhere in that destruction. She didn't know. She'd been betrayed.

Then the cameras were on Riffa. Villas were still standing, starkly clean of smoke and flames, purple bougainvillea bright and flowering as though just watered and pruned, but something was different. Another ragged young reporter appeared and it took me a few seconds to realize the peculiarity, the things in the picture that didn't belong. Scurrying black figures were barely visible on the rooftops behind the reporter, the

Kalashnikov's unmistakable silhouette an appendage on every body. *The people have taken over Riffa, home to foreign infidels, Western colonizers, and corrupt Sunni politicians,* the bespectacled reporter explained matter-of-factly. Embassy officers had gathered around the screens, gravely silent as though watching a massacre of their children. Jimmy joined the crowd, slack-jawed, his Whiskey Tango Foxtrot tee shirt stained with heavy perspiration.

The camera was inside one of the villas, trained on two masked men in Western fatigues standing on each side of a kneeling figure. *Minister of Interior,* the reporter's disembodied voice narrated. *Oppressor and torturer of the people.* The camera zoomed in as the two militia men slowly, methodically, began their work. Back and forth, back and forth, remarkably steady, the weapon oscillating with professional persistence. Covered in a black hood, the minister was strangely silent, almost cooperative, as though he'd been drugged. His executioners were equally silent. The crowd inside the room watched, transfixed, the whir of shredders and crumple of paper extinct. Down the hall someone laughed.

Back to a Riffa rooftop where a man was shouting in Arabic, his Kalashnikov pointed at the sky in triumph. The outline of a familiar villa was visible in the background. Someone began turning off the televisions; someone else turned to the crowd, barked instructions about what we could bring on the helicopters. Clipboard out, reading the first list of names: Tier One evacuees. Ambassador, Deputy Chief of Mission, Naval Attaché—already on board. *Whitney Alden Mitchell.* In the midst of murmurs and roll call, no one seemed to notice that Whitney was not making his way to the exit. Jordan was indifferent, distracted, tensely riffling through her purse. I glanced around, looking for his earnest face, stout frame, his mop of boyish hair. Maybe he'd caught a ride on a Navy bird. I patted my pockets, muttered to Jordan I'd forgotten my passport.

Crowds were still swarming the gates when I exited the building, the only one headed upstream, away from safety. Nervous twittering rippled through the throngs, the sun relentless, trying the sinners waiting for salvation. Amid the sea of casually comfortable garments—tee shirts, blue jeans, tank tops—a lone woman stood on the fringes in a bulky black abaya, all but her eyes covered. She wouldn't last long in the heat. By the time I'd begun to escape the hordes, the woman had plunged into the crowd, moving toward the embassy and a Marine distributing water bottles.

I was nearly to my car when I heard the blast, felt the heat, tiny pebbles pricking my skin, raining down on every surface. The wind had been half knocked out of me, force propelling me into the driver's side door. My eyes burned. I turned to see a story-high ball of flames consuming the crowd. What once were hundreds were now tens, staggering and limping along the edges of devastation. Watching from my distant perch, I experienced a kind of paralysis. Shouts, cries, shreds of unstaunched humans stumbling blindly, Marines rushing from the embassy, yelling orders. The privileged cadre up in smoke. Destroyed in a single explosion.

But I was safe; I was removed. Just an observer. Then I realized my arm was bleeding from a small piece of shrapnel. I ripped off a strip from a shirt in my suitcase, tied it around the wound. The sun was still beating down wretchedly. I climbed into the car, taking care not to look in my rearview mirror.

The city was wasted. Buildings were mutilated corpses, their glass eyes shattered or ripped from their sockets. A permanent shade of gray colored the air, trapping the heat like a heavy blanket. Masses of people were walking on every street in the direction of downtown. Somewhere in Bu Ashira I found a deserted lot where I could put on a pair of glasses, pour water onto my hair, comb it back. Change my clothes to an expensive but casual jacket, shirt, slacks. Discard my bloody rag and hope my wound wouldn't seep through. I checked my phone one last time for a message. No reception—the network down. Emptied my flask, got in my car. One final attempt. I had to.

But I only made it a few miles before I was forced to turn back. Barricades, fires, crowds, sloppy ersatz checkpoints manned by wild-eyed men who shot at the breeze. No way I could get to her. I would die trying. Hundreds of foreigners had just been blown to pieces; there was no chance they'd let me through.

Down Al Fatih Highway. Manama's tallest skyscrapers to the west still standing defiantly above the destruction, the Gulf to the east quiet. The sun had begun to set, splintering the smog with jagged honeyed rays. Few vehicles were on the road and my Lancer sped along, a lone survivor. When I reached Coral Bay, the buildings parted to give an unobstructed view of downtown. It was a sight I remember as clearly as the Admiral's farewell-to-summer party or the first time I saw Almaisa. Thousands of people, maybe tens of thousands, kneeling on the ground—in streets, fields, parking lots—facing west toward Mecca, synchronously praying. The haze broke momentarily and the setting sun turned the sea of human backs golden. A lilting stream of chanting was audible. It was a new country, a new land, the old already a charred memory.

The airport was a veritable fortress, a maze of Saraya Al-Mukhtar–controlled tanks apparently wrested from or surrendered by the Saudis, extending outward like an iron spiderweb. A few streetlights were shining their dim glow, which I took as a good omen: The airport still had electricity. Rather than press my luck navigating the layers of security, I parked at the nearby Mövenpick Hotel. Its new masters had not yet moved in, the only occupants a family of birds.

My first and, as it turned out, only checkpoint was outside the airport, where a young man in mismatched fatigues with a Kalashnikov on each shoulder eagerly demanded to know just where I thought I was going.

I showed him my passport. Tracer fire punctured the evening sky behind him, its footsteps galloping into the distance. "Just here on business." I smiled. "Heading home."

The boy looked at the document. His face showed the bewilderment of the young suddenly awarded too much power. The combined weight of the Kalashnikovs was surely more than his body. "Julian . . ."

"Schmid."

He studied the passport again. "You are Swiss?"

"Yes."

"You have ticket?"

"No. I'll buy one inside."

He stepped away, spoke in guttural Arabic into his radio. When he came back, he shook his head. "You cannot go. No planes going out."

I realized my breath stank of drink, wondered if this was contributing to the boy's belligerence, my habits suddenly not just an element of instability but a moral affront. I relaxed my face. "I'm sure there is at least one plane. Please let me through."

His eyes were growing mean, insecure, and he nervously drummed his fingers on one of the weapons. "I tell you no. No planes going out."

From my wallet, thinner and lighter now that it was empty of any shred of Shane Collins, case officer for the CIA, I withdrew a hefty sum in dinars. Blood from my wound was oozing beneath my sleeve, still hidden by my jacket, and I angled my body to hide the incriminating limb. The boy looked at the bills doubtfully, shifted from one foot to the other. As though suddenly remembering he owned my most valuable possession, he pocketed my passport. My vision was blurring. He was a young insouciant demon. I blinked hard.

"Maybe this is better." I pulled a wad of euros from my pocket. His expression grew sharp and indignant, and I realized plainly the revolution was irreversible; its soldiers couldn't be bought, wouldn't even tolerate an offer. He was turning back to his radio, angry now. Quickly I touched his arm. "*Habibi*. I am a friend of Rashid. You know Rashid?" The boy looked confused, and I added, "Ask your leader. You have someone in charge of the airport? He will know."

The boy flashed me a look of skepticism and disdain, ordered me to wait in place. He moved away again, spoke harshly into his radio.

When he returned, he handed me my passport. "You can go." I smiled graciously, pushed up my glasses, lifted my luggage with sweaty hands, and walked into the airport, now nothing more than a fuzzy mirage.

It was the last commercial flight out and a rocky one, the plane's ascent a dogged attempt to navigate the clouds of smoke and debris rising from the island. Isolated still-towering fires were visible through breaks in the smog, as were the prayerful masses—definitely in the tens of thousands, I could now see—spread across the downtown. A sea of belief: the last thing I saw before leaving Bahrain airspace.

When the seat belt sign turned off, I ordered a Bloody Mary, cleaned my eyeglasses. At Frankfurt we disembarked and I boarded my second flight. Clouds parted to reveal white sun and the placid blue waters of Lake Zurich. Land of neutrality.

At Zurich Airport I took my place in the line for EU and Swiss nationals. It was my last checkpoint, the one through which I would never return. My luggage felt pleasantly light. I started to remove my jacket but the pain in my arm reminded me to keep it on. Most likely English spoken at the gate; if not, I knew enough French to get by, my transplanted businessman cover story well rehearsed. Hanging televisions broadcast the latest international news in sliding red words. *Uprising in Bahrain. Bahraini King Murdered by Opposition. Arab Spring Takes Root in Tiny Gulf Island. Iran Declares Manama New Capital of Greater Iran in the Gulf. Tehran Hints at War with Saudi Arabia. Rebels Announce End to Colonialism, Use Barbaric Tactics Against Westerners and Sunnis. Hundreds Killed Outside American Embassy.*

And then Al Jazeera footage was on the screen: a captive kneeling on a filthy street, shrouded in a black hood and cape that nevertheless failed to disguise his portly awkward frame, the near-clichéd masked duo standing on either side. *Previously recorded*, the silent words assured viewers, alternated with *Warning: violent and disturbing images.* The other airport arrivals were uninterested, occasionally glancing up at the screens as though waiting for the climactic scene. I fingered my passport, felt the tiny threads of fabric unraveling along the edges, perfectly worn as only an authentic passport can be.

Lopsided buildings, a lone streetlight. Bani Jamrah Avenue in the background. He'd gone to get her. A rescue operation. Maybe just a farewell. Either way, he'd ventured into incendiary enemy territory, a near-certain dead end within the last twenty-four hours for any Westerner, much less a known intelligence officer. Sheer folly, recklessness that bordered on insanity. An undeniable act of bravery. At least an act of heroics. Lust, maybe. *American Intelligence Chief in Bahrain Slaughtered.*

But something was amiss. It was too neat, too cinematic, Whitney too perfect a catch for Saraya Al-Mukhtar's randomly cast net. The assassins had known who he was. Had expected him. They'd known everything about him—what he looked like, what car he drove, his routes and habits, that he would come get her. They'd cornered him before he had a fighting chance. They had the ultimate trump card: an inside source. It was a setup. She'd betrayed him just as she'd betrayed me.

The line was moving quickly, not nearly as much security as I'd feared, the Swiss apparently reassured by the homogeneity of our lot; no Arabs had been permitted exit from Bahrain and all my fellow passengers had been European. In the corner a K9 and his handler were content to watch from the sidelines. Whitney Alden Mitchell was dead. Betrayed. She'd duped us all, standing outside the embassy gates in her bulky black abaya. I'd almost failed to recognize her. Only when the plane had cleared the pollution, when we'd reached a safe altitude and a drink had restored my vision, did her face resurrect itself in all its beautiful assemblage. Two brilliant green eyes staring out at the crowd, scar on the left side. The scar. Unmistakable. Perhaps I'd even known in that moment, when I watched the masses ignite in flames. I just hadn't wanted to believe, maybe. Belief, after all, is a powerful thing.

None of it mattered now. I was at the pearly gates of Switzerland, bastion of neutrality or at least the appearance of neutrality, where one never has to take sides. Where the winters are long and the desert heat far away. The final threshold, the sopping ugliness of choices and decisions behind me. Approaching the uniformed man behind the glass, I felt a modicum of comfort she hadn't chosen me. She'd found someone better, more valuable, consequential. It hadn't been me waiting for her on Bani Jamrah Avenue. I'd been only the mule, Whitney the star, the prime target. Maybe in the end she'd even tried to save me. Spotted me at the embassy, waited until I'd left before detonating her smooth golden body, blowing it to pieces. Maybe she'd loved me after all.

Footage had shifted to Riffa and its candy-colored villas. *CIA House Overtaken*. The camera was giving a jerky tour of Whitney's home, already halfway converted to a local command center, suits strewn across the floor to make room for a closetful of weapons, an armed occupier boastfully explaining the secure phone Whitney had kept at his bedside. *We will decode this phone. We will learn all the secrets of the American CIA. How it helped our oppressors colonize the Middle East and try to make war with Iran*. The occupier held up Whitney's *Arabs at War*—American imperialist doctrine, he declared.

I stepped up to the glass, showed the Swiss official my passport. The news coverage had broken for a luxury watch commercial. *At Piaget, time is measured only in gold.*

"Welcome home, Mr. Schmid." The official handed back my passport.

56

December 2015
Zurich, Switzerland

When clients walk into my practice—a sparse room in my flat on Aargauerstrasse—it's almost always a case of suspected infidelity. Invariably the client's suspicions prove true, at which point I offer a few words of counsel. Temper your reaction. Temper expectations. Everyone cheats. Consider yourself lucky you found out. People cheat for different reasons, maybe not the ones you'd expect. It doesn't mean your partner doesn't love you. I'm a worm, I'll admit if the client needs further mollification, an underground digger, the mealiest, slimiest, most unsophisticated of creatures. The business of a private dick runs on jealousy, paranoia, worst assumptions. I look only at the evidence, not the story behind it. And usually there is a story, an explanation. At this, the client shifts uncomfortably in her seat, unsure whether she should appreciate or take umbrage at this extraneous moral counsel, especially from someone in a questionable profession. Then I rein it in and spill the evidence, fearful of losing business; my Swiss bank account will eventually run dry.

I haven't touched the reserve account Rashid set up for me. Not so much a moral compunction as distrust; who knows where the fingerprints will lead? There are always people watching, cameras in bank vaults, handwriting analysts, financial wire traces. Multiple avenues that could link him to me. Rashid, after all, has become something of a celebrity—deputy head of the newly formed Atomic Energy Organization of Bahrain, a fact I first learned watching television in a Zurich dive bar. Suddenly, his face was on the screen, gaunt and scraggly as ever, spouting his fiery rhetoric, staring through the camera as though watching me down my beer. Turned out his education abroad, background in engineering, and technical expertise made him perfect for the gig. His yearlong cooperation with the infidel American intelligence service was dead and buried, a secret that would never see the light of day. Now he was on top and I was on the bottom, handler and informant linked as ever, except now the relationship was inverted, or maybe just perverted, depending on how you looked at it.

When enough time has passed, I'll probably tap the reserve account. I fulfilled my duty.

It took several days after the revolution for papers and the world to understand how the king had died. Poison, good old-fashioned tried-and-true poison, fatal elixir for such other historical greats as Socrates, Cleopatra, Hitler. A cook had slipped poison into the royal breakfast and the king had died within hours, found sprawled on his bedroom floor in an inconspicuous villa south of Riffa, one of his rotating hideouts, his strategy of spending each night in a different covert place an apparent failure, his safe houses unsafe. By that time, all but one of his four wives (Fatima, it seemed, had natural immunity) and a sizable extension of the royal family had also been poisoned, their sumptuous morning victuals consisting of the same ingredients or the king's generous leftovers. It was a well-orchestrated operation leaving a trail of decomposing royalty spread across the island like an elegant opera, a string of victims felled by a singularly elite contagion. Based on the reported symptoms and circumstances, American toxicologists and intelligence analysts postulated that the poison was *Gelsemium elegans*, heartbreak grass, a potent fast-acting substance favored by Russia in assassinations, most likely smuggled into Bahrain in copious quantities, in ready-to-use liquid form. Russia, experts said, had recently been supporting Shiite revolutionaries to counter Saudi oil interests and regional dominance. The poison's vehicle of transport into the country, however, was anyone's guess.

Five unlabeled dropper bottles, milky brown glass filled with liquid, faint scent of jasmine. They were benign, I'd told myself. Maybe even an antidote—something that would boost the right side, save lives in the end. Or if not benign, something incapable of wreaking widespread havoc. A delusional revolutionary's genies-in-a-bottle. Rashid had asked if I wanted to know the contents and I'd said no. I was just a mule, after all. Nothing more than a mule.

But it wasn't just the dropper bottles. Like a mosaic, every finished work requires hundreds of pieces, some minute, some nondescript and unimportant, hardly worthy of notice. Without them, the piece is incomplete. It fails.

Sleight of hand. Mere brush of flesh. The prison officer didn't notice when I slipped the name of the bribed guard and the breakout signal into Junaid's hand at Dry Dock. A minimal transaction—wrapped in the brilliant cover story I'd fed to Whitney, that I sought to recruit the guy. For Junaid's escape only, Rashid had assured me. (And hadn't the ailing poet taken blame for a crime that rightfully belonged to me and Rashid?) No one in my position would have anticipated all the inmates would escape. No one would have anticipated that revolutionaries carrying assault weapons would storm the prison, aided by the bribed guard and the breakout signal—an abrupt outage arranged by workers at the Al Dur power plant—that within two hours the entire prison population would be loose on the streets. Some ninety guards killed, the flimsy armored barrier on Arad Highway easily overrun. I hadn't known that so many would be massacred. I hadn't known that two of the escaped inmates, if the propaganda is to be believed, would become Whitney's assassins.

I offer reassurances to myself. What I knew, when I knew it, that I was on the right side. I remind myself that the warehouse on the naval base was a fair target. I'd done my due diligence—at great effort and risk to myself—confirmed the weapons smuggled deep in the bowels of the *Eisenhower*, then stored in the warehouse. A clear violation of the still-intact embargo. In the tally sheets of history, I'm certain, Saraya Al-Mukhtar's mortar attack on the warehouse, the explosion so massive Jordan had heard it at the embassy, was justified. No need to rearrange the event in my mind until it's a pulpy mess. And all I'd done was mark the location with my puny handheld GPS.

I didn't know the truth about the bombs in Adliya. I resurrect Rashid's explanation in my mind—that the king's men were to blame—every detail, every hole filled, every box checked. A pyramid beautifully constructed. I convince myself all over again. Facts—no less than facts—had swayed me: Walid revealing a fifth intended bomb that morning; the Ministry's incomplete falsified investigation; violence that spooked Westerners just enough to get them angry but not flee. It wasn't until I'd been in Zurich nearly a month that I realized the truth. Watching stakeout footage in my office one evening, I abruptly recalled the video in Walid's office. The figure scampering in and out of Lilou's dumpster had hung from the rim with his left arm. Same slight frame, same uncanny agility as Rashid. A silhouette more familiar than my son. An engineer with the technical knowledge to build bombs. He'd beaten me at my own game.

Maybe we hadn't needed to kill the police officer in the slums that night. Was the guy really going to shoot me or just ask a few questions? I no longer know truth from lies, whether Rashid saved my life or deceived me. It's even possible my shot missed. Only one round was found in his body.

Sometimes I think about the afternoon I showed Rashid where Whitney lived. That wasn't my intention, of course. *The biggest villa on the street.* Had I hinted too strongly, too often, that my chief was no friend of the revolution? Saraya Al-Mukhtar had posthumously branded Whitney a key puppet of the monarchy, a collaborator and conduit for oppressive royal policies, consummate traitor to the Bahraini people. Had I planted those ideas in Rashid's head? (And did I believe them myself?) I'd like to think they were nothing more than assumptions inevitably attached to the senior intelligence officer on the ground. Chief of Station came with risks. A job title that grabbed headlines; glory had its price. I, after all, wasn't important enough to be slaughtered on international TV. And in the end hadn't Whitney caused his own death? He should never have been in the slums. Looking for Almaisa. She was the one who betrayed him.

Almaisa. The vortex of the storm. Her art undoubtedly favored by the king for the plainest of reasons: to keep an eye on the niece of a known dissident. But his eyes had failed to see. *You will never guess*, the Swede had said in Cambodia. *The guns, the explosives, the money—they are smuggled through a charity! Global Society. Very clever, I think. Nobody suspects humanitarian supplies, yah? And Customs never opens boxes from a charity.* But had I really needed Sven to tell me? Almaisa—who traveled abroad to solicit donations, returned with large sums of cash, supported herself on nothing more than mosaic sales, stockpiled supplies in every nook and cranny of the orphanage. Surely I knew before Sven told me. Yet even when, returning to that empty orphanage to search for Almaisa, I'd opened a box to find plastic jugs of purple liquid, I still wondered if it was a fluke, a mistake that the children's shelter was storing vast quantities of potassium permanganate. I'd known, of course, but maybe I hadn't. What is knowledge? The intricacies, the angles, the distinction between foreground and background—all just varied output of an artist, viewed differently from different pairs of eyes.

On June 25, 2013, a new world order began. Within a year the regime in Bahrain had become more brutal and repressive than any of its neighbors or its predecessor, limited only by its modest size and resources. All persons with even tenuous ties to the prior regime were imprisoned or killed, deals with cooperative Sunnis quickly abandoned. Even moderate Shia were purged. Altogether, a death toll estimated in the thousands. Junaid, rhetorical voice of the revolution, was permanently relegated to the sidelines. International sanctions, withdrawal of Saudi oil concessions, and insufficient funds from Tehran ensured that the vast majority of Shia lived in poverty more dire than before. Neglect and destruction of water purification and sewage treatment plants resulted in mass epidemics of typhoid and cholera. The people too had been betrayed.

Less than two months after the revolution, Tehran and its puppets in Manama called on their compatriots in Iraq, Azerbaijan, Yemen, and Lebanon to take up arms against their enemies, sparking fresh waves of sectarian violence around the globe. While Iran worked closely with the newly formed Atomic Energy Organization of Bahrain, Russia once again swapped sides, signing a deal to assist Saudi Arabia in developing a uranium enrichment facility. A new season that pundits termed the Arab Winter extended its icy fingers. Evicted from its home, the U.S. Navy resorted to a floating base for its Fifth Fleet, a distinctly handicapped perch from which to keep the peace and fend off its newly empowered foes.

Fitna—discord and fighting, sedition and trial. Destiny written in blood. The small inconsequential island that never made headlines had changed history. Five bottles of poison had released a genie that could never be returned.

Almaisa's suicide bombing was the worst attack on an American diplomatic facility in history—and the only one perpetrated by a woman—the death toll of 312, mostly Americans, handily surpassing the 1998 Nairobi and Dar es Salaam embassy bombings. Her attack became known as the Manama embassy bombing and was examined by experts as a case study in unexpected gender roles, the female bomber an example of the importance of anticipating unpredictability. Her identity remains unknown.

The Bakowskis made it out that day. Probably got a hero's welcome at Headquarters. No one, as far as I could tell, ever learned of Jimmy's sins—only Gretchen, who I now see had been trying in her timorous way to tell me at the opera, when all I could think about was Almaisa. Flipping through the pages of a *New York Times* one morning, I'd spotted a photo of an American-Indonesian bilateral naval exercise with a man in the background who looked an awful lot like Jimmy. He'd lost so much weight as to be almost unrecognizable—the kind of thin that comes from a serious illness or your wife leaving you or maybe just the buried knowledge of your own transgressions.

The Admiral escaped on one of the first birds. Not long after I'd disappeared into the streets of Zurich, an anonymous source tipped off the Navy, and a few months later the Admiral was summoned to a court-martial on charges of bribery, conspiracy, wire fraud, and disclosure of classified information. In an elaborate scheme involving confederates from flag-level officers to seamen to middlemen overseas, the Admiral was accused of arranging weapons deliveries to Bahrain on U.S. vessels in contravention of federal laws and policy, receiving substantial kickbacks from all sides—manufacturers, foreign fixers, the Jasri royal family. To inflate the perceived threat from Iran, demand for weapons, and his own indispensability, the

Admiral was also accused of tainting intelligence and fabricating sources. His corruption reached back years and spanned the globe, the world learned during the Admiral's trial, a mess of lucrative naval contracts in exchange for decadence and dens of iniquity, headlines revealing suckling pig feasts in Singapore, game hunting in Thailand, solid gold watches and cuff links, a Gucci fashion show with his wife, Cambodian prostitutes without his wife, a colorful cast of characters that included Prince Jamil, a prostitute named Dai-Tai, and a rotund Chinese man called Uncle Chang. Convicted and sentenced to eleven years at Terminal Island, the distinguished Naval Academy graduate hanged himself in his cell.

Whitney became one of those erudite legends born of international events—terrorist attacks, POW captures, hijackings. Long after he'd faded from the news, he was immortalized in books and on the wall of stars at Langley. His diplomat parents were surprisingly silent about the whole affair, a hint of grief to the press but not much more, polished terse platitudes about what an exemplary public servant he'd been. Hadn't spoken to his son in years, the father admitted, but Whitney's accomplishments and service record spoke for themselves. I cut out every article about Whitney, just as I cut out every article about the mysterious suicide bomber, shoved them to the back of my desk.

And then there's me, and as far as the world is concerned, Shane Collins is dead. A lifetime underground affords one a wealth of subterranean connections, access to the world of traceless sin, the black vacuum of inconsequence. Within days of my arrival in Zurich, my contact in the Philippines morgues had supplied my death certificate precisely as contracted, sent it to my son and Marlene. Not that it was particularly necessary; in the Manama embassy bombing and chaotic bloodshed of the revolution, scores of unidentified Westerners were killed, disappeared, or ultimately lost in the anonymous folds of history. Certainly no one was inquiring about my whereabouts. But I wanted to cover my bases just in case. I was a good spy.

I see Whitney sometimes—on a street, in a bar. He looks older, distinguished in a smart gray suit, purposefully clutching a briefcase or shopping bag. At the end of the road, I've learned, espionage turns into a profession of ghosts. The culmination of actions taken or not taken, ends swallowed by means. A place where, even after you disappear, you can't escape. Time loops back on itself, furls the days, keeps them from marching forward. Memories become lodged between fugue and reality, their existential quality unclear the way events in dreams sometimes have the texture of real life.

My skin and eyes grow yellower, my gut fatter and softer, my liver sicker. My hands tremble more. I listen to Sinatra and look at photos and stare at the curios on my desk purchased from a local antiques shop. I await my death. I am a survivor. I take my punishment as it falls: the knowledge that I'm the same man I always was, no better.

I do not miss Bahrain. I think about Bahrain sometimes. The scent of jasmine in the springtime Alps is sweet and chalky and slightly spiced like Almaisa's studio mixed with her perfume. I remember too the scent of the bougainvillea. Or the souq at noontime. Amber, musk, oud.

What is Manama? It is the *maghrib* call to prayer, the Gulf tide brushing its lips along the curves of the coast like a diffident lover, jumbled city blocks and nameless streets and burial mounds that have neither the beauty nor ugliness for memory, porcelain flamingos roosting only until the weather permits a return home. It is the slums. It is a woman. It is a place where spies go to die.

Acknowledgments

I'm deeply grateful to my agent, David McCormick, for being the first person to officially believe in me, for his invaluable guidance and wisdom, his steadfastness and encouragement. He was my sounding board for just about everything, graciously answering my myriad questions every step of the way. Thank you as well to Susan Hobson at McCormick Literary for all her efforts on my behalf. I'm incredibly thankful to my editor at Atria Books, Peter Borland, for taking a chance on me, understanding my work, making the process seamless and easy, and providing perfectly tuned feedback. Thank you to Sean deLone—whose incisive edits made me wonder whether he knew my book better than I did—and the entire Simon & Schuster/Atria team, including Sonja Singleton.

Nada Prouty and Ro Hamze provided superb Arabic translations, often at a moment's notice. Ahmad Halabi and Mo Halabi went above and beyond in supplying last-minute linguistic support. Mark Zaid, national security attorney extraordinaire and dear friend, never hesitated to provide a sanity check on legal issues, and was always only a text message away.

Donna Kaminski has been my shout-it-from-the-rooftops friend, and Julie Zachariadis has been the quiet support I needed. Ian Bryan (best initials ever) pulled off the road to read my draft pitches and summaries; his witticisms, sometimes at my expense, kept me grounded and going. Tai Hsia and Alex Sierra were always happy to listen, and provided much-needed comic relief.

A huge thank you to Ian Caldwell, fellow Thomas Jefferson High

School for Science and Technology alumnus, who gamely read an early draft of this book, providing input worthy of a college paper. Several other TJHSST folks deserve thanks: Dustin Thomason, for reading a writing sample, sagely advising me to "pick a lane" (fiction, it turned out, was the sweet spot), and championing "messy" characters; Yin Ly, for translating Khmer on demand; Patti Hinckley Young, Sean Young, and Sebastian Fønss, for translating and finding more meaning than I could have fathomed in my lone Swedish sentence; and Mrs. Feldman, my senior-year literature teacher, who suggested I drop everything and devote my life to writing. Likewise, Mrs. Rugel, my eighth-grade English teacher, who enrolled me in writing contests and pushed me to pursue all things literary. It took me a while to heed my teachers' advice.

I'm immeasurably grateful to my parents for raising me with love, in a house full of books, and cheering on my life choices, even when those choices dragged me to warzones and dark corners of the globe. Infinite love to my son, Zev, who jumped on the bed with delight when I was offered a book deal, and inspires me every day. And thank you, above all, to my husband Rick, my most valuable critic, who never hesitates to make room in his life for my writing (or anything else I set my heart on), who reassures me when no one else can, who is my rock and my love.

Lecture Notes in Mathematics

Edited by A. Dold and B. Eckmann

1394

T.L. Gill W.W. Zachary (Eds.)

Nonlinear Semigroups, Partial Differential Equations and Attractors

Proceedings of a Symposium
held in Washington, D.C., August 3–7, 1987

Springer-Verlag
Berlin Heidelberg New York London Paris Tokyo Hong Kong

PREFACE

This volume comprises the proceedings of the second Symposium on Nonlinear Semigroups, Partial Differential Equations, and Attractors held at Howard University in Washington, D.C. on August 3-7, 1987. The proceedings of the first symposium, held two years earlier, was published as volume 1248 of this Lecture Notes Series. The present Symposium was made possible by grant support from the following funding agencies: U.S. Air Force Office of Scientific Research, U.S. Army Research Office, U.S. Department of Energy, National Aeronautics and Space Administration, U.S. National Science Foundation, and the U.S. Office of Naval Research.

The local support committee consisted of James A. Donaldson (Howard University), Lawrence C. Evans (University of Maryland), James Sandefur and Andrew Vogt (Georgetown University), and Michael C. Reed (Duke University) whom we thank for their helpful advice.

The Symposium brought together a total of 76 distinguished researchers in the Mathematical, Physical, and Engineering sciences working on analytical, topological, and numerical aspects of a large variety of nonlinear partial differential equations. This multidisciplinary character of the Symposium attendees brought about a productive exchange of ideas on various approaches to current problems in applied mathematics.

In the past twenty or so years, there has been an increased interest in the study of nonlinear models of physical, chemical, biological, and engineering systems. The evolution of new analytical and topological methods for the study of infinite dimensional systems concurrently with the advent of large-scale computers and efficient algorithms has served to further stimulate research on problems that were considered impossible to attack just a few years ago.

There are many problems in the natural sciences which are naturally formulated in terms of nonlinear partial differential equations. Over the years, new methods and special techniques have evolved for the study of nonlinear problems. In addition, there has been a great deal of recent activity devoted to the study of stochastic ("chaotic") solutions to nonlinear differential equations in cases where the "conventional wisdom" physics leads us to believe that only deterministic solutions exist. Many of these studies have been numerical and confined to either maps or ordinary differential equations, which are more easily analyzed than are partial differential equations. Recently however, various methods have been developed for the study of partial differential equations which, because of the complicated nature of these equations, are a valued addition to the mathematical sciences.

A general method that has been very effective in the treatment of large classes of nonlinear partial differential equations makes use of the theory of nonlinear semigroups. Given appropriate conditions, these semigroups generate solutions to nonlinear evolution equations which may have a compact global attractor with finite Hausdorff dimension. This type of analysis applies to numerous nonlinear partial differential equations. Most of the papers contained in the present collection are concerned with nonlinear semigroups.

A major contribution to the multidisciplinary character of the Symposium is the existence of the Large Space Structures Institute at Howard University. This is a special institute devoted to the study of physical, engineering, and mathematical problems that arise in the development of large structures (space-stations) to support life in space. It is a joint effort of the departments of mathematics and of electrical, mechanical, and civil engineering. One afternoon session of the Symposium was devoted to the presentation and general discussion of new classes of nonlinear problems that model certain components of these structures. The rationale was to introduce direct interaction among the symposium participants and some of the research engineers concerned with analyses of these types of problems. We feel that this interaction among scientists with varying backgrounds and interests gave the symposium a distinctive flavor and provided a unique cross-fertilization of ideas.

Tepper L. Gill
W.W. Zachary
Washington, D.C.
November 1988

TABLE OF CONTENTS

Other contributions to the symposium:

M.E. Aluko, Controller-induced Bifurcations in a Distributed Particulate (crystallizer) non-isothermal System.

Anthony K. Amos, Nonlinear P.D.E. Issues for Space Structure Problems of Interest to AFOSR

Stuart Antman, Asymptotics of Quasilinear Equations of Viscoelasticity

Joel D. Avrin, The Semilinear Parabolic Equations of Electrophoretic Separation

Stavros A. Belbas, Parabolic Nonlinear Partial Differential Equations arising in Stochastic Game Theory

Melvyn S. Berger, Vortex motions in Mathematics and Fluids, their Bifurcation and Instabilities

Nam P. Bhatia, Separated Loops and an Extension of Sarkovskii's Theorem

Shui-Nee Chow, Bifurcation of Homoclinic Orbits

Michael G. Crandall, Hamilton-Jacobi Equations in Infinite Dimensions

Lawrence C. Evans, Hamilton-Jacobi Equations in Large Deviation

Tepper L. Gill, Time-Ordered Nonlinear Evolutions

Carlos Hardy, Generating Quantum Energy Bounds by the Moment method - a Linear Programming Approach

Christopher K.R.T. Jones, Behavior of the Nonlinear Wave Equation Near an Equilibrium Solution

Jack Lagnese, Infinite Horizon Linear-Quadratic Problems for Plates

John Mallet-Paret, Poincare-Bendixon Theory for Reaction Diffusion Equations

David V. McLaughlin, The Semiclassical Limit of a Nonlinear Schrodinger Equation

R. Mickens, Exact Solutions to a Nonlinear Advection Equation

Walter Miller, Dynamics of Periodically Forced Traveling Waves of the KDV Equation and Chaos

Mary E. Parrott, The Weak Solution of a Functional Differential Equation in a General Banach Space

Michael Polis, On issues Related to Stabilization of Hyperbolic Distributed-Parameter Systems

Michael C. Reed, Singular Solutions to Semilinear Equations

Robert Reiss, Optimization Criteria for Large Space Structures Modeled as Continuous Media

Joel C.W. Rogers, The Triangle Inequality for Classes of Functions in Function Spaces

George R. Sell, Melnikov Transformations, Bernouilli Bundles, and Almost Periodic Perturbations

P. Souganidis, A Geometrical Optics Approach to Certain Reaction Diffusion Equations

Robert Sternberg, Symmetry in Geometrical Optics

Walter Strauss, Global Existence in the Kinetic Theory of Plasmas

Michael Weinstein, Remarks on Stability, Instability, and Resonances

W.W. Zachary, Upper Bounds for the Dimension of Attracting Sets for a system of Equations Arising in Ferromagnetism

S. Zaidman, A Note on the well-posed Ultraweak Cauchy Problem

STATE-SPACE FORMULATION FOR FUNCTIONAL DIFFERENTIAL EQUATIONS OF NEUTRAL-TYPE

John A. Burns*
Terry L. Herdman**

Department of Mathematics
Virginia Polytechnic Institute and State University
Blacksburg, Virginia 24061

Janos Turi***

Department of Mathematical Sciences
Worcester Polytechnic Institute
Worcester, MA 01609

1. INTRODUCTION

In recent years various classes of functional differential equations (FDE) have been studied in the context of functional analytic semigroup theory (see e.g. [1], [2], [4], [10] and the references therein). The basic approach in this direction is to establish equivalence between the FDE and an abstract evolution equation (AEE) in some appropriate state space (i.e. space of initial data.) Furthermore if the associated AEE is well-posed, then the equivalence between the FDE and the AEE provides an excellent framework to study approximation techniques for systems governed by FDEs. Well-posedness is dependent on the choice of a state-space and the choice of an appropriate state-space is tied to the particular application. It was shown (see [1], [2] for retarded and [10], [12] for neutral functional differential equations) that certain classes of FDEs can be transformed into well-posed Cauchy problems in the product spaces $\mathbb{R}^n \times L_p$. The product space model also proved to be very useful

*The work of this author was supported in part by the Air Force Office of Scientific Research under grant AFOSR-85-0287, the Defense Advanced Research Projects Agency under grant F49620-87-C-0116 and SDIO under contract F49620-87-C-0088.

**The work of this author was supported in part by the Air Force Office of Scientific Research under grant AFOSR-84-0326 and Defense Advanced Research Projects Agency under contract F49620-87-C-0016.

***The work of this author was supported in part by the Air Force Office of Scientific Research under grant AFOSR-85-0287. Parts of this research were carried out while this author was a visitor at the Interdisciplinary Center for Applied Mathematics, VPI and SU, Blacksburg, VA and was supported by Defense Advanced Research Projects Agency under contract F49620-87-C-0016.

in investigating a variety of control and identification problems for problems governed by FDEs ([2], [3], [5], [10], [12]).

In this paper we extend previous results concerning the well-posedness of FDEs on the product spaces $\mathbb{R}^n \times L_p$. In particular we develop general necessary and sufficient conditions for the well-posedness of neutral systems to include non-atomic neutral equations and certain classes of singular integro-differential equations.

2. WELL-POSEDNESS OF FDEs ON $\mathbb{R}^n \times L_p$

We consider the FDE of neutral-type

$$\frac{d}{dt} Dx_t = Lx_t + f(t) \tag{1}$$

with initial data

$$Dx_0(\cdot) = \eta;\ x_0(s) = \varphi(s), \quad -r \leq s < 0 \tag{2}$$

where D and L are linear $\mathbb{R}^n$-valued operators with domains $\mathcal{D}(D)$ and $\mathcal{D}(L)$ subspaces of the Lebesgue-measurable $\mathbb{R}^n$-valued functions on $[-r,0]$. We assume that $W^{1,p} \subseteq \mathcal{D}(D) \cap \mathcal{D}(L)$, $(\eta,\varphi) \in \mathbb{R}^n \times L_p([-r,0],\mathbb{R}^n)$ (or shortly $\mathbb{R}^n \times L_p$), $f \in L_{p,loc}$, $1 \leq p < \infty$, $0 \leq r < \infty$ and n is a positive integer.

Define the linear operator $\mathcal{A}$ with domain

$$\mathcal{D}(\mathcal{A}) = \{(\eta,\varphi) \in \mathbb{R}^n \times L_p / \varphi \in W^{1,p}, D\varphi = \eta\} \tag{3}$$

by

$$\mathcal{A}(\eta,\varphi) = (L\varphi\ \dot{\varphi}) \tag{4}$$

and consider the AEE

$$\dot{z}(t) = \mathcal{A}z(t) + (f(t),0) \tag{5}$$

with

$$z(0) = z_0 = (\eta,\varphi). \tag{6}$$

The well-posedness of the FDE (1)-(2) and the AEE (5)-(6) has been studied extensively ([5], [10], [13]) assuming various continuity conditions on L and D. It is known (see [13]) that, if L and D belong to $\mathcal{B}(W^{1,p},\mathbb{R}^n)$, then the FDE (1)-(2) is well-posed if and only if the AEE (5)-(6) is well-posed (i.e., if $\mathcal{A}$ defined by (3)-(4) is the infinitesimal generator of a C_0-semigroup $\{S(t)\}_{t\geq 0}$ on $\mathbb{R}^n\times L_p$). It is also known (see [5]) that if $\mathcal{A}$ generates a C_0-semigroup on $\mathbb{R}^n\times L_p$, then it is necessary that i) $L \in \mathcal{B}(W^{1,p},\mathbb{R}^n)$ and $D \in \mathcal{B}(W^{1,p},\mathbb{R}^n)$ and ii) $D \notin \mathcal{B}(L_p,\mathbb{R}^n)$. Concerning the sufficiency conditions for the well-posedness of the FDE (1)-(2) it is known that i*) $L \in \mathcal{B}(W^{1,p}, \mathbb{R}^n)$; ii*), $D \in \mathcal{B}(C,\mathbb{R}^n)$ and D is atomic at zero imply well-posedness, but condition ii*) above is not necessary (see Remarks 2 and 3).

<u>Remark 1</u>: Observe that if D is defined on $W^{1,p}$ by $D\varphi = \varphi(0)$ (i.e., when the FDE is retarded), then $L \in \mathcal{B}(W^{1,p},\mathbb{R}^n)$ is necessary and sufficient condition for the well-posedness of the FDE (1)-(2), because $D \in \mathcal{B}(W^{1,p},\mathbb{R}^n)$, $D \notin \mathcal{B}(L_p,\mathbb{R}^n)$, D has a bounded extension to C, and D is atomic at zero.

<u>Remark 2</u>: Consider the scalar FDE of the form (1) with $L\varphi \equiv 0$ and $D\varphi \equiv \int_{-r}^{0} \varphi(s)|s|^{-\alpha}ds$; $0 < \alpha < 1$. It can be shown (see [5], [9]) that the FDE is well-posed on $\mathbb{R}\times L_p$ if and only if $p < 1/(1-\alpha)$. This example demonstrates that: I): $L \in \mathcal{B}(W^{1,p},\mathbb{R})$, $D\in\mathcal{B}(W^{1,p},\mathbb{R})$ and $D\notin\mathcal{B}(L_p,\mathbb{R})$ is not sufficient (i.e., consider $p = 1/(1-\alpha)$), and II): $L\in\mathcal{B}(W^{1,p},\mathbb{R})$, $D\in\mathcal{B}(C,\mathbb{R})$ and D is atomic at zero is not necessary (i.e., consider $p < 1/(1-\alpha)$) for the well-posedness of the FDE on $\mathbb{R}\times L_p$.

<u>Remark 3:</u> The authors studied (see [16]) a scalar equation of the form (1) with $L\varphi\equiv 0$ and $D\varphi \equiv \int_{-r}^{0} \dot{\varphi}(s)|s|^{-\alpha}ds$; $0 < \alpha < 1$, and established well-posedness of this equation on $\mathbb{R}\times L_p$ for $p > 1/(1-\alpha)$. Since D does not have a bounded extension to C, this example implies, that $D \in \mathcal{B}(C,\mathbb{R})$ is not necessary for well-posedness.

<u>Remark 4:</u> Kappel and Zhang [9] considered the problem (1)-(2) in the space $C([-r,0], \mathbb{R})$ under the assumptions that D belongs to $\mathcal{B}(C,\mathbb{R})$ and $L \equiv 0$. They proved that the well-posedness of the FDE (1)-(2) in

the state space C implies that D is weakly atomic at zero.

Remark 5: At this point the most general necessary condition for well-posedness is given in [16]. Assuming only that $L \in \mathcal{B}(W^{1,p},\mathbb{R}^n)$, $D \in \mathcal{B}(W^{1,p},\mathbb{R}^n)$ and $D \notin \mathcal{B}(L_p,\mathbb{R}^n)$ it was shown that $\mathcal{A}$ defined by (3)-(4) is the generator of a C_0-semigroup only if the $n \times n$ matrix valued function $D(e^{\lambda \cdot}I) = [D(e^{\lambda \cdot}e_1) \vdots D(e^{\lambda \cdot}e_2) \vdots \ldots \vdots D(e^{\lambda \cdot}e_n)]$ exhibits "certain" asymptotic behavior as $\lambda \to \infty$; $(\lambda \in \mathbb{R})$.

As the previous Remarks indicate it is not known if there is a set of conditions on L and D that are both necessary and sufficient for $\mathcal{A}$ to generate a C_0-semigroup on $\mathbb{R}^n \times L_p$.

In the next section we consider a relatively large class of nonatomic neutral equations (NNFDE)(i.e., D is not necessarily atomic at zero) and give conditions which imply the well-posedness of these equations on $\mathbb{R}^n \times L_p$ for certain values of p.

3. NONATOMIC NEUTRAL EQUATIONS (NNFDEs)

In this section we consider the class of neutral functional differential equations given by (1)-(2) and provide conditions on D and L that imply the well-posedness of these equations on the product spaces $\mathbb{R}^n \times L_p$. Our results extend the results of Burns, Herdman and Stech [5] in that we obtain the well-posedness of (1)-(2) without assuming that the operator D be atomic at zero.

Our approach is based on the fact that the FDE (1)-(2) is well-posed on $\mathbb{R}^n \times L_p$ provided that $\mathcal{A}$ defined by (3)-(4) is an infinitesimal generator of a C_0-semigroup on $\mathbb{R}^n \times L_p$. Thus, our main result will establish sufficient conditions on D and L implying that $\mathcal{A}$ generates a C_0-semigroup on $\mathbb{R}^n \times L_p$. We assume that $W^{1,p} \subseteq \mathcal{D}(D) \cap \mathcal{D}(L)$, where the operators L and D satisfy the following conditions:

(H1) The operator $D \in \mathcal{B}(C,\mathbb{R}^n)$ has representation

$$D\varphi = \int_{-r}^{0} [A d\beta(s) + d\mu(s)]\varphi(s) \tag{7}$$

where the $n \times n$ matrix functions μ, β and the nonsingular matrix A

satisfy: i) μ is of bounded variation on $[-r,0]$, $\mu(0) = 0$, μ is left continuous on $[-r,0]$ and $\lim_{\epsilon\to0^+} \mathrm{Var}_{[-\epsilon,0]}\mu = 0$; ii) β is a diagonal matrix and there exists an integer k; $0 \le k \le n$, such that the entries, β_{ii}, satisfy $\beta_{ii}(s) = -\rho(-s)$ for $i \le k$, $\beta_{ii}(s) = -(-s)^{1-\alpha_i}/(1-\alpha_i)$ for $i > k$, where $\rho:[0,r] \to \mathbb{R}$, $\rho(0) = 0$, $\rho(s) = 1$ for $s > 0$ and the constants α_i; $i > k$, satisfy $0 < \alpha_i < 1$; iii) A has the block matrix form $A = \mathrm{diag}(A_{11}, A_{22})$ where A_{11} and A_{22} are $k \times k$ and $\ell \times \ell$ matrices, respectively, with $k + \ell = n$.

(H2) The operator $L \in \mathscr{B}(W^{1,p},\mathbb{R}^n)$ has representation

$$L\varphi = B\varphi(0) + \int_{-r}^{0} B(s)\varphi(s)ds \tag{8}$$

where B is a $n \times n$ constant matrix and $B(\cdot)$ is a $n \times n$ matrix valued function having column vectors in L_q, $\frac{1}{p} + \frac{1}{q} = 1$.

(H3) The $n \times n$ matrix valued function α defined on $[-r,0]$ by

$$\alpha(s) = \mu(s) - \int_{0}^{s} B(u)du$$

has the representation $\alpha(s) = [\alpha_1(s) \mid \alpha_2(s)]^T$ where α_1 and α_2 are $k \times n$ and $\ell \times n$ matrix valued functions, α_2 is absolutely continuous and $\dot{\alpha}_2$ is of bounded variation on $[-r,0]$.

Remark 6: If (H1) holds, then we may assume, without loss of generality, that $A = I$. In the event that the original nonsingular matrix A is not the identity matrix, one can multiply (1)-(2) by A^{-1}, introduce $\hat{\mu} = A^{-1}\mu$; $\hat{B} = A^{-1}B$, $\hat{B}(\cdot) = A^{-1}B(\cdot)$ and reduce the original problem to the case of $A = I$.

Remark 7: The operators L and D defined in (H1) and (H2) belong to $\mathscr{B}(W^{1,p},\mathbb{R}^n)$. In the case $k = n$ (i.e., $\ell = 0$), the operator D is atomic at zero and the sufficiency result of Burns, Herdman and Stech [5, Theorem 2.3] yields the well-posedness of (1)-(2) on $\mathbb{R}^n \times L_p$. The case $k = 0$, $\ell = n = 1$, $\mu(\cdot) \equiv 0$, $L \equiv 0$ and $f \equiv 0$ was also considered in [5] and well-posedness of (1)-(2) on $\mathbb{R}\times L_p$ was established for $1 \leq p < 1/(1-\alpha_1)$.

In Theorem 1 below we establish the well-posedness of a large class of FDEs (1)-(2) on the product spaces $\mathbb{R}^n \times L_p$ for certain values of p.

Theorem 1: Let $\alpha_{min} = \min_{i>k} \{\alpha_i\}$, $1 \leq p < 1/(1-\alpha_{min})$ and $D \in \mathcal{B}(C, \mathbb{R}^n)$, $L \in \mathcal{B}(W^{1,p}, \mathbb{R}^n)$ have representations (7), (8), respectively. If conditions H1) - H3) are satisfied, then the system

$$y(t) = \eta + \int_0^t (Lx_u + f(u))du, \quad t > 0$$

$$Dx_t = y(t) \text{ a.e. on } [0,\infty)$$

with initial condition

$$x_0(s) = \varphi(s) \text{ a.e. on } [-r,0]$$

has a unique solution $y(t) = y(t;\eta,\varphi,f)$, $x(t) = x(t;\eta,\varphi,f)$ defined on $[0,\infty)$ and $[-r,\infty]$, respectively such that $y(\cdot)$ is continuous and $x_t(\cdot) \in L_p$. Moreover, for $t_1 > 0$ the mapping $(\eta,\varphi,f) \rightarrow (y(\cdot;\eta,\varphi,f), x(\cdot;\eta,\varphi,f))$ from $\mathbb{R}^n \times L_p([0,t_1], \mathbb{R}^n)$ into $C([0,t_1], \mathbb{R}^n) \times L_p([-r,t_1], \mathbb{R}^n)$ is continuous.

Proof: First we note that to prove the theorem it is sufficient to consider the problem

$$\begin{aligned} Dx_t &= \eta + \int_0^t [Lx_u + f(u)]du \quad \text{a.e. on } [0,\infty) \\ x(s) &= \varphi(s) \quad \text{a.e. on } [-r,0]. \end{aligned} \tag{9}$$

Using the representations (7) and (8) and changing the order of integration of the integral involving $B(s)$, (9) becomes

$$\begin{aligned} &\int_{-r}^0 [d\beta(s) + d\mu(s)]x(t+s) - \int_{-r}^0 B(s)x(t+s)ds \\ &-B\int_0^t x(u)du = \eta - \int_{-r}^0 B(s)\varphi(s)ds + \int_0^t f(u)du. \end{aligned} \tag{10}$$

For $0 < t \leq r$ we can rewrite (10) as

$$\int_0^t [d\bar{\beta}(s) + d\bar{\gamma}(s)]x(t-s) = g(t), \tag{11}$$

where $\bar{\beta}(s) = -\beta(-s)$, $\bar{\gamma}(s) = -\gamma(-s)$, $\gamma(s) = \alpha(s) - Bs$ and

$$g(t) = \eta - \int_{-r}^{0} B(s)\varphi(s)ds + \int_0^t f(u)du - \int_{-r}^{-t} [d\beta(s) + d\alpha(s)]\varphi(t+s) . \tag{12}$$

Note that $\bar{\alpha}, \bar{\gamma} \in NBV([0,r],\mathbb{R}^{nxn})$, where $NBV([0,r],\mathbb{R}^{nxn})$ denotes the space of nxn matrix-valued functions which are of bounded variation on $[0,r]$, right continuous for $0 < s < r$, and take the value 0 at $s = 0$.

Define $h(\cdot) \in NBV([0,r],\mathbb{R}^{nxn})$ by $h(\cdot) = [h_{ij}(\cdot)]$, $1 \leq i,j \leq n$, where for all $1 \leq j \leq n$

$$h_{ij}(s) = \begin{cases} \bar{\gamma}_{ij}(s) & i \leq k, \\ \frac{d}{ds}\left[\frac{\sin\alpha_i\pi}{\pi}\int_0^s (s-u)^{\alpha_i-1}\,\bar{\gamma}_{ij}(u)du\right] & i > k . \end{cases} \tag{13}$$

It can be shown (see [15] for details) that for $0 < t \leq r$, equation (11) is equivalent to

$$\int_0^t d\bar{\beta}\,(s)\, w(t-s) = g(t), \tag{14}$$

where

$$w(t) = x(t) + \int_0^t dh(u)x(t-u) . \tag{15}$$

Recall that (15) is a Volterra-Stieltjes integral equation. Our assumptions guarantee that $h \in NBV([0,r],\mathbb{R}^{nxn})$ and that h is continuous at 0 from the right, i.e.

$$\lim_{t\to 0^+} h(t) = 0 . \tag{16}$$

Note that (16) is a sufficient condition (see for example [12]) for the existence and uniqueness of the fundamental solution, $\zeta \in NBV([0,r],\mathbb{R}^{nxn})$, of equation (15) Moreover, if $x(\cdot)$ the unique solution of (15), then $X(\cdot)$ belongs to $L_p([0,r],\mathbb{R}^n)$ and has representation

$$x(t) = \int_0^t d\zeta(s)w(t-s). \tag{17}$$

Continuous dependence of x on w with respect to the L_p - norm is an immediate consequence of (17). In particular, for $0 < t_1 \leq r$, we have the estimate

$$||x||_{L_p([0,t_1],\ \mathbb{R}^n)} \leq Var_{[0,t_1]}(h)||w||_{L_p([0,t_1],\ \mathbb{R}^n)}. \tag{18}$$

Next we consider equation (14) in component form, i.e.

$$\int_0^t d\overline{\beta}_i(s)w_i(t-s) = g_i((t);\ 1 \leq i \leq n. \tag{19}$$

Using the special form of $\overline{\beta}(\cdot)$, equation (19) implies that

$$w_i(t) = g_i(t) \quad , \quad i \leq k \tag{20}$$

and

$$\int_0^t s^{-\alpha_i} w_i(t-s)ds = g_i(t) \quad , \quad i > k . \tag{21}$$

For $t \in (0,r]$ define G_i by

$$G_i(t) \equiv \int_0^t (t-s)^{\alpha_i - 1} g_i(s)ds \quad , \quad i > k . \tag{22}$$

Note that if $(\eta,\varphi) \in \mathbb{R}^n x L_p$, $f \in L_p([0,r],\ \mathbb{R}^n)$ and $1 \leq p < 1/(1-\alpha_{min})$, then $g_i \in L_p([0,r],\mathbb{R})$, $1 \leq i \leq k$, and $G_i \in W^{1,p}([0,r],\mathbb{R})$, $i > k$ (see [5], [9] or [15] for details). Therefore, w_i, the ith component of the unique L_p solution of (14), is given by

$$w_i(t) = \begin{cases} g_i(t) & , \text{ for } i \leq k \\ \frac{d}{dt}[\frac{\sin \alpha_i \pi}{\pi} G_i(t)], & \text{for } i > k . \end{cases} \tag{23}$$

Moreover, there exists a nonnegative, increasing function $M \in C([0,r],R)$ such that

$$||w||_{L_p([0,t],\mathbb{R}^n)} \leq M(t)||(\eta,\varphi,f)||_{\mathbb{R}^n \times L_p \times L_p([0,r],\mathbb{R}^n)} \tag{24}$$

for $t \in [0,r]$ (see[15]). Substituting (23) into (17) we get a representation for the unique, L_p-solution to (9) for $0 \leq t \leq r$. Continuity of the mapping $(\eta,\varphi,f) \to (y(\cdot;\eta,\varphi,f),x(\cdot;\eta,\varphi,f))$ from $\mathbb{R}^n \times L_p([0,t_1],\mathbb{R}^n)$ into $C([0,t_1],\mathbb{R}^n) \times L_p([-r,t_1],\mathbb{R}^n)$ is an easy consequence of the estimates (18) and (24) for $0 < t_1 \leq r$. The "method of steps" is employed to extend the above results to $[0,+\infty)$. □

As an immediate consequence of Theorem 1 and the equivalence of the FDE (1)-(2) and the AEE (5)-(6) we have the following sufficiency result.

Theorem 2: If (H1)-(H3) hold, $1 \leq p < 1/(1-\alpha_{min})$, and D and L have the representations (7) and (8), then $\mathcal{A}$ defined by (3)-(4) is the infinitesimal generator of a C_0-semigroup on $\mathbb{R}^n \times L_p$.

4. CONCLUSIONS:

We have extended earlier results concerning the well-posedness of FDEs on product spaces. In particular, we have presented sufficient conditions for the well-posedness of a large class of functional differential equations (NNFDE). This class contains the "standard" neutral and retarded functional differential equations and many weakly singular integro-differential equations. It appears that results in this paper can be applied to infinite delay problems by using proper weighting on the state-space.

REFERENCES

[1] H. T. Banks and J. A. Burns, Hereditary control problems: Numerical methods based on averaging approximations, SIAM J. Control and Optimization, 16 (1978), 169-208.

[2] H. T. Banks and J. A. Burns, An abstract framework for approximate solutions to optimal control problems governed by hereditary systems, International Conference on Differential Equations, H. A. Antosiewicz ed., Academic Press, New York, 1975, 10-25.

[3] H. T. Banks, J. A. Burns and E. M. Cliff, Parameter estimation and identification for systems with delays, SIAM J. Control and Optimization, 19 (1981), 791-828.

[4] H. T. Banks and F. Kappel, Spline approximations for functional differential equations, Journal Differential Equations, 34 (1978), 496-522.

[5] J. A. Burns, T. L. Herdman and H. W. Stech, Linear functional differential equations as semigroups on product spaces, SIAM J. Math. Anal., 14 (1983), 98-116.

[6] J. K. Hale, Theory of Functional Differential Equations, Springer-Verlag, New york, 1977.

[7] F. Kappel, Approximation of neutral functional differential equations in the state-space $\mathbb{R}^n \times L_2$, in Colloquia Mathematica Societatis Janos Bolyai,, 30. Qualitative Theory of Differential Equations, Vol. I" (M. Farkas, Ed.), pp. 463-506, Janos Bolyai Math. Soc. and North Holland Publ. Comp., Amsterdam 1982.

[8] F. Kappel and Kang pei Zhang, Equivalence of functional equations of neutral type and abstract Cauchy-problems, Monatsch Math. 101 (1986), 115-133.

[9] F. Kappel and Kang pei Zhang, On neutral functional differential equations with nonatomic difference operator, J. M. A. A., 113 (1986), 311-343.

[10] F. Kappel and D. Salamon, Spline approximation for retarded systems and the Riccati equation, MRC Technical Summary Report No. 2680, 1984.

[11] A. Pazy, Semigroups of Linear Operators and Applications to PDE's, Springer-Verlag, New York, 1984.

[12] D. Salamon, Control and Observation of Neutral Systems, Pitman, 1984.

[13] G. Tadmor, $\mathbb{R}^n \times L_2$ representation of linear functional differential equations of neutral type, Preprint.

[14] Kang pei Zhang, On a neutral equation with nonatomic D-operator, Ph.D. Thesis, Institute for Mathematics, University of Graz, 1983.

[15] J. Turi, Well-posedness questions and approximation schemes for a general class of functional differential equations, Ph.D. Thesis, VPI and SU, Blacksburg, VA, 1986.

[16] J. A. Burns, T. L. Herdman and J. Turi, Neutral Functional Integro-Differential Equations with Weakly Singular Kernels, (to appear in J. M. A. A.).

SOME REMARKS ON FORCED INTEGRABLE SYSTEMS

Robert Carroll
University of Illinois
Urbana, IL 61801

1. INTRODUCTION

The use of the inverse scattering transform in dealing with "classical" integrable evolution systems is well known. In particular, it provides an effective method of studying soliton dynamics etc. via the time evolution of spectral data. Less well understood is the situation when such a nonlinear evolution equation for (say) $u(x,t)$ is "forced" by an input of the form $u(0,t) = \mathcal{Q}(t)$ for example (with $u(x,0)$ also prescribed, $0 \leq x < \infty$). The question now is not one of existence-uniqueness (which can be studied separately), but rather to use an inverse scattering technique in such a way as to determine soliton behavior via spectral information. As we indicate later, following especially Kaup [21-31], such a procedure involves overdetermining the system by specifying $u_x(0,t) = \mathcal{P}(t)$ in order to obtain the time evolution of the spectral data. Many such problems have been studied (cf. [21-31;37;38]) but the theory is not yet complete. In this paper we give some preliminary results in the following direction. We take a very special nonlinear Schrödinger equation (NLS) of the form (*) $iu_t = u_{xx} - 2|u|^2u$ which can be related to a special AKNS system of the form

$$v_{1x} + i\zeta v_1 = uv_2; \; v_{2x} - i\zeta v_2 = \bar{u}v_1 \tag{1.1}$$

$$v_{1t} = Av_1 + Bv_2; \; v_{2t} = Cv_1 - Av_2 \tag{1.2}$$

$$A = i|u|^2 + 2i\zeta^2; \; B = -iu_x - 2\zeta u; \; C = i\bar{u}_x - 2\zeta\bar{u}. \tag{1.3}$$

Such a system (1.1) - (1.3) does not in fact involve solitons and thus one can concentrate on the behavior of maps involving continuous spectrum in an attempt to understand this aspect of the situation. This will allow us to provide explicit formulas for the composition of maps $(\mathcal{P},\mathcal{Q}) \to$ spectral data $\to u(x,t) \to \mathfrak{M}(\mathcal{P},\mathcal{Q}) = (\hat{\mathcal{P}},\hat{\mathcal{Q}}) = (u_x(0,t),u(0,t))$. One then hopes to find a fixed point theorem for $\mathfrak{M}$ in suitable spaces. The point of our approach is to give insofar as possible explicit expressions for this map $\mathfrak{M}$ in terms of $(\mathcal{P},\mathcal{Q})$ so that properties of $\mathfrak{M}$ can be determined more readily. In this direction the paper only serves to set the stage heuristically. The machinery encounters (not surprisingly) some technical restrictions which require further study; any limitations thereby discovered would be revealing (perhaps indicating conditions needed to preserve the form (1.1)- (1.3)). Perhaps most interesting, the development here suggests adopting a similar procedure for general forced systems, in particular for NLS with solitons not precluded (i.e., with $w = -\bar{u}$ in (2.1) - (2.2));

such work is in progress. The technical restrictions indicated above arise in part due to the attempt to display the results via formulas where the objects of interest occur explicitly. Thus, alternatively, one could use standard AKNS inversion procedures, solve (in some way) the Marčenko integral equations for say $K(x,y,t)$, and then study $u(x,t) = -2K_1(x,x,t)$ and $u_x(x,t) = -2K_1'(x,x,t)$; our spectralization of these formulas creates integrals which need more investigation. Let us remark that in a previous version of this paper we proceeded via half line systems studied in geophysics and transmission lines, connections of that to various inverse spectral transforms, etc. (cf. [4-17;19;20;33-37;44;45]); appropriate versions of (3.4)-(3.5) were developed in that context along with many other relations between full and half line spectral data. However, the presentation before was too long and flawed at times by having recourse to real potentials. Thus, we forgo the exhibition of a number of interesting formulas connecting spectral data and potentials (many of which are in fact implicit or explicit in [6-15]) and simply give here a fast and eminently generalizable development based on AKNS theory.

2. BACKGROUND INFORMATION

We give here a brief background sketch (cf. [1;2;5;13;40;41;43;46;48]) and introduce the forcing framework of [21-32]. The AKNS generalization of the Zakharov-Shabat (Z-S) systems involves (cf. [1;2;13])

$$v_{1x} + i\zeta v_1 = uv_2;\ v_{2x} - i\zeta v_2 = wv_1 \tag{2.1}$$

$$v_{1t} = Av_1 + Bv_2;\ v_{2t} = Cv_1 - Av_2. \tag{2.2}$$

The compatability conditions for (2.1)-(2.2) have the form

$$A_x = uC - wB;\ B_x + 2i\zeta B = u_t - 2Au;\ C_x - 2i\zeta C = w_t + 2Aw. \tag{2.3}$$

One defines generalized eigenfunctions of (2.1) via

$$\varphi \sim \binom{1}{0}e^{-i\zeta x} \text{ and } \hat{\varphi} \sim \binom{0}{-1}e^{i\zeta x}\ (x \to -\infty);\ \psi \sim \binom{0}{1}e^{i\zeta x} \text{ and } \hat{\psi} \sim \binom{1}{0}e^{-i\zeta x}\ (x \to \infty). \tag{2.4}$$

One checks easily that

$$\varphi = a\hat{\psi} + b\psi;\ \hat{\varphi} = -\hat{a}\psi + \hat{b}\hat{\psi};\ \psi = \hat{b}\varphi - a\hat{\varphi};\ \hat{\psi} = b\hat{\varphi} + \hat{a}\varphi \tag{2.5}$$

(note $a\hat{a} + b\hat{b} = 1$). We will only be concerned here with a particular situation involving a forced nonlinear Schrödinger equation of a very special type. This situation will allow quick access to formulas containing the relevant information from which one hopes to establish a suitable fixed point theorem. The formulation of this in the special context here should provide better understanding of how to proceed in general. Thus,

<u>EXAMPLE 2.1.</u> For $w = \mp \bar{u}$ and $A = iuw + 2i\zeta^2$, $B = -iu_x - 2\zeta u$; and $C = iw_x - 2\zeta w$ one has a NLS (♦) $iu_t = u_{xx} \pm 2|u|^2 u$. We will take $\bar{u} = w$ so (♦) becomes (★) $iu_t = u_{xx} -$

$-2|u|^2u$ and one has various symmetry properties for the $a,\hat{a},b,\hat{b}$; in particular, $\hat{a}(\zeta,t) = \bar{a}(\bar{\zeta},t)$ and $\hat{b}(\zeta,t) = -\bar{b}(\bar{\zeta},t)$. Moreover $\hat{\psi}_1(x,\zeta) = \bar{\psi}_2(x,\bar{\zeta})$, $\hat{\psi}_2(x,\zeta) = \bar{\psi}_1(x,\bar{\zeta})$, $\hat{\varphi}_1(x,\zeta) = -\bar{\varphi}_2(x,\bar{\zeta})$, and $\hat{\varphi}_2(x,\zeta) = -\bar{\varphi}_1(x,\bar{\zeta})$. This is the model problem we will examine. One notes that solitons are not to be expected (cf. [1] and Remark 3.4), but this will simplify the analysis (in the spirit of [9;20]) and allow us to concentrate on other matters.

Given the situation leading to (★) now in Example 2.1 ($w = \bar{u}$) we assume $u(x,0)$ is given along with $u(0,t) = \mathbb{Q}(t)$. In terms of existence-uniqueness for (★) suitable information of this kind would be sufficient, but in order to eventually study soliton dynamics (for more general situations) one wants to proceed via the inverse spectral transform (IST). We refer here to [21-32] for general forced integrable systems in this spirit and will comment below on some of Kaup's procedure and results. Thus one looks for the time evolution of suitable spectral data (e.g., $a,\hat{b}$) and constructs recovery formulas for $u(x,t)$ in terms of such data. Now $u(x,t) = 0$ for $x < 0$ and a simple argument yields

$$\psi(0,t) = \begin{pmatrix}\hat{b}(\zeta,t)\\ a(\zeta,t)\end{pmatrix} \tag{2.6}$$

(note $\varphi = \binom{1}{0}\exp(-i\zeta x)$ and $\hat{\varphi} = \binom{0}{-1}\exp(i\zeta x)$ for $x \le 0$ so from $\psi = \hat{b}\varphi - a\hat{\varphi}$ one gets (2.6)). The classical AKNS procedure for determining time evolution of spectral data has to be modified here. We assume u and $u_x \to 0$ at ∞ as rapidly as needed and set $A_+ = \lim A(x,t,\zeta)$ as $x \to \infty$. Hence $A_+ = 2i\zeta^2$ and, in a similar notation, $B_+ = C_+ = 0$ here. Next one notes that (2.2) will not admit time-independent asymptotic conditions of the form (2.4) (which would require $0 \sim A_+\varphi_1 \sim 2i\zeta^2\exp(-i\zeta x)$ in particular) so one considers so-called time-dependent eigenfunctions $\varphi^t = \varphi\exp(A_+t)$, $\psi^t = \psi\exp(-A_+t)$, $\hat{\varphi}^t = \hat{\varphi}\exp(-A_+t)$, and $\hat{\psi}^t = \hat{\psi}\exp(A_+t)$. Then one asks that the φ^t, ψ^t, etc. satisfy (2.2) (they automatically satisfy (2.1)). This requires e.g.,

$$\psi_t = \begin{pmatrix} A+A_+ & B \\ C & A_+-A \end{pmatrix}\psi \tag{2.7}$$

and is compatible with the asymptotic conditions (2.4) as $x \to \infty$.

Now look at A,B,C for $x = 0$ with $\mathbb{Q}(t) = u(0,t)$ and $\mathbb{P}(t) = u_x(0,t)$ (note $\mathbb{P}$ is not determined a priori by $\mathbb{Q}$ - it must be specified independently). Thus ($A_+ = 2i\zeta^2$)

$$A(0,t,\zeta) = i|\mathbb{Q}|^2 + 2i\zeta^2;\ B(0,t,\zeta) = -i\mathbb{P} - 2\zeta\mathbb{Q};\ C(0,t,\zeta) = i\bar{\mathbb{P}} - 2\zeta\bar{\mathbb{Q}}. \tag{2.8}$$

Then write down (2.7) at $x = 0$, using (2.6), to obtain

$$\hat{b}_t = (i|\mathbb{Q}|^2 + 4i\zeta^2)\hat{b} - (i\mathbb{P} + 2\zeta\mathbb{Q})a;\ a_t = (i\bar{\mathbb{P}} - 2\zeta\bar{\mathbb{Q}})\hat{b} - i|\mathbb{Q}|^2a. \tag{2.9}$$

This is essentially the same thing as in [32] but with a slightly different formulation (details below). It is perhaps worth symmetrizing (2.9) here by writing

$$\binom{\hat{b}}{a} = \binom{\beta}{\alpha}e^{2i\zeta^2t};\ \beta_t - 2i\zeta^2\beta = i|\mathbb{Q}|^2\beta - (i\mathbb{P}+2\zeta\mathbb{Q})\alpha;\ \alpha_t + 2i\zeta^2\alpha = -i|\mathbb{Q}|^2\alpha + (i\bar{\mathbb{P}} - 2\zeta\bar{\mathbb{Q}})\beta.$$

THEOREM 2.2. The time evolution of spectral data b,a for the forced (NLS) (*) with $\mathbb{Q}$ and $\mathbb{P}$ given independently is determined by (2.9).

REMARK 2.3. The new feature here, discussed by Kaup and others in [21-32], is that the problem must be overdetermined by assigning $\mathbb{P}$ in order to discover the time evolution of $(\hat{b},a)$. Many models and approximation schemes are developed in [21-32] and considerable understanding emerges. Our approach aims at a purely mathematical result (a fixed point theorem) and produces a framework which should generalize nicely. We will assume (a suitable) $\mathbb{Q}$ is given and assign independently (a suitable) $\mathbb{P}$; then we compute u and u_x directly by spectralizing the Marčenko kernel at x = 0. This yields $(\hat{\mathbb{P}},\hat{\mathbb{Q}}) = (u_x(0,t),u(0,t)) = \mathfrak{m}(\mathbb{P},\mathbb{Q})$ and we ask for a fixed point of the map $\mathfrak{m}$. The $\mathbb{P}$ determined in this manner then leads to u(x,t) for $x,t \geq 0$ via inverse scattering and in more complicated situations where solitons are present this should allow one to study soliton dynamics etc. Requirements on $\mathbb{P}$ and $\mathbb{Q}$ will emerge from the investigation of $\mathfrak{m}$.

REMARK 2.4. Following Kaup [21-32], one can also look at (2.9), for example, as another kind of eigenvalue problem for $\psi(0,t)$ with "principal" term ζ^2 instead of ζ in (2.1) - cf. here [18;30;31]. One is then concerned with matching up solutions having different regions of analyticity ($\mathrm{Im}\zeta > 0$ and $\mathrm{Im}\zeta^2 > 0$ for example) and this can be related to a Riemann-Hilbert problem (cf. [21]).

3. DEVELOPMENT OF $\mathfrak{m}$.

Consider a general AKNS system as in (2.1)-(2.3) with (cf. [1;2])

$$\psi = \binom{0}{1}e^{i\zeta x} + \int_x^\infty K(x,s)e^{i\zeta s}ds;\ \hat{\psi} = \binom{1}{0}e^{-i\zeta x} + \int_x^\infty \hat{K}(x,s)e^{-i\zeta s}ds. \tag{3.1}$$

Let C (resp. $\hat{C}$) be suitable contours relative to the zeros of a (resp. $\hat{a}$ - see e.g. [1]) and operate on $\hat{\psi},\psi$ respectively by $(1/2\pi)\int_C \exp(i\zeta y)d\zeta$ and $(1/2\pi)\int_{\hat{C}} \exp(-i\zeta y)d\zeta$ for y > x to obtain

$$(1/2\pi)\int_C \hat{\psi}(\zeta,x)e^{i\zeta y}d\zeta = \binom{1}{0}\delta(y-x) + \hat{K}(x,y); \tag{3.2}$$

$$(1/2\pi)\int_{\hat{C}} \psi(\zeta,x)e^{-i\zeta y}d\zeta = \binom{0}{1}\delta(y-x) + K(x,y)$$

(t is suppressed momentarily). Now from [1;2] $u = -2K_1(x,x)$ and $w = -2\hat{K}_2(x,x)$ so from (3.2) formally

$$w(x) = -(1/\pi)\int_C \hat{\psi}_2(\zeta,x)e^{ix\zeta}d\zeta;\ u(x) = -(1/\pi)\int_{\hat{C}} \psi_1(\zeta,x)e^{-ix\zeta}d\zeta. \tag{3.3}$$

One sees that the formulation here is very general and we are presently examining other situations beyond the NLS considered here. Now insert t as needed, take $w = \bar{u}$ with $u(0,t) = \mathbb{Q}(t)$ where u = 0 for x < 0, recall (2.6), and with $a,\hat{a}$ having no zeros take $C = \hat{C} = (-\infty,\infty)$; recall also for ζ real $\psi_1(0,\zeta,t) = \hat{b}(\zeta,t) = -\bar{b}(\zeta,t)$, $\hat{\psi}_2(0,\zeta,t) = \bar{\psi}_1(0,\zeta,t) = -b(\zeta,t)$, and $\hat{a}(\zeta,t) = \bar{a}(\zeta,t)$ (cf. also Remark 3.4). Now working with say (3.2b), for example, with $w = \bar{u}$ ((3.2a) would be equivalent) one obtains formally

$$\hat{Q} = -(1/\pi)\int_{\infty}^{\infty} \hat{b}(\zeta,t)d\zeta. \tag{3.4}$$

Next one has $u' = u_x = -(1/\pi)\int_{-\infty}^{\infty} [\psi_1' - i\zeta\psi_1]\exp(-i\zeta x)d\zeta$. But $\psi_1' = -i\zeta\psi_1 + u\psi_2$ so $\psi_1'(0,\zeta,t) = -i\zeta\hat{b} + \mathcal{Q}a$ and hence formally

$$u'(0,t) = \hat{P}(t) = -(1/\pi)\int_{\infty}^{\infty} [\mathcal{Q}a - 2i\zeta\hat{b}]d\zeta. \tag{3.5}$$

THEOREM 3.1. Given $w = \bar{u}$ the formulas (3.4)-(3.5) determine the output $\hat{P},\hat{Q}$ in terms of spectral data $\hat{b},a$ determined by input $\mathcal{P},\mathcal{Q}$ in (2.9) (formally - see below).

Clearly the procedure here will generalize to more general forced AKNS systems where $C,\hat{C}$ are more general contours etc., and this will be developed at another time.

REMARK 3.2. By way of technical restrictions one sees that (3.4)-(3.5) make sense if one has "classical" variation $\hat{b} \sim (c/|\zeta|)\exp(4i\zeta^2 t)$, with $(2i\zeta\hat{b} - \mathcal{Q}a) \sim \hat{b}_t/2\zeta$ in (3.5). In general one should go back to the origin (3.2)-(3.3) of (3.4)-(3.5), and for small x leave factors $\exp(\pm i\zeta x)$ and $\exp(-i\zeta y)$ in the integrands; then with some adjustment for δ functions one obtains $\hat{Q} = -(1/\pi)\lim_{x\downarrow 0}\int_{-\infty}^{\infty} \hat{b}(\zeta,t)\exp(-2i\zeta x)d\zeta$; $\hat{P} = (1/\pi)\lim_{x\downarrow 0}\int_{-\infty}^{\infty} [\mathcal{Q}(1-a) + 2i\zeta\hat{b} - \hat{Q}]\exp(-2i\zeta x)d\zeta$. This corresponds to a standard way of treating step functions with Fourier integrals and the integrals will make sense under the growth behavior of $\hat{b},a$ to be expected here.

REMARK 3.3. There should be a playoff between smoothness of u and growth of $\hat{b}$ as $\zeta \to \infty$; smoothness involves perhaps some extensions and limiting arguments. This will be examined in another paper (cf. [6;7;20;44]).

REMARK 3.4. Let us collect here some information about the preservation of form and the continued absence of solitons in (2.1). Thus, one assumes (2.1)-(2.3) and A,B,C as in Example 2.1 so (*) $iu_t = u_{xx} - 2|u|^2 u$ holds, etc. We recall $\varphi = a\hat{\psi} + b\psi$, $\hat{\varphi} = -\hat{a}\psi + \hat{b}\hat{\psi}$, (φ,ψ) as in (2.4), $a\hat{a} + b\hat{b} = 1$, $\hat{a}(\zeta,t) = \bar{a}(\zeta,t)$ for ζ real with $\hat{b}(\zeta,t) = -\bar{b}(\zeta,t)$, $W(u,v) = u_1v_2 - u_2v_1$, $a = W(\varphi,\psi)$, $\hat{a} = W(\hat{\varphi},\hat{\psi})$, etc. Now generally $\hat{a}(\zeta) = \bar{a}(\bar{\zeta})$ (and $\hat{b}(\zeta) = -\bar{b}(\bar{\zeta})$ - suppressing t for simplicity) so the roots ζ_k, where $a(\zeta_k) = 0$ are paired with $\bar{\zeta}_k$ (i.e. $\hat{a}(\bar{\zeta}_k) = \bar{a}(\zeta_k) = 0$); we will show there aren't any ζ_k. Write (2.1) as

$$Lv - Qv = \zeta v;\ L = D_x\begin{pmatrix} i & 0 \\ 0 & -i\end{pmatrix};\ Q = \begin{pmatrix} 0 & iu \\ -i\bar{u} & 0\end{pmatrix}. \tag{3.6}$$

Evidently L and Q are formally self adjoint on $(-\infty,\infty)$ ($D_x^* = -D_x$, etc.), so if ζ is an eigenvalue of $\tilde{L} = L - Q$ with $\tilde{L}u = \zeta u$, then $(\tilde{L}u,u) = \zeta\|u\|^2 = (u,\tilde{L}u) = \bar{\zeta}\|u\|^2$. Hence ζ would be real (and paired with ζ). Here we are identifying eigenvalues of $\tilde{L}$ with discrete spectra ζ_k, where $a(\zeta_k) = 0$. Note at bound states ζ_k, defined by $a(\zeta_k) = 0$, $\varphi = b\psi$ and $\hat{\varphi} = b\hat{\psi}$ (recall $\hat{\varphi}_1(\zeta) = -\bar{\varphi}_2(\bar{\zeta})$, $\hat{\varphi}_2(\zeta) = -\bar{\varphi}_1(\bar{\zeta})$, $\hat{\psi}_1(\zeta) = \bar{\psi}_2(\bar{\zeta})$, and $\hat{\psi}_2(\zeta) = \bar{\psi}_1(\bar{\zeta})$). Now one knows e.g., $\exp(i\zeta x)\varphi$ and $\exp(-i\zeta x)\psi$ are analytic and bounded for $\text{Im}\zeta > 0$ under reasonable hypotheses on u. Hence if $\varphi = b\psi$ with $\text{Im}\zeta > 0$, then for $x < 0$, $|\exp(i\zeta x)\varphi| \le c$ implies $|\varphi| \le |c\exp(-i\zeta x)| \in L^2$ near $-\infty$ and for $x > 0$, $|\exp(-i\zeta x)\varphi| \le |b||\exp(-i\zeta x)\psi| \le \tilde{c}$ so $\varphi \in L^2$ near ∞. Hence φ and $\psi \in L^2$. Thus any bound state with $\mathrm{Im}\zeta_k > 0$ corresponds to an L^2 eigenfunction φ. We have however excluded such eigenfunctions by selfadjointness and hence such ζ_k are excluded. Note also we can in fact deal with L^2 on $(-\infty,\infty)$ here (since on $[0,\infty)$ selfadjointness of $\widetilde{L}$ is not established). Indeed, φ and ψ are defined for all x and for $a(\zeta_k) = 0$ the argument above holds. As for real zeros, we see from $a\hat{a} + b\hat{b} = 1$ that if ζ is real with $a(\zeta) = 0$, then $\hat{b}(\zeta) = -\bar{b}(\zeta)$ implies $1 = -|b|^2$ which is impossible. Hence there are no real zeros.

REMARK 3.5. Writing $\beta = \hat{\beta}\exp(i\int_0^t |Q|^2 ds)$ and $\alpha = \hat{\alpha}\exp(-i\int_0^t |Q|^2 ds)$ in the α,β equations after (2.9), we obtain

$$\hat{\alpha}_t + 2i\zeta^2\hat{\alpha} = (i\bar{P}-2\zeta\bar{Q})e^{2i\int_0^t |Q|^2 ds}\hat{\beta};\quad \hat{\beta}_t - 2i\zeta^2\hat{\beta} = -(iP+2\zeta Q)e^{-2i\int_0^t |Q|^2 ds}\hat{\alpha}. \tag{3.7}$$

This can be related to [18;31] with $2\zeta^2 = \lambda^2$ and will be discussed elsewhere. Note also that the equations (22) in [32] are $v_{1t} + i(2z^2 + \tilde{r}\tilde{q})v_1 = (2z\tilde{q} + i\tilde{q}_x)v_2$ and $v_{2t} - i(2z^2 + \tilde{r}\tilde{q})v_2 = (2z\tilde{r} - i\tilde{r}_x)v_1$. Set $z = -\zeta$, $\tilde{q} = \bar{u}$, and $\tilde{r} = u$ to obtain (at $x = 0$) the α,β equations after (2.9) with $\alpha = v_1$ and $\beta = v_2$.

REFERENCES

1. M. Ablowitz and H. Segur, Solitons and the inverse scattering transform, SIAM Philadelphia, 1981.
2. M. Ablowitz, D. Kaup, A. Newell, and H. Segur, Studies Appl. Math., 53 (1974), 249-315.
3. M. Ablowitz and H. Segur, Jour. Math. Physics, 16 (1975), 1054-1056.
4. A. Bruckstein, B. Levy, and T. Kailath, SIAM Jour. Appl. Math., 45 (1985), 312-335.
5. F. Calogero and A. Degasperis, Spectral transform and solitons, North-Holland, Amsterdam, 1982.
6. R. Carroll, Transmutation, scattering theory, and special functions, North-Holland, Amsterdam, 1982.
7. R. Carroll, Transmutation theory and applications, North-Holland, Amsterdam, 1985.
8. R. Carroll, Oakland Conf. PDE and Appl. Math., Pitman Press, London, 1987, pp. 1-38.
9. R. Carroll, Some features of the maps from potential to spectral data, Applicable Anal., 1987, to appear.
10. R. Carroll, Applicable Anal., 22 (1986), 21-43.
11. R. Carroll, CR Royal Soc. Canada, 9 (1987), 237-242.
12. R. Carroll, Acta Applicandae Math., 6 (1986), 109-184.
13. R. Carroll, Mathematical physics, North-Holland, Amsterdam, 1988, to appear.
14. R. Carroll, Diff. Eqs. in Banach spaces, Springer Notes Math. 1223, 1986, pp. 25-36.
15. R. Carroll and S. Dolzycki, Applicable Anal., 23 (1986), 185-208.
16. K. Chadan and P. Sabatier, Inverse problems in quantum scattering theory, Springer, N.Y., 1977.
17. L. Fadeev, Uspekhi Mat. Nauk, 14 (1959), 57-119; Sov. Prob. Math., 31 (1974), 93-180.

18. V. Gerdzhikov, M. Ivanov, and P. Kulish, Teor. Mat. Fiz., 44 (1980), 342-357.

19. M. Howard, Geophys. Jour. Royal Astr. Soc., 65 (1981), 191-215.

20. T. Kappeler and E. Trubowitz, Comment. Math. Helv., 61 (1986), 442-480.

21. D. Kaup, Lect. Appl. Math., AMS, 1986, pp. 195-215.

22. D. Kaup, Physica 25D (1987), 361-368.

23. D. Kaup, Jour. Math. Physics, 25 (1984), 277-281.

24. D. Kaup and H. Neuberger, Jour. Math. Physics, 25 (1984), 282-284.

25. D. Kaup and A. Newell, Proc. Royal Soc. London, 361A (1978), 413-446.

26. D. Kaup, Wave phenomena, Elsevier, 1984, pp. 163-174.

27. D. Kaup, Advances in nonlinear waves, Pitman, 1984, 197-209.

28. D. Kaup and A. Newell, Advances Math., 31 (1979), 67-100.

29. D. Kaup, SIAM Jour. Appl. Math., 31 (1976), 121-133.

30. D. Kaup and A. Newell, Lett. Nuovo Cimento, 20 (1977), 325-331.

31. D. Kaup and A. Newell, Jour. Math. Physics, 19 (1978), 798-801.

32. D. Kaup and P. Hansen, Physica 18D (1986), 77-84; 25D (1987), 369-381.

33. I. Kay and H. Moses, Inverse scattering papers, 1955-1963, Math. Sci. Press, 1982.

34. B. Levitan, Inverse Sturm-Liouville problems, Moscow, 1984.

35. B. Levitan and I. Sargsyan, Introduction to spectral theory ..., Moscow, 1970.

36. B. Levy and A. Yagle, Acta Applicandae Math., 3 (1985), 255-284.

37. V. Marčenko, Sturm-Liouville operators and their applications, Kiev, 1977.

38. H. Moses, Studies Appl. Math., 58 (1978), 187-207.

39. H. Moses, Jour. Math. Physics, 17 (1976), 73-75.

40. A. Newell, Solitons in mathematics and physics, SIAM, Phila., 1985.

41. A. Newell, Solitons, Springer, N.Y., 1980, pp. 177-242.

42. R. Newton, Conf. Inverse Scattering, SIAM, Phila., 1983, pp. 1-74.

43. S. Novikov, S. Manakov, L. Pitaevskij, and V. Zakharov, Theory of solitons, Plenum, 1984.

44. J. Pöschel and E. Trubowitz, Inverse spectral theory, Academic Press, N.Y., 1987.

45. A. Shabat, Sel. Math. Sov., 4 (1985), 19-35.

46. V. Zakharov, Solitons, Springer, N.Y., 1980, pp. 243-285.

47. V. Zakharov and P. Shabat, Funkts. Anal. Priloz., 8 (1974), 226-235; 13 (1979), 166-174.

48. L. Fadeev and L. Takhtajan, Hamiltonian methods in the theory of solitons, Springer, N.Y., 1987.

SOME REMARKS ON THE NONLINEAR SCHRÖDINGER EQUATION IN THE CRITICAL CASE

Thierry Cazenave [1] and Fred B. Weissler [1][2]

[1] Analyse Numérique, Université Pierre et Marie Curie,

4, Place Jussieu, 75252 PARIS CEDEX 05, FRANCE.

[2] Department of Mathematics, Texas A&M University,

COLLEGE STATION, TX 77843-3368, USA

1. INTRODUCTION

We consider the Cauchy problem (initial value problem) for nonlinear Schrödinger equations in $\mathbf{R}^n$, of the form

$$iu_t + \Delta u = g(u) \quad , \quad u(0,\cdot) = \varphi(\cdot) \; . \tag{NLS}$$

Here u is a complex-valued function defined on $[0,T)\times\mathbf{R}^n$ for some $T>0$, φ is some initial condition defined on $\mathbf{R}^n$ and g is some nonlinear (local or non-local) mapping. In most of the examples that have been considered, g has some symmetry properties and is also the gradient of some functional G. Thus, at least formally, we have both conservation of charge and conservation of energy, that is

$$\int_{\mathbf{R}^n} |u(t,x)|^2\,dx = \int_{\mathbf{R}^n} |\varphi(x)|^2\,dx \;,$$

$$\frac{1}{2}\int_{\mathbf{R}^n} |\nabla u(t,x)|^2\,dx + G(u(t,\cdot)) = \frac{1}{2}\int_{\mathbf{R}^n} |\nabla\varphi(x)|^2\,dx + G(\varphi(\cdot)) \;.$$

Clearly, the charge and energy involve the L^2 and H^1 norms of the solution. Therefore, it is appropriate to solve the local Cauchy problem in the space $L^2(\mathbf{R}^n)$ or $H^1(\mathbf{R}^n)$. Indeed, when it is possible to solve the local Cauchy problem in $L^2(\mathbf{R}^n)$, then global existence follows immediately from the conservation of charge; and when it is possible to solve the local Cauchy problem in $H^1(\mathbf{R}^n)$, then global existence follows from the conservation laws if G satisfies certain conditions (for example $G\geq 0$). Obviously, in order to solve the local Cauchy problem in those spaces, there are some necessary requirements on g. In the applications, this will impose some "growth" conditions on g.

We consider here the model case $g(u)=\lambda|u|^\alpha u$, where $\alpha\geq 0$ and $\lambda\in\mathbf{R}$. It is known that the local Cauchy problem is well posed in $L^2(\mathbf{R}^n)$ for $1\leq\alpha<4/n$ (see Y. Tsutsumi [8]) and in $H^1(\mathbf{R}^n)$ for

$1 \le \alpha < 4/(n-2)$ (see J. Ginibre and G. Velo [4,5], T. Kato [7], T. Cazenave and F. B. Weissler [2]).

In this paper, we study the "critical" cases $\alpha=4/(n-2)$ (for the Cauchy problem in $H^1(\mathbf{R}^n)$), and $\alpha=4/n$ (for the Cauchy problem in $L^2(\mathbf{R}^n)$). We basically apply the previously known methods, based on sharp dispersive esitmates for the (linear) Schrödinger equation, to establish local well-posedness of the Cauchy problem. However, the local existence results that we obtain are nonclassical in the sense that a bound on the solution in $L^2(\mathbf{R}^n)$ or $H^1(\mathbf{R}^n)$ does not imply global existence. (See the remarks at the end of Sections 3 and 4).

Before stating our main results, we introduce some notation. We denote by $S(t)$ the group of isometries generated by $i\Delta$ in $H^s(\mathbf{R}^n)$, $s \in \mathbf{R}$. For $\alpha \ge 0$ and $\lambda \in \mathbf{R}$, we define the function g by

$$g(u)=\lambda|u|^\alpha u, \text{ for } u \in \mathbf{C}.$$

We shall consider the following integral formulation of (NLS), which is in general equivalent (see [7]).

$$u(t) = S(t)\varphi + \mathcal{F}(u)(t). \tag{1.1}$$

Here, φ is some given initial datum, and $\mathcal{F}$ is defined by

$$\mathcal{F}u(t) = -i\int_0^t S(t-s)\, g(u(s))\, ds, \tag{1.2}$$

whenever the right hand side of (1.2) makes sense. Finally, we define the energy E by

$$E(\psi) = \frac{1}{2}\int_{\mathbf{R}^n} |\nabla\psi|^2\, dx + \frac{\lambda}{\alpha+2}\int_{\mathbf{R}^n} |\psi|^{\alpha+2}\, dx, \text{ for } \psi \in H^1(\mathbf{R}^n).$$

As in the papers [2,5,7,8], our results and proofs make use of mixed spaces of the type $L^q(0,T,L^r(\mathbf{R}^n))$ and $L^q(0,T,W^{1,r}(\mathbf{R}^n))$, where q and r always satisfy the same conditions. Thus, we make the following definition.

DEFINITION 1. A pair (q,r) is <u>admissible</u>, if $r \in [2,2n/(n-2))$ ($r \in [2,\infty]$ if $n=1$ and $r \in [2,\infty)$ if $n=2$) and q satisfies $2/q=n(1/2-1/r)$.

Also, we will have occasion to use the spaces $C([0,T],X)$ with $X=L^2$ or H^1. In order to simplify the statements of our results, it will be understood that if $T=\infty$, then $C([0,T],X)$ means $C([0,\infty),X)$. Our main results are the following.

THEOREM 1. Assume $\alpha=4/n$. For every $\varphi \in L^2(\mathbf{R}^n)$, there exists a unique maximal solution $u \in C([0,T^*),L^2(\mathbf{R}^n)) \cap L^{\alpha+2}_{loc}([0,T^*),L^{\alpha+2})$ of (1.1). Furthermore,

(i) $u \in L^q(0,T,L^r(\mathbf{R}^n))$, for every $T<T^*$, and every admissible pair (q,r),

(ii) $\|u(t)\|_{L^2} = \|\varphi\|_{L^2}$, for every $t \in [0,T^*)$,

(iii) if $T^*<\infty$, then $\|u\|_{L^q(0,T^*,L^r)} = \infty$ for every admissible pair (q,r) with $r\geq\alpha+2$.

THEOREM 2. Assume $n\geq 3$ and $\alpha=4/(n-2)$. For every $\varphi\in H^1(\mathbf{R}^n)$, there exists a maximal solution $u\in C([0,T^*),H^1(\mathbf{R}^n))\cap C^1([0,T^*),H^{-1}(\mathbf{R}^n))$ of (1.1). Furthermore,

(i) $u\in L^q(0,T,W^{1,r}(\mathbf{R}^n))$, for every $T<T^*$, and every admissible pair (q,r),

(ii) u is unique in $L^\infty(0,T,H^1(\mathbf{R}^n))\cap L^q(0,T,W^{1,r}(\mathbf{R}^n))$, for $T<T^*$, and every admissible pair (q,r), with $r>2$,

(iii) $E(u(t)) = E(\varphi)$ and $\|u(t)\|_{L^2} = \|\varphi\|_{L^2}$, for every $t\in[0,T^*)$,

(iv) if $T^*<\infty$, then $\|u\|_{L^q(0,T^*,W^{1,r})} = \infty$ for every admissible pair (q,r), with $r>2$.

The paper is organized as follows. In Section 2 we collect some basic estimates, in Section 3 we study the L^2-critical case, and in Section 4 we study the H^1-critical case. Moreover, in Sections 3 and 4 we give precise results concerning continuous dependence of solutions and some sufficient conditions for global solutions.

2. SOME PRELIMINARY RESULTS.

In this section, we collect some basic estimates for the linear Schrödinger equation, as well as some useful inequalities. We first recall the following result. (See [5,11,7,2]).

LEMMA 1. Let (q,r) be an admissible pair. There exists C_1 depending only on n and r such that

$$\|S(\cdot)\varphi\|_{L^q(\mathbf{R},L^r(\mathbf{R}^n))} \leq C_1 \|\varphi\|_{L^2}, \text{ for every } \varphi\in L^2(\mathbf{R}^n).$$

Moreover, for every admissible pair (γ,ρ) there exists C_2 depending only on n,r and ρ such that

$$\Big\| \int_0^t S(t-s)\, f(s)\, ds \Big\|_{L^q(0,T,L^r)} \leq C_2 \|f\|_{L^{\gamma'}(0,T,L^{\rho'})}, \text{ for every } T\in(0,\infty] \text{ and } f\in L^{\gamma'}(0,T,L^{\rho'}(\mathbf{R}^n)).$$

The following estimate will be used in section 3.

LEMMA 2. Assume $\alpha=4/n$. Let $T\in(0,\infty]$, let $\sigma=\alpha+2$, and let (q,r) be an admissible pair. Then, whenever $u\in L^\sigma(0,T,L^\sigma(\mathbf{R}^n))$, it follows that $\mathcal{F}u\in C([0,T],H^{-1}(\mathbf{R}^n))\cap L^q(0,T,L^r(\mathbf{R}^n))$. Furthermore, there exists K, independent of T, such that

$$\|\mathcal{F}v - \mathcal{F}u\|_{L^q(0,T,L^r)} \leq K \left(\|u\|^{\alpha}_{L^\sigma(0,T,L^\sigma)} + \|v\|^{\alpha}_{L^\sigma(0,T,L^\sigma)} \right) \|v-u\|_{L^\sigma(0,T,L^\sigma)} \qquad (2.1)$$

for every $u,v\in L^\sigma(0,T,L^\sigma(\mathbf{R}^n))$.

PROOF. Observe that if $u\in L^{\sigma}(0,T,L^{\sigma}(\mathbf{R}^n))$, then $|u|^{\alpha}u\in L^{\sigma'}(0,T,L^{\sigma'}(\mathbf{R}^n))\subset L^1(0,T,H^{-1}(\mathbf{R}^n))$, since $L^{\sigma'}(\mathbf{R}^n)\subset H^{-1}(\mathbf{R}^n)$. Therefore $\mathcal{F}u\in C([0,T],H^{-1}(\mathbf{R}^n))$. Estimate (2.1) follows from Lemma 1 (applied with $\gamma=\rho=\sigma$), and Hölder's inequality.

For $\alpha=4/(n-2)$ (when $n\geq 3$), and $m\in \mathbf{N}$, we define g_m by

$$g_m(u) = g(u) \text{ if } |u|\leq m, \text{ and } g_m(u) = \lambda m^{\alpha}u \text{ if } |u|\geq m. \tag{2.2}$$

For convenience, we set $g_{\infty}=g$. We define $\mathcal{F}_m$, for $m\in \mathbf{N}$, by

$$\mathcal{F}_m u(t) = -i\int_0^t S(t-s)g_m(u(s))\,ds, \tag{2.3}$$

and we set $\mathcal{F}_{\infty}=\mathcal{F}$.

LEMMA 3. Assume $n\geq 3$ and $\alpha=4/(n-2)$. Let (γ,ρ) be an admissible pair, with $2<\rho<n$. Let $\rho^*=\rho n/(n-\rho)$ and let (a,b) be the admissible pair where $b>2$ is given by $1/b'=1/b+\alpha/\rho^*$. (It follows that a satisfies $1/a'=1/a+\alpha/\gamma$). Then, for every admissible pair (q,r), for every $m\in \mathbf{N}\cup\{\infty\}$, $T\in (0,\infty]$, and every $u\in L^{\gamma}(0,T,L^{\rho^*}(\mathbf{R}^n))\cap L^a(0,T,L^b(\mathbf{R}^n))$, it follows that $\mathcal{F}_m(u)\in L^q(0,T,L^r(\mathbf{R}^n))\cap C([0,T],H^{-1}(\mathbf{R}^n))$. Moreover, there exists a constant L, independent of m and T, such that

$$\|\mathcal{F}_m(v)-\mathcal{F}_m(u)\|_{L^q(0,T,L^r)} \leq L\,(\|v\|^{\alpha}_{L^{\gamma}(0,T,L^{\rho^*})}+\|u\|^{\alpha}_{L^{\gamma}(0,T,L^{\rho^*})})\,\|v-u\|_{L^a(0,T,L^b)} \tag{2.4}$$

for every $u,v\in L^{\gamma}(0,T,L^{\rho^*}(\mathbf{R}^n))\cap L^a(0,T,L^b(\mathbf{R}^n))$.

PROOF. Observe that we have

$$|g_m(y) - g_m(x)| \leq C\,(|y|^{\alpha} + |x|^{\alpha})\,|y-x|\,, \text{ for } x,y\in \mathbf{C}, \tag{2.5}$$

uniformly for $m\in \mathbf{N}\cup\{\infty\}$. Hölder's inequality and (2.5) imply

$$\|g_m(v)-g_m(u)\|_{L^{a'}(0,T,L^{b'})} \leq L\,(\|v\|^{\alpha}_{L^{\gamma}(0,T,L^{\rho^*})}+\|u\|^{\alpha}_{L^{\gamma}(0,T,L^{\rho^*})})\,\|v-u\|_{L^a(0,T,L^b)}.$$

Applying Lemma 1, we get (2.4). In addition, $g_m(u)\in L^1(0,T,H^{-1}(\mathbf{R}^n))$, since $L^{b'}(\mathbf{R}^n)\subset H^{-1}(\mathbf{R}^n)$; and so $\mathcal{F}_m(u)\in C([0,T],H^{-1}(\mathbf{R}^n))$.

COROLLARY 1. Under the hypotheses of Lemma 3, if $u\in L^a(0,T,L^b(\mathbf{R}^n))\cap L^{\gamma}(0,T,L^{\rho^*}(\mathbf{R}^n))$ and if $\nabla u\in L^a(0,T,L^b(\mathbf{R}^n))$, then for every admissible pair (q,r), $\mathcal{F}_m(u)\in L^q(0,T,W^{1,r}(\mathbf{R}^n))\cap C([0,T],H^1(\mathbf{R}^n))$, with

$$\|\nabla\mathcal{F}_m(u)\|_{L^q(0,T,L^r)} \leq 2L\,\|u\|^{\alpha}_{L^{\gamma}(0,T,L^{\rho^*})}\,\|\nabla u\|_{L^a(0,T,L^b)}. \tag{2.6}$$

PROOF. (2.6) is an immediate consequence of (2.4), since the gradient is the limit of the finite differences quotient. It remains to establish the continuity in H^1. First, observe that in Lemma 3, (q,r) is arbitrary. In particular, $\mathcal{F}_m(u)\in L^{\infty}(0,T,H^1(\mathbf{R}^n))$. Since also $\mathcal{F}_m(u)\in C([0,T],H^{-1}(\mathbf{R}^n))$, $\mathcal{F}_m(u)$ is weakly continuous in H^1. Let $t\in[0,T]$ and $h>0$ be such that $t+h\in[0,T]$. We have

$$\mathcal{F}_m(u)(t+h) - \mathcal{F}_m(u)(t) = (S(h)-I)\mathcal{F}_m(u)(t) + \mathcal{F}_m(u(\cdot+t))(h)\,. \tag{2.7}$$

Now, $\mathcal{F}_m(u)(t)\in H^1(\mathbf{R}^n)$, and so

$$(S(h)-I)\mathcal{F}_m(u)(t)\to 0 \text{ in } H^1(\mathbf{R}^n), \text{ as } h\to 0. \tag{2.8}$$

Using the weak continuity in H^1, and formulas (2.4) and (2.6), we see that

$$\|\mathcal{F}_m(u(\cdot+t))(h)\|_{H^1} = \|\mathcal{F}_m(u(\cdot+t))\|_{L^\infty(0,h,H^1)} \le 2L\, \|u\|^\alpha_{L^\gamma(t,t+h,L^{\rho^*})}\, \|u\|_{L^a(t,t+h,W^{1,b})}\,,$$

which converges to 0 as $h\to 0$. A similar argument works if $h<0$. Hence the H^1-continuity, by (2.7) and (2.8).

LEMMA 4. With the notation of Lemma 3, there exists M such that if $T<\infty$, $m\in\mathbf{N}$, and $u\in L^\infty(0,T,H^1(\mathbf{R}^n))\cap L^\gamma(0,T,L^{\rho^*}(\mathbf{R}^n))$, then

$$\|\mathcal{F}_m(u)-\mathcal{F}(u)\|_{L^q(0,T,L^r)} \le M\, T^{1/a'}\, m^{-(2^*-b)/b}\, \|u\|^\alpha_{L^\gamma(0,T,L^{\rho^*})}\, \|u\|^{2^*/b}_{L^\infty(0,T,H^1)}\,,$$

where $2^*=2n/(n-2)$.

PROOF. It is easily verified (see the proof of Lemma 3) that

$$\|g_m(u) - g(u)\|_{L^{a'}(0,T,L^{b'})} \le C\, \|u\|^\alpha_{L^\gamma(0,T,L^{\rho^*})}\, \|\chi u\|_{L^a(0,T,L^b)}\,,$$

where $\chi=\chi_E$ with $E=\{(t,x),\ |u(t,x)|\ge m\}$. From Sobolev and Hölder inequalities, we get

$$C\,T^{1/a'}\, \|u\|^{2^*/b}_{L^\infty(0,T,H^1)} \ge T^{1/a'}\, \|\chi u\|^{2^*/b}_{L^\infty(0,T,L^{2^*})} \ge T^{1/a'}\, m^{(2^*-b)/b}\, \|\chi u\|_{L^\infty(0,T,L^b)} \ge m^{(2^*-b)/b}\, \|\chi u\|_{L^a(0,T,L^b)}\,.$$

The lemma follows from the above two inequalites and Lemma 1.

3. THE L^2-CRITICAL CASE.

In all this section, we assume $\alpha=4/n$, and we set $\sigma=\alpha+2$. We begin with the following.

PROPOSITION 1. There exists $\delta>0$ with the following property. If $\varphi\in L^2(\mathbf{R}^n)$ and $T\in(0,\infty]$ are such that $\|S(\cdot)\varphi\|_{L^\sigma(0,T,L^\sigma)} < \delta$, there exists a unique solution $u\in C([0,T],L^2(\mathbf{R}^n))\cap L^\sigma(0,T,L^\sigma(\mathbf{R}^n))$ of (1.1). In addition, $u\in L^q(0,T,L^r(\mathbf{R}^n))$ for every admissible pair (q,r); and $\|u(t)\|_{L^2}=\|\varphi\|_{L^2}$ for $t\in[0,T]$. Finally, u depends continuously in $C([0,T],L^2(\mathbf{R}^n))\cap L^\sigma(0,T,L^\sigma(\mathbf{R}^n))$ on $\varphi\in L^2(\mathbf{R}^n)$. If $\varphi\in H^1(\mathbf{R}^n)$, then $u\in C([0,T],H^1(\mathbf{R}^n))$.

PROOF. Let $\delta>0$, to be chosen later, and let φ, T be as above. Consider the set

$$E = \{u\in L^\sigma(0,T,L^\sigma(\mathbf{R}^n)),\ \|u\|_{L^\sigma(0,T,L^\sigma)} \le 2\delta\,\}.$$

For $w\in E$, we define $\mathcal{H}w$ by

$$\mathcal{H}w(t) = S(t)\varphi + \mathcal{F}w(t).$$

It follows easily from (2.1) that if δ is small enough (independently of φ and T), $\mathcal{H}$ is a strict contraction on E. Thus $\mathcal{H}$ has a fixed point u, which is the unique solution of (1.1) in E. Applying again (2.1), we see that $u\in L^q(0,T,L^r(\mathbf{R}^n))$, for every admissible pair (q,r).

Consider now a sequence $\varphi^m\in L^2(\mathbf{R}^n)$, with $\varphi^m\to\varphi$, as $m\to\infty$. It follows from Lemma 1 that for m large enough, we have

$$\|S(\cdot)\varphi^m\|_{L^\sigma(0,T,L^\sigma)} < \delta .$$

We can apply the above argument to construct the solution $u^m\in L^\sigma(0,T,L^\sigma(\mathbf{R}^n))$ of (1.1) with initial datum φ^m. Applying again (2.1), we obtain that $u^m\to u$ in $L^\sigma(0,T,L^\sigma(\mathbf{R}^n))\cap L^\infty(0,T,L^2(\mathbf{R}^n))$, as $m\to\infty$, and in fact in $L^q(0,T,L^r(\mathbf{R}^n))$, for every admissible pair (q,r).

Assume now $\varphi\in H^1(\mathbf{R}^n)$. Since (1.1) is uniquely locally solvable in $H^1(\mathbf{R}^n)$ (see [7]), it follows that $u\in C([0,\tau],H^1(\mathbf{R}^n))\cap L^\sigma(0,T,W^{1,\sigma}(\mathbf{R}^n))$, for some $\tau\in(0,T]$. We claim that we can take $\tau=T$. Otherwise, we would have $\|u(t)\|_{H^1}\to\infty$ as $t\uparrow\tau$. But, from (2.1) and Lemma 1, we deduce with the method of proof of Corollary 1, that

$$\|u\|_{L^q(0,t,W^{1,r})} \le C_1\|\varphi\|_{H^1} + 2K\|u\|^{\alpha}_{L^\sigma(0,t,L^\sigma)}\|u\|_{L^\sigma(0,t,W^{1,\sigma})} \le C_1\|\varphi\|_{H^1} + 2K(2\delta)^\alpha\|u\|_{L^\sigma(0,t,W^{1,\sigma})}$$

for every $t\le\tau$. Choosing successively $(q,r)=(\sigma,\sigma)$ and $(q,r)=(\infty,2)$, and assuming that δ is small enough (still independently of φ and T), we get, by letting $t\uparrow\tau$,

$$\|u\|_{L^\infty(0,\tau,H^1)} \le C\|\varphi\|_{H^1} ,$$

which is a contradiction. Thus $\tau=T$. It follows (see [4,2]) that $\|u(t)\|_{L^2}=\|\varphi\|_{L^2}$, for every $t\in[0,T]$.

For an arbitrary $\varphi\in L^2(\mathbf{R}^n)$, let $\varphi^m\in H^1(\mathbf{R}^n)$, with $\varphi^m\to\varphi$ in $L^2(\mathbf{R}^n)$, as $m\to\infty$; and let u^m be the corresponding solutions of (2.1). From the continuous dependence, it follows that $u^m\to u$ in $L^\infty(0,T,L^2(\mathbf{R}^n))$, as $m\to\infty$. Thus $u\in C([0,T],L^2(\mathbf{R}^n))$ and $\|u(t)\|_{L^2}=\|\varphi\|_{L^2}$, for every $t\in[0,T]$. This completes the proof of Proposition 1.

PROOF OF THEOREM 1. Let $\varphi\in L^2(\mathbf{R}^n)$. Observe that $\|S(\cdot)\varphi\|_{L^\sigma(0,T,L^\sigma)}\to 0$, as $T\to 0$. Thus for sufficiently small T, the hypotheses of Proposition 1 are satisfied. Applying iteratively Proposition 1, we can construct the maximal solution $u\in C([0,T^*),L^2(\mathbf{R}^n))\cap L^\sigma_{loc}([0,T^*),L^\sigma(\mathbf{R}^n))$ of (1.1). It remains to establish property (iii) of the theorem. Thus, assume $T^*<\infty$, and suppose to the contrary that $\|u\|_{L^\sigma(0,T^*,L^\sigma)}<\infty$. Let $t\in[0,T^*)$. For every $s\in[0,T^*-t)$, we have

$$S(s)u(t) = u(t+s) - \mathcal{F}(u(t+\cdot))(s). \tag{3.1}$$

From (3.1) and (2.1), we obtain

$$\|S(\cdot)u(t)\|_{L^\sigma(0,T^*-t,L^\sigma)} \le \|u\|_{L^\sigma(t,T^*,L^\sigma)} + K(\|u\|_{L^\sigma(t,T^*,L^\sigma)})^{\alpha+1}.$$

Therefore, for t close enough to T*, it follows that

$$\|S(\cdot)u(t)\|_{L^\sigma(0,T^*-t,L^\sigma)} < \delta.$$

Applying Proposition 1, we find that u can be extended after T*, which contradicts the maximality. Let now (q,r) be an admissible pair, with $r \ge \sigma$. By Hölder's inequality, we have for $T<T^*$

$$\|u\|_{L^\sigma(0,T,L^\sigma)} \le \|u\|^{\mu}_{L^\infty(0,T,L^2)} \|u\|^{1-\mu}_{L^q(0,T,L^r)} \le \|\varphi\|^{\mu}_{L^2} \|u\|^{1-\mu}_{L^q(0,T,L^r)}, \text{ with } \mu = \frac{2(r-\sigma)}{\sigma(r-2)}.$$

Letting $T \to T^*$, we obtain $\|u\|_{L^q(0,T^*,L^r)} = \infty$. The proof of Theorem 1 is thereby completed.

REMARK 1. For initial data in $L^2(\mathbf{R}^n)$ but not in $H^1(\mathbf{R}^n)$, the energy is not well defined, thus we cannot expect conservation of energy.

REMARK 2. It follows from Proposition 1 that if $u \in L^\sigma(0,T,L^\sigma(\mathbf{R}^n))$ is a solution of (1.1) in the sense of Theorem 1, and if $\varphi \in H^1(\mathbf{R}^n)$, then $u \in C([0,T],H^1(\mathbf{R}^n))$. In other words, the solution exists in $H^1(\mathbf{R}^n)$ as long as it exists in $L^2(\mathbf{R}^n)$.

REMARK 3. With the argument of the proof of Proposition 1, it is not difficult to prove the following continuous dependence result. If $\varphi^m, \varphi \in L^2(\mathbf{R}^n)$, with $\varphi^m \to \varphi$, as $m \to \infty$, and if the corresponding sequence of solutions u^m of (1.1) is bounded in $L^\sigma(0,T,L^\sigma(\mathbf{R}^n))$ for some $T>0$, then the solution u of (1.1) with initial datum φ exists on [0,T] and $u^m \to u$ in $L^\sigma(0,T,L^\sigma(\mathbf{R}^n)) \cap C([0,T],L^2(\mathbf{R}^n))$, as $m \to \infty$. Moreover, one can show by these methods, that given $\varphi \in L^2(\mathbf{R}^n)$ and $T<T^*(\varphi)$, then $T^*(\psi)>T$ and $\|u_\psi\|_{L^\sigma(0,T,L^\sigma)} \le 2\,\|u_\varphi\|_{L^\sigma(0,T,L^\sigma)}$ for ψ in some L^2-neighborhood of φ. For a detailed proof, see [3], Theorem 1.2.

REMARK 4. If $\|\varphi\|_{L^2}$ is small enough, we can apply Proposition 1 with $T=\infty$ (see Lemma 1). In this case, we have $u \in L^q(0,\infty,L^r(\mathbf{R}^n))$, for every admissible pair (q,r). However, this does not hold in general for large data. Indeed, if $\lambda<0$, there exists nontrivial solutions (standing waves) of the form $u(t,x)=e^{i\omega t}\psi(x)$, with $\psi \in H^1(\mathbf{R}^n)$ (see [1]). These solutions obviously do not belong to $L^q(0,\infty,L^r(\mathbf{R}^n))$, if $q<\infty$. However, if $\lambda \ge 0$ and $\varphi \in H^1(\mathbf{R}^n)$, it is known that $T^*=\infty$ and $u \in L^q(0,\infty,L^r(\mathbf{R}^n))$ (see [6,9]).

REMARK 5. Despite of the conservation of charge, T* can be finite in some cases. For example, assume $\lambda<0$ and let $\psi \in H^1(\mathbf{R}^n)$ be a nontrivial solution of the equation $-\Delta\psi + \psi = g(\psi)$, and let u be given by

$$u(t,x) = (1-t)^{-n/2} \exp(i(4t-|x|^2)/4(1-t))\, \psi(x/(1-t)), \text{ for } x\in \mathbf{R}^n \text{ and } t\in [0,1).$$

Then u solves (1.1) on [0,1) (see [10]). However, if (q,r) is an admissible pair, we have

$$\|u(t)\|_{L^r} = C/(1-t)^{2/q}, \text{ for } t\in [0,1).$$

Therefore, $u\notin L^q(0,1,L^r(\mathbf{R}^n))$ if r>2, and so T*=1.

REMARK 6. We conjecture that if $\lambda\geq 0$, then $T^*=\infty$ for all $\varphi\in L^2(\mathbf{R}^n)$. However, we only have the following partial result. Assume $\lambda\geq 0$, and suppose that $\varphi\in L^2(\mathbf{R}^n)$ is such that $|x|\varphi(x)\in L^2(\mathbf{R}^n)$. Then $T^*=\infty$. To see this, consider a sequence $\varphi^m\in H^1(\mathbf{R}^n)$, with $\varphi^m\to\varphi$, as $m\to\infty$, and $|x|\varphi^m(x)$ bounded in $L^2(\mathbf{R}^n)$. The corresponding solutions satisfy $u^m\in C([0,\infty),H^1(\mathbf{R}^n))$ (see [4,2]), and from the pseudo-conformal conservation law (see[4]), we obtain

$$\|u^m(t)\|_{L^\sigma}^{\sigma} \leq \frac{C}{t^2}, \text{ for } t>0,$$

where C is independent of m. It now follows easily from Lemma 1 and Proposition 1 that $(u^m)_{m\in\mathbf{N}}$ is bounded in $L^\sigma(0,\infty,L^\sigma(\mathbf{R}^n))$. By Remark 3, we obtain $T^*=\infty$, and $u\in L^\sigma(0,\infty,L^\sigma(\mathbf{R}^n))$.

4. THE H¹-CRITICAL CASE.

Throughout this section, we assume $n\geq 3$ and $\alpha=4/(n-2)$. For convenience, we set

$$\rho = \frac{2n^2}{n^2-2n+4}, \quad \gamma = \frac{2n}{n-2}.$$

Then (γ,ρ) is an admissible pair, $\rho<n$, and by Lemma 3 and Sobolev's inequality, for every admissible pair (q,r), there exists C such that for every $T\in (0,\infty]$ and $m\in\mathbf{N}\cup\{\infty\}$,

$$\|\mathcal{F}_m(v)-\mathcal{F}_m(u)\|_{L^q(0,T,L^r)} \leq C\,(\|\nabla v\|^{\alpha}_{L^\gamma(0,T,L^\rho)} + \|\nabla u\|^{\alpha}_{L^\gamma(0,T,L^\rho)})\,\|v-u\|_{L^\gamma(0,T,L^\rho)}. \tag{4.1}$$

PROPOSITION 2. There exists $\delta>0$ with the following property. If $\varphi\in H^1(\mathbf{R}^n)$ and $T\in(0,\infty]$ are such that $\|S(\cdot)\nabla\varphi\|_{L^\gamma(0,T,L^\rho)} < \delta$, there exists a unique solution $u\in C([0,T],H^1(\mathbf{R}^n))\cap L^\gamma(0,T,W^{1,\rho}(\mathbf{R}^n))$ of (1.1). In addition, $u\in L^q(0,T,W^{1,r}(\mathbf{R}^n))$ for every admissible pair (q,r). Furthermore,

$$\|u(t)\|_{L^2} = \|\varphi\|_{L^2} \text{ and } E(u(t)) = E(\varphi), \text{ for every } t\in[0,T].$$

PROOF. Let $\delta>0$, to be chosen later, and let φ,T be as above. We use g_m and $\mathcal{F}_m$ as defined in (2.2) and (2.3), and for $m\in\mathbf{N}$, we consider the solution $u_m\in C([0,\infty),H^1(\mathbf{R}^n))$ of

$$u_m(t) = S(t)\varphi + \mathcal{F}_m(u_m)(t). \tag{4.2}$$

In particular, we have (see[4,2])

$$\|u_m(t)\|_{L^2} = \|\varphi\|_{L^2} \text{ and } E_m(u_m(t)) = E_m(\varphi), \text{ for every } t\geq 0, \tag{4.3}$$

where

$$E_m(w) = \frac{1}{2}\int_{\mathbf{R}^n} |\nabla w|^2 + \frac{\lambda}{\alpha+2}\int_{\mathbf{R}^n} G_m(w)\text{, for } w\in H^1(\mathbf{R}^n)\text{, and } G_m(x) = \int_0^x g_m(s)\, ds \text{ for } x\geq 0.$$

We shall first show that if δ is small enough (independently of φ and T), then

$$\|\nabla u_m\|_{L^\gamma(0,T,L^\rho)} \leq 2\,\delta\text{, for every } m\in \mathbf{N}. \tag{4.4}$$

To see this, note that from Lemma 1 and (2.6), we have

$$\|\nabla u_m\|_{L^\gamma(0,T,L^\rho)} \leq \delta + C\,\|\nabla u_m\|^{\alpha+1}_{L^\gamma(0,T,L^\rho)}.$$

These norms are finite, because of [7]. Thus (4.4) holds if $2C(2\delta)^\alpha < 1$. Lemma 1, (2.6) and (4.4) now imply that for every admissible pair (q,r), there exists C, independent of m (but depending on φ) such that

$$\|\nabla u_m\|_{L^q(0,T,L^r)} \leq C\text{, for every } m\in \mathbf{N}. \tag{4.5}$$

Next, we claim that, by choosing δ possibly smaller, we have

$$\|u_m\|_{L^\gamma(0,T,L^\rho)} \leq C\text{, for every } m\in \mathbf{N}, \tag{4.6}$$

where C is independent of m, but depends on φ. Indeed, by (4.1), (4.4) and Lemma 1, we have

$$\|u_m\|_{L^\gamma(0,T,L^\rho)} \leq C\,\|\varphi\|_{L^2} + C\,(2\delta)^\alpha\,\|u_m\|_{L^\gamma(0,T,L^\rho)}.$$

Thus (4.6) holds if $C(2\delta)^\alpha < 1$. (4.1), (4.4), (4.6) and Lemma 1, now imply that u_m is a bounded sequence in $L^q(0,T,L^r(\mathbf{R}^n))$, for every admissible pair (q,r); and combining this with (4.5), we get

$$u_m \text{ is a bounded sequence in } L^q(0,T,W^{1,r}(\mathbf{R}^n))\text{, for every admissible pair (q,r).} \tag{4.7}$$

For $j,m\in \mathbf{N}$, we have

$$u_m - u_j = [\mathcal{F}_m(u_m) - \mathcal{F}_m(u_j)] + [\mathcal{F}_m(u_j) - \mathcal{F}(u_j)] + [\mathcal{F}(u_j) - \mathcal{F}_j(u_j)]\,. \tag{4.8}$$

We now use (4.1) to estimate the first term on the right hand side of (4.8) and Lemma 4 to estimate the two other terms. Taking into account (4.4) and (4.7), we get, for every $\tau<\infty$, $\tau\leq T$,

$$\|u_m - u_j\|_{L^\gamma(0,\tau,L^\rho)} \leq C\,(2\delta)^\alpha\,\|u_m - u_j\|_{L^\gamma(0,\tau,L^\rho)} + \varepsilon_{j,m}\,,$$

where $\varepsilon_{j,m}\to 0$ as $j,m\to\infty$. Therefore, u_m is a Cauchy sequence in $L^\gamma(0,\tau,L^\rho(\mathbf{R}^n))$. Applying again (4.1), we see that u_m is a Cauchy sequence in $L^q(0,\tau,L^r(\mathbf{R}^n))$, for every admissible pair (q,r). Let u be its limit. We have $u\in L^q(0,T,L^r(\mathbf{R}^n))$, for every admissible pair (q,r), and

$$u_m \to u \text{ in } L^q(0,\tau,L^r(\mathbf{R}^n)), \text{ for every admissible pair } (q,r), \text{ and for } \tau<\infty,\ \tau\leq T. \tag{4.9}$$

It follows from (4.9) and (4.7) that

$$u \in L^q(0,T,W^{1,r}(\mathbf{R}^n)), \text{ for every admissible pair } (q,r). \tag{4.10}$$

Next, from (4.2), (4.9), (4.10), (2.4) and Lemma 4, we deduce that u solves (1.1). By Corollary 1, we have $u\in C([0,T],H^1(\mathbf{R}^n))$. Uniqueness follows from inequality (4.1).

It remains to establish the conservation laws. Conservation of charge follows from (4.3) and (4.9), applied with $(q,r)=(\infty,2)$. For the conservation of energy, the argument is more delicate. First, observe that by (4.9) (applied with $(q,r)=(\infty,2)$), (4.5), Sobolev and Hölder inequalities, we have,

$$u_m \to u \text{ in } L^q(0,\tau,L^{2n/(n-2)}(\mathbf{R}^n)), \text{ for every } \tau<\infty,\ \tau\leq T,\ 2<q<\infty. \tag{4.11}$$

It follows from (4.11) that there exists a subsequence, which we still denote by $(u_m)_{m\in\mathbf{N}}$, such that

$$u_m(t) \to u(t) \text{ in } L^{2n/(n-2)}(\mathbf{R}^n), \text{ a.e. in } (0,T).$$

Thus, $G_m(u_m)\to G(u)$, a.e. in $(0,T)$.Using lower semicontinuity in the gradient term, we can pass to the limit in (4.3) to get

$$E(u(t)) \leq E(\varphi), \text{ for a.e. } t\in(0,T).$$

Thus, from the continuity in $H^1(\mathbf{R}^n)$,

$$E(u(t)) \leq E(\varphi), \text{ for } t\in[0,T].$$

Considering the reverse equation, which enjoys the same properties, and in particular uniqueness, it is not difficult to show the converse inequality. This completes the proof of Proposition 2.

PROPOSITION 3. Let $T\in(0,\infty)$ and $\varphi\in H^1(\mathbf{R}^n)$. Let (q,r) be an admissible pair with $2<r<n$ and let $u\in L^q(0,T,L^{r^*}(\mathbf{R}^n))\cap L^\infty(0,T,H^1(\mathbf{R}^n))$ be a solution of (1.1). Then $u\in L^\mu(0,T,W^{1,\nu}(\mathbf{R}^n))$, for every admissible pair (μ,ν).

PROOF. Observe that by Lemma 3, $u\in C([0,T],H^{-1}(\mathbf{R}^n))$; and so u is weakly continuous in $H^1(\mathbf{R}^n)$. Thus, for every $t\in[0,T]$ and $s\in[0,T-t]$, the following holds in $H^1(\mathbf{R}^n)$.

$$u(t+s) = S(s)u(t) + \mathcal{F}(u(t+\cdot))(s). \tag{4.12}$$

Consider a family $(\nabla_h)_{h\in(0,1)}$ of difference quotients approximating ∇ and such that

$$\|\nabla_h w\|_{L^2} \leq C\, \|\nabla w\|_{L^2}, \text{ for every } w\in H^1(\mathbf{R}^n). \tag{4.13}$$

Let (a,b) the admissible pair associated with (q,r) by Lemma 3. It follows from (4.12), Lemma 1, (4.13) and (2.4) that

$$\|\nabla_h u\|_{L^a(t,t+s,L^b)} \leq C\, \|u\|_{L^\infty(0,T,H^1)} + C\, \|u\|^\alpha_{L^q(t,t+s,L^{r^*})} \|\nabla_h u\|_{L^a(t,t+s,L^b)}\ .$$

Therefore, there exists $\eta>0$ such that if $\|u\|_{L^q(t,t+s,L^{r^*})} \leq \eta$, then $\|\nabla_h u\|_{L^a(t,t+s,L^b)} \leq 2C\,\|u\|_{L^\infty(t,t+s,H^1)}$. Letting $h\to 0$, we get $\nabla u\in L^a(t,t+s,L^b(\mathbf{R}^n))$. Breaking $[0,T]$ into a finite number of intervals I_j such that $\|u\|_{L^q(I_j,L^{r^*})}\leq\eta$, we obtain $\nabla u\in L^a(0,T,L^b(\mathbf{R}^n))$. The result now follows from Lemma 1 and Corollary 1.

PROOF OF THEOREM 2. Uniqueness follows from Propositions 2 and 3.

Let now $\varphi\in H^1(\mathbf{R}^n)$. Applying iteratively Proposition 2, we can construct the maximal solution $u\in C([0,T^*),H^1(\mathbf{R}^n))\cap L^\gamma_{loc}([0,T^*),W^{1,\rho}(\mathbf{R}^n))$ of (1.1). Note that we also have the conservation laws and that $u\in L^q_{loc}([0,T^*),W^{1,r}(\mathbf{R}^n))$, for every admissible pair (q,r). Since (1.1) is equivalent to (NLS), we obtain $u\in C^1([0,T^*),H^{-1}(\mathbf{R}^n))$. It remains to establish property (iv) of the theorem. We argue by contradiction; and so we assume $T^*<\infty$ and $\|u\|_{L^q(0,T^*,W^{1,r})}<\infty$, for some admissible pair (q,r). By Proposition 3, it follows that $\|u\|_{L^\gamma(0,T^*,W^{1,\rho})}<\infty$. Let $t\in[0,T^*)$., we have

$$S(s)u(t) = u(t+s) - \mathcal{F}(u(t+\cdot))(s)\,,\ \text{for every } s\in[0,T^*-t). \tag{4.14}$$

From (4.14) and Corollary 1, we obtain

$$\|\nabla S(\cdot)u(t)\|_{L^\gamma(0,T^*-t,L^\rho)} \leq \|\nabla u\|_{L^\gamma(t,T^*,L^\rho)} + C(\|\nabla u\|_{L^\gamma(t,T^*,L^\rho)})^{\alpha+1}\,.$$

Therefore, for t close enough to T^*, we have

$$\|\nabla S(\cdot)u(t)\|_{L^\gamma(0,T^*-t,L^\rho)} < \delta.$$

Applying Proposition 2, we find that u can can be extended after T^*, which contradicts the maximality. The proof of Theorem 1 is thereby completed.

REMARK 7. It follows from Proposition 2 and Lemma 1 that if $\|\nabla\varphi\|_{L^2}$ is small enough, we have $T^*=\infty$, and $u\in L^q(0,T,W^{1,r}(\mathbf{R}^n))$ for every admissible pair (q,r). Observe that this remark includes initial data that can be large in $H^1(\mathbf{R}^n)$ since there is no assumption on $\|\varphi\|_{L^2}$. On the other hand, it is well known that if $\lambda<0$, there exists initial data for which $T^*<\infty$. We conjecture that if $\lambda\geq 0$, then $T^*=\infty$. The underlying problem for showing this is that a bound of the solution in $L^\infty(0,T,H^1(\mathbf{R}^n))$ does not immediately imply a bound in $L^q(0,T,W^{1,r}(\mathbf{R}^n))$.

REMARK 8. We could have treated the Cauchy problem in $H^1(\mathbf{R}^n)$ by a contraction method. This is the point of view that we adopt in [3]. However, there is at some stage an approximation argument, in order to prove conservation of energy.

REMARK 9. For results concerning continuous dependence of the solutions on the initial data, see [3], Theorem 1.2.

REFERENCES.

[1] H. BERESTYCKI and P.L. LIONS, Nonlinear scalar field equations. Arch. Rat. Mech. Anal., **82** (1983), 313-375.

[2] T. CAZENAVE and F. B. WEISSLER, The Cauchy problem for the nonlinear Schrödinger equation in H^1. Manuscripta Math., **61** (1988), 477-494.

[3]T. CAZENAVE and F. B. WEISSLER, The Cauchy problem for the critical nonlinear Schrödinger equation in H^s. Nonlinear Anal. T.M.A., to appear.

[4] J. GINIBRE and G. VELO, On a class of nonlinear Schrödinger equations. J. Funct. Anal., **32** (1979), 1-71.

[5] J. GINIBRE and G. VELO, The global Cauchy problem for the nonlinear Schrödinger equation revisited. Ann. Inst. Henri Poincaré, Analyse Non Linéaire, **2** (1985), 309-327.

[6] J. GINIBRE and G. VELO, Scattering theory in the energy space for a class of nonlinear Schrödinger equations. J. Math. Pures Appl., **64** (1985), 363-401.

[7] T. KATO, On nonlinear Schrödinger equations. Ann. Inst. Henri Poincaré, Physique Théorique, **46** (1987), 113-129.

[8] Y. TSUTSUMI, L^2-solutions for nonlinear Schrödinger equations and nonlinear groups. Funk. Ekva., **30** (1987), 115-125.

[9] Y. TSUTSUMI, Scattering problem for nonlinear Schrödinger equations. Ann. Inst. Henri Poincaré, Physique Théorique, **43** (1985), 321-347.

[10] M. WEINSTEIN, On the structure and formation of singularities of solutions to nonlinear dispersive evolution equations. Comm. PDE, **11** (1986), 545-565.

[11] K. YAJIMA, Existence of solutions for Schrödinger evolution equations. Comm. Math. Phys., **110** (1987), 415-426.

Research supported by NSF grants DMS 8201639 and DMS 8703096.

ON THE INTEGRABILITY OF NONLINEAR EVOLUTION EQUATIONS

H. H. Chen
Laboratory for Plasma and Fusion Energy Studies
University of Maryland. College Park, MD 20742, USA
and
J. E. Lin
Department of Mathematical Sciences
George Mason University. Fairfax, VA 22030, USA

Abstract

We propose a new method to find the spectral problem for the inverse scattering equations for integrable nonlinear evolution equations. This approach is able to construct the Miura transformation directly and utilize a factorization scheme to construct a linear recursion operator of gradients of constants of motion which can be taken as the spectral operator of a pair of Lax operators for the integrable equation. We have applied this method to the KdV, MKdV, and sine-Gordon equations. We have also obtained the Lax pair operators for their modified equations.

1. Introduction

It is now well known that a large class of nonlinear evolution equations such as the Korteweg- de Vries (KdV) equation,

$$u_t = 6uu_x + u_{xxx}, \tag{1}$$

is completely integrable [1-8]. Gardner, Greene, Kruskal, and Miura [1] first showed that Eq. (1) is the compatibility condition of two linear equations

$$L\phi = (\partial_x^2 + u)\phi = \lambda\phi \tag{2.1}$$

and

$$A\phi = (4\partial_x^3 + 6u\partial_x + 3u_x)\phi = \phi_t. \tag{2.2}$$

The so-called Lax pair [2], L and A, satisfies the operator equation

$$L_t = [A, L] = AL - LA. \tag{3}$$

Historically, Eq. (2.1) was found through the Miura [9] transformation,

$$u = M(v) = -(1/2)v_x - (1/4)v^2 + \lambda, \tag{4}$$

which relates the KdV equation (1) to the Modified KdV (MKdV) equation,

$$v_t = -(3/2)v^2 v_x + v_{xxx} + 6\lambda v_x, \tag{5}$$

and Eq. (2.1) was obtained from (4) by letting

$$v = 2\phi_x/\phi.$$

Actually, Miura [9] found this transformation by carefully examining the first few conservation laws that were known for both KdV and MKdV equations. This is certainly not the desired method to search the Lax pair for other integrable equations.

About ten years ago, one of the authors (HHC) together with Y. C. Lee and C. S. Liu [10] had proposed a method to systematically construct the infinite set of conservation laws directly from the linear variational equation (see Definition 3 below) of the original equation. They also used a linear mode coupling scheme to construct the spectral problem which together with the linear variational equation constitutes the Lax pair for the inverse scattering method. In this paper, we would like to propose a second method to find the spectral problem. The new approach is able to construct the Miura transformation directly and utilize a factorization scheme to construct the linear recursion operator of the symmetries (see Definition 3 below) which is also the spectral operator that we are seeking for. The adjoint of this pair consists of the eigenvalue problem of the linear recursion operator of gradients (see Definition 1 below) of constants of motion together with the adjoint of the linear variational equation.

Before we start, let us clarify some of the terminologies that are used heavily in this paper.

Definition 1. [2,11] Given a functional $I = \int_{-\infty}^{\infty} q(u)dx$ where $u = u(x,\alpha)$ is its argument function which depends on a parameter α as well as on the variable x and vanishes sufficiently fast at infinity. The gradient (functional derivative) of I with respect to u, $\delta I/\delta u$, is defined by

$$(d/d\alpha)I\{u\} = \int_{-\infty}^{\infty} (\delta I/\delta u)(\partial u/\partial\alpha)dx.$$

Definition 2. Given a function $F(u) = F(x,t,u,u_x,\ldots)$ which depends on u and its partial derivatives and possibly on the variables x and t,

$$F'[v] \equiv \partial F(u + \varepsilon v)/\partial\varepsilon|_{\varepsilon=0}$$

is the Gateaux derivative of F in the direction v with respect to u. Therefore, the operator F' which depends on u acts on the function v.

Definition 3. Given a nonlinear evolution equation,

$$u_t = K(u),$$

where K(u) depends on u and its partial derivatives and possibly on the variables x and t. A function $\sigma(x,t,u,u_x,\ldots)$ is called a symmetry of the equation if σ satisfies the linear variational equation of $u_t = K(u)$:

$$(\partial/\partial t)\sigma = K'[\sigma], \tag{6}$$

where $K'[\sigma]$ is defined as in Definition 2. These symmetries are the infinitesimal generators of one-parameter groups of invariant transformations of the equation. A classical theorem of Noether [12] states that constants of motion can usually be constructed from the symmetries of the Lagrangian.

We will denote the adjoint of an operator N by N*. All the solutions of equations in the following work are assumed to vanish sufficiently fast at infinity unless otherwise stated.

2. The new approach

(I) Background

Given an evolution equation $u_t = K(u)$, it is known [2] that the gradient ψ of the constants of motion of this equation satisfies the adjoint of the linear variational equation (6) of the original equation:

$$\psi_t = -K'^*(\psi). \tag{7}$$

On the other hand, it is known that an integrable equation possesses an infinite set of polynomial conservation laws, therefore, there must exist infinitely many polynomial solutions $\psi = \psi(x,t,u,u_x,\ldots)$ of (7). Taking Eq. (7) as one member of the Lax equations, it is proposed in [10] a method to construct infinitely many constants of motion for integrable nonlinear evolution equations. They used a WKB-type expansion of the solution ψ of Eq. (7),

$$\psi = \exp(kx + \omega t + \textstyle\int_{-\infty}^{x} T dx) \tag{8}$$

with $T = T(u) = \sum_{n=0}^{\infty} k^{-n} T_n$, where k and ω satisfy the linear dispersion relation. Take KdV equation (1) as an example, Eq. (7) becomes

$$\psi_t = 6u\psi_x + \psi_{xxx}. \tag{9}$$

Substituting (8) into (9) yields a recursive relation for T_n's:

$$\int_{-\infty}^{x} (T_n)_t dx = 6u\delta_{n+1,0} + 6uT_n + (T_n)_{xx} + 3(T_{n+1})_x + 3\sum_{j=0}^{n} T_j (T_{n-j})_x + \sum_{j+m=0}^{n} T_j T_m T_{n-j-m} + 3\sum_{j=0}^{n+1} T_j T_{n+1-j} + 3T_{n+2}, \tag{10}$$

where we use $\delta_{i,j}$ to denote the Kronecker delta. The above formula can be used to generate T_{n+2} (the last term on the right-hand side) from the knowledge of T_j's, j = 0, 1, 2,..., n+1. Assuming $T_{-2} = 0 = T_{-1}$, we have, from (10),

$$T_0 = 0,$$

$$T_1 = -2u,$$

$$T_2 = 2u_x,$$

$$T_3 = -2u^2 - 2u_{xx},$$

$$T_4 = (4u^2 + 2u_{xx})_x,$$

$$T_5 = -2u_{xxxx} - 10u_x^2 - 12uu_{xx} - 4u^3,$$

etc.. Note that if $T_0, T_1, \ldots, T_{n+2}$ are differential polynomials of u, then $\int_{-\infty}^{\infty} (T_n)_t dx = 0$ provided u vanishes sufficiently fast at $x = \pm\infty$ and we have a conservation law $\int_{-\infty}^{\infty} T_n dx$ = constant. If this process can go on forever, we then have an infinite set of conservation laws and whether this process can go on depends on whether $\int_{-\infty}^{x} T_n dx$ can be expressed in terms of differential polynomials of u. In fact, for general nonlinear evolution equations, this criterion can serve as a test to distinguish an integrable system from a nonintegrable one. For example, the quartic KdV equation,

$$w_t = 6w^3 w_x + w_{xxx},$$

would not pass this test. We can show immediately that

$$T_1 \text{ (quartic KdV equation)} = -2w^3,$$

but w^3 is not a conserved density at all. The recursive formula therefore is disrupted at n = 1. On the other hand, for the integrable system like the KdV equation (1), we can go on to find the higher-order members of the conserved densities without disruption. However, from Eq. (10) alone, we can not establish the fact that the recursive operation could go on forever. More information is needed to do that. If we can find a nontrivial linear recursion operator of the solutions of (9), which are gradients of constants of motion for the KdV equation (1), then this would be enough to establish the existence of infinitely many conservation laws provided there is at least one nontrivial polynomial conservation law for the equation. This recursion operator can also serve as the spectral operator for a pair of Lax operators (see Remark 1 and Remark 2 below). It is therefore our goal in this paper to construct this linear recursion operator.

Remark 1. The linear recursion operator R_a of gradients of constants of motion acts on ψ_n and generates ψ_{n+1} where ψ_n and ψ_{n+1} are solutions of Eq. (7), that is to say,

$$\psi_{n+1} = R_a\psi_n, \tag{11a}$$

$$(\partial/\partial t)\psi_n = -K'^*(\Psi_n), \text{ and} \tag{11b}$$

$$(\partial/\partial t)\psi_{n+1} = -K'^*(\psi_{n+1}). \tag{11c}$$

The consistency of Eqs. (11a), (11b), and (11c) requires that

$$(\partial/\partial t)R_a = [-K'^*, R_a] = -K'^*R_a - R_a(-K'^*)$$

which is exactly the Lax condition (3) for the integrability of nonlinear evolution equations. Thus the linear recursion operator R_a plays the role of the spectral operator while $-K'^*$ plays the role of the temporal evolution operator.

Remark 2. There is also a recursion operator R_s acting on the symmetries of the equation $u_t = K(u)$ and this operator together with K' also form a Lax pair,

$$(\partial/\partial t)R_s = [K', R_s] = K'R_s - R_sK'.$$

Note that R_s plays the role of the spectral operator and K' plays the role of the temporal evolution operator.

Remark 3. The existences and applications of recursion operators for the nonlinear integrable evolution equations with one-spatial dimension have been discussed in many literatures, see, for example, [10, 13-17].

(II) A new twist

We would like to extend Eq. (8), re-interprete it, and construct from it a linear recursion operator of the gradients of constants of motion for the KdV equation (1). Note first that Eq. (8) implies that

$$\psi_x - p\psi = 0$$

where $p = k + T$. We now propose to add a new term ξ to the above equation for ψ,

$$\psi_x - p\psi = \xi. \tag{12}$$

Proposition: ξ is a solution of the adjoint of the linear variational equation of another nonlinear evolution equation.

Demanding that Eq. (12) be compatible with (9), we obtain

$$p_t = 6u_xp + 6up_x + p_{xxx} + 3p_{xx}p + 3p^2p_x + 3p_x^2 \tag{13}$$

and

$$\xi_t = \xi_{xxx} + (6u + 3p_x)\xi_x + (6u_x + 3p_{xx} + 3p_xp)\xi. \tag{14}$$

We now assume that Eq. (14) is the adjoint of the linear variational equation of a nonlinear evolution equation (see the proposition above), say,

$$v_t = m(v,v_x,\ldots). \tag{15}$$

Of course, $u = u(v,v_x,\ldots)$ and $p = p(v,v_x,\ldots)$. Taking the adjoint of (14), we get

$$\eta_t = \eta_{xxx} + (6u + 3p_x)\eta_x - 3p_xp\eta$$

which should be the linear variational equation of (15). Thus we have

$$v_t = v_{xxx} + F(v,v_x)$$

for some function F which depends only on v and v_x. We immediately see that

$$\begin{cases} \partial F/\partial v = -3pp_x & (16.1)\\ \partial F/\partial v_x = 6u + 3p_x. & (16.2)\end{cases}$$

It follows that $u = u(v,v_x)$ depends only on v and v_x and $p = p(v)$ depends only on v. The compatibility of (16.1) and (16.2) now yields

$$-3pp' = 6(\partial u/\partial v) + 3p''v_x,$$

where $p' = dp/dv$, $p'' = d^2p/dv^2$, and, in general, $p^{(n)} = d^np/dv^n$. Hence,

$$u = -p^2/4 - p_x/2 + f(v_x) \tag{17}$$

for some function f which depends only on v_x. Also from (16.1), we have

$$F = -(3/2)p^2v_x + g(v_x)$$

for some function g which depends only on v_x. From (16.2) and (17), we have

$$-(3/2)p^2 + dg/dv_x = 6u + 3p_x = -(3/2)p^2 + 6f(v_x).$$

Hence

$$dg/dv_x = 6f(v_x). \tag{18}$$

Recall that the equation for p, given in (13),

$$\begin{aligned} p_t &= 6u_xp + 6up_x + p_{xxx} + 3p_{xx}p + 3p_x^2 + 3p^2p_x \\ &= -(3/2)p^2p_x + p_{xxx} + 6(pf)_x, \end{aligned} \tag{19}$$

has to be consistent with

$$v_t - v_{xxx} = F(v,v_x) = -(3/2)p^2 v_x + g(v_x). \tag{20}$$

Substituting $p = p(v)$ into (19), we get

$$v_t - v_{xxx} = -(3/2)p^2 v_x + (p'''/p')v_x^3 + 6v_x f$$
$$+ 3((p''/p')v_x + 2(p/p')(df/dv_x))v_{xx}. \tag{21}$$

Hence, we have from (20) and (21)

$$\begin{cases} g(v_x) = (p'''/p')v_x^3 + 6v_x f & (22.1) \\ p''v_x + 2(df/dv_x)p = 0. & (22.2) \end{cases}$$

Now, (18) and (22.1) imply that

$$3(p'''/p')v_x^2 + 6v_x(df/dv_x) = 0,$$

hence, from (22.2),

$$p''p' = pp'''. \tag{23}$$

On the other hand, (22.2) implies

$$p''/p = -(2/v_x)(df/dv_x) = c, \tag{24}$$

where c is a constant, since the left-hand side and the middle term of (24) depend only on v and v_x, respectively. Note that (24) implies (23). Thus we have

$$f = -(c/4)v_x^2 + \lambda$$

for some constant λ. Eq. (17) and Eq. (21) then imply that

$$u = -(p^2/4) - (p_x/2) - (c/4)v_x^2 + \lambda \tag{25}$$

and

$$v_t - v_{xxx} = -(3/2)p^2 v_x - (c/2)v_x^3 + 6\lambda v_x, \tag{26}$$

respectively. In order for (25) to be a Miura transformation relating (1) to (26), that is to say,

$$u_t - 6uu_x - u_{xxx} = M'(v_t - v_{xxx} + (3/2)p^2 v_x + (c/2)v_x^3 - 6\lambda v_x),$$

where

$$M' = -(1/2)pp' - (1/2)p''v_x - (1/2)p'\partial_x - (c/2)v_x\partial_x,$$

we need $c = 0$. Hence, we have $p'' = 0$, that is to say,

$$p = av + b,$$

for some constants a and b. Therefore, (25) becomes the Miura transformation (4),

$$u = -(1/4)(av + b)^2 - (1/2)av_x + \lambda,$$

and (26) becomes the MKdV equation (5),

$$v_t - v_{xxx} = -(3/2)(av + b)^2 v_x + 6\lambda v_x.$$

(III) A construction of Lax-pair operators

Lemma 1. Suppose that L, A, and B are operators depending on t and related by $L_t = BL - LA$. Then $L\eta$ satisfies $\xi_t = B\xi$ if η satisfies $\eta_t = A\eta$.

Proof. $(L\eta)_t = L_t\eta + L\eta_t = (BL - LA)\eta + LA\eta = BL\eta$.

Lemma 2. Suppose that two evolution equations $u_t = K(u)$ and $v_t = G(v)$ are related by a transformation $u = M(v)$, then

(a) M' maps the symmetries of $v_t = G(v)$ to those of $u_t = K(u)$ and

(b) M'^* maps the gradients of constants of motion of $u_t = K(u)$ to those of $v_t = G(v)$.

Proof. (a) Let $u \to u + \varepsilon\sigma$ and $v \to v + \varepsilon\theta$ be two infinitesimal transformations of the solutions of $u_t = K(u)$ and $v_t = G(v)$, respectively. Then, from $u = M(v)$, we obtain

$$\sigma = M'\theta. \tag{27}$$

Note that σ and θ are symmetries (Definition 3) of $u_t = K(u)$ and $v_t = G(v)$, respectively. Thus we have obtained (a).

(b) Differentiating with respect to t on both sides of (27), we get

$$\sigma_t = (M'\theta)_t = (M')_t\theta + M'\theta_t.$$

Now, by the definition of symmetries, σ and θ satisfy $\sigma_t = K'\sigma$ and $\theta_t = G'\theta$, respectively. Hence

$$K'M'\theta = K'\sigma = (M')_t\theta + M'G'\theta,$$

that is to say,

$$K'M' = (M')_t + M'G'.$$

Taking adjoints on both sides, we obtain

$$(M'^*)_t = M'^*K'^* - G'^*M'^* = (-G'^*)M'^* - M'^*(-K'^*).$$

By Lemma 1, $M'^*\eta$ satisfies $\xi_t = -G'^*\xi$ if η satisfies $\eta_t = -K'^*\eta$. By a result of Lax [2], $\eta = \eta(x,t,u,u_x,\ldots)$ and $\xi = \xi(x,t,v,v_x,\ldots)$ are the gradients of constants of motion of $u_t = K(u)$ and $v_t = G(v)$, respectively. Thus (b) is proved.

Theorem. In addition to the assumption of Lemma 2, if equations $u_t = K(u)$ and $v_t = G(v)$ have independent Hamiltonian structures (i.e., none of them can be derived from the other by the relation $u = M(v)$), then we can construct the linear recursion operators of symmetries and gradients of constants of motion for both equations. In fact, let

J_1 and J_2 be the Hamiltonian operators of these two independent† Hamiltonian systems for $u_t = K(u)$ and $v_t = G(v)$, respectively, then

$$R_s = M'J_2M'^*J_1^{-1}$$

and

$$U_s = J_2M'^*J_1^{-1}M'$$

are linear recursion operators of symmetries of equations $u_t = K(u)$ and $v_t = G(v)$, respectively, and

$$R_a = J_1^{-1}M'J_2M'^*$$

and

$$U_a = M'^*J_1^{-1}M'J_2$$

are linear recursion operators of gradients of constants of motion of equations $u_t = K(u)$ and $v_t = G(v)$, respectively.

(† This implies that $J_1 \neq M'J_2M'^*$. Thus R_s, U_s, R_a, and U_a are not the identity operators).

Proof. Since the Hamiltonian operator maps the gradients of constants of motion of the equation to its symmetries, we see from Lemma 2 that symmetries of $u_t = K(u)$ are mapped by J_1^{-1} to the gradients of constants of motion of $u_t = K(u)$, which are then mapped by M'^* to the gradients of constants of motion of $v_t = G(v)$, which are then mapped by J_2 to the symmetries of $v_t = G(v)$, which are then mapped by M' to the symmetries of $u_t = K(u)$. Thus $R_s = M'J_2M'^*J_1^{-1}$ maps symmetries of $u_t = K(u)$ to those of $u_t = K(u)$, that is to say, R_s, which is obviously linear, is a recursion operator of symmetries for $u_t = K(u)$. This proves a part of the assertion. The rest of the assertion can be proved similarly.

Remark 4. As mentioned in Remark 1 and Remark 2, we now have the following two pairs of Lax operators (3) for $u_t = K(u)$:

R_s with K'

and

R_a with $-K'^*$.

Likewise,

U_s with G'

and

U_a with $-G'^*$

are two pairs of Lax operators for $v_t = G(v)$.

Now, the KdV and MKdV equations (1) and (5) are related by the Miura transformation (4) and, moreover, they have the Hamiltonian structures $u_t = J_1(\delta H_1/\delta u)$ and $v_t = J_2(\delta H_2/\delta v)$, respectively, where the Hamiltonian operators

$$J_1 = \partial_x = J_2 \tag{28}$$

and the Hamiltonians

$$H_1 = \int_{-\infty}^{\infty} (u^3 - (1/2)u_x^2)dx$$

and

$$H_2 = \int_{-\infty}^{\infty} (-(1/8)v^4 - (1/2)v_x^2 + 3\lambda v^2)dx.$$

Therefore, from the result of the above theorem, we have

Corollary 1. $R_s = M'J_2M'^*J_1^{-1} = -(1/4)(\partial_x^2 + 2u_x\partial_x^{-1} + 4u) + \lambda$, where $M' = -(1/2)(\partial_x + v)$ and is derived from (4), is a linear recursion operator of symmetries of the KdV equation (1).

Corollary 2. $R_a = J_1^{-1}M'J_2M'^* = -(1/4)(\partial_x^2 - 2\partial_x^{-1}\cdot u_x + 4u) + \lambda$ is a recursion operator of gradients of constants of motion of the KdV equation (1).

Corollary 3. Similarly, $U_s = J_2M'^*J_1^{-1}M' = -(1/4)(\partial_x^2 - v_x\partial_x^{-1}\cdot v - v^2)$ and $U_a = M'^*J_1^{-1}M'J_2 = -(1/4)(\partial_x^2 + v\partial_x^{-1}\cdot v_x - v^2)$ are linear recursion operators of symmetries and gradients of constants of motion of the MKdV equation (5), respectively.

Remark 5. From Remark 4, we now have the following two pairs of Lax operators for the KdV equation (1):

(i) R_s with $A = 6u_x + 6u\partial_x + \partial_x^3$ and

(ii) R_a with $B = -A^* = 6u\partial_x + \partial_x^3$.

Likewise, we also have the following two pairs of Lax operators for the MKdV equation (5):

(iii) U_s with $A_m = -3vv_x - (3/2)v^2\partial_x + \partial_x^3 + 6\lambda\partial_x$ and

(iv) U_a with $B_m = -A_m^* = -(3/2)v^2\partial_x + \partial_x^3 + 6\lambda\partial_x$.

Remark 6. The set of the inverse scattering equations (2.1) and (2.2) associated with the Lax operators in the Remark 5 can be reduced to the more familiar set of the inverse scattering equations [1-8]. For example, the set of the inverse scattering equations associated with (i) of the Remark 5,

$$\begin{cases} \phi_{xx} + 2u_x\partial_x^{-1}\phi + 4u\phi = 4\lambda\phi \\ \phi_t = \phi_{xxx} + 6u\phi_x + 6u_x\phi, \end{cases}$$

is related to the more familiar set [1,2],

$$\begin{cases} \psi_{xx} + u\psi = \lambda\psi \\ \psi_t = 4\psi_{xxx} + 6u\psi_x + 3u_x\psi, \end{cases}$$

by

$$\phi = \psi\psi_x.$$

3. Other integrable equations

We can apply the above method to other integrable equations such as MKdV equation and sine-Gordon equation.

(A) MKdV equation (5). We shall set $\lambda = 0$ in (5) for simplicity. Applying the above method, we find that MKdV equation and Modified MKdV (MMKdV) equation [18],

$$r_t = (3/2)\alpha^2 r_x \sin^2 r + (1/2)r_x^3 + r_{xxx}, \tag{29}$$

are related by

$$v = N(r) = i(\alpha \sin r + r_x) \tag{30}$$

where α is a constant.

Again by noting that the MMKdV equation (29) has the Hamiltonian structure

$$r_t = J(\delta/\delta r)H,$$

where

$$J = \partial_x + r_x\partial_x^{-1}\cdot r_x \tag{31}$$

and

$$H = -(1/2)\int_{-\infty}^{\infty} (\alpha^2 \sin^2 r + r_x^2 + 2\alpha r_x \sin r + 2\alpha^2(\cos r - 1))dx,$$

we have

Remark 7. $R_{sm} = N'JN'^*J_2^{-1} = \partial_x^2 - v^2 - v_x\partial_x^{-1}\cdot v - \alpha^2,$

where $N' = i(\alpha\cos r + \partial_x)$ is from (30), J is from (31), and J_2 is from (28), and

$$\begin{aligned} U_{smm} &= JN'^*J_2^{-1}N' \\ &= \partial_x^2 + r_x\partial_x^{-1}\cdot r_x\partial_x - \alpha^2\cos^2 r + \alpha^2 r_x\partial_x^{-1}\cdot \sin r \cos r \end{aligned}$$

are linear recursion operators of symmetries and

$$R_{am} = J_2^{-1}N'JN'^*$$

$$= \partial_x^2 - v^2 + v\partial_x^{-1}\cdot v_x - \alpha^2$$

and

$$U_{amm} = N'{}^*J_2^{-1}N'J$$
$$= \partial_x^2 + r_{xx}\partial_x^{-1}\cdot r_x + r_x^2 - \alpha^2\cos^2 r + (\alpha^2\cos r \sin r)\partial_x^{-1}\cdot r_x$$

are linear recursion operators of gradients of constants of motion for the MKdV equation (5) and MMKdV equation (29), respectively.

Remark 8. We now have the following two pairs of Lax operators for the MMKdV equation (29):

$$U_{smm} \text{ with } A_{mm} = (3/2)\alpha^2(\sin^2 r)\partial_x + 3\alpha^2 r_x \sin r \cos r + (3/2)r_x^2\partial_x + \partial_x^3$$

and

$$U_{amm} \text{ with } -A_{mm}^* = (3/2)\alpha^2(\sin^2 r)\partial_x + 3r_x r_{xx} + (3/2)r_x^2\partial_x + \partial_x^3.$$

(B) Sine-Gordon (SG) equation,

$$q_{xt} = \sin q. \tag{32}$$

Applying the above method, we find that the sine-Gordon equation (32) and the Modified sine-Gordon (MSG) equation [18],

$$\begin{cases} w_t = \sin z \\ z_x = \sin w, \end{cases} \tag{33}$$

are related by

$$q = Q(w) = w + \partial_x^{-1}\sin w. \tag{34}$$

Again by noting that the SG equation (32) and the MSG equation (33) have the following Hamiltonian structures

$$q_t = J_s(\delta/\delta q)H_s$$

and

$$w_t = J_m(\delta/\delta w)H_m,$$

respectively, where

$$J_s = \partial_x^{-1}, \tag{35}$$

$$H_s = \int_{-\infty}^{\infty} (1 - \cos q)dx,$$

$$J_m = \partial_x + w_x\partial_x^{-1}\cdot w_x, \tag{36}$$

and

$$H_m = \int_{-\infty}^{\infty} (\cos q - \cos w)dx,$$

we have

Remark 9. $R_{ssg} = Q'J_mQ'^*J_s^{-1}$

$$= \partial_x^2 + q_x^2 - q_x\partial_x^{-1}\cdot q_{xx} - 1$$

and

$$U_{sms} = J_mQ'^*J_s^{-1}Q'$$

$$= \partial_x^2 + w_x\partial_x^{-1}\cdot w_x\partial_x - \cos^2 w + w_x\partial_x^{-1}\cdot \sin w \cos w$$

are linear recursion operators of symmetries for SG equation(32) and MSG equation (33), respectively, and

$$R_{asg} = J_s^{-1}Q'J_mQ'^*$$

$$= \partial_x^2 + q_x^2 + q_{xx}\partial_x^{-1}\cdot q_x - 1$$

and

$$U_{ams} = Q'^*J_s^{-1}Q'J_m$$

$$= \partial_x^2 + w_{xx}\partial_x^{-1}\cdot w_x + w_x^2 - \cos^2 w - (\cos w \ \sin w)\partial_x^{-1}\cdot w_x$$

are linear recursion operators of gradients of constants of motion for the SG equation (32) and the MSG equation (33), respectively, where J_s is from (35), J_m is from (36), and $Q' = 1 + \partial_x^{-1}\cdot \cos w$ is from (34).

Remark 10. We now have the following two pairs of Lax operators for the SG equation (32):

$$R_{ssg} \text{ with } A_s = \partial_x^{-1}\cdot \cos q$$

and

$$R_{asg} \text{ with } -A_s^* = (\cos q)\partial_x^{-1}$$

Likewise, we have the following two pairs of Lax operators for the MSG equation (33):

$$U_{sms} \text{ with } A_{sm} = (\cos(\partial_x^{-1}\sin w))\partial_x^{-1}\cdot \cos w$$

and

$$U_{ams} \text{ with } -A_{sm}^* = (\cos w)\partial_x^{-1}\cdot(\cos(\partial_x^{-1}\sin w)).$$

Acknowledgments

The work of H. H. Chen is supported in part by NSF.

References

1. Gardner, C., Greene, J., Kruskal, M. and Miura, R., Commun. Pure Appl. Math. 27, 97-133 (1974).

2. Lax, P., Commun. Pure Appl. Math. 21, 467-490 (1968).

3. Scott, A. C., Chu, F. Y. F. and McLaughlin, D. W., Proc. IEEE, 61, 1443-1483 (1973).

4. Ablowitz, M. J., Kaup, D. J., Newell, A. C., and Segur, H., Studies in Appl. Math. 53, 249-315 (1974).

5. Miura, R. M., SIAM Review 18, 412-459 (1976).

6. Ablowitz, M. J. and Segur, H., Solitons and the Inverse Scattering Transform (SIAM, Philadelphia 1981).

7. Newell, A. C., Solitons in Mathematics and Physics (SIAM, Philadelphia 1985).

8. Novikov, S., Manakov, S., Pitaevskii, L. and Zakharov, V. E., Theory of Solitons (Consultants Bureau, New York 1984).

9. Miura, R., J. Math. Phys. 9, 1202-1204 (1968).

10. Chen, H. H., Lee, Y. C. and Liu, C. Ş., Physica Scripta 20, 490-492 (1979).

11. Gardner, C. S., J. Math. Phys. 12, 1548-1551 (1971).

12. Arnold, V. I., Mathematical Methods of Classical Mechanics (Springer, New York 1978).

13. Lax, P., SIAM Review 18, 351-375 (1976).

14. Chen, H. H., Lee, Y. C. and Lin, J. E., Pysica 26D, 165-170 (1987).

15. Fuchssteiner, B. and Fokas, A. S., Physica 4D, 47-66 (1981).

16. McKean, H. P., in Topics in Functional Analysis (Academic Press, New York 1978) 217-226.

17. Zhu, G. C. and Chen, H. H., J. Math. Phys. 27, 100-103 (1986).

18. Chen, H. H., in Backlund Transformations (Lecture Notes in Mathematics, Vol. 515. Springer-Verlag, New York 1976) 241-252.

On Quasilinear Hyperbolic Integrodifferential Equations in Unbounded Domains

Hans Engler
Department of Mathematics
Georgetown University
Washington, D.C. 20057, U.S.A.

1. Introduction

In this note, we study the existence and some asymptotic properties of weak solutions for quasilinear integrodifferential equations of the form

(1.a)
$$\partial_t^2 u^i(x,t) = div_x \sigma^i(x,t) + f^i(x,t) = \sum_j \partial_{x_j} \sigma_j^i(x,t) + f^i(x,t) \quad ,$$

(1.b)
$$\sigma_j^i(x,t) = g_j^i(\nabla u(x,t)) + \int_0^\infty a(s) h_j^i(\nabla u(x,t), \nabla u(x,t-s))\, ds$$

for $(x,t) \in \Omega \times [0,T]$, where $\Omega \subset \mathbf{R}^n$ is some unbounded domain. Here $u : \Omega \times [0,T] \to \mathbf{R}^k$ $(k \geq 1)$ is the unknown function, $g : \mathbf{R}^{nk} \to \mathbf{R}^{nk}$, $h : \mathbf{R}^{2nk} \to \mathbf{R}^{nk}$, the kernel $a : (0,\infty) \to \mathbf{R}^+$ is integrable, and $f : \Omega \times [0,T] \to \mathbf{R}^k$ is a given function. We give homogeneous boundary conditions

(2)
$$u^i(x,t) = 0 \quad \text{on} \quad \partial\Omega \times [0,T]$$

and prescribe initial data

(3)
$$\begin{aligned} u^i(x,t) &= U_o^i(x,t) \quad (x \in \Omega \times (-\infty, 0)) \\ u^i(x,0) &= U_{oo}^i(x) \quad (x \in \Omega), \\ \partial_t u^i(x,0) &= U_1^i(x) \quad (x \in \Omega). \end{aligned}$$

Equations of this form occur in the description of certain motions of viscoelastic materials. Consider e. g. a onedimensional viscoelastic bar occupying an interval $I \subset \mathbf{R}$, and let $u(x,t) \in \mathbf{R}$ denote the position of the particle x at time t. Assuming unit density, the motion is governed by the equation

(4)
$$\partial_t^2 u(x,t) = \partial_x \sigma(x,t) + f(x,t)$$

where σ is the stress and f denotes body forces. A general constitutive assumption is

(5)
$$\begin{aligned} \sigma(x,t) &= p(\partial_x u(x,t)) - \int_0^\infty a(s) q(\partial_x u(x,t-s))\, ds \\ &= (p - Aq)(\partial_x u(x,t)) + \int_0^\infty a(s)\left(q(\partial_x u(x,t)) - q(\partial_x u(x,s))\right) ds \end{aligned}$$

with $A = \int_0^\infty a(s)\, ds$. Then (4) and (5) reduce to (1).

The equation also arises in the description of shear motions of viscoelastic materials. Let $\xi \in \mathbf{Q} \subset \mathbf{R}^3$ denote the position of a material particle in a reference configuration and let $y(\xi, t)$ denote the position of this particle at time t. For incompressible elastic liquids of K – BKZ – type (cf. [1]), shear motions of the form

$$(y^1, y^2, y^3) = (\xi^1 + u^1(\xi^3, t), \xi^2 + u^2(\xi^3, t), \xi^3) \tag{6}$$

lead to a sytem of the form (1) with $n = 1, k = 2$ (ξ^3 is the spatial variable). Also, antiplane shear motions of the form

$$(y^1, y^2, y^3) = (\xi^1, \xi^2, \xi^3 + u(\xi^1, \xi^2, t)), \tag{6}$$

lead to this form with $n = 2, k = 1$, if some additional assumptions concerning the constitutive equation hold. In this case, $(\xi^1, \xi^2) \in \Omega \subset \mathbf{R}^2$, and $\mathbf{Q} = \Omega \times \mathbf{R}$ is an infinite column. For details of the derivation we refer to [7] . It should be noted that no further assumption has to be made when the full three-dimensional equations of motion for the material are reduced to (1). However, in spite of a superficial similarity, the general equations of motion for a viscoelastic material of integral type are in general not of the form (1).

For previous work on integrodifferential equations in unbounded domains, we refer to [7] and the bibliography given there. Most of these results are concerned with global smooth solutions for small smooth data. In [6], a global existence theorem in one space dimension is shown that uses compensated compactness and holds for equations of a somewhat more special structure, but without assuming the singularity of the kernel. Results similar to the ones presented in this note were given in [2] for the case of bounded domains; there it was also shown that the support of perturbations of rest states must spread with finite speed, and that the singularity of the kernel has a regularizing effect on the data. General boundary value problems for this equation were discussed in [3].

The main purpose of this note is to show that a kernel $a(\cdot)$ with an integrable singularity at zero allows one to find global weak solutions of (1) for arbitrary initial data (Theorem 1). In addition, these solutions satisfy a natural entropy inequality which can be used to show that solutions will tend to a rest state as time goes to infinity and that some exponential spatial decay must hold for the amplitude of these perturbations, if their support was initially confined to a compact subset of Ω (Theorem 2). We overcome the lack of compactness for certain imbeddings of Sobolev spaces of functions on unbounded domains, a property that was crucial in the arguments in [2] and [3]. We actually show the existence of weak solutions with infinite L^2 - energy. Thus travelling wave solutions such as those constructed in [5] fall into the class considered here.

Below, we introduce some basic notation. The main results are then given in the following section 2, together with comments and examples. The proofs are outlined in section 3.

Throughout, $\Omega \subset \mathbf{R}^n$ is assumed to be open, and the boundaries of the sets $\Omega_M = \Omega \cap \mathbf{B}_M(0)$ are to satisfy uniform Lipschitz conditions (cf. [4]). Here $\mathbf{B}_M(0)$ is the ball about 0 with radius M. Let $\rho \in C^1(\overline{\Omega}, \mathbf{R}^+)$ be a function that is bounded away from zero and that satisfies

$$|\nabla \rho(x)| \leq K \cdot \rho(x) \quad (x \in \Omega) \tag{8}$$

for some $K \geq 0$. We then define the norms

$$\|u\|_{0,\rho} := \left(\int_\Omega \rho \cdot |u|^2 \, dx \right)^{1/2} \tag{9}$$

and

$$\|u\|_{1,\rho} := \left(\|\nabla u\|_{0,\rho}^2 + \|u\|_{0,\rho}^2\right)^{1/2} \tag{10}$$

for $u \in C_o^\infty(\Omega, \mathbf{R}^k)$ and denote by $L_\rho^2(\Omega)$ resp. $H_\rho^1(\Omega)$ the closure of $C_o^\infty(\Omega)$ with respect to these norms.

Also, the dual space of $H_\rho^1(\Omega)$ can be viewed as the space of distributions

$$H_\rho^{-1}(\Omega) = \{v \mid |v(\phi)| \leq C \cdot \|\phi\|_{1,\rho} \quad \text{for all} \quad \phi \in C_o^\infty(\Omega, \mathbf{R}^n)\},$$

with its natural norm $\|\cdot\|_{-1,\rho}$. We then have natural imbeddings from $H_\rho^1(\Omega)$ into $L_\rho^2(\Omega)$ and from $L_\rho^2(\Omega)$ into $H_\rho^{-1}(\Omega)$, the latter being defined by $i(u)(v) = \int_\Omega \rho uv$,, if $u \in L_\rho^2(\Omega)$ and $v \in H_\rho^{-1}(\Omega)$.

Generic constants that depend only on given fixed quantities are denoted by the same letter C; their value may change from line to line. We also suppress superscripts, wherever no confusion can arise.

2. Main Results

Throughout, it will be assumed that

$$g_j^i(\xi) = \frac{\partial}{\partial \xi_j^i} G(\xi), \quad h_j^i(\xi, \zeta) = \frac{\partial}{\partial \xi_j^i} H(\xi, \zeta) \tag{11}$$

for some $G \in C^2(\mathbf{R}^{nk}, \mathbf{R}), H \in C^2(\mathbf{R}^{nk} \times \mathbf{R}^{nk}, \mathbf{R})$, and that g_j^i and h_j^i are globally Lipschitz continuous. Without loss of generality, we assume that

$$G(0) = 0, \quad h(0,0) = g(0) = 0, \quad H(\xi, \xi) = 0 \quad (\xi \in \mathbf{R}^{nk}).$$

Let $A = \int_0^\infty a(s)\,ds > 0$, then we assume in addition that

$$z_l^j \cdot \frac{\partial}{\partial \xi_l^j} \left(g_r^i(\xi) + A \cdot h_r^i(\xi, \zeta)\right) \cdot z_r^i \geq \lambda \cdot |z|^2 \quad (\xi, \zeta, z \in \mathbf{R}^{nk}) \tag{12}$$

and

$$H(\xi, \zeta) \geq \lambda \cdot |\xi - \zeta|^2 \quad (\xi, \zeta \in \mathbf{R}^{nk}) \tag{13}$$

for some $\lambda > 0$. The ellipticity condition (12) is needed to make (1) locally well posed. The kernel a will be assumed to be locally absolutely integrable on $(0, \infty)$,

$$a' = b - c, \quad b \in L^1(0, \infty; \mathbf{R}^+), \quad c \in L^1(\delta, \infty; \mathbf{R}^+) \quad \text{for all } \delta > 0 \tag{14}$$

where the negative part c of a' satisfies

$$\int_0^1 c(s)\,ds = \infty. \tag{15}$$

Thus a must go to ∞ as $t \downarrow 0$.

THEOREM 1. *Let (11) - (15) hold. Let* $\rho \in C^1(\overline{\Omega}, \mathbf{R}^+)$ *be a given function with (8), and let* $U_o \in L^\infty(-\infty, 0; H^1_\rho(\Omega))$, $U_{oo} \in H^1_\rho(\Omega), U_1 \in L^2_\rho(\Omega), f \in L^1(0,T; L^2_\rho(\Omega)) \cap L^2(0,T; H^{-1}_\rho(\Omega))$ *be given. Then there exist functions* $\sigma \in L^\infty(0,T; L^2_\rho(\Omega))$ *and* $u \in L^\infty(0,T; H^1_\rho(\Omega)) \cap W^{1,\infty}([0,T], L^2_\rho(\Omega))$ *for which (1.a) holds in the sense of distributions and (1.b) holds almost everywhere, and such that* $u(\cdot,t) \to U_{oo}$ *weakly in* $H^1_\rho(\Omega)$, $\partial_t u(\cdot,t) \to U_1$ *weakly in* $L^2_\rho(\Omega)$, *as* $t \downarrow 0$. *In addition, u can be found such that*

$$\partial_t \eta + div_x q + \Sigma' \leq \partial_t u \cdot f \tag{16}$$

in the sense of distributions and $q \cdot \nu = 0$ *in the trace sense on* $\partial\Omega \times [0,T]$, *where*

$$\eta(x,t) = \frac{1}{2}|\partial_t u(x,t)|^2 + G(\nabla u(x,t)) + \int_0^\infty a(s) H(\nabla u(x,t), \nabla u(x,t-s))\, ds, \tag{17}$$

$$q(x,t) = -\partial_t u(x,t) \cdot \sigma(x,t), \tag{18}$$

$$\Sigma'(x,t) = -\int_0^\infty a'(s) H(\nabla u(x,t), \nabla u(x,t-s))\, ds \tag{19}$$

and ν *is the outer unit normal vector field on* $\partial\Omega$.

Property (16) is an entropy inequality that should hold for physically meaningful solutions. The three terms on the left hand side of (16) are kinetic and potential energy η, energy flux q, and a dissipation term $\sum'$. Due to the singularity of a, the fact that $\sum'$ is integrable is non-trivial. Various additional properties of solutions of (1) can be deduced from (16); in [2], it is shown to imply that the support of perturbations of rest states must spread with finite speed.

It is not clear whether the condition (8) for the weight function ρ can be dropped. For linear equations, it is certainly unnecessary, since due to the finite speed of propagation and the superposition principle solutions can be localized. However, even so the result allows one to construct weak solutions which have only locally integrable $H^1(\Omega)$ - norms.

We also show that under suitable assumptions solutions must tend to rest states in certain weighted spaces in an exponential manner. We make the following assumption on the domain Ω:

$$\text{dist}(x, \partial\Omega) \leq C \quad \text{for all} \quad x \in \Omega$$

for some constant C. This is equivalent to the existence of a constant $\gamma > 0$ such that for all $u \in C^\infty_o(\Omega, \mathbf{R})$

$$\int_\Omega |\nabla u|^2\, dx \geq \gamma \cdot \int_\Omega |u|^2\, dx. \tag{20}$$

Additional assumptions are

$$a'(t) + \delta a(t) \leq 0 \quad \text{on} \quad (0, \infty) \tag{21}$$

for some $\delta > 0$, and for all $\xi \in \mathbf{R}^{n,k}$ either

$$G(\xi) = 0 \tag{22}$$

or

$$g(\xi) \cdot \xi \geq \epsilon |\xi|^2 \tag{23}$$

for some $\epsilon > 0$.

If (22) holds, then any t - independent function is a rest state of (1), since (12) and (13) imply that $h(\xi, \xi) = 0$ for all ξ. This assumption corresponds to a liquid - like behavior for a viscoelastic material. If on the other hand (23) holds, then only the zero solution will be a rest state; this is a property for solids.

THEOREM 2. *Let assumptions (11) - (15), together with (20) and (21) hold.*
a) If in addition (22) holds, then there exist $\beta, \kappa > 0$ such that for all $\rho \in C^1(\overline{\Omega}, \mathbf{R}^+)$ with

$$|\nabla \rho| \leq \kappa |\rho| \quad \text{on} \quad \Omega, \tag{24}$$

all $U_o \in L^\infty(-\infty, 0; H^1_\rho(\Omega))$, $U_{oo} \in H^1_\rho(\Omega)$, $U_1 \in L^2_\rho(\Omega)$ and $f \in L^1(0,\infty; L^2_\rho(\Omega))$ any solution constructed in Theorem 2 must satisfy

$$\|\partial_t u(\cdot,t)\|_{o,\rho} + \|u(\cdot,t) - u_\infty\|_{1,\rho} \leq e^{-\beta t}(C_o + \int_0^t e^{\beta s}\|f(\cdot,s)\|_{o,\rho}\, ds) \tag{25}$$

for some C_o depending on the initial data and some undetermined $u_\infty \in H^1_\rho(\Omega)$.
b) If instead (23) holds, then the same conclusion as in a) is true, with $u_\infty = 0$.

If $g(\xi) \cdot \xi$ is not bounded from below, then stable rest states cannot be expected. The first conclusion of the result again reflects the behavior of liquids for which no preferred rest state exists.

Theorem 2 remains true for any weak solution of (1) for which the entropy inequality (16) holds. It has two types of applications:

Firstly, by choosing ρ to be function that decays as $x \to \infty$, it allows one to show that even solutions with infinite $H^1(\Omega)$- norms must still tend to rest states . Some restriction of the form (24) (i.e. some growth constraint for the solution) must be assumed, as is shown by the following example.

Example 1. Consider the equation (1) in $\Omega = (0,\pi) \times (-\infty,\infty) \subset \mathbf{R}^2$, with $k = 1$ and

$$\sigma(x,t) = \int_{-\infty}^t e^{s-t}(\nabla u(x,t) - \nabla u(x,s))\, ds \tag{26}$$

and $U_o = 0, f = 0$. Then the system (1) is equivalent to the equation

$$\partial_t^2 u(x,t) + \partial_t u(x,t) - \Delta u(x,t) = U_1(x)\,.$$

This equation has solutions of the form

$$u(x,t) = e^{\alpha t} \cdot \sin x \cdot \cosh \lambda y$$

with $\alpha > 0$, if $\lambda \geq \frac{\sqrt{3}}{2}$, implying that in (24) , κ must be less than $\frac{\sqrt{3}}{2}$.

Secondly, Theorem 2 can be applied to study the asymptotic spatial behavior of solutions. For compactly supported initial data, the support of the solution will grow at a finite speed (cf. [2]), and its amplitude will decay exponentially, if the kernel $a(\cdot)$ has certain decay properties, as is shown for linear model equations on the real axis in [7]. By choosing the weight function ρ to be exponentially growing, Theorem 2 can be used to guarantee this latter effect in domains for which the Poincaré's inequality holds and for general kernels.

Example 2. Consider an equation of the form (1) in the spiral - like domain

$$\Omega = \{(r,\theta)\,|\,\theta > 0, 1 + \frac{\theta}{\pi} < r < 2 + \frac{\theta}{\pi}\} \quad \subset \quad \mathbf{R}^2$$

where polar coordinates are used. Then the support of any compactly supported perturbations of the rest state 0 must spread around the spiral, and its amplitude shuold be expected to decay in proportion to the distance that it has travelled. We can then use the weight function

$$\rho(r,\theta) = e^{\alpha\theta^2}$$

for a suitably small α, to verify this heuristic consideration by means of Theorem 2.

3. Proofs

Since the results of the previous section have been proved in detail in [2] in the case of bounded domains, only outlines will be given, and the main difficulties due to the unboundedness of the domain will be pointed out.

For the proof of Theorem 1, we need several lemmas.

Lemma 1. *Let $\rho, \rho' : \overline{\Omega} \to \mathbf{R}^+$ be given as above, with $\frac{\rho'(x)}{\rho(x)} \to 0$ as $|x| \to \infty$ in Ω. Then the imbeddings $i : H^1_\rho(\Omega) \to L^2_{\rho'}(\Omega)$ and $j : L^2_\rho(\Omega) \to H^{-1}_{\rho'}(\Omega)$ are compact.*

Proof: Let $(u_n)_{n\geq 0}$ be a sequence in $H^1_\rho(\Omega)$ that converges weakly to zero. For $\epsilon > 0$, pick $M > 0$ such that $\rho'(x) \leq \epsilon\rho(x)$ for $|x| \geq M$. Setting as before $\Omega_M = \Omega \cap B_M(0)$, we then have

$$\int_{\Omega_M} \rho' |u_n|^2 \, dx \to 0$$

by Rellich's theorem, and

$$\int_{\Omega - \Omega_M} \rho' |u_n|^2 \leq \epsilon \|u_n\|^2_{1,\rho}$$

can be made as small as we want. This proves the compactness of i; the compactness of j follows in a similar manner.

Lemma 2. *Let $b : (0,T) \to \mathbf{R}^+$ be given such that $b \in L^1(\epsilon, T)$ for all $\epsilon > 0$ and $\int_0^1 b(s)\, ds = \infty$. Let X and Y be Hilbert spaces with compact imbedding from Y into X. Then the set*

$$\mathbf{C} = \{u \in L^2(0,T;X) \mid \sup_t \| \int_0^t u(s)\, ds\|_Y \leq 1, \int_0^T \int_0^t b(t-s)\|u(t) - u(s)\|^2_X \, ds\, dt \leq 1\}$$

is relatively compact in $L^2(0,T;X)$.

Proof: One easily sees that $\mathbf{C}$ is bounded in $L^2(0,T;X)$. Let $(u_n)_{n\geq 1}$ be a sequence in $\mathbf{C}$; we can then extract a subsequence that converges weakly in this space, and without loss of generality its limit is zero.

For $0 < h < T$, we next define $0 < \delta(h) < h$ by the condition

$$\int_{\delta(h)}^h b(s)\, ds = \frac{1}{h}$$

Set $b_h(\cdot) = b(\cdot)\mathbf{I}_{[\delta(h),h]}(\cdot)$, where $\mathbf{I}_J$ is the indicator function of the set J, and then choose compactly supported smooth nonnegative functions ψ_h on $(0,T)$ such that

$$\int_0^T \psi_h(t)\, dt = 1 \quad , \quad \int_0^T |\psi_h(t) - h \cdot b_h(t)|\, dt \leq h.$$

Now set

$$u_{n,h}(t) = \int_0^t u_n(s)\psi_h(t-s) ds = \int_0^t \int_0^s u_n(\tau)\, d\tau\, \psi_h'(s)\, ds \quad ,$$

where u_n is continued as zero outside the interval $[0,T]$. Then for any fixed h, the $u_{n,h}$ are equicontinuous (due to the smoothness of the ψ_h) and have values in some bounded set in Y (due to the first property of elements of $\mathbf{C}$) ; thus by the theorem of Arzela and Ascoli, $u_{n,h} \to 0$ as $n \to \infty$ in $L^\infty(0,T;X)$. A straight forward calculation, using the second property of $\mathbf{C}$, shows that

$$\|u_n - u_{n,h}\|_{L^2(0,T;X)} \leq C \cdot h$$

for some constant $C > 0$. This implies that $u_n \to 0$ strongly, which proves the lemma.

Let a, g, h, ρ satisfy the general assumptions (11) – (15) of section 2. For smooth compactly supported functions u, v we continue u by U_o for $t < 0$, compute σ as in (1.b), and form the scalar product of $div_x \sigma$ with $v \cdot \rho$. The result is the expression

$$(27)\ \mathbf{B}(u)(v) = -\int_0^T \int_\Omega \Big(g(\nabla u(x,t)) + \\ + \int_{-\infty}^t a(t-s) h(\nabla u(x,t), \nabla u(x,s))\, ds\Big) (\nabla v(x,t)\rho(x) + v(x,t)\nabla\rho(x))\, dx\, dt.$$

This expression is still meaningful if $u, v \in H^1_\rho(\Omega)$, and we thus define an operator

$$\mathbf{B} : L^2(0,T; H^1_\rho(\Omega)) \to L^2(0,T; H^{-1}_\rho(\Omega)) \quad .$$

LEMMA 3. *For all* $u, v \in L^2(0,T; H^1_\rho(\Omega))$

$$(28) \qquad (\mathbf{B}(u) - \mathbf{B}(v))(u - v) \geq c\|u - v\|^2_{1,\rho} - C\|u - v\|_{0,\rho}$$

for some $c, C > 0$.

The proof is straight forward, using the ellipticity condition (12) and the global Lipschitz continuity of g and h. Details can be found in [2].

LEMMA 4. *Let* $\Omega \subset \mathbf{R}^n$ *be such that*

$$\int_\Omega |\nabla u|^2\, dx \geq \gamma \int_\Omega |u|^2\, dx$$

for all $u \in C^\infty_o(\Omega, \mathbf{R})$, *with some* $\gamma > 0$. *Then*

$$\int_\Omega \rho |\nabla u|^2\, dx \geq (\sqrt{\gamma} - \frac{\kappa}{2})^2 \int_\Omega \rho |u|^2\, dx$$

for all $u \in H^1_\rho(\Omega)$, *if* $|\nabla\rho| \leq \kappa\rho$ *on* $\overline{\Omega}$ *and* $\kappa^2 < 4\gamma$.

PROOF: Let $u \in C^\infty_o(\Omega, \mathbf{R})$ and set $v = \sqrt{\rho} u$, thus

$$\begin{aligned} \rho|\nabla u|^2 &= |\nabla v - \frac{\nabla\rho}{2\rho} v|^2 \\ &\geq |\nabla v|^2 - \frac{|\nabla\rho|}{\rho}|\nabla v||v| + \frac{|\nabla\rho|^2}{4\rho^2}|v|^2 \\ &\geq (1-s)|\nabla v|^2 + (1 - \frac{1}{s})\frac{|\nabla\rho|^2}{4\rho^2}|v|^2 \\ &\geq (1-s)|\nabla v|^2 + (1 - \frac{1}{s})\frac{\kappa^2}{4}|v|^2 \end{aligned}$$

for all $0 < s < 1$. Setting $s = \frac{\kappa}{2\sqrt{\gamma}}$, integrating, and using the assumption gives the desired result.

PROOF OF THEOREM 1: Without loss of generality, we assume that $A = \int_0^\infty a(t)\,dt = 1$. For $M > 0$, let $V_M = H_o^1(\Omega_M, \mathbf{R}^k) = H_o^1(\Omega \cap B_M(0), \mathbf{R}^k)$ and $H_M = L^2(\Omega_M, \mathbf{R}^k)$. The dual space $H^{-1}(\Omega_M, \mathbf{R}^k)$ will be denoted by V_M^*.

Let $\phi_M : \mathbf{R}^n \to \mathbf{R}^+$ be a smooth function with $\phi(x) = 0$ for $|x| \geq M$ and $\phi(x) = 1$ for $|x| \leq M-1$. We define operators $\mathbf{B}_M : L^2(0,T;V_M) \to L^2(0,T;V_M^*)$ as in (28), with $\rho = 1$, where u is to be continued as $\phi_M \cdot U_o$ on $\Omega \times (-\infty, 0)$. Then each $\mathbf{B}_M$ is globally Lipschitz continous.

Approximate solutions. For $M \geq 1$, consider the integral equation

$$(29) \qquad \partial_t w(x,t) - M^{-1}\Delta w(x,t) + \int_0^t \mathbf{B}_M(w)(x,s)\,ds = \phi_M(x)\left(U_1(x) + \int_0^t f(x,s)\,ds\right)$$

in $L^2(0,T;V_M^*)$, with initial data $w(\cdot,0) = \phi_M \cdot U_{oo}$. By well-known results on weak solutions of the linear heat equation ([4])and a standard fixed point argument, a unique solution $u_M \in W^{1,2}([0,T],V_M) \cap C^1([0,T],H_M) \cap W^{2,2}([0,T],V_M^*)$ can be found. We view u_M as an element of $W^{1,2}([0,T],H_\rho^1(\Omega))$ etc. .

A priori estimates. The equation (29) for u_M can be differentiated with respect to t and the inner product in H_M resp. in (V_M, V_M^*) with $\partial_t u_M \cdot \psi$ can be formed, where $\psi \in C_o^\infty(\overline{\Omega_M} \times [0,T), \mathbf{R})$ is an arbitrary test function. The regularity properties of u_M permit this operation. After integrating over $\Omega_M \times [0,T)$ and dropping temporarily the subscript M in u_M, the result is the identity

$$(30) \qquad \begin{aligned} &\int_0^T \int_\Omega (-\partial_t\psi\eta - \nabla\psi(q + M^{-1}\partial_t u \nabla\partial_t u) + \psi(M^{-1}|\nabla\partial_t u|^2 + \Sigma'))\,dx\,dt \\ &= \int_\Omega \psi(\cdot,0)\eta(\cdot,0)\,dx + \int_0^T \int_\Omega \partial_t u \psi_M f\,dx\,dt \end{aligned}$$

where η, q, and Σ' are defined as in (17 - (19). Taking in particular $\psi(x,s) = \rho(x)\chi_t(s)$, where χ_t is some smooth approximation of the indicator function $\mathbf{I}_{[0,t]}$ for arbitrary t, we get the estimate

$$\begin{aligned} &\frac{1}{2}\|\partial_t u(\cdot,t)\|_{o,\rho}^2 + \int_\Omega \rho(x)\Sigma(x,t)\,dx + M^{-1}\int_0^t \|\nabla\partial_t u(\cdot,s)\|_{o,\rho}^2\,ds \\ &\quad + \int_0^t \int_\Omega \rho(x)\Sigma'(x,s)\,dx\,ds \\ &\leq \frac{1}{2}\|U_1\|_{0,\rho}^2 + \int_\Omega \rho(x)\Sigma(x,0)\,dx + \int_0^t \int_\Omega \|\partial_t u(\cdot,s)\|_{o,\rho}\|f(\cdot,s)\|_{o,\rho}\,ds \\ &\quad + C\int_0^t \|\partial_t u(\cdot,s)\|_{o,\rho}\left(\|\sigma(\cdot,s)\|_{o,\rho} + M^{-1}\|\nabla\partial_t u(\cdot,s)\|_{o,\rho}\right)\,ds \quad , \end{aligned}$$

where

$$\Sigma(x,t) = G(\nabla u(x,t)) + \int_{-\infty}^t a(t-s)H(\nabla u(x,t), \nabla u(x,s))\,ds$$

and where σ is defined as in (1.b). Noting that (12), (13), and the Lipschitz bounds for g and h

imply the estimates

$$\int_\Omega \rho(x) H(\nabla v(x), \nabla w(x))\,dx \ge c\|v-w\|_{1,\rho}^2 - C\|v-w\|_{0,\rho}^2,$$

$$\int_\Omega \rho(x)\left(G(\nabla v(x)) + H(\nabla v(x), \nabla w(x))\right)dx \ge c\|v\|_{1,\rho}^2 - C\|w\|_{1,\rho}^2,$$

$$\int_0^t\int_\Omega |\nabla\rho(x)||\partial_t u_M(x,s)||\sigma(x,s)|\,dx\,ds \le C\cdot\left(\int_0^t \left(\|\partial_t u_M(\cdot,s)\|_{0,\rho}^2 + \|u_M(\cdot,s)\|_{1,\rho}^2\right)ds\right),$$

$$\int_0^t\int_\Omega |\nabla\rho(x)||\partial_t u_M(x,s)||\nabla\partial_t u_M(x,s)|\,dx\,ds \le C\int_0^t \|\partial_t u_M(\cdot,s)\|_{1,\rho}^2\,ds,$$

and using Gronwall's inequality implies the estimates

$$\sup_{[0,T]}\left(\|\partial_t u_M(\cdot,t)\|_{0,\rho} + \|u_M(\cdot,t)\|_{1,\rho}\right) \le C_1 \tag{31}$$

$$M^{-1}\int_0^T \|\partial_t u_M(\cdot,t)\|_{1,\rho}^2\,dt \le C_1, \tag{32}$$

$$\int_0^T\int_0^t c(t-s)\|u_M(\cdot,t) - u_M(\cdot,s)\|_{1,\rho}^2\,ds\,dt \le C_1, \tag{33}$$

with C_1 not depending on M. Let $w_M = \mathbf{B}(u_M) \in L^2(0,T;H_\rho^{-1}(\Omega))$, then (33) implies that additionally

$$\int_0^T \|w_M(\cdot,t)\|_{-1,\rho}^2\,dt \le C_1\,, \tag{34}$$

$$\int_0^T\int_0^t c(t-s)\|w_M(\cdot,t) - w_M(\cdot,s)\|_{-1,\rho}^2\,ds\,dt \le C_1\,. \tag{35}$$

Here $c(\cdot)$ is the negative part of a' as in (14). The details of the last argument can be found in [2].

Passing to the limit. We can choose subsequences of the $(u_M)_{M\ge 1}$, labelled in the same way, such that

$$u_M \rightharpoonup u \quad \text{weak-* in} \quad L^\infty(0,T;H_\rho^1(\Omega)), \tag{36}$$

$$\partial_t u_M \rightharpoonup \partial_t u \quad \text{weak-* in} \quad L^\infty(0,T;L_\rho^2(\Omega)), \tag{37}$$

$$w_M \rightharpoonup w \quad \text{weak-* in} \quad L^2(0,T;H_\rho^{-1}(\Omega))\,. \tag{38}$$

By Lemma 1 and the Theorem of Arzela and Ascoli,

$$u_M \to u \quad \text{strongly in} \quad L^\infty(0,T;L_\rho^2(\Omega))\,. \tag{39}$$

In addition, since $M^{-1}\Delta\partial_t u_M \to 0$ strongly in $L^2(0,T;H_\rho^{-1}(\Omega))$ due to (32), the limit w must equal $\partial_t^2 u - f$ in $L^2(0,T;H_\rho^{-1}(\Omega))$.

It remains to show that the equation holds in the weak sense, i. e. that $w = \mathbf{B}(u)$ in $L^2(0,T;H_\rho^{-1}(\Omega))$. For this purpose, let ρ' be some weight function that satisfies (8) and for which

$\rho'(x) = o(\rho(x))$ as $|x| \to \infty$. We then use (32), (33) and Lemma 2 to deduce that actually $w_M \to w$ strongly in $L^2(0,T;H_{\rho'}^{-1}(\Omega))$. Next, using Lemma 3, we obtain

$$
\begin{aligned}
(40) \qquad c \cdot \int_0^T \|u_M - u_N\|_{1,\rho'}^2 \, dt &\le \int_0^T (\mathbf{B}(u_M) - \mathbf{B}(u_N))(u_M - u_N) \, dt \\
&\quad + C \int_0^T \|u_M - u_N\|_{0,\rho'}^2 \, dt \\
&= \int_0^T (w_M - w_N)(u_M - u_N) \, dt + C \int_0^T \|u_M - u_N\|_{0,\rho'}^2 \, dt \\
&\le \frac{1}{2c} \int_0^T \|w_M - w_N\|_{-1,\rho'}^2 \, dt + \frac{c}{2} \int_0^T \|u_M - u_N\|_{1,\rho'}^2 \, dt \\
&\quad + C \int_0^T \|u_M - u_N\|_{0,\rho'}^2 \, dt \, ,
\end{aligned}
$$

where all inner products are between $H_{\rho'}^1(\Omega)$ and its dual $H_{\rho'}^{-1}(\Omega)$. This implies that $(u_M)_{M\geq 1}$ must be a Cauchy sequence in $L^2(0,T;H_{\rho'}^1(\Omega))$. The equation thus holds in the weak sense and also in $L^2(0,T;H_\rho^1(\Omega))$. Passing to the limit in (30) (in the sense of distribution) for a non-negative test function ψ implies also the entropy inequality (16).

Proof of Theorem 2: The proof is similar to an argument given in [2]. Let $\gamma > 0$ be as in (20) and $\delta > 0$ as in (21). Choose $\rho \in C^1(\overline{\Omega}, \mathbf{R}^+)$ with $|\nabla\rho| \le \kappa\rho$, where $0 < \kappa < 2\sqrt{\gamma}$ is to be fixed below, and let u be a solution of (1) that satisfies the entropy inequality (16). We define the energy components

$$(41) \qquad E_1(t) = \frac{1}{2} \int_\Omega \rho(x) |\partial_t u(x,t)|^2 \, dx$$

$$(42) \qquad E_2(t) = \int_\Omega \rho(x) \int_0^\infty a(s) H(\nabla u(x,t), \nabla u(x,t-s)) \, ds \, dx$$

$$(43) \qquad E_3(t) = \int_\Omega \rho(x) G(\nabla u(x,t)) \, dx \quad ,$$

where E_3 is understood to vanish if (22) holds, and

$$(44) \qquad D(t) = - \int_\Omega \rho(x) \int_0^\infty a'(s) H(\nabla u(x,t), \nabla u(x,t-s)) \, ds \, dx \quad .$$

Then $D(t) \ge \delta E_2(t)$ by (21). We also define

$$F_1(t) = \int_\Omega \rho(x) \partial_t u(x,t) \int_0^\infty b(s) \left(u(x,t) - u(x,t-s)\right) \, ds \, dx$$

and

$$F_2(t) = \int_\Omega \rho(x) \partial_t u(x,t) u(x,t) \, dx$$

where $b(t) = \min\{1, a(t)\}$. Set $\beta = \int_0^\infty b(s)ds > 0$. Then

$$
\begin{aligned}
(45) \qquad |F_1(t)| &\leq C \cdot \left(E_1(t) + \int_0^\infty b(s) \int_\Omega |u(x,t) - u(x,t-s))|^2 \, dx \, ds \right) \\
&\leq C \cdot (E_1(t) + E_2(t)) \quad ,
\end{aligned}
$$

where (13) and Lemma 4 have been used, and similarly

$$
(46) \qquad |F_2(t)| \leq C \cdot (E_1(t) + E_3(t)) \quad ,
$$

if (23) holds. Also, (16) implies that in the sense of distributions

$$
\begin{aligned}
\frac{d}{dt}(E_1(t) + E_2(t) + E_3(t)) &\leq -D(t) + \sqrt{E_1(t)} \|f(\cdot,t)\|_\rho + \int_\Omega |\nabla \rho(x,t)| |\partial_t u(x,t)| |\sigma(x,t)| \, dx \\
&\leq -D(t) + c \cdot \kappa \cdot (E_1(t) + E_2(t) + E_3(t)) + \sqrt{E_1(t)} \|f(\cdot,t)\|_{o,\rho} \quad ,
\end{aligned}
$$

where c is some constant.

We next produce some differential inequalities for the F_i. This can be done by differentiating their definitions and inserting the equation, since all occuring terms are well-defined. After some standard manipulations and estimates the results are for a. e. $t \in [0, \infty)$

$$
F_1'(t) \geq \beta E_1(t) - C \cdot D(t) - \left(\frac{C}{\mu} + \kappa \right) E_2(t) - (\mu + \kappa) E_3(t) - C \cdot \sqrt{E_2(t)} \|f(\cdot,t)\|_{o,\rho} \quad ,
$$

where C is some constant and $\mu > 0$ is arbitrary, to be chosen later; $\mu = 1$, if $E_3 = 0$. Similarly, if (22) holds,

$$
F_2'(t) \leq 2E_1(t) + CE_2(t) + (C\kappa - \eta) E_3(t) + \sqrt{E_3(t)} \|f(\cdot,t)\|_{o,\rho}
$$

for some $\eta > 0$. Now set

$$
(47) \qquad E(t) = E_1(t) + E_2(t) + E_3(t) + \alpha(\beta F_2(t) - 3F_1(t)) \quad ,
$$

where $\alpha > 0$ is such that

$$
(48) \qquad \frac{1}{2}(E_1(t) + E_2(t) + E_3(t)) \leq E(t) \leq 2(E_1(t) + E_2(t) + E_3(t)) \quad ;
$$

by (45) and (46) such a choice is possible. If (23) holds, we then obtain in the sense of distributions

$$
\begin{aligned}
E'(t) \leq (3C\alpha - 1) D(t) - \alpha \beta E_1(t) + \alpha \left(C\beta + 3\frac{C}{\mu} + 3\kappa \right) E_2(t) \\
+ \alpha (C\beta\kappa - \beta\gamma + \mu + \kappa) E_3(t) + C \|f(\cdot,t)\|_{o,\rho} \sqrt{E(t)} \quad .
\end{aligned}
$$

Now choose first μ and κ so small that the coefficient of E_3 is negative. Since $D(t) \geq \delta E_2(t)$, α can be picked sufficiently small to make the coefficient of E_2 negative, and we obtain an estimate

$$
(49) \qquad E'(t) \leq -2\omega E(t) + C \|f(\cdot, t\|_\rho \sqrt{E(t)}
$$

for some $\omega > 0$. Integrating this differential inequality gives

$$
(50) \qquad E(t) \leq Ce^{-2\omega t} \left(1 + \int_0^t e^{-\omega s} \|f(\cdot,s)\|_{0,\rho} \, ds \right)^2
$$

for some constant C. In case a), $E(t)$ dominates $\|u(\cdot,t)\|_{1,\rho}^2 + \|\partial_t u(\cdot,t)\|_{0,\rho}^2$, and the first part of the theorem follows.

In the proof of the second part, i.e. when E_3 vanishes identically, we drop F_2 in the definition of E and obtain again (49) and (50) by first choosing κ and then α sufficiently small. Now set $a_0(t) = t_0^{-1}\mathbf{I}_{[0,t_0]}$, where t_0 is so small that $a(t) \geq 1$ for all $t \in (0, t_0)$; then we have

$$\|u(\cdot,t) - \int_{-\infty}^{t} a_0(t-s)u(\cdot,s)\,ds\|_{1,\rho}^2 \leq t_0^{-1} \int_{-\infty}^{t} a(t-s)\|u(\cdot,t) - u(\cdot,s)\|_{1,\rho}^2\,ds \leq CE(t)$$

and also

$$\|\partial_t u(\cdot,t)\|_{0,\rho}^2 \leq CE(t) \quad .$$

This shows that $\partial_t u(\cdot,t)$ decays as in the statement of the theorem, and $u(\cdot,t)$ must satisfy a linear Volterra integral equation

$$u(\cdot,t) = \int_{-\infty}^{t} a_0(t-s)u(\cdot,s)\,ds = f_1(\cdot,t)$$

with a right hand side f_1 that behaves as (50) in $H_\rho^1(\Omega)$. By a standard resolvent argument for linear Volterra equations, this implies the existence of a limit for u as $t \to \infty$. Details of this argument can again be found in [2].

This work was supported by the National Science Foundation under grant # DMS 8601762.

References

1. R. B. Bird, R. C. Armstrong, O. Hassager, *Dynamics of Polymeric Liquids.* John Wiley, New York 1987.
2. H. Engler , Weak Solutions of a Class of Quasilinear Hyperbolic Integrodifferential Equations Describing Viscoelastic Materials. Arch. Rat. Mech. Anal., to appear.
3. H. Engler, Boundary Value Problems for Weak Solutions of a Class of Quasilinear Hyperbolic Integrodifferential Equations. In G. Da Prato, M. Ianelli (ed.), *Volterra Integro-Differential Equations in Banach Spaces and Applications*, Longman Group, London, to appear.
4. O.A. Ladyženskaja, V.A. Solonnikov, N.N. Ural'ceva, *Linear and Quasilinear Equations of Parabolic Type.* Amer. Math. Soc., Providence, R.I. ,1968.
5. J. A. Nohel, R. C. Rogers, A. Tzavaras, Weak Solutions for a Nonlinear System in Viscoelasticity, M. R. C. Technical Summary Report # 2976, Madison, WI, 1987.
6. M. Renardy, W. J. Hrusa, J. A. Nohel, *Mathematical Problems in Viscoelasticity.* Longman Group, London, 1987.
7. T. P. Liu, Nonlinear Waves for Viscoelasticity with Fading Memory. Preprint, University of Maryland, College Park, MD, 1987.

POSITIVE SOLUTIONS FOR SEMILINEAR ELLIPTIC SYSTEMS

W. E. Fitzgibbon
Department of Mathematics
University of Houston
Houston, Texas 77004 U.S.A.

and

J.J. Morgan
Department of Mathematics
Texas A & M University
College Station, Texas 77843 U.S.A.

1. Introduction

We shall be concerned with the existence of solutions to boundary value problems for semilinear elliptic systems. Such systems describe steady state solutions for systems of reaction diffusion equations. Specifically we consider systems of the form:

$$-\Delta u_i(x)=f_i(u(x)) \qquad x\in\Omega;\ i=1 \text{ to } m \qquad (1.1a)$$
$$u_i(x)=a_i(x) \qquad x\in\partial\Omega;\ i=1 \text{ to } m \qquad (1.1b)$$

where Ω is a smooth bounded Lipschitz domain in $\mathbf{R}^n$ with C^{2+d} boundary for some $\alpha\in(0,1)$.

Speaking in the roughest possible terms we employ the properties of a convex function H to control the growth of the reaction vector field f. This in turn allows the use of a scalar comparison function to dominate the summation of the components of the system and thereby provides a-priori bounds for the system. Classical arguments then establish the existence of a solution.

Convex functions, similar to H have been abstracted and successfully used by Morgan in [18], [19] to provide generalized Lyapunov functions which are used to obtain a-priori bounds, global the case at hand we work with a time independent problem. However we can exploit the structure provided by the convex function or generalized Lyapunov function. This paper is related to several appearing in the literature. Groger [8], [9] makes use of a particular convex function which is related to the rate of chemical dissipativity. Closely related but conceptually distinct are the invariance methods of Bates [3], Haedeler, Rothe and Vogt [10], Schmidt [22] and Weinberger [26].

2. Discussion of the Main Result

We begin by delineating the hypotheses required to guarantee solutions to (1.1a-b). We denote $\mathbf{R}^m_+ = \{u \mid u \in \mathbf{R}^m$ and $u_i \geq 0$ for $i=1$ to $m\}$. If $A \subseteq \mathbf{R}^m$ then int(A) will denote the interior of A and cl(A) will denote the closure of A. The vectors in $\mathbf{R}^m$ whose components are all zero or all 1 will be denoted by Θ_m and 1_m respectively. If $u \in \mathbf{R}^m$ then u^+ is the vector whose components are given by $(u^+)_i = u_i$ if $u_i \geq 0$ and $(u^+)_i = 0$ if $u_i < 0$; we now let $u^- = (-u)^+$.

The constant $\lambda_0 > 0$ will always denote the principal eigenvalue of $-\Delta$ on Ω with homogeneous Dirichlet boundary data. Other notation will be standard or will be subsequently developed.

We place the following restriction on our boundary data $a=(a_i)$.

<u>B</u> $\quad a \in C^{2+\alpha}(\partial\Omega, \mathbf{R}^m_+)$ for some $0<\alpha<1$

The vector field $f=(f_i)$: $\mathbf{R}^m_+ \to \mathbf{R}^m$ is required to be locally Lipschitz. Moreover we assume:

<u>P</u> $\quad$ For $1 \leq i \leq m$ and $u \in M$ $f_i(u) \geq 0$ whenever $u_i = 0$

We remark that this condition requires the ith component of f to be nonnegative on the coordinate hyperplane $u_i = 0$.

We now introduce what we term a generalized Lyapunov structure for the vector field f. Namely we postulate the existence of $H \in C^2(\text{int}(\mathbf{R}^m_+)\mathbf{R}_+) \cap C(M, \mathbf{R}_+)$ which satisfies:

<u>H1</u> There exists z_0 M such that $H(u)=0$ if and only if $u=z_0$

<u>H2</u> The Hessian $\partial^2 H(u)$ is nonnegative definite for all $u \in int(\mathbf{R}^m_+)$

<u>H3</u> $H(u) \to \infty$ as $|u| \to \infty$, $u \in \mathbf{R}^m_+$

<u>H4</u> If B is a bounded subset of $\mathbf{R}^m_+$ then

$$\lim_{\epsilon \to 0} \sup_{u \in B} |\partial H(u + 1_m)| \epsilon = 0$$

<u>H5</u> There exists a $\lambda (0 \leq \lambda \leq \lambda_0)$ and $K>0$ so that for all $u \in int(\mathbf{R}^m_+)$ one has $\partial H(u) f(u) \leq \lambda H(u) + K$

We note that <u>H2</u>-<u>H3</u> imply that H is a nonnegative convex function mapping $\mathbf{R}^m_+$ to $\mathbf{R}_+$. At the risk of belaboring the obvious we point out that the multiplication of <u>H5</u> is that of the m×1 row vector $\partial H(u)$ by the 1×m column vector $f(u)$. We shall frequently make use of this type of multiplication. If $\lambda = K = 0$, then <u>H5</u> implies that the reaction vector field points inward along level curves of H. Thus we may give <u>H5</u> the geometric interpretation as providing a limitation of the growth of the vector field across level curves of the function H.

We state our main result.

<u>Theorem 2.1</u>: If conditions <u>B</u>, <u>P</u> and <u>H1</u>-<u>H5</u> are satisfied then there exists $u=((u_i) \in C^2(\Omega, M) \cap C^1(cl(\Omega, \mathbf{R}^m_+))$ which solves (1.1a-b). Furthermore, if $v \in C^2(\Omega, \mathbf{R}_+ \cap C^1(cl(\Omega), \mathbf{R}_+)$ is a nonnegative solution of

$$-\Delta v(x) = \lambda v(x) + K \qquad x \in \Omega \qquad (2.2a)$$

$$v(x) = H(a(x)) \qquad x \in \partial\Omega \qquad (2.2b)$$

then $H(u(x)) \leq v(x)$ for all $x \in cl(\Omega)$.

Before outlining a proof of Theorem 2.1 we provide an example to illustrate our notion of a nonnegative convex function H. In [8], [9] Groger considers a class of systems modelling dissipative chemical reactions where f satisfies <u>P</u> and an additional polynomial growth restriction. He requires that there exists $e=(e_i) \in int(\mathbf{R}^m_+)$ and a function $g\colon \mathbf{R}^m_+ \to \mathbf{R}_+$ so that

$$\underset{i=1}{} f_i(u)\ln \frac{}{e_i} + g(u)\ \ 0 \ \text{ for all } u\ \ \text{int}(\mathbf{R}_+) \qquad (2.3)$$

and

$$\lim_{\substack{|u|\to\infty \\ u\,\in\,\text{int}(\mathbf{R}^m_+)}} \frac{g(u)}{|u|}=0 \qquad (2.4)$$

We note that if we define

$$H(u)=\sum_{i=1}^{m}\{(u)\ln\left(\frac{u_i}{e_i}\right)-u+e_i\} \text{ for all } u\in\text{int}(\mathbf{R}^m_+)$$

then H can be extended continuously to a nonnegative function on $\mathbf{R}^m_+$ (2.3) may be rewritten as $\partial H(u)f(u)\leq g(u)$. From (2.4) we see that if $0<\lambda<\lambda_0$ then there exists a $K_\lambda>0$ such that for all $u\in\mathbf{R}^m_+$

$$g(u)\leq\lambda\sum_{i=1}^{m}u_i+K_\lambda$$
$$\leq \lambda H(u) + \lambda\exp(2\sum_{i=1}^{m}e_i)$$

Setting $z_0=e$ we easily verify that H1-H5 hold. Consequently condition (2.3) and (2.4) are subsumed by H1-H5.

Our approach to the question of existence is standard in partial differential equations. We modify our reaction vector field by cutting it off outside of a region. We then use comparison principles to establish that the solution to the modified system lies in the region where the modified vector field and the original vector field agree and consequently a solution to the modified system is also a solution to the original system.

We introduce our cutoff function. If $v\in C^2(\Omega,\mathbf{R}_+)\cap C^1(\text{cl}(\Omega),\mathbf{R}_+)$ is a nonnegative function satisfying (2.2a-b) we let $k\in C^\infty(M,[0,1])$ be such that

$$k(u)=1 \text{ for } u\in M \text{ and } H(u)\leq\|v\|_{\infty,\Omega} \qquad (2.5a)$$

$$k(u)=0 \text{ for } u\in M \text{ and } H(u)\geq 2\|v\|_{\infty,\Omega} \qquad (2.5b)$$

We define a modified reaction vector field $F=(F_i)$: $\mathbf{R}^m\to\mathbf{R}^m$ as follows:

$$F(u)=k(u^+)f(u^+) \text{ for } u\in\mathbf{R}^m \text{ and } 1\leq i\leq m \qquad (2.6)$$

The following lemma which establishes the existence of solutions to a modified version of

(1.1a-b) is established via a Schauder Fixed Point Theorem argument.

Lemma 2.7: If F is defined via (2.6), then there exists $u\in C^2(\Omega,\mathbf{R}^m)\cap C^1(cl(\Omega),\mathbf{R}^m)$ which satisfies:

$$-\Delta u_i(x)=F_i(u(x)) \qquad x\in\Omega,\ i=1 \text{ to } m \qquad (2.8a)$$
$$u_i(x)=a_i(x) \qquad x\in\partial\Omega,\ i=1 \text{ to } m \qquad (2.8b)$$

A proof appears in [5].

Solutions to (2.8a-b) are confined to $\mathbf{R}^m_+$. The following positivity lemma is established in [5].

Lemma 2.9: If u is a solution to (3.4a-b) then u∈M. Consequently u satisfies

$$-\Delta u_i(x)=k(u(x))f(u_i(x)) \qquad x\in\Omega,\ i=1 \text{ to } m \qquad (2.10a)$$
$$u_i(x)=a_i(x) \qquad x\in\partial\Omega,\ i=1 \text{ to } m \qquad (2.10b)$$

We show that the solution provided by Lemmae 2.7 and 2.9 is indeed a solution to (1.1a-b).

Proof of Theorem 2.1: Let $\epsilon>0$ and u satisfy (2.10 a-b); we consider $H(u+\epsilon 1_m)$. Using H2 and H5 we compute,

$$\begin{aligned} -\Delta H(u+\epsilon 1_m) &\le \partial H(u+\epsilon 1_m)k(u)f(u) \qquad (2.11)\\ &=k(u)\partial H(u+\epsilon 1_m)f(u+\epsilon 1_m)\\ &\qquad +k(u)\partial H(u+\epsilon 1_m)[g(u)-F(u+\epsilon 1_m)]\\ &=\lambda H(u+\epsilon 1_m)+K+E(u,\epsilon) \end{aligned}$$

where $E(u,\epsilon)=|\partial H(u+\epsilon 1_m)||f(u)-f(u+\epsilon 1_m)|$. We recall f is locally Lipschitz and u is bounded on $cl(\Omega)$. Thus we are assured the existence of an $L>0$ so that $E(u,\epsilon)\le|\partial H(u+\epsilon 1_m)|L\epsilon$. Thus from H4, $\|E(u(\cdot),\epsilon\|_{\infty,\Omega}\to\infty$ as $\epsilon\to 0$. For each $\epsilon>0$ we set $E_\epsilon=\|E(u(\cdot),\epsilon\|_{\infty,\Omega}$.

Because λ 0 is less than the principal eigenvalue of $-\Delta$ on Ω we are guaranteed the existence of a unique nonnegative solution of

$$-\Delta v_\epsilon(x)=\lambda v_\epsilon(x)+K+E_\epsilon \qquad x\in\Omega \qquad (2.12a)$$

$$v_\epsilon(x)=H(a(x)+\epsilon 1_m \qquad x\in\partial\Omega \qquad (2.12b)$$

Classical continous dependence results via maximum principles (cf. [18], [22]) insure that

$$\lim_{\epsilon\to 0}\ \sup_{x\in\Omega}(v_\epsilon(x)-v(x))\leq 0$$

The comparison principle implies $H(u(x)+\epsilon 1_m)\leq v_\epsilon(x)$ for all $x\in\Omega$ and $\epsilon>0$. Hence from the definition of k we have $k\equiv 1$. Consequently $F(u)=f(u)$ and Lemma 2.7 provides the desired solution to (1.1a-b).

3. Examples

In this section, we illustrate various aspects of our theory by examining some semilinear elliptic systems.

Our first example is a three component system;

$$-\ d_1\Delta u=w-uv \qquad \Omega \qquad (3.1a)$$

$$-\ d_2\Delta v=w-uv \qquad \Omega \qquad (3.1b)$$

$$-\ d_3\Delta w=uv-w \qquad \Omega \qquad (3.1c)$$

$$u=a_1\ ;\ v=a_2\ ;\ w=a_3 \qquad \Omega \qquad (3.1d)$$

We assume $M=\mathbf{R}^3_+$, the boundary data satisfies B, and $d_1,d_2,d_3>0$. This system represents a steady state of a parabolic system proposed by Rothe [21] to model a chemical reation U+V W involving reactants U,V and product W.

We rewrite the system by multiplying each component by $1/d_i$ and observe that condition P is met. Our H-function is defined as $H(u,v,w) = d_1u+d_2v+2d3w$ for all $(u,v,w)\epsilon M$. It is immediate that H1-H4 are satisfied. Moreover one readily verifies that

$$\partial H(u,v,w)f(u,v,w)=0$$

and hence H5 is satisfied with constant $\lambda=0$. Thus we may use a comparison function z which satisfies

$$-\Delta z=0 \qquad \Omega$$
$$z=H(a_1,a_2,a_3) \qquad \partial\Omega$$

and thereby guarantee solutions to (3.1a-d) via Theorem 2.1. These solutions are nonnegative in each component and satisfy an a-priori bound given by

$$H(u(x),v(x),w(x))\leq z(x) \qquad \text{for all } x\epsilon\Omega$$

In fact, one can readily verify that solutions of (3.1a-d) must satisfy

$$H(u(x),v(x),w(x))=z(x) \qquad \text{for all } x\epsilon\Omega.$$

We now turn our attention to the system:

$$-\ \Delta u_1=c_1[\lambda(1-\rho)u_1-f(\rho,\beta)u_2] \qquad \Omega \qquad (3.2a)$$
$$-\ \Delta u_2=c_2[f(\rho,\beta)u_1+\lambda(1-\rho)u_2] \qquad \Omega \qquad (3.2b)$$
$$u_1=a_1\ ;\ u_2=a_2 \qquad \Omega \qquad (3.2c)$$

Here we assume the boundary data satisfies B and that c_1,c_2,λ are positive constants. The nonlinearity f is assume f to be 2π periodic in β, bounded for bounded ρ and locally Lipschitz for $\rho>0$ uniformly in β. The pair (ρ,β) represent polar coordinates for (u_1,u_2) in phase space. Although β is not defined at $\rho=0$ the functions f_1u_1 and f_2u_2 can be extended to have value zero at $\rho=0$ and thus we may assume that the vector field

$$F(u_1,u_2)=(c_1[\lambda(1-\rho)u_1-f(\rho,\beta)u_2],\ c_2[f(\rho,\beta)u_1+\lambda(1-\rho)u_2]T \qquad (3.3)$$

is locally Lipschitz.

This system represents the steady state of a semilinear parabolic system proposed by Lasry [15] to model neural conduction. A treatment of a generalization of the Lasry model is given by Barrow and Bates [2].

We let $M=\mathbf{R}$ and set $H(u_1,u_2)=u_1\ /2c_1+u_2\ /2c_2$ for $(u_1,u_2)\epsilon M$. We observe

$$\partial H(u_1,u_2)F(u_1,u_2)=\lambda(\rho^2-\rho^3)\leq 4\lambda/27. \qquad (3.4)$$

Clearly H1-H5 are satisfied. Hence if v is the nonnegative solution of

$$-\Delta v=4\lambda/27 \qquad \Omega \qquad (3.5a)$$

$$v=H(a_1,a_2) \qquad \partial\Omega \qquad (3.5b)$$

then Theorem 2.1 insures the existence of a classical solution (u_1,u_2) to (3.2a-c) with $H(u_1(x),u_2(x))\leq v(x)$ for all $x\in\Omega$.

4. Further Considerations

Two questions arise from Theorem 2.1. First, can condition H1-H5 be stated locally rather than globally and second can Theorem 2.1 be weakened to permit more growth of the reaction vector field. The answer to both questions is affirmative and it will be addressed in a forthcoming paper, [5]. Here we are able to localize our consideration to a trust region $A\subseteq\mathbf{R}^m_+$ where we expect to find solutions to our system. We control the growth of our reaction vector field by postulating the existence of a nondecreasing function $g(x)$, so that,

$$\partial H(u)f(u)\leq g(H(u)).$$

Instead of (2.2a-b), our scalar comparison function takes the form

$$-\Delta v=g(v) \qquad x\in\Omega$$

$$v(x)\geq H(a(x)) \qquad x\in\partial\Omega.$$

Furthermore, it is possible to consider convex functions H() which are defined on all of $\mathbf{R}^m$.

We point out that an analog of Theorem 2.1 holds in greater generality, namely we can apply our techniques to systems of the form

$$-L(u_i(x))=f_i(u(x)) \qquad x\in\Omega$$

$$u_i(x)=a_i(x) \qquad x\in\partial\Omega$$

where L is a strictly positive elliptic operator with smooth coefficients written in divergence form.

Here the scalar comparison function will be a solution to

$$-L(v(x))=\lambda v(x)+K$$

It is possible to adjust our hypotheses and consider (1.1a) subject to Robin-Neuman type boundary conditions of the form.

$$\partial u_i/\partial n=b_i(a_i-u_i)$$

Finally, we remark that one may use similar techniques, [6], to consider advection diffusion systems of the form

$$-\Delta u_i+\sum_{j=1}^{n}c_{ij}(x)\partial x_j u_i=f_i(u) \qquad x\in\Omega,\ i=1 \text{ to } m$$

$$u_i(x)=a_i(x) \qquad x\in\partial\Omega,\ i=1 \text{ to } m$$

In this paper we also obtain wave-like solutions to parabolic systems on radially symmetric domains which arise in response to periodic forcing on the boundary.

Bibliography

1. H. Amann, "Parabolic Evolution Equations with Nonlinear Boundary Conditions", Proc. Sym. Pure Math., Vol. 45 Part I, Amer. Math. Soc., Proficence, R.I. 1986, 17-27.

2. D. Barrow and P. Bates, "Bifurcation and Stability of Periodic Traveling Waves for a Reaction Diffusion System", J. Diff. Equations 50, 1983, 218-233.

3. P. Bates, "Containment for Weakly Coupled Parabolic Systems", Houston J. of Math., Vo. 11, 1985, 151-158.

4. P. Bates and K.J. Brown, "Convergence to Equilibrium in a Reaction Diffusion System", Nonlinear Anal. TMA, 8, 1984, 227-235.

5. W.E. Fitzgibbon and J.J. Morgan, "Existence of Solutions for a class of weakly coupled semilinear elliptic systems", (to appear).

6. W.E. Fitzgibbon and J.J. Morgan "Steady state solutions for certain reaction diffusion systems", (to appear).

7. D. Gilbarg and N. Trudinger, Elliptic Partial Differential Equations of Second Order , Springer - Verlag, Berlin, 1977.

8. K. Groger, "On the Existence of Steady States for a Class of Diffusion Reaction Equations", Math. Nachr. 112 (1985), 183-185.

9. K. Groger, "On the Existence of Steady States of Reaction Diffusion Equations, Arch. Rat., Mech, and Anal., Vol 1, 1986, 297-306.

10. K. Hadeler, F. Rothe and H. Vogt, "Stationary Solutions of Reaction Diffusion Equations", Meth. in Appl. Sci., 1979, 418-431.

11. D. Henry, Geometric Theory of Parabolic Equations, Lecture Notes in Mathematics, 840, Springer - Verlag, Berlin, 1981.

12. S.L. Hollis, R. H. Martin and M. Pierre, "Global Existence and Boundedness in Reaction Diffusion Systems", (to appear).

13. O.A. Ladyshenkaya, Boundary Value Problems of Mathematics Physics, Springer-Verlag, Berlin. 1984.

14. O.A. Ladyshenkaya, V.A. Solonnikov, and N.N. Uracleva, Linear and Quasilinear Equations of Parabolic Type, Transl. Amer. Math. Soc. Monographs, 33, Amer. Math. Soc., Providence, R.I., 1968.

15. J.M. Lasry, "International Working Paper of the Mathematics Research Center", Ceremade, University of Paris - Dauphine.

16. R.H. Martin, Nonlinear Operators and Differential Equations in Banach Spaces, John Wiley and Sons, New York, 1976.

17. R.H. Martin, "Nonlinear Pertubations of Coupled Systems of Elliptic Operators", Math. ann., 211, 1974, 155-169.

18. J.J. Morgan, "Global Existence, Boundedness and Decay for Solution of Semilinear Parabolic Systems of Partial Differential Equations", Dissertation, University of Houston, 1986.

19. J.J. Morgan, "Global Existence for Solutions of Semilinear Parabolic Systems", (to appear).

20. M. Protter and H. Weinberger, Maximum Principles in Differential Equations, Prentice Hall, Englewood Cliffs, N.J., 1967.

21. F. Rothe, Global Solutions of Reaction Diffusion Equations, Lecture Notes in Math., 1072, Springer - Verlag, Berlin, 1984.

22. K. Schmidt, Boundary Value Problems for Quasilinear Second Order Elliptic Equations, Nonlinear Anal. TMA, 2, 1978, 263-309.

23. R. Sperb, Maximum Principles and Their Applications, Academic Press, New York, 1981.

24. J. Smoller, Shock Waves and Reaction Diffusion, Springer - Verlag, Berlin, 1978.

25. F. Treves, Basic Linear Partial Differential Equations, Academic Press, New York, 1975.

26. H.F. Weinberger, "Invariant Sets for Weakly Coupled Parabolic and Elliptic Systems", Rendicontii di Matematica, 8, Serie, 295-310.

RECENT RIGOROUS RESULTS IN THOMAS-FERMI THEORY

Jerome A. Goldstein
Department of Mathematics
and Quantum Theory Group
Tulane University
New Orleans, LA 70118

Gisèle Ruiz Rieder
Department of Mathematics
Louisiana State University
Baton Rouge, LA 70803

0. OVERVIEW

This is a survey of mathematically rigorous results in Thomas-Fermi theory. Since the physical background may be unfamiliar to many mathematicians and since the mathematical tools used are quite diverse, we have decided to emphasize the heuristics and avoid most of the technical details. This approach will, we hope, highlight and make clear the key conceptual ideas in the theory.

We discuss the results of E. Lieb and B. Simon [17], [18], [16], utilizing the approach of Ph. Bénilan and H. Brezis [1], [4], [5]. Then we discuss our work, emphasizing our recent paper [13] and our new preprint [14]. Included will be a few new results which supplement [13] and [14].

1. INTRODUCTION TO GROUND STATE ELECTRON DENSITIES

Consider a quantum mechanical system of N electrons in $\mathbb{R}^3$. The Hamiltonian describing this system is

$$H = -\frac{\Delta}{2m} - \frac{1}{2} \sum_{\substack{i \neq j \\ 1 < i,j \leq N}} |x_i - x_j|^{-1} + \sum_{j=1}^{N} V(x_j) \tag{1}$$

acting on the Hilbert space $\mathcal{H} = L_a^2(\mathbb{R}^{3N})$ of square integrable antisymmetric functions of $(x_1,\ldots,x_N) \in (\mathbb{R}^3)^N$. The potential

$V \in L^1_{loc}(\mathbb{R}^1,\mathbb{R})$ is arbitrary at this point, but the most important cases are

an atom: $$V(x) = -Z/|x| \ , \tag{2}$$

a molecule: $$V(x) = -\sum_{j=1}^{M} Z_j/|x - R_j| \ , \tag{3}$$

where $R_1, \ldots, R_m$ are the locations of M fixed nuclei with charge Z_j at R_j. (In an atom, $M = 1$, $Z_1 = Z$, and $R_1 = 0$.) The *ground state wave function* Ψ_0 and the *ground state energy* E_0 are defined by

$$E_0 = \inf\{\langle H\Psi,\Psi\rangle : \Psi \in \mathrm{Dom}(H) \subset \mathcal{H} \ , \ \|\Psi\| = 1\} = \langle H\Psi_0,\Psi_0\rangle$$

and $\|\Psi_0\| = 1$.

In the case of bulk matter, N is enormously large ($N \approx 10^{26}$), and it is a hopeless task to find Ψ_0, an antisymmetric function of $3N$ variables. Rather than seeking the wave function Ψ_0 we instead try to find the ground state density ρ_0, a function on $\mathbb{R}^3$.

Any wave function (i.e. unit vector Ψ in $\mathcal{H}$) determines an N-electron density ρ by the formula

$$\rho(x) = N \int_{\mathbb{R}^{3(N-1)}} |\Psi(x,x_2,\ldots,x_N)|^2 dx_2 \ldots dx_N \ . \tag{4}$$

We have $\rho(x) \geq 0$, $\int_{\mathbb{R}^3} \rho(x)dx = N$, and for Γ any Borel set in $\mathbb{R}^3$, $\int_\Gamma \rho(x)dx$ is the expected number of electrons in Γ of our quantum mechanical system in the state Ψ.

The idea of *density functional theory* is to express the energy $\langle H\Psi,\Psi\rangle$ in the state Ψ as a functional $E(\rho)$ of the density ρ given by (4). One then minimizes $E(\rho)$ as ρ varies over the domain $\mathrm{Dom}(E)$ of E, in particular, $\rho(x) \geq 0$ a.e. and $\int_{\mathbb{R}^3} \rho(x)dx = N$. Thus the strategy is the dispense with the wave function Ψ and work instead with the density ρ; the symbol for this is

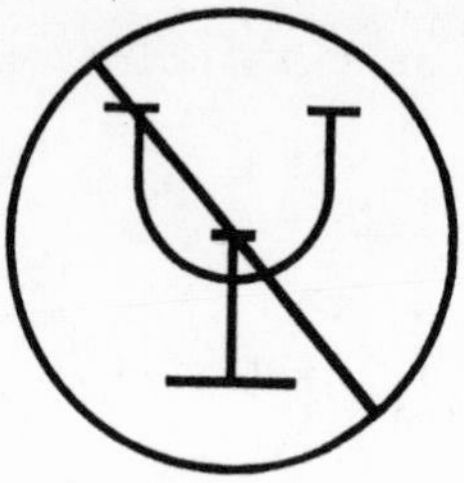

(cf. [22][†]).

The map $\Psi \to \rho$ (see (4)) is not injective, so we cannot write $\langle H\Psi, \Psi\rangle$ as $E(\rho)$. The next best thing is to find a functional $\mathcal{E}(\rho)$ which approximates $\langle H\Psi, \Psi\rangle$. This was first done (independently) by Thomas [25] and Fermi [6]. Their ideas were as follows:

$\mathcal{E}(\rho)$ consists of three terms, corresponding to the three terms in $\langle H\Psi, \Psi\rangle$ coming from (1). The electron-nuclear attraction term is exactly

$$\langle \sum_{j=1}^{N} V(x_j)\Psi, \Psi\rangle = \int_{\mathbb{R}^3} V(x)\rho(x)dx \ .$$

The quantum electron-electron repulsion term is approximated by the classical Coubomb repulsion energy given by

$$\frac{1}{2} c_{ee} \int_{\mathbb{R}^3} \int_{\mathbb{R}^3} \frac{\rho(x)\rho(y)}{|x - y|} dx\, dy \ .$$

Here it is traditional to take the constant c_{ee} to be one, but Fermi and Amaldi [7] suggested $c_{ee} = (N - 1)/N$; this c_{ee} approximates one for large N , but vanishes for $N = 1$ (when there is no electron-electron repulsion).

The kinetic energy term $(2m)^{-1}\langle -\Delta\Psi, \Psi\rangle$ has no obvious expression in terms of ρ. Thomas and Fermi suggested the expression

$$c \int_{\mathbb{R}^3} \rho(x)^{5/3} dx \ .$$

"Derivations" of this expression can be found in, for example, Landau and Lifschitz [15], Bethe and Jackiw [3], and Lieb and Simon [18].

---

† We learned this symbol from John Perdew, who attributes it to Doug Allen.

We sketch two formal supporting arguments here; the reader is urged not to take them too seriously.

In a system's (minimum energy) ground state, electrons can have all allowable momenta $p \in \mathbb{R}^3$ with $|p| \leq k_F$, where k_F defines the *Fermi level*. The kinetic energy of an electron with mass m and momentum p is $(2m)^{-1}|p|^2 \leq (2m)^{-1}k_F{}^2$. According to the Heisenberg uncertainly principle (in the form "$\Delta x \cdot \Delta p \geq 1$"), the number of electrons in a unit cube near x is proportional to

$$2 \sum_{|p| \leq k_F} 1 \approx 2 \int_{|p| \leq k_F} dp = c_1 k_F{}^3 .$$

The factor 2 allows for spin up and spin down electrons; thus, ignoring spin, the Pauli exclusion principle permits (at most) two electrons to occupy the same spatial state. Consequently, the density $\rho(x)$ is proportional to $k_F{}^3$ (which is x dependent). The kinetic energy in a small region is the number of electrons there times the kinetic energy per electron. Summing gives, since $k_F = c_2 \rho^{1/3}$,

$$\int_{\mathbb{R}^3} (2m)^{-1} k_F(x)^2 \rho(x)\, dx = c_3 \int_{\mathbb{R}^3} \rho(x)^{5/3} dx .$$

Next we give a scaling argument. For Ψ a unit vector in H and $\lambda > 0$, the wave function Ψ *scaled by* λ is the unit vector

$$\Psi_\lambda(x) = \lambda^{3N/2}\, \Psi(\lambda x) , \quad x \in \mathbb{R}^{3N} .$$

Since $\langle -\Delta\Psi_\lambda, \Psi_\lambda \rangle = \lambda^2 \langle -\Delta\Psi, \Psi \rangle$, one says that the kinetic energy *scales like* λ^2. Let

$$T(\rho) = c \int_{\mathbb{R}^3} \rho(x)^p\, dx$$

be a candidate for expressing kinetic energy as a functional of the density. Then the scaled density ρ_λ corresponding to $\rho \geq 0$, $\int_{\mathbb{R}^3} \rho(x)dx = N$ is

$$\rho_\lambda(x) = \lambda^3 \rho(\lambda x) , \quad x \in \mathbb{R}^3$$

(so that $\int_{\mathbb{R}^3} \rho_\lambda(x)dx = N$). Since

$$T(\rho_\lambda) = c \int_{\mathbb{R}^3} \lambda^{3p} \rho(\lambda x)^p dx = \lambda^{3(p-1)} c \int_{\mathbb{R}^3} \rho(y)^p dy ,$$

the kinetic energy functional T scales like $\lambda^{3(p-1)}$; this becomes λ^2 precisely when $p = 5/3$.

2. THE MINIMIZATION PROBLEM AND THE EULER-LAGRANGE EQUATION

Thus we take

$$E(\rho) = c \int_{\mathbb{R}^3} \rho(x)^{5/3} dx + \frac{1}{2} c_{ee} \int_{\mathbb{R}^3} \int_{\mathbb{R}^3} \frac{\rho(x)\rho(y)}{|x - y|} dx\, dy$$
$$+ \int_{\mathbb{R}^3} V(x)\rho(x) dx , \tag{5}$$

where c and c_{ee} are positive constants and, for simplicity and concreteness, V is given by (2) or (3). Besides allowing more general choices of V we could take a kinetic energy term of the form $\int_{\mathbb{R}^3} \rho(x)^p w(x) dx$ where w is a nonnegative weight function and $p > 1$, but we stick with (5) for ease of exposition.

Let $\mathrm{Dom}(E;N)$ consist of all $0 \leq \rho \in L^1(\mathbb{R}^3)$ for which $\int_{\mathbb{R}^3} \rho(x) dx = N$ and each term on the right hand side of (5) is finite. The basic problem of Thomas-Fermi theory is the *Minimization Problem*:

$$\text{minimize}\{E(\rho) : \rho \in \mathrm{Dom}(E;N)\} ;$$

more precisely, find $\rho_0 \in \mathrm{Dom}(E;N)$ such that

$$E(\rho_0) \leq E(\rho) \quad \text{for all } \rho \in \mathrm{Dom}(E;N) .$$

This problem was solved by Lieb and Simon [17], [18]. They made a frontal attack on the problem, using the direct methods of the calculus of variations.

In the late 1970s, Benilan and Brezis approached this problem using the modern theory of elliptic partial differential equations [1]. Since $E(\rho)$ is a strictly convex function of ρ , its minimum should be its only critical point, thus it sufficies to solve the Euler-Lagrange problem which formally takes the form $dE/d\rho = 0$. We "derive" this "equation" now and sketch its reduction to a nonlinear elliptic problem.

Let ρ_0 be a minimizer for E on $\mathrm{Dom}(E;N)$. Let $\sigma \in L^1(\mathbb{R}^3)$ satisfy $\int_{\mathbb{R}^3} \sigma(x)dx = 0$ and suppose $\rho_0 + \varepsilon\sigma \geq 0$ for ε real and small. Then $E(\rho_0 + \varepsilon\sigma) \geq E(\rho_0)$. Thus $(d/d\varepsilon)E(\rho_0 + \varepsilon\sigma)|_{\varepsilon=0} = 0$, which formally means $E'(\rho_0) = 0$ since σ is arbitrary. Formally,

$$E(\rho_0 + \eta) = E(\rho_0) + E'(\rho_0)\cdot\eta + o(\eta) ,$$

thus $E'(\rho_0)$ acts on η as a linear functional (i.e., $E'(\rho_0)$ is a function F on $\mathbb{R}^3$ and $E'(\rho_0)\cdot\eta$ means $\int_{\mathbb{R}^3} F(x)\eta(x)dx$)

Applying this formal procedure to (5) gives

$$E'(\rho)(x) = c_1\rho^{2/3}(x) + c_{ee}\int_{\mathbb{R}^3} \frac{\rho(y)}{|x - y|}\,dy + V(x) \tag{6}$$

with $c_1 = (5/3)c$. Concerning the integral term, let

$$B\rho(x) = \int_{\mathbb{R}^3} \frac{\rho(y)}{|x - y|}\,dy ;$$

then from Newtonian potential theory we know that $B = (4\pi)(-\Delta)^{-1}$. Since B is a self-adjoint operator on $L^2(\mathbb{R}^3)$, the derivative of

$$\rho \to \int_{\mathbb{R}^3} \rho(x)B\rho(x)dx = \langle\rho,B\rho\rangle = \langle B\rho,\rho\rangle$$

is $2B\rho$.

Next, to account for the constraint $\int_{\mathbb{R}^3} \rho(x)dx = N$, to $E(\rho)$ we add the term $\lambda(\int_{\mathbb{R}^3} \rho(x)dx - N)$, where λ is a Lagrange multiplier. This entails adding the constant λ to the right hand side of (6). There is one more subtle point to consider. When ρ is perturbed to become $\rho + \sigma$, where $\int_{\mathbb{R}^3} \sigma(x)dx = 0$, σ must be nonnegative on $\{x \in \mathbb{R}^3 : \rho(x) = 0\}$ since we demand $\rho \geq 0$ and $\rho + \sigma \geq 0$. The end result is as follows:

EULER-LAGRANGE PROBLEM. *Find* $\rho_0 \in \mathrm{Dom}(E;N)$ *and* $\lambda \in \mathbb{R}$ *such that*

$$c_1\rho^{2/3} + c_{ee}B\rho + V + \lambda = 0 \quad \textit{a.e. on} \quad \{x \in \mathbb{R}^3 : \rho(x) > 0\} , \tag{7}$$

$$c_1\rho^{2/3} + c_{ee}B\rho + V + \lambda \geq 0 \quad \textit{a.e. on} \quad \{x \in \mathbb{R}^3 : \rho(x) = 0\} . \tag{8}$$

For a class of potentials V including those given by (2) or (3), the Euler-Lagrange problem and the minimization problem are equivalent. Since $V(x) \to 0$ as $|x| \to \infty$ and since $\rho \in L^1(\mathbb{R}^3) \cap L^{5/3}(\mathbb{R}^3)$, it follows that $\lambda \geq 0$. Thus we may replace "$\lambda \in \mathbb{R}$" by "$\lambda \geq 0$" in the above problem.

Next let

$$u = -c_{ee} B\rho - V .$$

Applying the Laplacian to both sides gives

$$\Delta u = 4\pi c_{ee}\rho - \Delta V .$$

The function $s \to c_1 s^{2/3}$ is monotone increasing from $[0,\infty)$ to $[0,\infty)$; let γ be its inverse. Extend γ by defining $\gamma(t) = 0$ for $t < 0$. Thus $\gamma(t) = c_1^{-3/2}(t_+)^{3/2}$. Consequently (7), (8) collapse to the single equation

$$\rho = \gamma(u - \lambda) \quad \text{a.e. on } \mathbb{R}^3 .$$

Thus the minimization problem reduces to solving

$$\Delta u = 4\pi c_{ee}\gamma(u - \lambda) - \Delta V \quad \text{on } \mathbb{R}^3 , \tag{9}$$

$$\int_{\mathbb{R}^3} \gamma(u(x) - \lambda)dx = N . \tag{10}$$

With (9) we associate the boundary condition that u vanishes at infinity (at least in some weak sense). When V is given by (2) or (3), ΔV is a finite measure on $\mathbb{R}^3$, so the nonlinear elliptic equation (9) involves a measure as an inhomogeneous term.

For each $\lambda \geq 0$, problem (9) (together with "$u(\infty) = 0$") can be solved uniquely thanks to a result of Benilan, Brezis, and Crandall [2]. Denoting the solution by u_λ, let

$$I(\lambda) = \int_{\mathbb{R}^3} \gamma(u_\lambda(x) - \lambda)dx . \tag{11}$$

Then $I \in C([0,\infty),[0,\infty))$, I is strictly decreasing on the set where it is positive, $I(\lambda) \to 0$ as $\lambda \to \infty$, and $I(0) = c_{ee}Z$ where Z is given by (2) or by $Z = \sum_{j=1}^{M} Z_j$ if V is given by (3).

The conclusion is that the minimization problem has no solution for $N > c_{ee}^{-1} Z$ and one and only one solution for $0 < N \leq c_{ee}^{-1} Z$. The maximum allowable N is $N_{max} = c_{ee}^{-1} Z$, and this maximal N

corresponds to $\lambda = 0$. This reduces to $N_{max} = Z$ when $c_{ee} = 1$, while for $c_{ee} = (N - 1)/N$, $N_{max} = Z + 1$. Thus negative ions $(N > Z)$ do not exist in ordinary Thomas-Fermi theory (with $c_{ee} = 1$), while in the Fermi-Amaldi modification of Thomas-Fermi theory, singly negative ions exist (but not doubly negative ions).

Recall that $N < N_{max}$ iff $\lambda > 0$. The formula

$$\rho(x) = \gamma(u_\lambda(x) - \lambda)$$

together with $u_\lambda(\infty) = 0$ suggests that ρ has compact support when $N < N_{max}$ while $\operatorname{supp} \rho = \mathbb{R}^3$ for $N = N_{max}$. These conclusions are valid. Thus when the number of electrons is less than maximal, the compact support conclusion means that the electrons are very tightly bound to the nuclei.

3. THE NUCLEAR CUSP CONDITION

Our original interest in Thomas-Fermi theory was purely mathematical and was inspired by a lecture of Haim Brezis. Our papers [10], [11], [12], [24] all deal with mathematical questions growing out of the Lieb-Simon-Benilan-Brezis results described above. Thus we replaced the ambient space $\mathbb{R}^3$ by $\mathbb{R}^n$ for $n \geq 3$, the kinetic energy term $c \int_{\mathbb{R}^3} \rho(x)^{5/3} dx$ was replaced by $\int_{\mathbb{R}^3} J(\rho(x)) w(x) dx$ for J a convex function and w a weight function, etc. In [11] we argued that the electron-electron repulsion potential and the electron-proton attraction potential, which both have the form $\text{const}(\text{distance})^{-1}$ in $\mathbb{R}^3$, have different forms in $\mathbb{R}^n$ for $n > 3$. In [12] we studied a class of singular ordinary differential equations arising from generalized Thomas-Fermi Theory.

But our ultimate goal was to establish physically significant results in a rigorous way. We have two positive results to report along this line [13], [14]. We hope eventually to obtain further results. In describing these results we shall continue to emphasize the heuristics, keeping the technical details to a minimum.

Our first physically significant result concerns the *nuclear cusp condition*. In a molecule, the (true) electron density is continuous at a nucleus but has a cusp there. Slightly more precisely, near a nucleus with Z_j protons positioned at R_j, the density is approximately a radial function of $x - R_j$ given by

$\rho(x) = c_0 \exp\{-Z_j|x - R_j|\} + o(1)$. Thus ρ is continuous at R_j but $\nabla\rho$ need not be. For simplicity we consider an atom with nucleus at the origin, so that V is given by (2). Near the origin, $\rho(x)$ is positive, so that according to (7),

$$c_1 \rho(x)^{2/3} + c_{ee} B\rho(x) - Z/|x| + \lambda = 0 .$$

Since $B\rho$ and λ are bounded near 0, it follows that

$$\rho(x) \sim \text{const}|x|^{-3/2}$$

near $x = 0$. In particular, ρ is unbounded, which is physically unreasonable.

In a very interesting note, R. Parr and S. Ghosh [20] proposed a variant of Thomas-Fermi theory which formally yields a continuous density satisfying the nuclear cusp condition at each nucleus. Moreover, computer simulations indicated that their results gave improved ground state energies. In [13] we provided a rigorous mathematical framework for the work of Parr and Ghosh, and we proved the relevant theorems. Recently, Pratt, Hoffman, and Harris [23] found a different "statistical" approach to the problem of building the nuclear cusp condition into an approximate electron density.

The intuitive idea behind [20], [13] is that $\nabla\rho$ can have a discontinuity at each nucleus but elsewhere is a nice function which in some sense vanishes at infinity. Thus $\Delta\rho$ should be a signed Radon measure on $\mathbb{R}^3$. More precisely, we shrink the domain $\mathrm{Dom}(E;N)$ to $\tilde{D}_k(E;N)$ by requiring the extra constraint that $\rho \in W^{2,1}_{loc}(\mathbb{R}^3)$ and $\int_{\mathbb{R}^3} e^{-2k|x|} \Delta\rho(x)dx = M \in \mathbb{R}$. The constant $k > 0$ is fixed but the constant M is allowed to vary.

In deriving the Euler-Lagrange problem for our energy functional E, we introduce another Lagrange multiplier μ through a term of the form

$$\mu\left(\int_{\mathbb{R}^3} e^{-2k|x|} \Delta\rho(x)dx - M\right) .$$

An integration by parts (together with the assumption that $\nabla\rho(\infty) = 0$) converts this term to

$$\mu\left(\int_{\mathbb{R}^3} (4k^2 - 4k|x|^{-1})e^{-2k|x|} \rho(x)dx - M\right) .$$

Thus in the Euler-Lagrange equation (and inequality, see (7), (8)) $1/|x|$ appears with the coefficient $-Z + Ze^{-2k|x|}$ if we choose $\mu = Z/4k$. The point is that this term, $Z(e^{-2k|x|} - 1)/|x|$, is not singular at the origin.

The same theory holds for this modified Thomas-Fermi theory, that is, there exists a unique solution density ρ_0 if and only if $0 < N \leq N_{max}$ where N_{max} is Z or $Z + 1$, according as c_{ee} is 1 or $(N - 1)/N$; support(ρ_0) is compact if $N < N_{max}$; etc. Moreover, as $|x| \to 0$ (or, more generally, near a nucleus) $\rho_0(x) \sim \text{const}\ e^{-2Z|x|}$. Moreover, ρ_0 is bounded and smooth away from the nuclei. The connection between the constant $k > 0$ and other ingredients of the problem is

$$k = [(5/9)c\rho_0(0)^{2/3}]^{1/2} ,$$

where c is the coefficient appearing in the kinetic energy term. More details can be found in [13].

We end this section be stating one new result in this context. *In the variant of Thomas-Fermi theory incorporating the nuclear cusp condition, the ground state electron density of an atom is a radially decreasing function of the distance from the nucleus.* That is, when V is given by (2), the solution satisfies $\rho_0(x) = \rho_0(|x|)$ and $r \to \rho_0(r)$ is a strictly decreasing function on $\{r \geq 0 : \rho_0(r) > 0\}$. This interval coincides with $[0,\infty)$ if and only if $N = N_{max}$.

The idea of the proof is as follows. The solution ρ_0 is given by $\rho_0 = \gamma(u - \lambda)$ where $\Delta u = \gamma_1(u) - \Delta V$ for a suitable nondecreasing function γ_1. This problem for u (with $u(\infty) = 0$) has a unique solution. But one can reduce this to an ordinary differential equation for $u = u(r)$, $r = |x|$. This can be shown to have a decreasing solution. By uniqueness, the solution u of our PDE is radially decreasing.

A detailed proof of the above theorem in the general setting of [13] is based on a result of Gallouët and Morel [9]; the details are similar to those in Section 5 of [14].

4. SPIN POLARIZED THOMAS-FERMI THEORY

No mention of spin has been made thus far. Let now $\rho_1(x)$ [resp. $\rho_2(x)$] be the density of spin up [resp. spin down] electrons in our (molecular) system, and let $\rho = \rho_1 + \rho_2$ be the total electron density. Let $N_j = \int_{\mathbb{R}^3} \rho_j(x)dx$ be the total number of spin up or spin down electrons, according as $j = 1,2$, and let $N = N_1 + N_2$ be the total number of electrons. A *spin polarized theory* is one in which $N_1 \neq N_2$. In *spin polarized Thomas-Fermi theory* we specify $N_j = \int_{\mathbb{R}^3} \rho_j(x)dx$ in advance and seek ground state densities $\rho_1, \rho_2 \geq 0$ to minimize the energy functional

$$E(\rho_1,\rho_2) = \sum_{j=1}^{2} c_0 \int_{\mathbb{R}^3} \rho_j(x)^{5/3}dx + \frac{1}{2}\int_{\mathbb{R}^3}\int_{\mathbb{R}^3} \frac{\rho(x)\rho(y)}{|x-y|}dxdy + \int_{\mathbb{R}^3} V(x)\rho(x)dx . \tag{12}$$

The possibility of studying this problem rigorously was mentioned by Lieb and Simon [18, p.34]. Gadiyak and Lozovik [8] and Pathak [21] formally obtained some results for spin polarized Thomas-Fermi theory. In [14] we established rigorous results in this context.

Minimization of (12) over all ρ_1, ρ_2 subject to $\rho_j \geq 0$, $\int_{\mathbb{R}^3} \rho_j(x)dx = N_j$, and $N_1 + N_2 = N$ (with N_1, N_2 variable) gives the minimization problem of Section 2. The unique solution (for $N \leq N_{max}$) is $\rho_1(x) = \rho_2(x) = 2^{-1}\rho_0(x)$, so that half the electrons are spin up while half are spin down.

Suppose that, in the sequel, the index α ranges over a complete set of one electron quantum numbers, excluding spin. The spin index will be j, with $j = 1$ [resp. 2] corresponding to spin up [resp. down]. If V_j is a self-consistent one electron potential and

$$(-(2m)^{-1}\Delta + V_j(x))\psi_{\alpha j}(x) = E_{\alpha j}\psi_{\alpha j}(x), \quad x \in \mathbb{R}^3, \tag{13}$$

then the corresponding density is

$$\rho_j(x) = \sum_{\alpha} n_{\alpha j} |\psi_{\alpha j}(x)|^2 , \tag{14}$$

where the occupation numbers $n_{\alpha j}$ are given by $n_{\alpha j} = 1$ or 0, according as $E_{\alpha j} \leq \mu_F$ or $E_{\alpha j} > \mu_F$, where μ_F is the Fermi energy level. (This is analogous to some of the heuristic discussion in Section 1.)

Varying the kinetic energy as a functional of the density leads to the definition

$$T(\rho_1,\rho_2) = \sum_{\alpha,j} n_{\alpha,j}\langle -(2m)^{-1}\Delta\psi_{\alpha j},\psi_{\alpha j}\rangle .$$

It follows that

$$T(\rho_1,\rho_2) = T(0,\rho_2) + T(\rho_1,0) .$$

For a spin unpolarized system with $\rho_1 = \rho_2 = \rho/2$,

$$T(\rho_1,\rho_2) = T(\rho/2,0) + T(0,\rho/2) = 2T(\rho/2,0) ,$$

which for a fully polarized system with $\rho_1 = \rho$, $\rho_2 = 0$,

$$T(\rho_1,\rho_2) = T(\rho,0) .$$

If $T(\rho,0) = k \int_{\mathbb{R}^3} \rho(x)^{5/3}\, dx$ for some positive constant k , then

$$2T(\rho/2,0) = 2^{-2/3}T(\rho,0) ,$$

whence the fully polarized system has a higher kinetic energy by a factor of $2^{2/3}$. For the hydrogen atom and other systems, this leads to a more accurate ground state energy. (Cf. Oliver and Perdew [19], to whom the above calculation is due; further references are given in [14].)

To simplify the mathematics and the exposition we shall avoid discussing the Fermi-Amaldi modification in the spin polarized theory. In the sequel we shall proceed at a brisk pace.

Thus we begin with (12). To this expression for $E(\rho_1,\rho_2)$ we add $\sum_{j=1}^{2} \lambda_j(\int_{\mathbb{R}^3} \rho_j(x)dx - N_j)$, where λ_1 , λ_2 are Lagrange multipliers. The Euler-Lagrange equations come from setting $(\partial/\partial\rho_j)E(\rho_1,\rho_2) = 0$ for $j = 1,2$. The precise formulation of the Euler-Lagrange problem is as follows:

For $j = 1,2$ *find* $\lambda_j \geq 0$ *and* $0 \leq \rho_j \in L^1(\mathbb{R}^3)$ *such that* $\int_{\mathbb{R}^3} \rho_j(x)dx = N_j$ *and*

$$c_2\rho_j(x)^{2/3} + B\rho(x) + V + \lambda_j = 0 \quad a.e. \quad on \quad [\rho_j > 0] ,$$
$$c_2\rho_j(x)^{2/3} + B\rho(x) + V + \lambda_j \geq 0 \quad a.e. \quad on \quad [\rho_j = 0]$$

where $\rho = \rho_1 + \rho_2$.

This problem involves solving *two* equations and *two* inequalities. Let $u_j = -B\rho_j - V/2$, and let $\gamma(s) = 0$ on $(-\infty,0]$ and $\gamma(s)$ be the inverse function of $t \to c_2 t^{2/3}$ on $[0,\infty)$. The Euler-Lagrange problem thus reduces to $\rho_j = \gamma(u - \lambda_j)$,

$$\Delta u_j = 4\pi\gamma(u - \lambda_j) - \Delta V/2 .$$

For $u = u_1 + u_2$ and $\lambda_1 , \lambda_2 \geq 0$ we can find a unique solution of the single equation

$$\Delta u = 4\pi \sum_{j=1}^{2} \gamma(u - \lambda_j) - \Delta V ,$$

$u(\infty) = 0$. Call this u as u_λ for $\lambda = (\lambda_1,\lambda_2)$ and define

$$I_j(\lambda_1,\lambda_2) = \int_{\mathbb{R}^3} \gamma(u_\lambda(x) - \lambda_j)dx$$

for $j = 1,2$. The main lemma in [14] established that (I_1,I_2) is a bijection:

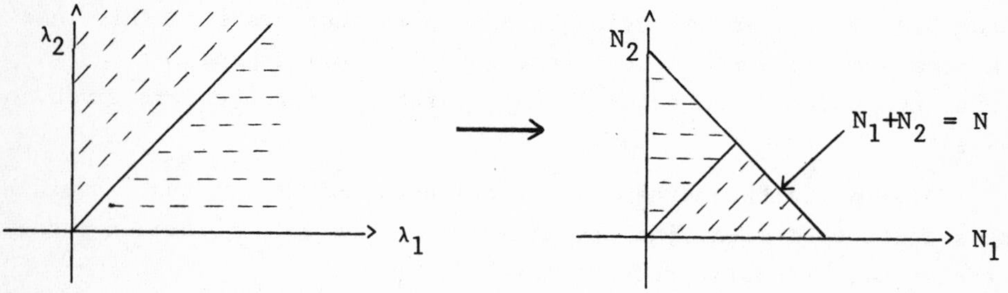

Thus, given $N_1 , N_2 > 0$, we can choose $\lambda_1 \geq 0$, $\lambda_2 \geq 0$ so that $I_j(\lambda_1,\lambda_2) = N_j$ for $j = 1,2$ if and only if $N_1 + N_2 = N \leq N_{max} = Z$.

Moreover, ρ_j has compact support whenever $\lambda_j > 0$; and $\lambda_1 = \lambda_2$ if and only if $N_1 = N_2$. Thus ρ_1 and ρ_2 both have compact support for $N < N_{max}$, while ρ_2 has compact support but ρ_1 has full support whenever $N_1 + N_2 = N_{max}$ and $N_1 > N_2$. In this case, in an electrically neutral system with more spin up than spin down electrons, the spin down electrons are very tightly bound to the nuclei. Finally, both ρ_1 and ρ_2 are radially decreasing functions in the case of an atom.

ACKNOWLEDGEMENTS

At various stages in our work on Thomas-Fermi theory we have profited greatly from the generous advice and help of Haim Brezis, Mel Levy, Elliott Lieb, Robert Parr, Rajeev Pathak, and John Perdew. This work partially supported by NSF grant DMS-86-20148 (JG) and the Louisiana Education Quality Support Fund, contract number 86-LBR-016-04 (GR).

REFERENCES

1. Bénilan, Ph. and H. Brezis, The Thomas-Fermi problem, in preparation.

2. Bénilan, Ph., H. Brezis, and M. G. Crandall, A semilinear elliptic equation in $L^1(\mathbb{R}^N)$, *Ann. Scuola Norm. Sup. Pisa* 2 (1975), 523-555.

3. Bethe, H. and R. W. Jackiw, *Intermediate Quantum Mechanics*, Benjamin, New York, 1968.

4. Brezis, H., Nonlinear problems related to the Thomas-Fermi equation, in *Contemporary Developments in Continuum Mechanics and Partial Differential Equations* (ed. by G. M. de la Penha and L. A. Medeiros), North-Holland, Amsterdam (1978), 81-89.

5. Brezis, H., Some variational problems of Thomas-Fermi type, in *Variational Inequalities and Complementary Problems: Theory and Applications* (ed. by R. W. Cottle, F. Giannessi and J. L. Lions), Wiley, New York (1980), 53-73.

6. Fermi, E., Un methodo statistico per la determinazione di alcune prioretà dell' atome, *Rend. Acad. Naz. Lincei* (1927), 602-607.

7. Fermi, E. and E. Amaldi, Le orbite ∞s degli elementi, *Mem. Accad. d'Italia* 6 (1934), 119-149.

8. Gadiyak, G. V. and Yu. E. Lozovik, Many-electron atoms in high magnetic fields, *J. Phys. B, Molec. Phys.* 13 (1980), 1531-1535.

9. Gallouët, Th. and J.-M. Morel, On some properties of the solution of the Thomas-Fermi problem, in *Nonlinear Analysis, Theory, Methods and Applications*, 7 (1983), 971-979.

10. Goldstein, J. A. and G. R. Rieder, Some extensions of Thomas-Fermi theory, in *Proceedings of the Conference on Differential Equations in Banach Spaces* (ed. by A. Favini and E. Obrecht), Springer Lecture Notes in Mathamatics 1223, Berlin (1986), 110-121.

11. Goldstein, J. A. and G. R. Rieder, The Coulomb potential in higher dimensions, in *Proceedings of the International Conference on Differential Equations and Mathematical Physics* (ed. by I. W. Knowles and Y. Saito), Springer Lecture Notes in Mathematics 1285, New York (1987), 143-149.

12. Goldstein, J. A. and G. R. Rieder, A class of ordinary differential equations connected with Thomas-Fermi theory, *Matem. Aplic. Comp.* 6 (1987), 57-68.

13. Goldstein, J. A. and G. R. Rieder, A rigorous modified Thomas-Fermi theory for atomic systems, *J. Math. Phys.* 28 (1987), 1198-1202.

14. Goldstein, J. A. and G. R. Rieder, Spin polarized Thomas-Fermi theory, *J. Math. Phys.* 29 (1988), 709-716.

15. Landau, L. D. and E. M. Lifschitz, *Quantum Mechanics. Non-Relativistic Theory* (2nd ed.), Pergamon, Oxford and New York, 1965.

16. Lieb, E. H., Thomas-Fermi theory and related theories of atoms and molecules, *Rev. Mod. Phys.* 53 (1981), 603-641.

17. Lieb, E. H. and B. Simon, Thomas-Fermi theory revisited, *Phys. Rev. Lett.* 33 (1973), 681-683.

18. Lieb, E. H. and B. Simon, The Thomas-Fermi theory of atoms, molecules, and solids, *Adv. Math.* 23 (1977), 22-116.

19. Oliver, G. L. and J. P. Perdew, Spin-density gradient expansion for the kinetic energy, *Phys. Rev. A* 20 (1979), 397-403.

20. Parr, R. G. and S. K. Ghosh, Thomas-Fermi theory for atomic systems, *Proc. Nat. Acad. Sci. USA* 83 (1983), 3577-3579.

21. Pathak, R., *Some Investigations on Atoms and Molecules within the Density Functional Formalism*, Ph.D. Thesis, Univ. of Poona, Pune, India, 1982.

22. Perdew, J. P. and Yue Wang, Electron density functionals from the gradient expansion of the density matrix: The trouble with long range iterations, in *Mathematics Applied to Science* (ed. by J. A. Goldstein, S. I. Rosencrans, and G. A. Sod), Academic Press, Boston (1988), 187-210.

23. Pratt, L. R., G. G. Hoffman, and R. A. Harris, The statistical theory of electron densities, *J. Chem. Phys.* 88 (1988), 1818-1823.

24. Rieder, G. R., Mathematical contributions to Thomas-Fermi theory, *Houston J. Math.*, in press.

25. Thomas, L. H., The calculation of atomic fields, *Proc. Cambridge Phil. Soc.* 23 (1927), 542-548.

Methods of Computing Fractal Dimensions

Fern Hunt*
Center for Computing and Applied Mathematics,
National Bureau of Standards
Gaithersburg, Maryland 20899

Francis Sullivan
Center for Computing and Applied Mathematics,
National Bureau of Standards
Gaithersburg, Maryland 20899

Abstract

We discuss two frequently calculated fractal dimensions, the capacity and information dimension and present efficient methods for their computation for sets embedded in $\mathbb{R}^n$. In particular we show how Monte Carlo calculation of areas and volumes can be used to compute the capacity using fewer box counting operations than straightforward box counting, and we discuss an efficient implementation of the method using a very fast one-dimensional sorting algorithm. Sets embedded in $\mathbb{R}^2$ and $\mathbb{R}^3$ are mapped to [0,1] using a folding map φ with the property that $n \bullet \dim(\varphi(X)) = \dim(X)$ where $X \subset \mathbb{R}^n$ and dim is either the Hausdorff or the capacity dimension. Thus the problem of calculating the capacity dimension or the information dimension (in the case it coincides with the Hausdorff dimension) is reduced to computing these quantities for $\varphi(X)$ in [0,1].

Also Department of Mathematics, Howard University, Washington, D.C. 20059.
Partially supported by NSF grant DMS-8603703.

Introduction.

In recent years fractal dimensions have become a leading tool for characterizing chaotic attractors. The Pesin and conjectured Kaplan-Yorke equalities suggest that these exponents play an important role in linking up the dynamics of a chaotic system to the invariant ergodic measure of the attractor. Our purpose is to discuss two aspects of the computation of dimension - convergence and efficiency of computations. In Section 1 we define two common exponents, the capacity and information dimension. It is not hard to show that the commonly used modifications of these definitions can also be used to calculate these numbers. This is done in Section 1 (also see appendix).

If the number of points representing the attractor, that is the point set size is large, it is well known [1] that the most direct means of evaluating the limits (see 1.1) in the definition of capacity, the so-called box counting methods, are expensive in time and storage. In Section 2 we present an alternative approach based on the Monte Carlo method. It has the advantage of requiring box counting only for a set of random test points - a set much smaller than the set of attractor points. The actual counting can be handled efficiently using sorting algorithms, and we briefly describe how this can be done for sets embedded in the unit interval. To make the method suitable for multidimensional sets, a map $\varphi: X \subset \mathbb{R}^n \to [0,1]$ is introduced with the property that $\dim(X) = n \bullet \dim(\varphi(X))$ for the capacity and Hausdorff dimensions. If X is such that the Hausdorff and information dimension (see (1.4)) coincide, this means equality holds for the information dimension as well. Such sets have been studied by Billingsley [11],[12] among others. The proof of this relation can be found in Section 3. Although all of the multidimensional arguments in this paper are in $\mathbb{R}^2$ they extend to higher dimensions with little or no change.

Eckmann and Ruelle [8, p. 647] and others [2] have observed that there is a finite range of box sizes where calculation of the information dimension can be carried out. This is also true for the capacity dimension. The length of this region increases as the point set size N increases, but outside of this region sampling error can significantly affect the validity of the least square technique used to obtain dimension estimates. This is particularly true of attractors that have large areas of low probability. In this paper we will neglect such finite sample effects; in our calculations heuristic methods were used to locate the region. One way around this problem is to extrapolate in N, to obtain results for an infinite sample size. This has been done for the Lorenz attractor by McGuiness [3] among others. In a forthcoming paper we will discuss the convergence of box counting calculations for finite samples of a specific attractor. We believe such conditions automatically define an asymptotic region.

Section 1.

To begin suppose the attractor is a compact set X embedded in $\mathbb{R}^2$. Without loss we assume $X \subset S$ the unit square. The capacity dimension is defined as follows. Let $N(\epsilon)$ be the minimum number of boxes of side ϵ that cover X. The capacity or box dimension is

(1.1) $$\lim_{\epsilon \to 0} \frac{\ln N(\epsilon)}{\ln 1/\epsilon} = D_c \text{ (where the limit is}$$

assumed to exist). From this it can be shown that [5]:

(1.2) $$\lim_{\epsilon\to 0} \frac{\ln V(\epsilon)}{\ln 1/\epsilon} = D_c - 2$$

where $V(\epsilon)$ is the area of the set $\{x \in \mathbb{R}^2: \inf_{y\in A} d(x,y) < \epsilon\}$, d is the Euclidean metric in $\mathbb{R}^2$. In the general multidimensional case $V(\epsilon)$ is the volume, and the right hand side of (1.2) is $D_c - D$ where D is the embedding dimension.

From a computational point of view it is convenient to subdivide S into a grid of dyadic boxes of side 2^{-n}. Let M_n equal the number of such boxes containing points of A. Then it can be shown that

(1.1') $$D_c = \lim_{n\to\infty} \frac{\log_2 M_n}{n}$$

(1.2') $$D_c - 2 = \lim_{n\to\infty} \frac{\log_2 A_n}{n}$$ (see appendix for proof)

where A_n is the area of the set $\bigcup_{x\in X} Z_n(x)$, and $Z_n(x)$ is the grid box containing x. A straightforward calculation of M_n, i.e., box counting, requires the examination of 2^{nD} (in general ϵ^{-D}) boxes. For large n and $D \geq 2$, this requires prohibitively large time and storage and consequently the calculation of A_n in (1.2') becomes an attractive alternative. This approach is not new [5],[6]. The novelty here lies in our use of Monte Carlo methods for estimating A_n which can be written as

$$A(2^{-n}) = A_n = \int_S f_{2^{-n}}(x)dx,$$

where

$$f_{2^{-n}}(r) = \begin{cases} 1 \text{ if } r \text{ is in } \bigcup_{x\in A} Z_n(x) \\ 0 \text{ otherwise} \end{cases}.$$

Let $\{r_i\}_{i=1}^M$ be a set of points in S chosen at random with a uniform distribution. The integral $A(\epsilon)$ can be estimated by

$$I_M(\epsilon) = \sum_{i=1}^{M} f_\epsilon(r_i)/M.$$

$I_M(\epsilon)$ is a random variable with mean $\mu = A(\epsilon)$ and variance $\sigma^2 = A(\epsilon)(1 - A(\epsilon))/M$. The strong law of large numbers implies that for large enough M, $I_M(\epsilon)$ is close to $A(\epsilon)$ for almost all selections of random points. Let

δ be the error inestimating $A(\epsilon)$ as measured by $\delta = \sigma/\mu$. Then (1.1) and (1.2') implies that $M \sim \left[\frac{\epsilon^{-D}}{\delta^2 N(\epsilon)}\right]$ as $\epsilon \to 0$. Thus for fixed δ, the number of evaluations of f_ϵ is much smaller than the number of boxes that need to be examined in straight-forward box counting. One can also modify this basic procedure by first subdividing the cube S into subcubes and then selecting a point at random from each cube rather than from all of S. This is called stratified sampling and has been shown to produce an estimate whose variance is no bigger than that of $I_M(\epsilon)$ (the mean is the same). Therefore one can expect to obtain a smaller error.

If a log-log plot of $A(\epsilon)$ as a function of ϵ is made then, assuming $N(\epsilon)\epsilon^{-D_C} = O(1)$ is slowly varying as $\epsilon \to 0$, we have

$$\log A(\epsilon) = (D - d_C)\log \epsilon + \zeta \tag{1.3}$$

where $\zeta = O(1)$ is slowly varying. For a finite point set this relation holds for a range of values ϵ where the plot appears linear. d_C is estimated by calculating the slope of the least square line. The method is easily implemented and reasonable results can be obtained with modest computation.

The information dimension is a second frequently calculated dimension. It is often estimated by the quantity

$$D_I = \lim_{\epsilon\to0} \frac{I(\epsilon)}{\ln 1/\epsilon}, \quad I(\epsilon) = \sum_{i=1}^{N(\epsilon)} \bar{p}_i \ln \bar{p}_i \tag{1.4}$$

where $\bar{p}_i$ is the probability of finding a point of X in the ith box of a minimal ϵ-covering of X. For the purpose of calculation we suppose that $I(\epsilon)$ may be replaced by $J(\epsilon) = \sum_{j=1}^{M(\epsilon)} p_j \ln p_j$ where p_j is the probability of finding a point of A in the jth box of a grid of size ϵ that covers X and $M(\epsilon)$ is the number of grid boxes. We then calculate

$$D_I = \lim_{\epsilon\to0} \frac{J(\epsilon)}{\ln 1/\epsilon}. \tag{1.5}$$

Our use of (1.5) is justified if X supports a probability measure P such that

$$\lim_{n\to\infty} \frac{\log P[Z_n(x)]}{n \log 2} = \alpha, \quad \text{a.e.}P, \tag{1.6}$$

where $Z_n(x)$ is the grid box of edge 2^{-n} that contains x and $\alpha = \dim_H P$ is the information dimension. The information dimension is defined as the minimum Hausdorff dimension of the sets which have P measure 1.

In what follows we assume that $\log P(z_n(x))$ is a P-measurable function of x.

Lemma: Let X be a set with probability measure P that is contained in S and suppose that (1.6) holds. Then the limit in (1.5) exists and $D_I = \alpha$, the information dimension.

Proof: Since $\frac{\log P(Z_n(x))}{n \log 2}$ converges a.e. P, $\left|\frac{\log P(Z_n(x))}{n \log 2}\right|$ is bounded a.e. P by $|\alpha| + 1$ for all sufficiently large n. By the bounded convergence theorem,

$$\lim_n \int_{S'} \frac{\log P(Z_n(x))}{n \log 2} P(dx) = \alpha.$$

But

$$\int_{S'} \frac{\log P(Z_n(x))}{n \log 2} P(dx) = \sum_{j=1}^{M_n} \frac{p_j \log p_j}{n \log 2},$$

where M_n is the number of grid boxes containing points of X and S' is the set of occupied grid boxes. Q.E.D.

Remark. L.S. Young in [7] showed that if P is a probability measure on a finite-dimensional manifold such that

(1.7) $$\lim_{r\to 0} \frac{\log P[B_r(x)]}{\log r} = \alpha \quad \text{a.e.P}$$

($B_r(x)$ is the ball of radius r and center x), then α is the information dimension. This condition is indeed satisfied for two-dimensional manifolds when P is the invariant measure of a $C^{1+\alpha}$ diffeomorphism with non-zero Lyapunoff exponents. Our assumption in (1.6) is obtained by replacing $B_r(x)$ in (1.7) by $Z_n(x)$. We have been unable to establish the equivalence of (1.6) and (1.7) in general, but conjecture that this is the case for Sinai-Ruelle-Bowen measures.

Section 2.

Calculation of both the capacity and information dimension as well as other Rényi exponents can be accomplished with greater speed and efficiency through the use of tree data structures, an idea due to F. Varosi [9]. To see this suppose $X \subset [0,1]$. The interval can be represented as a tree whose first node represents the entire interval. The intervals [0,1/2), [1/2,1] are represented by two nodes on the second level of the tree and so on. On the kth level, each of the 2^k subintervals of [0,1] is associated with a node. Each point $x \in X$ corresponds to a path in the tree whose nodes correspond to the dyadic subintervals to which x belongs. Box counting is accomplished by computing at each level n, the number of nodes through which a path passes; this is just M_n. The information dimension is calculated by computing the frequency of each occupied node. For large point sets this is close to the probability of an occupied box, so the frequencies can be used to estimate $J(2^{-n})$. The Monte Carlo technique can also be implemented with this scheme. Paths are assigned to each test point r_i $(i = 1,..M)$, and for each n, $f_{2^{-n}}(r_i) = 1$ if the path for r_i hits an occupied node on the nth level.

We can apply the method to sets embedded in higher dimension by first mapping them to the unit interval. For example, suppose without loss that X is embedded in the unit square in $\mathbb{R}^2$. If the binary expansions of the coordinates of a point $P = (x,y) \in S$ are $x = \sum_{i\geq 1} x_i 2^{-i}$, $y = \sum_{i\geq 1} y_i 2^{-i}$ the map $\phi: S \to I = [0,1]$ is defined as $q = \phi(P) = \sum_{i\geq 1} q_i 4^{-i}$ where $q_i = 2x_i + y_i$ is the ith digit in the 4-adic expansion of $\phi(P)$. ϕ is well defined if non-terminating expansions are used for non-zero x and y. Unfortunately, the tree data structure uses a great deal of storage since it has $O(\epsilon^{-D_c})$ nodes. Moreover, direct implementation of the tree leads to an inefficient non-vectorizable code for the CYBER 205. Our way out of these difficulties is explained again in terms of $X \subset \mathbb{R}^2$. After mapping X to [0,1] using the map ϕ defined above, the numbers are sorted using the Diamond sort method, a vectorized sorting algorithm [10]. The Monte Carlo method then consists of taking a set of random points in the plane $\{(r_i,\bar{r}_i)\}_{i=1}^M$ and calculating a corresponding set of test points $\{s_i\}_{i=1}^M$, $s_i = \phi(r_i,\bar{r}_i)$. Then a fast search method is used to locate each test point s in the sorted array $\{q(k)\}_{k=1}^N$. If $q(k) < s < q(k + 1)$ and $s - q(k) < 1/4^j$ or $q(k + 1) - s < 1/4^j$ for some j, then $f_\epsilon(s) = 1$ for $\epsilon = 1/4, 1/4^2,\dots,1/4^j$. Thus the information required for the Monte Carlo averages is accumulated. Box counting and the computation of the information dimension can also be implemented using this scheme. Table 1 displays the results of computations on the CYBER 205. All of these calculations took less than three minutes.

Section 3.

The justification for the use of the map is a consequence of the

Theorem 1. Let X be a set contained in the unit square, and let $\phi: S \to I$ be the map defined above. Then

$$\dim(X) = 2 \bullet \dim(\phi(X))$$

$$\text{h-}\dim(X) = 2 \bullet \text{h-}\dim(\phi(X))$$

where dim(X), h-dim(X) are the capacity and Hausdorff dimensions respectively. Beyers [16] proved the second part of this theorem for an equivalent map. Our proof relies on the work of Billingsley and Cajar.

Let $I_n(k)$ be the kth 4-adic interval where $[0, 1/4^n]$ if k = 0 and $(\frac{k-1}{4^n}, k/4^n]$ for $k = 1,\dots 4^n$. Let $\{S_n(k)\}_{k=1}^{4^n}$ $(n = 1,2,\dots)$ be the boxes in S whose sides are the dyadic intervals, $[0, 1/2^n]$, $(k - 1/2^n, k/2^n]$, $k = 1,\dots 2^n$.

In what follows, the sets $\{I_n(k)\}$ and $\{S_n(k)\}$ will be called cylinders. Each point in the cylinder set $S_n(k)$ is uniquely associated with an infinite sequence $\{x_i,y_i\}_{i\geq 1}$ where $\{x_i,y_i\}_{i=1}^n$ is determined by $S_n(k)$. Similarly, each point in the cylinder set $I_n(k)$ is uniquely associated with the infinite sequence $\{q_i\}_{i=1}^{\infty}$. Since the map $(x_i,y_i) \to q_i$ defining ϕ is a bijection of $\{0,1\} \times \{0,1\}$ onto $\{0,1,2,3\}$, ϕ itself is a bijection of $S_n(k)$ onto $I_n(k)$. It follows that

Lemma 1. $\phi(S_n(k)) = I_n(k)$, $\phi^{-1}(I_n(k)) = S_n(k)$.

The first statement in the theorem is an immediate consequence.

Proposition. $\dim(X) = 2 \bullet \dim(\phi(X))$.

Proof. If $x \in S_n(k)$ then $\phi(x) \in I_n(k)$, and vice versa. Thus there is a one to one correspondence between cylinders containing points of X and cylinders of I containing points of $\phi(X)$. If M_n is the number of occupied boxes in S, the dimension of $\phi(X)$ is

$$\lim_{n\to\infty} \frac{\log_2 M_n}{\log_2 4^n} = \frac{1}{2} \lim_{n\to\infty} \frac{\log_2 M_n}{n} = \frac{1}{2} \dim X. \quad \text{Q.E.D.}$$

A second immediate result of Lemma 1 is that ϕ is measure preserving on the class of partition cylinders, and therefore preserves the measure of sets that are disjoint unions of cylinders. Since this class of sets form a field that generates the Borel sets, by a well known theorem in measure theory, ϕ is a measure preserving map of (S,B_2,Λ) onto (I,B_1,λ) where B_1,B_2 are the Borel sets in I and S respectively, and λ,Λ are one and two-dimensional Lebesgue measure respectively. It is interesting to note that both ϕ and ϕ^{-1} are almost everywhere continuous (ϕ is discontinuous on the boundaries of the grid) so ϕ is almost a homeomorphism.

Although ϕ compresses <u>topological</u> and capacity dimension, it preserves another dimension which is related to the Hausdorff dimension and the second part of the theorem.

Let $(X,\underline{X},\mu)$ be a probability space where $\underline{X}$ is a σ-algebra of μ-measurable subsets of X and $\mu(X) = 1$. Our attention will be focused on the probability spaces (I,B_1,λ) and (S,B_2,Λ) with cylinder sets $\{I_n(k)\}_{k=1}^{4^n}$ and $\{S_n(k)\}_{k=1}^{4^n}$ $n = 1,2,\ldots$ respectively.

<u>Definition</u>. A μ - δ covering of a subset M of X is a denumerable collection of cylinders $\{B_i\}_{i\geq 1}$ that cover M and that satisfy $\mu(B_i) < \delta$.

For any $\alpha > 0$, let $L_\mu(M,\alpha,\delta) = \inf\{ \sum_{i\geq 1} \mu(B_i)^\alpha : \{B_i\}_{i\geq 1}$ is a μ-δ covering of $M\}$.
We then set $L_\mu(M,\alpha) = \lim_{\delta\to 0} L_\mu(M,\alpha,\delta)$.

Definition: The Billingsley dimension of M is defined as

$$\text{b-dim}_{\mu}(M) = \sup\{\alpha: 0 < \alpha \leq 1, L_{\mu}(M,\alpha) = \infty\}.$$

If $M \subset S$, M has finite outer measure, so $L_{\Lambda}(M,1) < \infty$ and it easily follows that $L_{\lambda}(\varphi(M),1) < \infty$. Therefore it is always the case that $\text{b-dim}_{\Lambda}(M) \leq 1$, $\text{b-dim}_{\lambda}(\varphi(M)) \leq 1$ even if sup were extended to $0 < \alpha < \infty$. Billingsley introduced this dimension in [11], for $X = I$, $\mu = \lambda$, and proved that the Billingsley and Hausdorff dimensions coincide. A clear discussion of these and related dimensions can be found in [13].

Let $\{B_i\}_{i\geq 1}$ be a set of cylinders of the form $S_n(k)$ for some n,k, and suppose they are a Λ - δ cover for $M \subset S$. If $x \in M \cap B_i$ for some i, then by Lemma 1, $\phi(x) \in \phi(M) \cap b_i$ where $b_i = \phi(B_i)$. Thus $\{b_i\}_{i\geq 1}$ is a λ - δ cover of $\phi(M)$. Now $\sum_{i\geq 1} \Lambda(B_i)^{\alpha} = \sum_{i\geq 1} \lambda(b_i)^{\alpha}$ so $L_{\Lambda}(M,\alpha,\delta) = L_{\lambda}(\phi(M),\alpha,\delta)$ for every δ. Given any $\epsilon > 0$ and for all sufficiently small $\delta > 0$ there is a Λ - δ covering $\{B_i\}$ satisfying $\sum_i \Lambda(B_i)^{\alpha+\epsilon} < \infty$, $\Lambda(B_i) < \delta$. Thus if $b_i = \phi(B_i)$, then $\sum_i \lambda(b_i)^{\alpha+\epsilon} < \infty$, where $\lambda(b_i) < \delta$, so $\text{b-dim}_{\lambda}\phi(M) < \alpha + \epsilon$. On the other hand for any $\epsilon > 0$, for all sufficiently small $\delta > 0$ there is a Λ - δ covering such that $\sum_i \Lambda(B_i)^{\alpha-\epsilon} > T$ for any pre-assigned $T > 0$. Thus, reasoning as before, $L_{\lambda}(\phi(M), \alpha - \epsilon) = \infty$ or $\text{b-dim}_{\lambda}\phi(M) > \alpha - \epsilon$. Since $\epsilon > 0$ was arbitrary, we can state that

Proposition 1: If $M \subset S$, $\text{b-dim}_{\Lambda}(M) = \text{b-dim}_{\lambda}(\phi(M))$.

The equality of the Billingsley dimension and Hausdorff dimension for sets in [0,1] was proved by Billingsley [11] using cylinders that were r-adic intervals. Wegmann and then Cajar [13] extended the result to more general cylinder sets in [0,1]. We will use a result proved by Cajar to prove the second part of the theorem.

Definition. A dimension system $(X,\{Z_n\}_{n\in N})$ is a non-empty set X together with a sequence of denumerable decompositions of X where $Z_0 = X$ and each Z_n is a refinement of Z_{n-1}. Let $Z_n(x)$ denote the cylinder of order n (in Z_n) containing x. We can now state Cajar's theorem.

Theorem 2. ([12]). Let $([0,1], \{Z_n\})$ be a dimension system on the unit interval, let $M \subset [0,1]$ and assume the following conditions:

(1) Each cylinder is an interval,

(2) $\lim_{n\to\infty} \lambda(Z_n(x)) = 0 \quad \forall x \in M$

(3) $\lim_{n\to\infty} \dfrac{\log \lambda(Z_n(x))}{\log \lambda(Z_{n+1}(x))} = 1 \quad \forall x \in M$.

Then h-dim(M) = b-dim$_\lambda$(M).

Now (S, $\{S_n(k)\}_{k=1}^{4^n}$, n = 1,2,...) is a dimension system. The interior of each $S_n(k)$ is an open sphere in the maximum metric, and these sets satisfy (2) and (3). This suggests that we can modify the arguments used by Cajar and prove

Proposition 2: Given the dimension system (S,$\{S_n(k)\}_{k=1,4^n}$, n = 1,2,...) and M ⊂ S, then

$$\text{h-dim}(M) = 2 \bullet \text{b-dim}_\Lambda(M).$$

This result will imply the second part of Theorem 1 for, by Propositions 1 and 2, h-dim(M) = 2 • b-dim$_\lambda$(φ(M)), and Theorem 2 implies that h-dim(M) = 2 • h-dim(φ(M)). It is interesting to note that this is a well known property of the Peano space-filling curve which φ^{-1} resembles [14] (see figure).

The proof of Proposition 2 depends on the following lemma.

Lemma: Every box O ⊂ S with sides parallel to those of S can be covered by F cylinders $\{B_i\}_{i=1}^F$ (of equal size) satisfying $\Lambda(B_i) < \Lambda(O) < F \bullet \Lambda(B_i)$ where F = 64 or 256.

Proof of Lemma. Using an argument of Billingsley [11] it can be shown that every interval in [0,1] can be covered by four 1-dimensional cylinders of length $1/2^n$, the length of the largest cylinder v, contained in the interval. The covering consists of the cylinder of length $1/2^{n-1}$ containing v, and the cylinder of the length $1/2^{n-1}$ containing the cylinder of length $1/2^n$ that is adjacent to v. Suppose we are given then an open square O of side ℓ with sides parallel to S. Define proj$_x$(O), proj$_y$(O) to be the projections onto (any one of) the horizontal and vertical sides of O respectively.

Let c be the length of the largest one-dimensional cylinder contained in proj$_y$(O) regarded as a subset of [0,1]. There is a two-dimensional cylinder of side c that meets O. Call this cylinder S_1. If the length(proj$_x$(S_1) ∩ proj$_x$(O)) < c/2 then there is a cylinder S_2, that is, the reflection of S_1 in the horizontal direction towards the interior of O such that length (proj$_x$(S_2) ∩ proj$_x$(O)) ≥ c/2. To see this, suppose this was not the case. Then S_2 could not be completely contained in O and thus (*) length(proj$_x$(S_2) ∩ proj$_x$(O)) + length(proj$_x$(S_1) ∩ proj$_x$(O)) ≥ ℓ. On the other hand the left hand side of the inequality is < c by our supposition. This contradicts the fact that c ≤ ℓ. Now if S_2 ⊂ O then the largest one-dimensional cylinder in proj$_x$(O) is at most 2c in length. For since length (proj$_x$(O)) = length (proj$_y$(O)), some translate, call it U of proj$_x$(O), contains a largest cylinder u of length c. Now if proj$_x$(O) contained a cylinder of length greater than 2c then length $(\text{proj}_x(0)) \geq 4c$, the length of the next largest cylinder. But as in Billingsley's one dimensional argment u in U and three other cylinders of length c form an interval which must contain U. Thus length (U) < 4c; in contradiction to the fact that length (U) = $\text{length}(\text{proj}_x(0))$. We can again apply the one-dimensional covering argument to $\text{proj}_x(0)$, using this cylinder. So there is a horizontal row of 8 cylinders (obtained by reflecting S_2) whose x-projections cover the horizontal sides of 0 and therefore every horizontal line in 0 of 0. This row can be reflected in the vertical direction to obtain a set of 64 cylinders in which every vertical line and horizontal line in 0 is covered. Thus 0 is covered. If $S_2 \cap 0 \neq \emptyset$ but $S_2 \subset 0$, then since $\text{length}(\text{proj}_x(S_2) \cap \text{proj}_x(0)) \geq c/2$, there is a subcylinder of S_2 with side of length c/2 that is completely contained in 0. By reflecting this cylinder horizontally we obtain a row of 16 cylinders whose x-projection covers every horizontal line. Thus the set of 256 squares obtained by reflecting the row, is a covering of 0.

Proof of Proposition 2: Let $\alpha = \text{b-dim}_\Lambda(M)$. For any $\epsilon > 0$, and given $\eta > 0$, there exists a $\delta > 0$ such that $\sum_i \Lambda(B_i)^{\alpha+\epsilon/2} < \eta$ where $\{B_i\}_{i\geq 1}$ is a denumerable collection of cylinders that cover M and satisfy $\Lambda(B_i) < \delta^2$. For such a covering it is also true that $\sum_i (\text{diam}(B_i))^{2\alpha+\epsilon} < 4\eta$ where $\text{diam}(B_i) < 2\delta$. Therefore $\text{h-dim}(M) < 2\alpha + \epsilon$, and if $\epsilon \to 0$ we can conclude that $\text{h-dim}(M) \leq 2\text{b-dim}_\Lambda(M)$. To obtain the reverse inequality consider $s > \text{h-dim}(M)$. Then for any $t > 0$, there is a $\delta > 0$ and a denumerable set of open boxes $\{0_i\}_{i\geq 1}$ that cover M and satisfy

$$\sum_i [\text{diam}(0_i)]^s < t, \ \text{diam}(0_i) < 4\delta.$$

Let $\{B_{ij}\}_{j\geq 1}^{F_i}$ be the cylinders covering 0_i as in the lemma. $\bigcup_i \bigcup_j B_{ij}$ is a denumerable covering of M and

$$\sum_{i,j} \Lambda(B_{ij})^{s/2} \leq \sum_i F\,\Lambda(0_i)^{s/2} < \text{const.} \bullet 2^{-s} \bullet \sum_i \text{diam}(0_i)^s$$

$< \text{const} \bullet t$ where $\Lambda(B_{ij}) < \delta^2$. It follows then that $2\text{b-dim}(M) < s$ for any $s > \text{h-dim}(M)$. Allowing $s \downarrow \text{h-dim}(M)$ we obtain the desired inequality.

Acknowledgement: The first author wishes to thank the Scientific Computing Division of the National Bureau of Standards for their generous hospitality during her stay.

BIBLIOGRAPHY

1. H. Greenside, A. Wolf, J. Swift, T. Pignataro, "The Impracticality of a Box Counting Algorithm for Calculating the Dimensionality of Strange Attractors", Phys. Rev. A25 (1982) 3453.

2. A. Cohen, I. Procaccia, "Computing the Kolmogorov entropy from time signals of dissipative and conservative dynamical systems," Phys. Rev. A31, (1985) 1872.

3. M. McGuiness, "A Computation of the Limit Capacity of the Lorenz Attractor", Physica 16D (1985), 265-275.

4. R. Badii, A. Politi, "Statistical Description of Chaotic Attractors: The Dimension Function", J. Stat. Phys., 40 (1985) 725-750.

5. E. Ott, J. Yorke, E. Yorke, "A Scaling Law: How an Attractor's Volume depends on Noise Level", preprint.

6. D.A. Russell, J. Hanson, E. Ott, Phys. Rev. Lett. 45 (1980), 1175.

7. L.S. Young, "Dimension entropy, and Lyapunov exponents", Erg. Th. & Dyn. Sys. 2 (1982), 109-124.

8. J.P. Eckmann, D. Ruelle, "Ergodic Theory of Chaos and Strange Attractors" Rev. Mod. Phys. 57, (1985), 617-655.

9. F. Varosi (to appear).

10. B. Brooks, H. Brock, F. Sullivan, "Diamond: A Sorting Method for Vector Machines", BIT 21 , (1981)142-152.

11. P. Billingsley, "Hausdorff Dimension in Probability Theory" Ill. J. Math 4 (1960), 187-209.

12. P. Billingsley, Ergodic Theory and Information, John Wiley and Sons, New York, 1965.

13. H. Cajar, Billingsley Dimension in Probability Spaces, Lecture Notes in Mathematics no. 892, Springer-Verlag, 1981.

14. F.M. Dekking, "Variations on Peano", O. Arch. v. Wisk 3 (1980), 275-281.

15. F. Hunt, F. Sullivan, "Efficient Algorithms for Computing Fractal Dimensions", in Dimensions and Entropies in Chaotic Systems, ed. G. Mayer-Kress, Springer Verlag, 1985.

16. W.A. Beyer, "Hausdorff dimension of level sets of some Rademacher series", Pac. J. Math. 12 (1962), 35-46 .

17. J.D. Farmer, E. Ott, J. Yorke, "The Dimension of Chaotic Attractors"; Physica 7D (1983), 153-180.

Appendix

Proof of (1.1') and (1.2'). The validity of (1.2') will be demonstrated first. Define $A_n = \text{area}\{\bigcup_{x\in X} Z_n(x)\}$ and $B(n,X)$ = set of points for which $d(y,X) < 2^{-n}$ where $d(y,x)$ is the max-norm distance. If B_n is the area of $B(n,X)$, we claim that

$$A_n \leq B_n \leq 9A_n.$$

First note that $\bigcup_{x\in X} Z_n(x) \subset B(n,X)$ since any y in the first set is in a grid box with a point of X. On the other hand, if y is in $B(n,X)$ then there exists an x_0 such that $d(x_0,y) < 2^{-n}$. Thus y is either on $Z_n(x_0)$ or in one of the 8 adjacent squares. The area of this region is at most $9A_n$. Thus

$$\frac{B_n}{9} \leq A_n \leq B_n .$$

Now suppose (A.1) $\lim_n \frac{\ln B_n}{n \ln 2} = \beta$. Then

$$\lim_n \frac{\ln A_n}{n \ln 2} = \beta.$$

We now show that (A.1) holds with $\beta = D_c - 2$. Let $U(\epsilon)$ be the set of points within ϵ of X in the Euclidean sense, i.e., $\inf_{x\in X} \rho(y,x) < \epsilon$ iff $y \in U(\epsilon)$. If $y \in U(2^{-(n+1)})$, then: $\exists\, x_0 \in X$ such that $\rho(x_0,y) < 2^{-(n+1)}$, therefore $d(x_0,y) < 2^{(-n+1)}$ so $y \in B(n,X)$. If $y \in B(n,X)$ then $d(x_0,y) < 2^{-n}$ for some $x_0 \in X$ thus $\rho(x_0,y) < 2^{-n}\sqrt{2}$. Thus $U(2^{-(n+1)}) \subset B(n,X) \subset U(2^{-n}\sqrt{2})$. This implies that the respective areas satisfy $A(U(2^{-(n+1)})) \leq B_n \leq A(U(2^{-n}\sqrt{2}))$. Since (see [5])

$$\text{(A.2)} \qquad \lim_{\epsilon\to 0} \frac{\ln A(U(\epsilon))}{\ln 1/\epsilon} = 2 - D_c \quad \text{and}$$

$$\frac{\ln A(U(2^{-(n+1)})}{n \log 2} \leq \frac{\ln B_n}{n \ln 2} \leq \frac{\ln A(U(2^{-n}\sqrt{2}))}{n \log 2}$$

on taking the limit as $n \to \infty$ and applying (A.2) for $\epsilon = 2^{-n}$ we obtain (A.1) with $\beta = 2 - D_c$. Equation (1.1') may be obtained from the relation $A_n = M_n 4^{-n}$ and applying (1.2').

TABLE 1

	Monte Carlo	Cell Count
Baker	1.635	1.603
Henon	1.383	1.304
Zaslavskij	-	1.762
Lorenz	2.21	2.04

Results are reported for four cases: the Baker's transformation, the Henon map, the Zaslavskij map and the Lorenz attractor. The columns headed Monte Carlo and Cell Count show calculations of the capacity dimension using the Monte Carlo and box counting methods described in the paper.

Asymptotic Behavior of Solutions to Quasimonotone Parabolic Systems

Robert H. Martin, Jr.
Department of Mathematics
North Carolina State University
Raleigh, North Carolina 27695-8205

Let Ω be a bounded domain in $\mathbf{R}^N$ with $\partial\Omega$ smooth and let Δ be the Laplacian operator on Ω. Also, let m be a positive integer, $m \geq 2$, and let $f = (f_i)_1^m$ be a C^2 function from $\mathbf{R}^m$ into $\mathbf{R}^m$. The purpose of this paper is to develop abstract results which apply to the study of the asymptotic behavior of solutions to reaction diffusion systems of the form

$$\begin{cases} \text{(a)} \quad \partial_t u_i = d_i \Delta u_i + f_i(u_1, \ldots, u_m), & t > 0,\ \sigma \in \Omega, \quad i = 1, \ldots, m \\ \text{(b)} \quad \alpha_i u_i + k_i \partial_n u_i = 0 & t > 0,\ \sigma \in \partial\Omega,\ i = 1, \ldots, m \\ \text{(c)} \quad u_i = x_i & t = 0,\ \sigma \in \Omega, \quad i = 1, \ldots, m \end{cases}$$

where $d_i > 0$, $\alpha_i \geq 0$, $k_i \in \{0, 1\}$, and ∂_n is the outward normal derivative on $\partial\Omega$. Also, if $k_i = 0$ it is assumed $\alpha_i = 1$ and if $\alpha_i = 0$ it is assumed that $k_i = 1$. The initial functions x_i are assumed nonnegative and continuous on $\overline{\Omega}$.

Under special assumptions on f, the results of this paper give precise information on the asymptotic behavior of solutions to (RD) as $t \to \infty$. Furthermore, these results are quite general and include earlier techniques by the author [**5**], [**6**], Capasso and Maddalena [**1**] and Messia and De Mottoni [**8**]. In particular, by using the concept of u_o-positive operators in Krasnoselskii [**3**], we are able to extend many techniques that require the underlying cone to have nonempty interior.

§1. **THE BASIC ABSTRACT SYSTEM.** Suppose X_i, $i = 1, \ldots, m$, is a real Banach space with norm denoted by $|\cdot|$, and let $X = \prod_{i=1}^m X_i$ with norm $|\cdot|$ defined on X by

$$|x| = \sum_{i=1}^m |x_i| \text{ for all } x = (x_i)_1^m \in X$$

For $C_i \subset X_i$ and $x_i \in X_i$ define $d_i(x_i; C_i) = \inf\{|x_i - y_i| : y_i \in C_i\}$ and for $C \subset X$ and $x \in X$ define $d(x; C) = \inf\{|x - y| : y \in C\}$. Observe that

$$(1.1) \qquad d(x; \prod_{i=1}^m C_i) = \sum_{i=1}^m d_i(x_i; C_i) \text{ for all } x = (x_i)_1^m \in X,\ C_i \subset X_i$$

Suppose also that X_i^+ is a positive cone in X_i with the property that the induced partial ordering "$\geq$" is a Banach lattice:

$$\begin{cases} \text{(a)} \quad \text{if } x_i, y_i \in X_i \text{ then } z_i = \sup\{x_1, y_i\} \text{ exists -- that is} \\ \qquad\quad z_i \geq x_i,\ z_i \geq y_i \text{ and if } w_i \geq x_i,\ w_i \geq y_i \text{ then } w_i \geq z_i \\ \text{(b)} \quad \text{if } 0 \leq x_i \leq y_i \text{ then } |x_i| \leq |y_i| \end{cases}$$

Remark 1.1. Property (b) is not precisely the same as the corresponding property in a Banach lattice given in Vulikh [**12**, p. 173]. However, it follows easily from [**12**, Theorem VII.1.4, p. 176] that the norm on X_i is equivalent to one having the property of monotonicity, and this is all that needed for our results.

For each $i = 1, \ldots, m$, we suppose that $T_i(t)$, $t \geq 0$, is an analytic semigroup of bounded linear operators on X_i with generator A_i (see Goldstein [**2**] or Pazy [**10**]). Suppose also that T_i is positive: $T_i(t)x_i \geq 0$ whenever $t \geq 0$ and $x_i \geq 0$. Furthermore, let ζ_i be either $+\infty$ or a member of X_i^+ and set $[0, \zeta_i] = X_i^+$ if $\zeta_i = +\infty$ and $[0, \zeta_i] = \{x_i : 0 \leq x_i \leq \zeta_i\}$ if $\zeta \in X_i^+$. Also define $D^+ \subset X$ by

$$D^+ = \prod_{i=1}^{m} [0, \zeta_i]$$

Now suppose that F_i is a continuous function from D^+ into X_i and consider the following system of integral equations:

$$u_i(t) = T_i(t)z_i + \int_o^t T_i(t-r)F_i(u(r))dr, \ t \geq 0, \ i = 1, \ldots, m \tag{1.2}$$
$$\text{where } z = (z_i)_i^m \in D^+ \text{ and } u = (u_i)_1^m.$$

A function $u = (u_i)_1^m$ from $[0, b) \to X$ is said to be a solution to (1.2) on $[0, b)$ if u is continuous, $u(t) \in D^+$ for all $t \in [0, b)$, and (1.2) is satisfied for $t \in [0, b)$. If each u_i is differentiable on $(0, b)$ then a solution u to (1.2) satisfies the system of differential equations

$$u_i'(t) = A_i u_i(t) + F_i(u(t)), \ u_i(0) = z_i, \ t \geq 0, \ i = 1, \ldots, m. \tag{1.3}$$

Define $T(t)x = (T_i(t)x_i)_1^m$ and $F(x) = (F_i(x))_1^m$ and consider

$$u(t) = T(t - T_o)z + \int_{t_o}^t T(t-r)F(u(r))dr, \ t \geq t_o \geq 0 \tag{1.4}$$

Note that the semigroup property of T [i.e., $T(t+s) = T(t)T(s)$ for $t \geq s \geq 0$] implies that a solution u to (1.2) on $[0, b)$ satisfies (1.4) for any $0 \leq t_o \leq t < b$. In addition to the aforementioned properties of T_i and F_i it is assumed that each of the following also hold:

(C1) For each $R > 0$ there is an $L(R) > 0$ so that
$|F(x) - F(y)| \leq L(R)|x - y|$ for $x, y \in D^+$ with $|x|, |y| \leq R$

(C2) $\lim_{h \to 0+} \frac{1}{h} d(T(h)x + hF(x); D^+) = 0$ for all $x \in D^+$.

(C3) $\lim_{h \to 0+} \frac{1}{h} d(x - y + h[F(x) - F(y)]; X^+) = 0$ for all $x, y \in D^+$ with $x \geq y$.

Remark 1.2. Notice that (1.1) implies (C3) may be written

(C3)′ $\lim_{h \to 0+} \frac{1}{h} d_i(x_i - y_i + h[F_i(x) - F_i(y)]; X^+) = 0$ for
$i = 1, \ldots, m$ whenever $x, y \in D^+$ with $x \geq y$.

Such a function F satisfying (C3) is called quasimonotone.

Theorem 1. If (C1) - (C3) hold and $z \in D^+$ then (1.2) has a unique noncontinuable solution $u = u(\cdot; z)$ on $[0, b_z)$ where $0 < b_z \leq +\infty$. Furthermore, if $U(t) = (U_i(t))_1^m$ is defined by

$$U(t)z = u(t,z) \text{ for } z \in D^+ \text{ and } 0 \leq t < b_z$$

then U is a (local) C_0 semigroup of nonlinear operators and the following properties are valid:

(i) $U(0)z = z$ and $0 \leq s < b_z$, $0 \leq t < b_{U(s)z}$ implies $t + s < b_z$ and $U(t+s)z = U(t)U(s)z$;

(ii) $(t,z) \to U(t)z$ is continuous on $\{(t,x) : z \in D^+,\ 0 < t < b_z\}$.

(iii) $|U(t)z| \to \infty$ as $t \to b_z -$ if $b_z < \infty$.

(iv) if $x, y \in D^+$ with $x \geq y$, then $b_y \geq b_x$ and $U(t)x \geq U(t)y$ for all $0 \leq t < b_x$.

In addition, $u(\cdot, z)$ is C^1 on $(0, b_z)$ satisfies the differential equation (1.3) on $(0, b_z)$.

Remark 1.3. This theorem follows from the results in [4, Chapter 8], for example. A convenient criteria to check (C2) is the following which places conditions on T and F separately:

$$(1.5) \qquad \begin{cases} \text{(C2) holds if } T(t) : D^+ \to D^+ \text{ for all } t \geq 0 \text{ and} \\ \lim\limits_{h \to o+} h^{-1} d_i(x_i + hF_i(x); [0, \zeta_i]) = 0 \text{ whenever} \\ x = (x_j)_1^m \in D^+ \text{ and } i = 1, \dots, m. \end{cases}$$

(see [4, Lemma 1.3, p. 326]). Observe, for example, that if $X = \mathbf{R}^m$, $X_i^+ = [0, \infty)$ and $D^+ = \prod_{i=1}^m [0, \zeta_i]$ where $0 < \zeta_i \leq \infty$, then the limit in (1.5) holds if and only if

$$(1.6) \qquad \begin{cases} \text{for each } k \in \{1, \dots, m\} \text{ and } x = (x_i)_1^m \in \prod_{i=1}^m [0, \zeta_i], \text{ if} \\ x_k = 0 \text{ then } F_k(x) \geq 0 \text{ and if } x_k = \zeta_k \text{ then } F_k(x) \leq 0. \end{cases}$$

Our next result uses positive linear functions to obtain estimates between solutions to (1.2) that are related to strict inequalities. In order to develop these techniques we assume a condition on $F = (F_i)_1^m$ that seems slightly more restrictive than quasi-monotonicity:

$$(1.7) \qquad \begin{cases} \text{For each compact } K \subset D^+ \text{ there is an } L(K) > 0 \text{ so that} \\ \qquad F_i(x) - F_i(y) \geq -L(K)(x_i - y_i) \\ \text{for all } i \in \{1, \dots, m\} \text{ and } x, y \in K \text{ with } x \geq y. \end{cases}$$

It is easy to see that (1.7) implies F is quasi-monotone. For if $K = \{x, y\}$ where $x \geq y$, then

$$x_i - y_i + h[F_i(x) - F_i(y)] \geq (1 - hL(K))(x_i - y_i) \geq 0$$

if $hL(K) < 1$, and it is immediate that (C3)′ in Remark 1.2 holds. A crucial property implied by (1.7) is given by the following lemma:

Lemma 1.1 Suppose (1.7) holds, $z, w \in D^+$ with $z \geq w$, and $0 \leq t_o < c < b_z$. Then there is an $M = M(c, z, w) > 0$ such that

$$U_i(t)z - U_i(t)w \geq e^{-M(t-t_o)}T_i(t-t_o)[U_i(t_o)z - U_i(t_o)w] \tag{1.8}$$

for all $t_o \leq t \leq c$ and $i = 1, \dots, m$.

Proof. Set $M = L(K)$ where L is as in (1.7) and $K = \{U(t)z, U(t)w : 0 \leq t \leq c\}$. Since each T_i is positive, $x_i \geq y_i$ implies $T_i(t) \geq T_i(t)y_i$, we have from (1.7) and (1.4) that

$$\begin{aligned} U_i(t)z - U_i(t)w &= T_i(t-t_o)[U_i(t_o)z - U_i(t_o)w] \\ &\quad + \int_{t_o}^{t} T_i(t-r)[F_i(U_i(r)z) - F_i(U_i(r)w)]dr \\ &\geq T_i(t-t_o)[U_i(t_o)z - U_i(t_o)w] \\ &\quad + \int_{t_o}^{t} T_i(t-r)[-M(U_i(r)z) - U_i(r)w)]dr \end{aligned}$$

Thus, $U_i(t)z - U_i(t)w \geq v_i(t)$ where v_i satisfies

$$v_i(t) = T_i(t-t_o)[U_i(t_o)z - U_i(t_o)w] - M\int_{t_o}^{t} T_i(t-r)v_i(r)dr$$

(see, e.g., [4, Proposition 6.1, p. 367]). One may verify directly that $v_i(t)$ is given by the right hand side of the inequality in (1.8).

In order to establish the main results on strict inequalities we use a class of positive linear functionals on each X_i and we follow the ideas introduced in Martin and Smith [7]. So let X_i^* denote the dual of X_i and let P_i^* be the class of positive members of $X_i^* : \Phi_i \in P_i^* \Leftrightarrow \Phi_i \in X_i^*$ and $\Phi(x_i) \geq 0$ for all $x_i \geq 0$. For each $i \in \{1, \dots, m\}$ select a subset Q_i^* of P_i^* and assume $0 \notin Q_i^*$ [Hence $\Phi_i \in Q_i^*$ implies $\Phi_i(x_i) > 0$ for some $x_i \in X_i^+$, since $X_i = X_i^+ - X_i^+$ is part of the definition of a Banach lattice]. The fundamental property for the linear semigroup T relative to the classes Q_i^* is the following assumption:

$$\left\{\begin{array}{l} \text{if } k \in \{1, \dots, m\} \text{ and } x_k \in X_k^+ \text{ with } \Phi_k(x_k) > 0 \text{ for} \\ \text{some } \Phi_k \in Q_k^*, \text{ then } \Phi_k(T_k(t)x_k) > 0 \text{ for all } t > 0 \text{ and all } \Phi_k \in Q_k^* \end{array}\right. \tag{1.9}$$

This assumption corresponds to the strong maximum principle when T_k is generated by the heat equation. For the nonlinear term $F = (F_i)_1^m$ we use the following assumption:

$$\left\{\begin{array}{l} \text{if } \Sigma \text{ is a proper, nonempty subset of } \{1, \dots, m\}, \Sigma^c \\ \text{the complement of } \Sigma, \text{ and } z, w \in D^+ \text{ with} \\ \quad \text{(a) } z \geq w \\ \quad \text{(b) } \Phi_j(z_j) = \Phi_j(w_j) \text{ for all } j \in \Sigma^c,\ \Phi_j \in Q_j^* \\ \quad \text{(c) } \Phi_j(z_j) > \Phi_j(w_j) \text{ for all } j \in \Sigma,\ \Phi_j \in Q_j^* \\ \qquad \text{then there is a } k \in \Sigma^c \text{ and a } \Phi_k \in Q_k^* \text{ such that} \\ \qquad\qquad \Phi_k(F_k(z)) > \Phi_k(F_k(w)). \end{array}\right. \tag{1.10}$$

Remark 1.4. Since F is quasi-monotone [see (C3) and (C3)′] we have that if $z \geq w$ and $k \in \{1, \ldots, m\}$ then

$$z_k - w_k + h[F_k(z) - F_k(w)] = p_h + o(h)$$

where $p_h \geq 0$ and $h^{-1}|o(h)| \to 0$ as $h \to 0+$. If $\Phi_k \in Q_k^*$ and $\Phi_k(z_k) = \Phi_k(w_k)$, then

$$h\Phi_k[F_k(z) - F_k(w)] = \Phi_k(p_h) + \Phi_k(o(h)) \geq \Phi_k(o(h))$$

and hence $\Phi_k(F_k(z)) \geq \Phi_k(F_k(w))$. Therefore, the crucial point in (1.10) is that a strict inequality holds for some $k \in \Sigma^c$ and some $\Phi_k \in Q_k^*$.

Theorem 2. Suppose that (C1) – (C3), (1.7), (1.9) and (1.10) are satisfied and that $z, w \in D^+$ with $z \geq w$. If $\Phi_n(z_n) > \Phi_n(w_n)$ for some $n \in \{1, \ldots, m\}$ and some $\Phi_n \in Q_n^*$, then

(1.11) $\qquad \Phi_j(U_j(t)z) > \Phi_j(U_j(t)w)$ for all $0 < t < b_z$, all $\Phi_j \in Q_j^*$ and all $j = 1, \ldots, m$

Proof. Consider the statement

(1.12) $\qquad \Phi_k(U_k(t_o)z) > \Phi_k(U_k(t_o)w)$ for some $k \in \{1, \ldots, m\}$, $\Phi_k \in Q_k^*$ and $0 \leq t_o < b_z$,

and define

$$\Gamma(t) = \{i : \Phi_i(U_i(t)z) > \Phi_i(U_k(t)w) \text{ for all } \Phi_i \in Q_i^*\}$$

It follows from (1.8) and assumption (1.9) that if (1.12) holds then $k \in \Gamma(t)$ for all $t_o < t < b_z$. Since (1.12) holds with $t_o = 0$ and $k = n$ by assumption, we have $\Gamma(t) \supset \{n\}$ for all $t > 0$. Suppose, for contradiction, that for some interval $[t_o, t_o + \epsilon] \subset (0, b_z)$, we have $\Gamma(t) \equiv \Sigma$ where Σ is a proper subset of $\{1, \ldots, m\}$. Replacing z by $U(t_o)z$ and w by $U(t_o)w$ in property (1.10) implies there is a $k \in \Sigma^c$ and a $\Phi_k \in Q_k^*$ such that

$$\Phi_k(F_k(U(r)z)) > \Phi_k(F_k(U(r)w) \text{ if } r = t_o$$

By continuity there is a $\delta > 0$ such that this inequality holds for all $t_o \leq r \leq t_0 + \delta$. Therefore,

$$\Phi_k(T_k(t-r)[F_k(U(r)z - F_k(U(r)w)] > 0 \text{ for } t_o < r < t < t_o + \delta$$

by (1.9), and hence

$$\Phi_k\left(\int_{t_o}^t T_k(t-r)F_k(U(r)z)dr - \int_{t_0}^t T_k(t-r)F_k(U(r)w)dr\right) > 0$$

for $t_o < t \leq t_0 + \delta$. Since

$$\Phi_k(T_k(t-t_o)U_k(t_o)z - T_k(t-t_0)U_k(t_0)w) \geq 0$$

for $t_o \leq t \leq t_o + \delta$, it follows from (1.2), with 0 replaced by $t_o \geq 0$, that

$$\Phi_k(U_k(t)z - \Phi_k(U_k(t)w) > 0 \text{ if } t_o < t \leq t_o + \delta.$$

Since $k \in \Sigma^c$ this contradicts the assumption $\Gamma(t) \equiv \Sigma$ on $[t_o, t_o + \epsilon]$. Thus $\Gamma(t)$ is constant on an open interval only in case $\Gamma(t) \equiv \{1, \ldots, m\}$ on this interval. From this it is obvious that $\Gamma(t) \equiv \{1, \ldots, m\}$ for all $t > 0$, and the proof is complete.

§2. **ESTIMATES USING POSITIVE EIGENVECTORS.** Very effective techniques are developed in this section that combine those using positive functionals from the preceding section with those using eigenvectors in the positive cone. In this section we assume throughout that the following hold:

$$(2.1)\quad \begin{cases} \text{(a)} & \text{The suppositions of Theorems 1 and 2 in §1 are satisfied.} \\ \text{(b)} & T_i(t) \text{ is compact for each } t > 0 \text{ and } i = 1, \ldots, m. \\ \text{(c)} & \text{If } i \in \{1, \ldots, m\} \text{ and } x_i \in X_i^+ \text{ with } x_i \neq 0, \text{ then} \\ & \text{there is a } \Phi_i \in Q_i^* \text{ so that } \Phi_i(x_i) > 0. \end{cases}$$

Therefore, the (local) C_o semigroup $U(t)z$, $0 < t < b_z$, defined as the solution $u(t) = u(t, z)$ to (1.2) satisfies each of the properties listed in Theorem 1 as well as the two additional properties

$$(2.2)\quad \begin{cases} \text{(a)} & z, w \in D^+ \text{ with } z \geq w \text{ and } z \neq w \text{ implies} \\ & \phi_i(U_i(t)z) > \Phi_i(U_i(t)w) \text{ for all } t > 0, \text{ all } \Phi_i \in Q_i^* \\ & \text{and all } i = 1, \ldots, m. \\ \text{(b)} & \text{if } C \text{ is a bounded subset of } D^+ \text{ and } |U(t)z| \leq R \\ & \text{for } z \in C,\ 0 \leq t \leq b < b_z \text{ and some } R > 0, \text{ then there} \\ & \text{is a compact subset } K(\epsilon, b, R, C) \text{ such that} \\ & U(t)z \in K(\epsilon, b, R, C) \text{ for all } 0 < \epsilon \leq t \leq b \text{ and } z \in C. \end{cases}$$

Assertion (2.2a) follows from Theorem 2 and assumption (2.1c), and (2.2b) follows in a standard manner from (2.1b) and the assumption (C1) on F, which implies F is bounded on bounded sets.

Our fundamental assumption is that each $T_i(t)$, $t > 0$, has a positive eigenvector corresponding to a dominant real eigenvalue:

$$(2.3)\quad \begin{cases} \text{(a)} & \text{For each } i \in \{1, \ldots, m\} \text{ there is a } \mu_i \in \mathbf{R} \text{ and a} \\ & p_i \in X_i^+,\ p_i \neq 0, \text{ such that } T_i(t)p_i = e^{\mu_i t} p_i \text{ for all } t \geq 0. \\ \text{(b)} & \text{For each } i \in \{1, \ldots, m\} \text{ and } (t, x_i) \in (0, \infty) \times X_i^+ \text{ with} \\ & x_i \neq 0, \text{ there are numbers } a_i(t, x_i),\ b_i(t, x_i) > 0 \text{ such} \\ & \text{that } a_i(t, x_i)p_i \leq T_i(t)x_i \leq b_i(t, x_i)p_i. \\ \text{(c)} & \text{If } A_i \text{ is the generator of } T_i \text{ and } x_i \in D(A_i) \cap X_i^+, \\ & \text{then } x_i \leq \gamma_i(x_i)p_i \text{ for some } \gamma_i(x_i) > 0. \end{cases}$$

It is immediate from (a) that $p_i \in D(A_i)$ and $A_i p_i = \mu_i p_i$. Combining (2.3a) and (2.3b) shows that

$$T_i(t)x_i = T_i(t-\epsilon)T_i(\epsilon)x_i \leq T_i(t-\epsilon)(b_i(\epsilon, x_i)p_i) \text{ and}$$

$$T_i(t)x_i = T_i(t-\epsilon)T_i(\epsilon)x_i \geq T_i(t-\epsilon)(a_i(\epsilon, x_i)p_i)$$

for all $t \geq \epsilon > 0$, and hence by linearity

$$(2.4)\quad \begin{cases} \left[a_i(\epsilon, x_i)e^{-\mu_i \epsilon}\right] p_i \leq T_i(t)x_i \leq \left[b_i(\epsilon, x_i)e^{-\mu_i \epsilon}\right] e^{\mu_i t} p_i \\ \text{for all } t \geq \epsilon > 0,\ i = 1, \ldots, m \text{ and } x_i \in X_i^+ \text{ with } x_i \neq 0. \end{cases}$$

Under these assumptions U possesses the following basic property:

Lemma 2.1. Suppose (2.1) and (2.3) are satisfied and $z, w \in D^+$ with $z \geq w$. Then for each $t > 0$ there are numbers $\overline{b}_i(t,z,w), \overline{a}_i(t,z,w) > 0$ such that

$$\overline{a}_i(t,z,w)p_i \leq U_i(t)z - U_i(t)w \leq \overline{b}_i(t,z,w)p_i \quad \text{for } i = 1,\dots,m \tag{2.5}$$

Proof. Since T_i is analytic and F_i is Lipschitz continuous, it follows from Pazy [**10**, p. 196], for example, that $U_i(t)z \in D(A_i)$, and hence such a function $\overline{b}$ exists by assumption (2.3c). If $t_o > 0$ then $U_i(t_o)z \neq U_i(t_o)w$ for all i by Theorem 2 and (2.1c), so it follows from (1.8) and (2.3b) that

$$U_i(t)z - U_i(t)w \geq e^{-M(t-t_o)}a_i(t-t_o, U_i(t_o)z - U_i(t_o)w)p_i$$

for all $t_o < t < b_z$ and all $i = 1,\dots,m$. Thus (2.5) holds with

$$\overline{a}_i(t,z,w) = e^{-M(t-t_o)}a_i(t-t_o, U_i(t_o)z - U_i(t_o)w)$$

where $t_o = t/2$, for example.

Under these assumptions we have the following basic result for linear perturbations:

Theorem 3. Suppose that (2.1) and (2.3) are satisfied and there is a bounded linear operator B on X such that $F(x) = Bx$ for all $x \in D^+ \equiv X^+$. Then $U(t)z \equiv V(t)z$ for all $t \geq 0$ and $z \in X^+$ where V is a positive analytic semigroup of compact linear operators on X with generator $A + B$. Furthermore, there is a $\nu \in \mathbf{R}$ and a $q = (q_i)_1^m \in X^+$ such that

(i) $\overline{\epsilon}p_i \leq q_i \leq \overline{\epsilon}^{-1}p_i$ for some $\overline{\epsilon} > 0$ and all $i = 1,\dots,m$
(ii) $V(t)q = e^{\nu t}q$ for all $t \geq 0$
(iii) for each $x \in X^+$, $x \neq 0$, and $t > 0$ there are $\tilde{a}(t,x)$, $\tilde{b}(t,x) > 0$ such that $\tilde{a}(t,x)q \leq V(t)x \leq \tilde{b}(t,x)q$.

Proof. Since B is bounded it is well-known that $A + B$ is the generator of a compact, analytic semigroup V and if $u(t) \equiv V(t)z$ for $t \geq 0$, then u is the solution to (1.2) (see Pazy [**10**, p. 80]). Thus $U(t)z \equiv V(t)z$ for $t \geq 0$ and $z \geq 0$, and so it remains to show that (i) – (iv) hold true. By (2.5) in Lemma 2.1 (with $U = V$ and $w = 0$) we have

$$\overline{a}(t,z)p \leq V(t)z \leq \overline{b}(t,z)p \text{ if } t > 0,\ z \in X^+,\ z \neq 0 \tag{2.5$'$}$$

where $\overline{a}(t,z) \equiv \min\{\overline{a}_i(t,z,0) : i = 1,\dots,m\}$ and $\overline{b}(t,z) \equiv \max\{\overline{b}_i(t,z,0)\ i = 1,\dots,m\}$. In the terminology of Krasnoselskii [**3**], $V(t)$ is p-positive for each $t > 0$. In particular by [**3**, Th 2.5, p. 67] for each $t > 0$ there is a $w(t) \in X^+$, $|w(t)| = 1$, and a $\lambda(t) \in (0,\infty)$ such that

$$V(t)w(t) = \lambda(t)w(t)$$

Furthermore, each $w(t)$ is the unique eigenvector of norm 1 in the cone X^+ for $V(t)$ [**3**, Th 2.11, p. 78], and $|\lambda| < \lambda(t)$ for all remaining eigenvalues of $V(t)$ [**3**, Th 2.13, p. 81]. Set $q = w(1)$ and note that the semigroup property implies

$$V(1)w(n^{-1}) = V(n^{-1})^n w(n^{-1}) = V(n^{-1})^{n-1}\lambda(n^{-1})w(n^{-1})$$
$$= \ldots = \lambda(n^{-1})^n w(n^{-1})$$

By uniqueness in X^+ we see that $w(n^{-1}) = q$ and $\lambda(n^{-1})^n = \lambda(1)$ for all $n = 1, 2, \ldots$. Therefore, if $\mu = \ell n(\lambda(1))$,

$$V(n^{-1})q = \lambda(n^{-1})q = e^{\mu/n}q$$

and by continuity and the denseness of the rationals we see that (ii) holds. Part (i) is immediate from (ii) by setting $z = q$ in (2.5)$'$. Part (iii) follows from (i) and (2.5)$'$ and part (iv) follows from (iii) and (ii) [see the derivation of (2.4)]. This completes the proof.

As our final abstract result we apply these ideas to study of the existence and stability properties of steady-state solutions to (1.2) when the nonlinear term F is concave. A point $z^* \in D^+$ is an <u>equilibrium</u> for (1.2) if $b_{z^*} = \infty$ and $U(t)z^* = z^*$ for all $t \geq 0$. In particular, from (1.3) we see that

(2.6) $\quad z^* \in D^*$ is an equilibrium for (1.2) only in case $z^* \in D(A) \cap D^+$ and $Az^* + Fz^* = 0$.

In addition to (2.1) and (2.3) we assume that $F(0) = 0$ and that F has a derivative at 0 in the following sense:

(2.7) $\quad$ there is a bounded linear operator $F'(0)$ on X satisfying each of the properties of B in Theorem 3 such that

- (a) $F'(0)x = \lim_{h \to o+} \frac{1}{h} F(hx)$ for all $x \in D^+$.
- (b) If $x \in D^+$ with $\hat{\alpha}p \leq x \leq \hat{\beta}p$ where $\hat{\alpha}, \hat{\beta} > 0$, and $\epsilon > 0$, then there is a $\delta_o = \delta_o(x, \epsilon)$ so that $F(hx) - F'(0)(hx) \geq -h\epsilon p \quad (0 < h < \delta_o)$

<u>Remark 2.1</u>. Observe that (2.7a) holds if $F'(0)$ is the (right) Gateaux derivative of F at 0. The condition in (2.7b) is a type of (one-sided) order convergence and places additional restrictions on F which are crucial to these techniques.

<u>Theorem 4</u>. In addition to (2.1), (2.3) and (2.7), suppose that F is p-concave:

(2.8) $\quad$ if $x \in D^+$ and $\hat{\alpha}p \leq x \leq \hat{\beta}p$ where $\hat{\alpha}, \hat{\beta} > 0$, then $F(hx) \geq hF(x)$ and $F(hx) \neq hF(x)$ for $0 < h < 1$.

Also, by assumption (2.7), let V, ν and q be as in Theorem 3 with $B = F(0)$.

(i) If $\nu \leq 0$, then $b_z = +\infty$ and $U(t)z \to 0$ as $t \to \infty$ for all $z \in D^+$

(ii) If $\nu > 0$, then $b_z = +\infty$ for all $z \geq 0$ and either $|U(t)z| \to \infty$ as $t \to \infty$ for all $z \geq 0$, $z \neq 0$, or there is a unique nontrivial equilibrium $z^* \in D^+$ such that $\hat{a}p \leq z^* \leq \hat{b}q$ where $\hat{a}, \hat{b} > 0$, $b_z = +\infty$, and $U(t)z \to z^*$ as $t \to \infty$ for all $z \geq 0$, $z \neq 0$.

The proof of this theorem is given by a sequence of lemmas which are assumed to be under the hypothesis of Theorem 4.

Lemma 2.2. Suppose that $w \in D^+$.

(i) If either $U(t)w \leq w$ for all $0 \leq t < b_w$ or $w \in D(A)$ and $Aw + F(w) \leq 0$, then $b_w = \infty$, $w \geq U(s)w \geq U(t)w$ for $t \geq s \geq 0$, and $U(t)w \to z^*$ as $t \to \infty$ where z^* is an equilibrium for (1.2).

(ii) If either $U(t)w \geq w$ for all $0 \leq t < b_w$ or $w \in D(A)$ and $Aw + F(w) \geq 0$, then $w \leq U(s)w \leq U(t)w$ for $t \geq s \geq 0$ and either $|U(t)w| \to \infty$ as $t \to b_w^-$ or $b_w = \infty$ and $U(t)w \to z^*$ as $t \to \infty$ where z^* in an equilibrium for (1.2).

Proof. Suppose first that $w \in D(A)$ and $Aw + F(w) \leq 0$. Then $T(t-r) : X^+ \to X^+$ and it follows that

$$\int_{t_o}^{t} T(t-r)Awdr + \int_{t_o}^{t} T(t-r)F(w)dr \leq 0 \quad \text{for } t \geq t_o \geq 0.$$

Since $\int_{t_o}^{t} T(t-r)Awdr = T(t-t_o)w - w$ we have

$$v^+(t) \geq T(t-t_o)v^+(t) + \int_{t_o}^{t} T(t-r)F(w)dr \quad \text{for } t \geq t_o \geq 0.$$

where $v^+(t) \equiv w$. Since F is quasi-monotone it follows that $0 \leq U(t)x \leq w$ for all $0 \leq t < b_w$ and $0 \leq x \leq w$ (see, e.g., [7, Proposition 1 and Corollary 1]). By the semigroup and order preserving properties of U we have

$$U(t)w = U(s)U(t-s)w \leq U(s)w \quad \text{for } 0 \leq s \leq t < b_w$$

Thus U is monotone decreasing and bounded, so it follows from (2.2b) that $b_w = \infty$ and $U(t)w \to z^* \in D^+$ as $t \to \infty$. By continuity and the semigroup property,

$$U(t)z^* = \lim_{S \to \infty} U(t)U(s)w = \lim_{S \to \infty} U(t+s)w = z^*$$

and hence z^* is an equilibrium for (1.3). Since the last part of this argument is valid whenever $U(t)w \leq w$ [and hence $0 \leq U(t)x \leq U(t)w$ for $0 \leq x \leq w$] we see that (i) is true. In case (ii), the fact that $U(t)w \geq U(s)w \geq w$ for $t \geq s \geq 0$ follows as in the proof of (i), and so either $|U(t)w| \to \infty$ as $t \to \infty$ or $|U(t)w| \leq M < \infty$ for all $t \geq 0$. In the latter case $\{U(t)w : t \geq 0\}$ must lie in a compact set by (2.2b), and hence $U(t)w \to z^*$ as $t \to \infty$ where z^* is an equilibrium. This complete the proof.

Lemma 2.3. Suppose that $x \in X^+$. Then $F(x) \le F(0)x$ and if $\hat{a}p \le x \le \hat{b}p$ where $\hat{a}, \hat{b} > 0$, then $F(x) \ne F(0)x$.

Proof. If $\hat{a}p \le x \le \hat{b}p$ then (2.8) implies

$$F(0)x = \lim_{h \to o+} \frac{1}{h} F(hx) \ge F(x)$$

The fact that $F'(0)y \ge F(y)$ for all $y \ge 0$ follows from continuity and the fact that $y = \lim_{n\to\infty} x_n$ where $\hat{a}_n p \le x_n \le \hat{b}_n p$ [e.g., if $y \ne 0$ take $x_n = V\left(\frac{1}{n}\right) y$ – see (iii) in Theorem 4 – and if $y = 0$ take $x_n = \frac{1}{n}p$]. Furthermore, since $\frac{1}{2}\hat{a}p \le \frac{1}{2}x \le \frac{1}{2}\hat{b}p$ we also have from (2.8) that

$$F(x) = F\left(2 \cdot \frac{1}{2}x\right) \ge 2F\left(\frac{1}{2}x\right) \text{ and } F(x) \ne 2F\left(\frac{1}{2}x\right).$$

Thus, by the first part of this lemma

$$2F\left(\frac{1}{2}x\right) \ge 2F'(0)\left(\frac{1}{2}x\right) = F'(0)x$$

This shows that the lemma is true.

Lemma 2.4. If $x \in D^+$ then $b_x = \infty$ and $0 \le U(t)x \le V(t)x$ for all $t \ge 0$.

Proof. Lemma 2.3 and the positivity of T and U implies

$$\begin{aligned} U(t)x &= T(t)x + \int_o^t T(t-r)F(U(r)x)dr \\ &\le T(t)x + \int_o^t T(t-r)F'(0)U(r)xdr \end{aligned}$$

Since $F'(0)$ is quasimonotone we have $U(t)x \le v(t)$ where

$$v(t) = T(t)x + \int_o^t T(t-r)F'(0)v(r)xdr$$

Since $v(t) \equiv V(t)x$ we see that $b_x = \infty$ and the lemma is established.

Lemma 2.5. Suppose that $0 \le z_1^* \le z_2^*$ are equilibria for U. If $z_1^* \ne z_2^*$ then $z_1^* = 0$.

Proof. Assume for contradiction that $z_1^* \ne 0$ and $z_1^* \ne z_2^*$. By (2.5) we have

$$\begin{aligned} \overline{a}p &\le U(t)z_2^* - U(t)z_1^* = z_2^* - z_1^* \le \overline{b}p, \\ \overline{a}p &\le U(t)z_1^* - U(t)0 = z_1^* \le \overline{b}p, \end{aligned}$$

and hence

(2.9) $$\overline{a}p \le z_1^* \le z_2^* \le \overline{b}p \text{ and } \overline{a}p \le z_2^* - z_1^* \le \overline{b}p$$

where $\overline{a}, \overline{b} > 0$. Now define

$$\rho = \sup\{\alpha \geq 0 : \alpha z_2^* \leq z_1^*\}$$

and observe that $0 < \rho$ by (2.9) and $\rho < 1$ since $z_2^* \geq z_1^*$ and $z_2^* \neq z_1^*$. Thus $\rho z_2^* \in D(A) \cap D^+$ and (2.8) implies

$$A(\rho z_2^*) + F(\rho z_2^*) \geq \rho[Az_2^* + F(z_2^*)] = 0 \quad \text{and} \quad A(\rho z_2^*) + F(\rho z_2^*) \neq 0$$

Hence $\rho z_2^* \neq z_1^*$ and $U(t)(\rho z_2^*) \geq \rho z_2^*$ by (ii) Lemma 2.2. But

$$z_1^* - U(t)(\rho z_2^*) = U(t)(z_1^*) - U(t)(\rho z_2^*) \geq \hat{a}p$$

by (2.5), and so $z_1^* - \hat{a}p \geq U(t)(\rho z_2^*) \geq \rho z_2^*$. Thus if $\epsilon z_2^* \leq \hat{a}p$ then

$$(\rho + \epsilon)z_2^* \leq (z_1^* - \hat{a}p) + \epsilon z_2^* \leq z_1^*$$

which contradicts the definition of ρ. This shows that either $z_1^* = 0$ or $z_1^* = z_2^*$ and completes the proof.

Lemma 2.6. Suppose that $z^* \in D^+$ is an equilibrium for U and $z^* \neq 0$. Then

(2.10) $$U(t)(\rho z^*) \downarrow z^* \text{ as } t \to \infty \text{ for all } \rho \geq 1.$$

Proof. This is obvious if $\rho = 1$ so suppose $\rho > 1$. Then $z^* \in D(A)$ and by (2.5)

$$\overline{a}p \leq U(t)z^* - U(t)0 = z^* \leq \overline{b}p$$

Thus $\overline{a}p \leq z^* \leq \overline{b}p$ for some $\overline{a}, \overline{b} > 0$ and it follows from (2.8) that

$$F(z^*) = F(\rho^{-1}\rho z^*) \geq \rho^{-1}F(\rho z^*)$$

Therefore,

$$A(\rho z^*) + F(\rho z^*) \leq \rho[Az^* + F(z^*)] = 0$$

and it follows from (i) of Lemma 2.3 that $U(t)(\rho z^*) \downarrow z_o^*$ where z_o^* is an equilibrium for U. But $\rho z^* \geq z^*$ implies

$$U(t)(\rho z^*) \geq U(t)z^* = z^* \text{ for all } t \geq 0$$

and hence $z_o^* \geq z^*$. But $z^* \neq 0$ and we conclude from Lemma 2.5 that $z_o^* = z^*$. This completes the proof.

Lemma 2.7. Suppose that q and ν are as in Theorem 3 with $B \equiv F'(0)$ and $\epsilon > 0$. Then there is a $\delta_o = \delta_o(\epsilon) > 0$ such that

$$A(\delta q) + F(\delta q) \geq (\gamma - \epsilon)\delta q \quad \text{for all } 0 < \delta \leq \delta_o.$$

Proof. From (ii) in Theorem 3 it is immediate that q is an eigenvector corresponding to the eigenvalue ν for $A+F'(0)$:

$$A(\delta q)+F'(0)(\delta q)=\delta\nu q \quad \text{for all } \delta>0 \tag{2.11}$$

From the p-differentiability assumption (2.7b) along with the fact that $\overline{\epsilon}p \le q \le \overline{\epsilon}^{-1}p$ [see (i) in Theorem 3] it follows that there is a $\overline{\delta}=\overline{\delta}(\epsilon,q)>0$ such that

$$F(\delta q)-F'(0)(\delta q)\ge -\delta\epsilon q$$

Combining this with (2.11) establishes the lemma.

Proof of Theorem 4 when $\nu<0$. Applying Lemma 2.4 and (iv) of Theorem 3 it follows that

$$0\le U(t)z\le V(t)z\le \hat{b}(1,z)e^{\nu t}q \quad \text{for all } z\ge 0,\ t\ge 1.$$

Hence $U(t)z\to 0$ as $t\to\infty$ for all $z\ge 0$ in the case $\nu<0$.

Proof of Theorem 4 when $\nu=0$. By Lemma 2.1 and the order preserving property of $U(t)$ and (i) of Theorem 3 we have

$$U(t)z=U(t-1)U(1)z\le U(t-1)(\overline{b}p)\le U(t-1)(\overline{b}\overline{\epsilon}^{-1}q)$$

for all $t\ge 1$. Therefore, it suffices to show $U(t)(\rho q)\to 0$ as $t\to\infty$ for each $\rho>0$. So let $\rho>0$ be given and since $\nu=0$ note that

$$0\le U(t)(\rho q)\le V(t)(\rho q)=\rho q$$

by Lemma 2.4 and (ii) of Theorem 3. Part (i) of Lemma 2.2 now implies $U(t)(\rho q)\to z^*$ as $t\to\infty$ where $z^*\ge 0$ is an equilibrium for (1.2). If $z^*\ne 0$ then Lemma 2.1 implies

$$z^*=U(t)z^*-U(t)z\ge \overline{a}p \quad (\overline{a}\ge 0)$$

and hence $z^*\ge \hat{a}q$ for some $\hat{a}>0$ by (i) of Theorem 3. Thus if

$$\overline{\sigma}\equiv\min\{\sigma>0: z^*\le\sigma q\}$$

Then $\overline{\sigma}$ is well defined and $\hat{a}\le\overline{\sigma}\le\rho$. Also, from Lemma 2.3, (2.11) and the assumption $\nu=0$,

$$A(\overline{\sigma}q)+F(\overline{\sigma}q)\le\overline{\sigma}[A(q)+F(0)q]=0 \text{ and } A(\overline{\sigma}q)+F(\overline{\sigma}q)\ne 0.$$

Thus, $z^*\ne\overline{\sigma}q$ and since $\overline{\sigma}q\ge z^*$ we have from Lemmas 2.4 and 2.1 along with (i) in Theorem 3 that

$$\overline{\sigma}q-z^*=V(t)(\overline{\sigma}q)-U(t)z^*\ge U(t)(\overline{\sigma}q)-U(t)z^*\ge\overline{a}p\ge\overline{a\epsilon}q$$

But this implies $z^* \le (\overline{\sigma} - \overline{a}\,\overline{\epsilon})q$ where $\overline{a}\,\overline{\epsilon} > 0$, contradicting the definition of $\overline{\sigma}$. Thus $z^* = 0$ and it follows that $U(t)(\rho q) \to 0$ as $t \to \infty$ for all $\rho > 0$. This completes the proof when $\nu = 0$.

Proof of Theorem 4 when $\nu > 0$. Taking $\epsilon = \nu$ in Lemma 2.7 show that there is a $\delta_o > 0$ such that

$$A(\delta q) + F(\delta q) \ge 0 \quad \text{for } 0 < \delta \le \delta_o$$

Applying (ii) of Lemma 2.2 shows exactly one of the following must hold:

$$(2.12) \qquad \begin{cases} \text{(a)} & |U(t)(\delta q)| \uparrow \infty \text{ as } t \to \infty \text{ for all } 0 < \delta \le \delta_o \\ \text{(b)} & U(t)(\delta_1 q) \uparrow z^* \text{ as } t \to \infty \text{ for some } 0 < \delta_1 \le \delta_o \end{cases}$$

where z^* is an equilibrium for (1.2). Suppose first that (2.12a) holds and $z \in D^+$, $z \ne 0$. Then $U(1)z \ge \overline{a}p \ge \overline{a}\overline{\epsilon}q$ where $\overline{a}\,\overline{\epsilon} > 0$. Without loss we may also assume $\overline{a}\,\overline{\epsilon} \le \delta_o$, and hence

$$U(t)z = U(t-1)U(1)z \ge U(t-1)(\overline{a}\,\overline{\epsilon}q) \quad \text{for } t \ge 1$$

and it is immediate that $|U(t)z| \to \infty$ as $t \to \infty$. Thus if (2.12a) holds then $|U(t)z| \to \infty$ as $t \to \infty$ for all $z \in D^+$, $z \ne 0$. Now suppose (2.12b) holds. If $0 < \delta \le \delta_1$ then

$$U(t)(\delta q) \le U(t)(\delta_1 q) \le z^* \quad \text{for all } t \ge 0,$$

and it follows that $U(t)(\delta q) \uparrow z^*_\delta$ as $t \to \infty$, where z^*_δ is an equilibrium for (1.2). But $z^*_\delta \le \delta q \ne 0$, and $z^*_\delta \le z^*$, so Lemma 2.5 implies $z^*_\delta = z^*$ for all $0 < \delta \le \delta_1$. Therefore,

$$(2.13) \qquad U(t)(\delta q) \uparrow z^* \quad \text{as } t \to \infty \quad \text{for all } 0 < \delta \le \delta_1.$$

Furthermore, by Lemma 2.6,

$$(2.14) \qquad U(t)(\rho z^*) \downarrow z^* \quad \text{as } t \to \infty \text{ for all } \rho > 1.$$

Now let $z \in D^+$, $z \ne 0$, be given. Then Lemma 2.1 implies

$$\overline{a}p \le U(1)z \le \overline{b}p \quad \text{where } \overline{a}, \overline{b} > 0$$

and as there exists δ, ρ such that $0 < \delta \le \delta_1$ and $\rho > 0$ with

$$\delta q \le U(1)z \le \rho z^*$$

Thus, for $t \ge 0$ $U(t+1)z = U(t)U(1)z$ and

$$U(t)(\delta q) \le U(t+1)z \le U(t)(\rho z^*)$$

Properties (2.13) and (2.14) imply immediately that $U(t+1)z \to z^*$ as $t \to \infty$ and we see that each of the assertions in Theorem 4 are established.

§3. SEMILINEAR PARABOLIC SYSTEMS. Throughout Ω is assumed to be a bounded domain in $\mathbf{R}^N$ with smooth boundary $\partial\Omega$, ∇ is the gradient operator on Ω, and ∂_n devotes the (outward) normal derivative operator on $\partial\Omega$. Let $\mathcal{C}(\overline{\Omega})$ be the space of continuous real-valued functions on $\overline{\Omega}$ with the maximum norm and for each $n \geq 1$ let $C^n(\overline{\Omega})$ denote the class of n-times continuously differentiable functions on $\overline{\Omega}$. For each $i = 1, \ldots, m$ define the operators L_i on $\mathcal{C}^2(\overline{\Omega})$ by

$$(3.1) \qquad \begin{cases} L_i v_i & \equiv \nabla \cdot (d_i \nabla v_i) - d_i \vec{a}_i \cdot \nabla v_i - c_i v_i \text{ for } v_i \in D(L_i) \\ D(L_i) & \equiv \{v_i \in \mathcal{C}^2(\overline{\Omega}) : \alpha_i(\sigma) v_i(\sigma) + k_i \partial_n v_i(\sigma) = 0 \text{ for } \sigma \in \partial\Omega\} \end{cases}$$

where it is assumed the following conditions are satisfied:

$$(3.2) \qquad \begin{cases} \text{(a)} & d_i \in \mathcal{C}^1(\overline{\Omega}) \text{ and } d_i(\sigma) > 0 \text{ for all } \sigma \in \overline{\Omega} \\ \text{(b)} & \vec{a}_i = (a_i^1, \ldots, a_i^N) \text{ and each } a_i^j \in \mathcal{C}^1(\overline{\Omega}) \\ \text{(c)} & c \in \mathcal{C}(\overline{\Omega}) \text{ and } c(\sigma) \geq 0 \text{ for all } \sigma \in \overline{\Omega} \\ \text{(d)} & k_i \in \{0,1\},\ \alpha_i \in \mathcal{C}^1(\overline{\Omega}),\ \alpha_i(\sigma) \geq 0 \text{ on } \partial\Omega, \\ & \text{and } \alpha_i \equiv 1 \text{ if } k_i = 0. \end{cases}$$

Now define $X_i = \mathcal{C}(\overline{\Omega})$ if $k_i = 1$ and $X_i = \mathcal{C}(\overline{\Omega})_o \equiv \{y \in \mathcal{C}(\overline{\Omega}) : y(\sigma) = 0 \text{ for } \sigma \in \partial\Omega\}$ if $k_i = 0$ (and hence $\alpha_i \equiv 1$). Then L_i is closable and densely defined in X_i, and if A_i is the closure of L_i then A_i is the generator of a compact, analytic semigroup $T_i = \{T_i(t) : t \geq 0\}$ of bounded linear operators on X_i. In particular, for each $y_i \in X_i$, $v_i(\cdot, t) \equiv T_i(t) y_i$ is the (classical) solution to the linear parabolic equation

$$\begin{cases} \partial_t v_i = \nabla \cdot (d_i \nabla v_i) - d_i \vec{a}_i \nabla v_i - c_i v_i & t > 0,\ \sigma \in \Omega \\ \alpha_i v_i + k_i \partial_n v_i = 0 & t > 0,\ \sigma \in \partial\Omega \\ v_i = y_i & t = 0,\ \sigma \in \Omega \end{cases}$$

Furthermore, if

$$X_i^+ = \{y_i \in X_i : y_i(\sigma) \geq 0 \text{ for all } \sigma \in \overline{\Omega}\}$$

then $T_i(t) : X_i^+ \to X_i^+$ by the maximum principle. In addition, if $\sigma(A_i)$ is the spectrum of A_i, then $\mu_i \equiv \sup\{\text{Re}\lambda : \lambda \in \sigma(A_i)\}$ is an eigenvalue of $A_i, \mu_i \leq 0$, and there is a corresponding eigenfunction p_i that is nonnegative on $\overline{\Omega}$:

$$(3.3) \qquad \begin{cases} \nabla \cdot (d_i \nabla p_i) - d_i \overline{a}_i \nabla p_i - c_i p_i = \mu_i p_i, & \sigma \in \Omega \\ \alpha_i p_i + k_i \alpha_n p_i = 0, & \sigma \in \partial\Omega, \\ p_i \geq 0, & \sigma \in \overline{\Omega}. \end{cases}$$

In particular, $T_i(t) p_i = e^{\mu_i t} p_i$ for all $t \geq 0$, so we see that (2.3a) is satisfied. Applying the strong maximum principle (see Protter and Weinberger [**11**, pp. 64 and 67]) shows that

$$(3.4) \qquad \begin{cases} \text{(a)} & p_i(\sigma) > 0 \text{ for all } \sigma \in \Omega \text{ and if } k_i = 1 \text{ then} \\ & p_i(\sigma) > 0 \text{ for all } \sigma \in \overline{\Omega} \\ \text{(b)} & \text{if } k_i = 0 \text{ then } \partial_n p_i(\sigma) > 0 \text{ for all } \sigma \in \partial\Omega \\ \text{(c)} & \mu_i \leq 0 \text{ and } \mu_i = 0 \text{ only in case } k_i = 1, \\ & \alpha_i \equiv 0 \text{ and } c_i \equiv 0. \end{cases}$$

Since the underlying space X_i is contained in $L^p(\Omega)$ for all $p > 1$, it follows that $D(A_i) \subset W_p^2(\Omega)$ for all $p > 1$, when $W_p^2(\Omega)$ in the usual Sobolev space. Hence members of $D(A_i)$ are at least in $C^1(\overline{\Omega})$ and this along with (3.4a,b) implies that (2.3c) is satisfied. The strong maximum principle for parabolic equations ([**11**, p. 170]) can be used to show that T_i also satisfies (2.3b), and hence we have that (2.3) is fulfilled in this case.

Define the classes $Q_i^* \subset X_i^*$ as follows:

$$(3.5) \qquad Q_i^* = \begin{cases} \{\phi_i^\rho : \rho \in \overline{\Omega},\ \phi_i^\rho(y_i) = y_i(\rho) \text{ for } y_i \in X_i\} \text{ if } X_i = C(\overline{\Omega}) \\ \{\phi_i^\rho : \rho \in \Omega,\ \phi_i^\rho(y_i) = y_i(\rho) \text{ for } y_i \in X_i\} \text{ if } X_i = C(\overline{\Omega})_o \end{cases}$$

Observing that the strong maximum principle implies that (1.9) holds we see that $T = (T_i)_1^m$ satisfies all of the assumptions on T in Theorems 1–4 with $D^+ = X^+$.

The concern now is to give properties for $f = (f_i)_1^m$ in order that these theorems apply to the reaction-diffusion system

$$(3.6) \qquad \begin{cases} \partial_t u_i = \nabla \cdot (d_i \nabla u_i) - d_i \vec{a}_i \nabla u_i - c_i u_i + f_i(\sigma, u_1, \dots, u_m) & t > 0,\ \sigma \in \Omega \\ \alpha_i u_i + k_i \partial_n u_i = 0 & t > 0,\ \sigma \in \partial\Omega \\ u_i = y_i & t = 0,\ \sigma \in \Omega \end{cases}$$

We assume $f = (f_i)^m : \overline{\Omega} \times [0, \infty)^m \to \mathbf{R}^m$ is continuous and satisfies the following:

$$(3.7) \qquad \begin{cases} \text{(a) for each } R > 0 \text{ there is an } L_R \geq 0 \text{ such that} \\ \qquad |f_i(\sigma, \xi_1, \dots, \xi_m) - f_i(\sigma, \eta_1, \dots, \eta_m)| \leq L_R \sum_{j=1}^m |\xi_j - \eta_j| \\ \qquad \text{for all } \sigma \in \overline{\Omega},\ (\xi_j)_1^m,\ (\eta_j)_1^m \in [0, R]^m \text{ and } i = 1, \dots, m. \\ \text{(b) } f \text{ is quasimonotone: if } (\xi_i)_1^m,\ (\eta_i)_1^m \in [0,\infty)^m \text{ with} \\ \qquad \xi_i \geq \eta_i \text{ for all } i = 1, \dots, m, \text{ then } \xi_k = \eta_k \text{ for some } k \in \{1, \dots, m\} \\ \qquad \text{implies } f_k(\sigma, \xi_i, \dots, \xi_m) \geq f_k(\sigma, \eta_1, \dots, \eta_m) \text{ for all } \sigma \in \overline{\Omega}. \\ \text{(c) } f \text{ is quasipositive: if } (\xi_i)_1^m \in [0,\infty)^m \text{ and } \xi_k = 0 \\ \qquad \text{for some } k \in \{1, \dots, m\}, \text{ then } f_k(\sigma, \xi_1, \dots, \xi_m) \geq 0 \text{ for all } \sigma \in \overline{\Omega}. \\ \text{(d) if } X_i = C(\overline{\Omega})_o \text{ then } f_i(\sigma, 0) = 0 \text{ for all } \sigma \in \partial\Omega. \end{cases}$$

Notice that if (3.7b) holds then (3.7c) holds only in case $f_i(\sigma, 0) \geq 0$ for all $\sigma \in \overline{\Omega}$ and $i = 1, \dots, m$. Furthermore, (3.7a) and (3.7b) imply

$$(3.8) \qquad \begin{cases} f_i(\sigma, \eta_i, \dots, \eta_m) - f_i(\sigma, \xi_1, \dots, \xi_m) \geq -L_R(\eta_i - \xi_i) \\ \text{for all } \sigma \in \overline{\Omega},\ i = 1, \dots, m \text{ and } (\xi_j)_1^m,\ (\eta_j)_1^m \in [0, R]^m \text{ with} \\ \xi_j \leq \eta_j \text{ for } j = 1, \dots, m. \end{cases}$$

For if $\xi, \eta \in [0, R]^m$ with $\xi \leq \eta$ and $i \in \{1, \dots, m\}$, define $\overline{\xi} = (\overline{\xi}_j)_1^m$ by $\overline{\xi}_i = \xi_i$ and $\overline{\xi}_j = \eta_j$ for $j \neq 1$. Then $\overline{\xi} \in [0, R]^m$ and $f_i(\sigma, \xi) \leq f_i(\sigma, \overline{\xi})$ by (3.7b). Hence (3.7a) implies

$$\begin{aligned} f_i(\sigma, \eta_1, \dots, \eta_m) - f_i(\sigma, \xi_1, \dots, \xi_m) &\geq f_i(\sigma, \eta_1, \dots, \eta_m) - f_i(\sigma, \overline{\xi}_1, \dots, \overline{\xi}_m) \\ &\geq -L_R \sum_{j=1}^m (\eta_j - \overline{\xi}_j) \\ &= -L_R(\eta_i - \xi_i) \end{aligned}$$

Define the map $F = (F_i)_1^m$ from $X^+ \equiv \prod_{i=1}^m X_i^+$ into $X = \prod_{i=1}^m X_i$ by

$$[F_i y](\sigma) = f_i(\sigma, y(\sigma)) \text{ for all } \sigma \in \overline{\Omega},\ y = (y_j)_1^m \in X^+ \text{ and } i = 1, \ldots, m \tag{3.9}$$

Clearly $F_i y \in X_i$ whenever $y \in X_i^+$ [see (3.7d)] and F is continuous from X^+ into X. Also assumptions (3.7) on f imply that F satisfies (C1) – (C3) and (1.7) in Section 1. It is immediate that (3.7a) implies (C1) and one can routinely show that (3.7b) implies (C3). Since (3.7c) implies

$$\lim_{h \to o+} \frac{1}{h} d_i(\xi_i + h f_i(\sigma, \xi); [0, \infty)) = 0 \text{ for } i = 1, \ldots, m,\ \sigma \in \overline{\Omega} \text{ and } \xi = (\xi_j)_1^m \in [0, \infty)^m$$

It is also routine to show that (1.5) in Remark 1.3 is valid, and hence (C2) is satisfied as well. The assumption (1.7) on F is immediate from (3.8). Applying Theorem 1 in Section 1 we see that for each $y = (y_i)_1^m$ in X^+, equation (3.6) has a unique nonnegative classical solution $u(\cdot; y) = (u_i(\cdot; y))_1^m$ defined on $\overline{\Omega} \times [0, b_y)$. Furthermore,

$$\begin{cases} 0 \le u_i(\sigma_1 t; y) \le u_i(\sigma, t; z) \text{ for all } (\sigma, t) \in \overline{\Omega} \times (0, b_z), \text{ and } i = 1, \ldots, m, \\ \text{whenever } y, z \in X^+ \text{ with } y(\sigma) \le z(\sigma) \text{ for } \sigma \in \overline{\Omega}. \end{cases} \tag{3.10}$$

and if $y \in X^+$ with $b_y < \infty$, then

$$\sum_{i=1}^m |u_i(\cdot, t; y)|_\infty \to \infty \text{ as } t \to b_y - .$$

In order to apply Theorem 2 to equation (3.6), we need a supposition on f which guarantees that the corresponding F satisfies condition (1.10). Such a property is a direct extension of the concept of an irreducible matrix and is characterized in the following manner:

$$\begin{cases} \text{if } \Sigma \text{ is a nonempty, proper subset of } \{1, \ldots, m\} \\ \text{and } (\xi_j)_1^m,\ (\eta)_1^m \in [0, \infty)^m \text{ with} \\ \qquad \text{(a) } \xi_j > \eta_j \text{ for all } j \in \Sigma \\ \qquad \text{(b) } \xi_j = \eta_j \text{ for all } j \in \Sigma^c \\ \text{then there is a } k \in \Sigma^c \text{ and a } \sigma_o \in \Omega \text{ such that } f_k(\sigma_o, \xi) > f_k(\sigma_o, \eta). \end{cases} \tag{3.11}$$

With Q_i^* defined by (3.5) it is easy to check that (3.11) implies that F satisfies (1.10). In particular if (3.11) holds then assertion (3.10) may be strengthened to read as follows:

$$\begin{cases} \text{if } y \text{ and } z \text{ are as in (3.10) and } y_i(\sigma_o) < z_i(\sigma_o) \\ \text{for some } i \in \{1, \ldots, m\} \text{ and some } \sigma_o \in \Omega, \text{ then} \\ \text{(a) } u_j(\sigma, t; y) < u_j(\sigma, t; z) \text{ for all } (\sigma, t) \in \overline{\Omega} \times (0, b_z) \\ \qquad \text{and all } j \in \{1, \ldots, m\} \text{ such that } X_j = C(\overline{\Omega}); \text{ and} \\ \text{(b) } u_j(\sigma, t; y) < u_j(\sigma, t; z) \text{ for all } (\sigma, t) \in \Omega \times (0, b_z) \\ \qquad \text{and all } j \in \{1, \ldots, m\} \text{ such that } X_j = C(\overline{\Omega})_o. \end{cases} \tag{3.12}$$

Of course (3.12) is a direct consequence of Theorem 2 with Q_i^* defined by (3.5).

The results in §2 may also be applied to system (3.6). Suppose that for each $i, j \in \{1, \dots, m\}$ $b_{ij} : \overline{\Omega} \to \mathbf{R}$ is continuous and the following holds:

$$(3.13) \qquad \begin{cases} \text{(a) } b_{ij}(\sigma) \geq 0 \text{ for all } \sigma \in \overline{\Omega} \text{ and } i \neq j. \\ \text{(b) if } \Sigma \text{ is a proper, nonempty subset of } \{1, \dots, m\}, \text{ then} \\ \qquad \text{there is a } k \in \Sigma^c \text{ and a } \sigma_o \in \Omega \text{ such that } \sum_{j \in \Sigma} b_{kj}(\sigma_o) > 0 \end{cases}$$

If the functions b_{ij} are constant, then property (3.13) is equivalent to the matrix $B = (b_{ij})$ being irreducible. In general, since (3.13b) is equivalent to (3.11) with

$$f_i(\sigma, \xi) \equiv \sum_{j=1}^{m} b_{ij}(\sigma)\xi_j \text{ for } \sigma \in \overline{\Omega},\ i = 1, \dots, m$$

we may apply Theorem 3 to the system

$$(3.14) \qquad \begin{cases} \partial_t v_i = \nabla \cdot (d_i \nabla v_i) - d_i \vec{a} \nabla v_i - c_k v_1 + \sum_{j=1}^{m} b_{ij} v_j, & t > 0,\ \sigma \in \Omega \\ \alpha_i v_i + k_i \partial_n v_i = 0 & t > 0,\ \sigma \in \Omega \\ v_i = y_i & t > 0,\ \sigma \in \Omega \end{cases}$$

In particular there is a $\nu \in \mathbf{R}$ and $q = (q_i)_1^m \in X^+$ so that

$$(3.15) \qquad \begin{cases} \text{(a) } \overline{\epsilon}\, p_i(\sigma) \leq q_i(\sigma) \leq \overline{\epsilon}^{-1} p_i(\sigma) \text{ for all } \sigma \in \overline{\Omega},\ i = 1, \dots, m. \\ \text{(b) } \nabla \cdot (d_i \nabla q_i) - d_i \vec{a}_i \nabla q_i - c_i q_i + \sum_{j=1}^{m} b_{ij} q_j = \nu q_j \text{ on } \Omega \\ \qquad \text{and } \alpha_i q_i + k_i \partial_n q_i = 0 \text{ on } \partial\Omega \text{ for } i = 1, \dots, m. \end{cases}$$

<u>Remark 3.1</u>. If $f = (f_i)_1^m$ is C^1 for $\xi \geq 0$, then (3.13) can be applied to the Jacobian matrix of f to determine if (3.11) is valid. In particular, (3.11) is valid under the following assumptions: if

$$b_{ij}(\sigma, \eta) \equiv \frac{\partial f_i(\sigma, \eta)}{\partial \xi_j} \text{ for } \sigma \in \overline{\Omega},\ \eta \in [0, \infty]^m$$

then f satisfied (3.11) if for each proper nonempty subset Σ of $\{1, \dots, m\}$ there is a $k \in \Sigma^c$ and a $\sigma_o \in \Omega$ such that

$$\sum_{j \in \Sigma} b_{kj}(\sigma_o, \xi) > 0$$

for all but an at most countable number of $\xi \in [0, \infty)^m$. (The proof of this fact follows as in [7, Remark 4.5]).

Finally, we show the implications of Theorem 4 relative to (3.6). Certainly, if f is assumed to have continuous first partial derivatives with respect to each ξ_i on $\overline{\Omega} \times [0, \infty)^m$, then F will satisfy (2.7a). Moreover, if f has the property

$$(3.16) \qquad \begin{cases} \xi_i > 0 \text{ for all } i = 1, \dots, m, \text{ implies} \\ \qquad f_i(\sigma, h\xi) \geq h f_i(\sigma, \xi) \text{ for all } \sigma \in \overline{\Omega},\ 0 < h < 1,\ i = 1, \dots, m \\ \text{and for each } 0 < h < 1 \text{ there is a } \sigma_h \in \Omega \text{ and } k_h \in \{1, \dots, m\} \\ \text{such that } f_{k_h}(\sigma_h, h\xi) > h f_{k_h}(\sigma_h, \xi). \end{cases}$$

then it is easy to see that (2.8) in Theorem 4 holds. Therefore, under assumption (3.16), we see that Theorem 4 applies to (3.6) whenever F also satisfies (2.7b).

Our aim now is to indicate the type of conditions on f that are needed in order to show that (2.7b) holds. So assume f has continuous first partial derivatives with respect to each ξ_i on $\overline{\Omega} \times [0,\infty)^m$ and define Γ to be the set of $i \in \{1,\dots,m\}$ such that $X_i = C(\overline{\Omega})_o$ (i.e., $i \in \Gamma^c$ only in case $k_i = 1$). Consider the following hypothesis for f:

(3.17) $$\frac{\partial}{\partial \xi_j} f_i(\sigma,\xi) \text{ is nondecreasing in } \xi_j \text{ for all } \sigma \in \overline{\Omega},\ i \in \Gamma \text{ and } j \in \Gamma^c.$$

Of course, (3.17) is vacuously satisfied if Γ is empty or Γ is all of $\{1,\dots,m\}$.

<u>Lemma 3.1</u>. Under the suppositions and notations in this section, if $f(\sigma,0) \equiv 0$ for $\sigma \in \Omega$ and (3.17) is satisfied, then F satisfies (2.7b).

<u>Proof</u>. Suppose $y = (y_i)_1^m \in X^+$. With $\hat{\alpha} p_i(\sigma) \le y_i(\sigma) \le \hat{\beta} p_i(\sigma)$ for all $\sigma \in \overline{\Omega}$ and $i = 1,\dots,m$, where $\hat{\alpha}, \hat{\beta} > 0$, and let $\epsilon > 0$ be given. Let i be in $\{1,\dots,m\}$ and apply the mean value theorem to show the existence of a $\delta_i(\sigma,h) \in (0,1)$ such that

$$f_i(\sigma,hy(\sigma)) = \sum_{j=1}^m \frac{\partial}{\partial \xi_j} f_i(\sigma,\delta_i(\sigma,h)hy(\sigma))hy_j(\sigma)$$

Then

$$\begin{aligned}
&[F_i(hy)](\sigma) - [F_i'(0)(hy)](\sigma) \\
&\quad = \sum_{j=1}^m \left[\frac{\partial}{\partial \xi_j} f_i(\sigma,\delta_i(\sigma,h)hy(\sigma)) - \frac{\partial}{\partial \xi_j} f_i(\sigma,0)\right] hy_j(\sigma) \\
&\quad = \sum_{j=1}^m [-\epsilon_j(h,\sigma)]\, hy_j(\sigma)
\end{aligned}$$

where

$$-\epsilon_j(h,\sigma) \equiv \frac{\partial}{\partial \xi_j} f_i(\sigma,\delta_i(\sigma,h)hy(\sigma)) - \frac{\partial}{\partial \xi_j} f_i(\sigma,0)$$

In particular, $|\epsilon_j(h,\sigma)| \to 0$ as $h \to 0+$ uniformly for $\sigma \in \overline{\Omega}$. Assume first that $i \in \Gamma^c$. Then $k_i = 1$ and it follows from (3.4a) that

$$\overline{\mu} \equiv \min\{p_i(\sigma) : \sigma \in \overline{\Omega}\} > 0.$$

If $\overline{\nu} = \max\{y_j(\sigma) : \sigma \in \overline{\Omega},\ j = 1,\dots,m\}$ and $h_o > 0$ is such that $|\epsilon_j(h,\sigma)| \le \epsilon\overline{\mu}(m\overline{\nu})^{-1}$ for $0 < h \le h_o$ and $\sigma \in \overline{\Omega}$, then $0 < h \le h_o$ implies

$$\begin{aligned}
[F_i(hy)](\sigma) - [F_i'(0)(hy)](\sigma) &= \sum_{j=1}^m [-\epsilon_j(h,\sigma)]\, y_j(\sigma) \\
&\le -\sum_{j=1}^m \epsilon\overline{\mu}(m\overline{\nu})^{-1} h\overline{\nu} = -h\epsilon\overline{\mu} \le -h\epsilon p_i(\sigma)
\end{aligned}$$

and we see that (2.7b) holds for all components of $i \in \Gamma^c$. Now assume $i \in \Gamma$. Then $-\epsilon_j(h,\sigma) \geq 0$ for all $j \in \Gamma^c$ by hypothesis (3.17), and it follows that

$$[F_i(hy)](\sigma) - [F_i'(0)(hy)](\sigma) \geq -\sum_{j\in\Gamma} \epsilon_j(h,\sigma)hy_j(\sigma)$$

However, if $j \in \Gamma$ then $p_j(\sigma) = 0$ and $\partial_n p_j(\sigma) > 0$ for all $\sigma \in \partial\Omega$, and it follows that there is a $\overline{\delta} > 0$ such that

$$\overline{\delta} p_j(\sigma) \leq p_i(\sigma) \text{ for all } \sigma \in \overline{\Omega} \text{ and } j \in \Sigma$$

Therefore, if $h_o > 0$ is chosen so that $|\epsilon_j(h,\sigma)| \leq \epsilon\overline{\delta}(\hat{\beta}m)^{-1}$ for $0 < h \leq h_o$ and $\sigma \in \overline{\Omega}$, then $y_j(\sigma) \leq \hat{\beta} p_j(\sigma) \leq \hat{\beta}\overline{\delta}^{-1} p_i(\sigma)$ for $j \in \Gamma$, $\sigma \in \overline{\Omega}$, and it follows that

$$[F_i(hy)](\sigma) - [F_i'(0)(hy)](\sigma) \geq -\sum_{j\in\Gamma} \epsilon\overline{\delta}(\overline{\beta}m)^{-1} h\hat{\beta}\overline{\delta}^{-1} p_i(\sigma) \geq \epsilon h p_i(\sigma)$$

for all $\sigma \in \overline{\Omega}$ and $0 < h \leq h_o$. This shows that (2.7b) also holds for all components $i \in \Gamma$ and completes the proof of the lemma.

Remark 3.2. Since assumption (3.16) is a type of concave condition on f, it is often the case that $\frac{\partial}{\partial \xi_j} f_i(\sigma,\xi)$ is nonincreasing in ξ_j. Thus, in many situations, assumption (3.17) reduces to $f_i(\sigma,\xi)$ being linear in ξ_j for each $\sigma \in \overline{\Omega}$, $i \in \Gamma$ and $j \in \Gamma^c$.

As a specific example to illustrate further these ideas, consider the following model of a cellular control process with positive feedback:

$$(3.18) \qquad \begin{cases} \partial_t u_1 = d_1 \Delta u_1 - c_1 u_1 + g(u_m) \\ \partial_t u_2 = d_2 \Delta u_2 - c_2 u_2 + u_1 \\ \qquad\qquad\qquad\qquad\qquad\qquad (t > 0,\ \sigma \in \Omega) \\ \cdots\cdots\cdots\cdots \\ \partial_t u_m = d_m \Delta u_m - c_m u_m + u_{m-1} \end{cases}$$

subject to the boundary condition

$$(3.19) \qquad \alpha_i u_i + k_i \partial_n u_i = 0 \quad (t > 0,\ \sigma \in \partial\Omega,\ i = 1, \ldots, m).$$

where d_i and c_i are positive constants and $g : [0,\infty) \to [0,\infty)$ is C^1 and strictly increasing with $g(0) = 0$, $g'(0) > 0$, and $hg(r) < g(hr)$ for all $0 < h < 1$ and $r > 0$ [e.g., $g(r) \equiv r(1+r)^{-1}$ for $r \geq 0$]. Then, according to (3.17), Theorem 4 applies to (3.18) whenever $k_1 = 1$ or if $k_i = 0$ for all $i = 1, \ldots, m$. In this case the linearization about the trivial equilibrium is

$$(3.20) \qquad \begin{cases} \partial_t v_1 = d_1 \Delta v_1 - c_1 v_1 + g'(0) v_m \\ \partial_t v_2 = d_2 \Delta v_2 - c_2 v_2 + v_1 \\ \qquad\qquad\qquad\qquad\qquad\qquad (t > 0,\ \sigma \in \Omega) \\ \cdots\cdots\cdots\cdots \\ \partial_t v_m = d_m \Delta v_m - c_m v_m + v_{m-1} \end{cases}$$

where each v_i is subject to the same boundary conditions as u_i in (3.19). In particular, the stability properties of the linear system (3.20) dictate those of (3.18) as described in Theorem 4. This example indicates that these methods extend those in Martin [**5**], [**6**], Capasso and Maddalena [**1**] and Messia and Mottoni [**8**]. Roughly speaking the results in [**5**] and [**6**] correspond to α_i and k_i being independent of i in (3.19), the results in [**1**] correspond to $k_i = 1$ for all i, and the results in [**8**] correspond to $k_1 = 1$ and $k_i = 0$ for $i = 2, \ldots, m$.

REFERENCES

1. Capasso, V. and Maddalana, L., "Convergence to equilibrium states for reaction-diffusion system modeling the spatial spread of a class of bacterial and viral diseases," *J. Math. Bio.* **13**(1981), 173–184.

2. Goldstein, J. A., *Semigroups of Linear Operators and Applications*, Oxford University Press, New York, 1985.

3. Krasnoselskii, M. A., *Positive Solutions of Operator Equations*, P. Noordhoff Ltd., Groningen, 1964.

4. Martin, R. H., *Nonlinear Operators and Differential Equations in Banach Spaces*, Wiley-Interscience New York, 1976.

5. Martin, R. H., "Asymptotic stability and critical points for nonlinear quasimonotone parabolic systems," *J. Diff. Eq.* **30**(1978), 391–423.

6. Martin, R. H., "Asymptotic behavior of solutions to a class of nonlinear parabolic systems," in *Nonlinear Partial Differential Equations and Their Applications*, Vol. 1, H. Brezis and J. L. Lions, Editors, *Research Notes in Math.* #53, Pitman, Boston, 1980.

7. Martin, R. H. and Smith, H., "Abstract functional differential equations and reaction-diffusion systems," (preprint).

8. Messia, M. E. and De Mottoni, P., "On some positive feedback systems with different boundary conditions," *J. Math. Anal. Appl.*, **103**(1984), 58–66.

9. De Mottoni, P. and Schiaffino, A., "Bifurcation of periodic solutions of periodic evolution equations in a cone," *J. Diff. Eq.* **45**(1982), 408–430.

10. Pazy, A., *Semigroups of Linear Operators and Applications to Partial Differential Equations*, Springer-Verlag, New York, 1983.

11. Protter, M. H. and Weinberger, H. F., *Maximum Principles in Differential Equations*, Prentice-Hall, Englewood Cliffs, NJ, 1967.

12. Vulikh, B. Z., *Introduction to the Theory of Partially Ordered Spaces*, Wolters-Noordhoff, Groningen, 1967.

GLOBAL EXISTENCE FOR SEMILINEAR PARABOLIC SYSTEMS VIA LYAPUNOV TYPE METHODS

JEFF MORGAN
DEPARTMENT OF MATHEMATICS
TEXAS A&M UNIVERSITY
COLLEGE STATION, TEXAS 77843
U.S.A.

ABSTRACT

We consider semilinear parabolic systems of partial differential equations of the form

$$u_t(t,x) = D\Delta u(t,x) + f(u(t,x)) \qquad t>0\ ,\ x\in\Omega \tag{1}$$

with bounded initial data and homogeneous Neumann boundary conditions, where D is an m by m diagonal positive definite matrix, Ω is a smooth bounded region in $\mathbf{R}^n$ and $f:\mathbf{R}^m \to \mathbf{R}^m$ is locally Lipschitz. We prove that if the vector field f satisfies a generalized Lyapunov type condition then either at least two components of the solution of (1) becomes unbounded in finite time or the solution exists for all $t > 0$. Our result generalizes a recent result of Hollis, Martin, and Pierre [4], and the proof given is considerably simpler.

1. INTRODUCTION AND NOTATION

Until recent years most global existence results for semilinear parabolic systems of partial differential equations have fallen into one of two types; one either assumes the existence of a bounded invariant region for the system or assumes that certain a priori bounds can be obtained for solutions of the sytem. Of these two approaches, generally only the first considers the vector field involved as anything more than an algebraic expression. Consequently, since many systems do not have invariant regions, the geometry of the vector field involved is often ignored. Recently however, Alikakos [1], Groger [3], Hollis, Martin, and Pierre [4], Masuda [9], and others have begun to exploit this geometry via Lyapunov type structures. In these works, the systems considered are essentially of the form

$$(1)\qquad \begin{aligned} &u_t(t,x) = D\Delta u(t,x) + f(u(t,x)) && t>0,\ x\in\Omega \\ &\partial u(t,x)/\partial\eta = 0 && t>0,\ x\in\partial\Omega \\ &u(0,x) = u_o(x) && x\in\Omega \end{aligned}$$

where

(A1) D is a diagonal m by m matrix with entries $d_i > 0$ on the diagonal,

(A2) Ω is a bounded region in $\mathbf{R}^n$ with smooth boundary $\partial\Omega$,

(A3) $f : \mathbf{R}^m \to \mathbf{R}^m$ is locally Lipschitz,

(A4) η denotes the unit outward normal on $\partial\Omega$,

(A5) $u_o : \Omega \to \mathbf{R}^m$ is bounded and measurable.

Furthermore, it is assumed there exists some invariant unbounded m-rectangle $M = M_1 \times \cdots \times M_m$ for (1) with faces parallel to the coordinate hyperplanes and a smooth function $H : M \to \mathbf{R}^+$ which satisfies:

(H1) there exists $z \in M$ such that $H(z) = 0$ and if $y \in M$, $y \neq z$, then $H(y) > 0$

(H2) $H(z) \to \infty$ as $|z| \to \infty$ in M

(H3) $\partial^2 H(z)$ is nonnegative definite for all $z \in M$

(H4) there exist $K, L \geq 0$ such that for all $z \in M$, $\partial H(z) f(z) \leq L_1 H(z) + L_2$.

That is, H is a nonnegative convex coercive functional and the vector field f has a linearly restricted growth rate across level curves of H (this is the geometric exploitation of f). In addition, if L_1, L_2 = 0 then (H1) and (H4) imply that H is a Lyapunov function for the ordinary differential equation

$$v' = f(v) \qquad t \in \mathbf{R}.$$

Hence, we refer to this H-structure as a generalized Lyapunov structure.

In the works of Alikikos, Hollis et al, and Masuda, M is the positive orthant

$$P^m = \{x : x=(x_i) \in \mathbf{R}^m \text{ and } x_i \geq 0 \text{ for all } 1 \leq i \leq m\},$$

$m = 2$, and the H-structure is obtained through $H(z_1,z_2) = z_1 + z_2$. We see that (H1) - (H3) are easily satisfied and that (H4) becomes $f_1(z) + f_2(z) \leq L_1(z_1 + z_2) + L_2$ for all $z = (z_i) \in \mathbf{P}^2$, i.e. , there is a "balancing" of higher order terms in f_1 and f_2. In the work of Groger $M = \text{int}(\mathbf{P}^m)$, there is no restriction on m and the H-structure is essentially given by

$$H(z) = \sum_{i=1}^{m} \left(z_i \ln(z_i) - z_i\right) + m .$$

Again (H1) - (H3) are easily satisfied and (H4) becomes a generalization of Groger's dissipativity condition (see [2], [3], and [4] for a physical justification of Groger's assumption).

The work in Hollis et al generalized the work of both Alikikos and Masuda. Basically Hollis proves that if his H-structure is present and $|f(z)|$ is polynomially bounded, then either at least two components of the solution u of (1) become unbounded in finite time or the solution u exists for all time. We generalize this result below. The results in Groger's paper are generalized in Morgan [10].

2. STATEMENT AND PROOF OF THE MAIN RESULT

Before stating and proving our generalization of Hollis' result we state the following well-known result (cf. Hollis et al [4]).

Theorem 2.1 : Suppose that (A1) - (A5) are satisfied. Then there exists $T_{max} > 0$ and $N = (N_i) \in C([0,T_{max}), \mathbf{R}^m)$ such that

(i) (1) has a unique, classical, noncontinuable solution $u(t,x)$ on $cl(\Omega)\times[0,T_{max})$,

and (ii) $| u_i(\cdot,t) |_{\infty,\Omega} \leq N_i(t)$ for all $1 \leq i \leq m$, $0 \leq t < T_{max}$.

Moreover, if $T_{max} < \infty$, then $| u_i(\cdot,t) |_{\infty,\Omega} \to \infty$ as $t \to T_{max}^-$ for some $1 \leq i \leq m$.

We now state our result as

Theorem 2.2 : Assume that (A1) - (A5) and (H1) - (H4) are satisfied and $u_0 : \Omega \to M$. Suppose

(H5) there exist $h_i : M_i \to \mathbf{R}^+$ for all $1 \leq i \leq m$ such that

$$H(z) = \sum_{i=1}^{m} h_i(z_i) \quad \text{for all } z \in M$$

and

(A6) there exist $q, K_1, K_2 \geq 0$ such that for all $1 \leq i \leq m$,

$$h_i'(z_i) f_i(z) \leq K_1 [H(z)]^q + K_2 \quad \text{for all } z \in M.$$

If $T_{max} < \infty$ then there exist $1 \leq j < k \leq m$ such that $| u_i(\cdot,t) |_{\infty,\Omega} \to \infty$ as $t \to T_{max}^-$ whenever $i = j, k$.

Proof of Theorem 2.2 : Suppose (by way of contradiction) that $T_{max} < \infty$ and there exists a unique $1 \leq i \leq m$ such that $| u_i(\cdot,t) |_{\infty,\Omega} \to \infty$ as $t \to T_{max}^-$. Without loss of generality assume that $i = m$. Let $0 < \tau < T_{max}$. We make the following simplifying definitions:

$$C = \max_{1 \leq i \leq m} \left\{ \frac{d_m L_1}{d_i} \right\},$$

$$F(t,x) = \sum_{i=1}^{m} h_i(u_i(\tau,x)) + \sum_{i=1}^{m-1} \left(\frac{d_i}{d_m} - 1 \right) h_i(u_i(t,x)) \quad \text{for all } (t,x) \in [\tau, T_{max}) \times \Omega,$$

$$g(t,x) = \sum_{i=1}^{m} d_i h_i''(u_i(t,x)) |\nabla u_i(t,x)|^2 \quad \text{for all } (t,x) \in [\tau, T_{max}) \times \Omega,$$

$$f_{m+1}(t,x,y) = C\Big[\sum_{i=1}^{m} \frac{d_i}{d_m} h_i(u_i(t,x)) + y\Big] + L_2 - \partial H(u(t,x))f(u(t,x)) \quad \text{for all } (t,x,y) \in [\tau,T_{max})\times\Omega\times R .$$

Furthermore, we let v be the unique solution of

$$\begin{aligned} &v_t(t,x) = d_m\Delta v(t,x) + g(t,x) + f_{m+1}(t,x,v(t,x)) && \tau < t < T_{max}, x \in \Omega , \\ &\partial v(t,x)/\partial\eta = 0 && \tau < t < T_{max}, x \in \partial\Omega , \\ &v(\tau,x) = 0 && x \in \Omega . \end{aligned}$$

Then since $g(t,x) + f_{m+1}(t,x,0) \geq 0$ for all $\tau < t < T_{max}$, $x \in \Omega$, we have $v \geq 0$ (cf. Lightbourne and Martin [8]). Finally, we define

$$w(t,x) = \int_{\tau}^{t} \Big[\sum_{i=1}^{m} \frac{d_i}{d_m} h_i(u_i(s,x)) + v(s,x)\Big] ds \quad \text{for all } (t,x) \in [\tau,T_{max})\times\Omega .$$

Then a simple calculation shows

$$\begin{aligned} &w_t(t,x) = d_m\Delta w(t,x) + F(t,x) + Cw(t,x) + L_2 t , && \tau < t < T_{max}, x \in \Omega , \\ &\partial w(t,x)/\partial\eta = 0 , && \tau < t < T_{max}, x \in \partial\Omega, \\ &w(\tau,x) = 0, && x \in \Omega, \end{aligned}$$

and since $F(t,x) + L_2 t$ is uniformly bounded on $[\tau,T_{max})\times\Omega$, we have $w(t,x)$ and hence $F(t,x) + Cw(t,x) + L_2 t$ uniformly bounded on $[\tau,T_{max})\times\Omega$. Thus, Theorem 9.1 in Ladyzenskaja [7] implies

$$w_t(\,,\,) \in L^p\big((\tau,T_{max})\big) \quad \text{for all } 1 \leq p < \infty$$

and consequently it follows that

$$H(u(\,,\,)) \in L^p\big((0,T_{max})\times\Omega\big) \quad \text{for all } 1 \leq p < \infty .$$

Now, if we let z be the unique solution of

$$\begin{aligned} &z_t(t,x) = d_{m+1}\Delta z(t,x) + K_1[H(u(t,x))]^q + K_2 && \tau < t < T_{max}, x \in \Omega, \\ &\partial z(t,x)/\partial\eta = 0 , && \tau < t < T_{max}, x \in \partial\Omega , \\ &z(\tau,x) = A , && x \in \Omega, \end{aligned}$$

where $A = \| h_m(u_m(0,\,)) \|_{\infty,\Omega}$, then (H3), (H5), (A6) and classical maximum principles (cf. Sperb [11]) imply that $z(t,x) \geq h_m(u_m(t,x))$ for all $\tau < t < T_{max}$, $x \in \Omega$. Therefore, if we apply our L^p bounds on $H(u(\,,\,))$ along with Lemma 3.3 and Theorem 9.1 in Ladyzenskaja [7], then there exists $N > 0$ such that $\| z(t,\,) \|_{\infty,\Omega} < N$ for all $0 \leq t < T_{max}$. That is, $\| h_m(u_m(t,\,)) \|_{\infty,\Omega} < N$

for all $0 \le t < T_{max}$. Consequently (H2) implies that $u_m(t,x)$ is uniformly bounded on $[0,T_{max})\times\Omega$. This contradicts $T_{max} < \infty$. The result follows.

REFERENCES

1. N. D. Alikakos, *L_p - bounds of solutions of reaction-diffusion equations* , Comm. Partial Diff. Eq. 4, No. 8 (1979) , 827 - 868.
2. M. Feinberg, *Complex balancing in general kinetic systems* , Arch Rational Mech. Anal. 49 (1972), 187 - 194.
3. K. Groger, *On the existence of steady states of certain reaction-diffusion systems* , Archive Rational Mechanics and Analysis, 1986.
4. S. Hollis, R. Martin, and M. Pierre, *Global existence and boundedness in reaction-diffusion systems* , (preprint).
5. F. Horn & R. Jackson, *General mass action kinetics* , Arch. Rational Mech. Anal. 47 (1972), 81 - 116.
6. F. Horn, *Necessary and sufficient conditions for complex balancing in chemical kinetics* , Arch. Rational Mech. Anal. 49 (1972), 172 - 186.
7. O. A. Ladyzenskaja, V. A. Solonnikov, and N. N. Uralceva, *Linear and Quasilinear Equations of Parabolic Type* . Transl. Math. Monographs Vol. 33, Amer. Math. Soc. , Providence, Rhode Island, 1968.
8. J. H. Lightbourne and R. H. Martin, *Relatively continuous nonlinear perturbations of analytic semigroups* , Nonlinear Anal. - T. M. A., 1(1977), 277 - 292.
9. K. Masuda, *On the global existence and asymptotic behavior of solutions of reaction-diffusion equations* , Hokkaido Math. J. , XII (1982) , 360 - 370.
10. J. Morgan, *Global existence for semilinear parabolic systems,* (preprint).
11. R. Sperb, *Maximum Principles and Their Applications* , Mathematics in Science and Engineering, Vol. 157, Academic Press, New York, 1981.

A Difference Inclusion

Esteban I. Poffald

Department of Mathematics
Wabash College
Crawfordsville, IN 47933

and

Simeon Reich

Department of Mathematics
University of Southern California
Los Angeles, CA 90089

Department of Mathematics
The Technion-Israel Institute of Technology
Haifa 32000, Israel

Our purpose in this paper is to study the difference inclusion

$$(1)\qquad \begin{cases} u_{i+1} - 2u_i + u_{i-1} \in c_i A u_i + f_i, & i = 1, 2, \ldots \\ u_0 = x, \\ \sup\{|u_i| : i \geq 0\} < \infty, \end{cases}$$

where A is a nonlinear (possibly discontinuous and set-valued) m-accretive operator in a Banach space $(X, |\cdot|)$, $\{c_i\}$ is a given sequence of positive numbers, and $\{f_i\}$ is a given sequence X.

This problem is of interest because it is the discrete analog of the quasi-autonomous incomplete Cauchy problem

$$(2)\qquad \begin{cases} u''(t) \in Au(t) + f(t), & 0 < t < \infty \\ u(0) = x \\ \sup\{|u(t)| : t \geq 0\} < \infty, \end{cases}$$

the solutions of which have several remarkable properties [**9**]. We obtain existence and uniqueness results for problem (1) and for the related boundary value problem (3), as well as results on the existence of periodic solutions and the asymptotic behavior of solutions to (1). We remark in passing that second-order difference equations with a single initial condition arise in monetary models [**2**], and that the asymptotic behavior of the solutions to such equations is of interest even on the real line [**4**].

We begin with the following boundary value problem

$$(3)\qquad \begin{cases} u_{i+1} - 2u_i + u_{i-1} \in c_i A u_i + f_i, & 1 \leq i \leq n \\ u_0 = x, \ u_{n+1} = y, \end{cases}$$

where n is a positive integer, $\{c_i : 1 \leq i \leq n\}$ is a given finite sequence of positive numbers, and $\{f_i : 1 \leq i \leq n\}$ is a given finite sequence of points in X. We denote by X^n the product space consisting of all n-tuples $u = (u_1, u_2, \ldots, u_n)$ with u_i in X for all $1 \leq i \leq n$, equipped with the norm

$$\|u\| = \left(\sum_{i=1}^{n} |u_i|^2 \right)^{1/2} .$$

Proposition 1. Let X be a Banach space and $A \subset X \times X$ an m-accretive operator. Then for each x and y in X and $\{f_i : 1 \leq i \leq n\}$ in X^n, the problem (3) has a unique solution in X^n.

Proof. Consider the operators $\mathcal{A} \subset X^n \times X^n$ and $\mathcal{B} : X^n \to X^n$ defined by

$$\mathcal{A}u = \{(c_1 v_1, c_2 v_2, \ldots, c_n v_n) : v_i \in Au_i\} + (x, 0, \ldots, 0, y)$$

and

$$\mathcal{B}u = (2u_1 - u_2, -u_1 + 2u_2 - u_3, \ldots, -u_{n-2} + 2u_{n-1} - u_n, -u_{n-1} + 2u_n).$$

It is known [10, p. 536] that $\mathcal{A}$ is m-accretive and that $\mathcal{B}$ is everywhere defined, continuous and strongly accretive. Therefore $\mathcal{A} + \mathcal{B}$ is also m-accretive and strongly accretive, hence surjective. The result now follows because (3) is equivalent to the inclusion

$$-(f_1, f_2, \ldots, f_n) \in \mathcal{A}u + \mathcal{B}u.$$

Another proof of Proposition 1 can be based on a judicious application of Banach's fixed point theorem [1].

It is clear that in general the difference inclusion (1) has no solution even if $A = 0$ and $\{f_i\} \in \ell^2(X)$. It turns out however, that if (1) has a solution for one point x in X, then it has a unique solution for all x in X.

Recall that the duality map J from X into the family of nonempty subsets of its dual X^* is always monotone. It is single-valued if and only if X is smooth. In this case it is said to be strongly monotone if there is a positive constant M such that

$$(4) \qquad (x - y, Jx - Jy) \geq M|x - y|^2$$

for all x and y in X. It is known [10, p. 528] that a smooth Banach space has a strongly monotone duality map if and only if it is uniformly convex with a modulus of convexity of power type 2. This is the case, for example, when X is one of the Lebesgue spaces L^p, $1 < p \leq 2$.

Theorem 2. Let X be a Banach space with a strongly monotone duality map and let $A \subset X \times X$ be an m-accretive operator. If problem (1) has a solution for some x in X, then it has a unique solution for all x in X.

Proof. Let $w = \{w_i : i = 0, 1, 2, \ldots\}$ be a solution to (1) with $w_0 = z$ and $\sup\{|w_i| : i \geq 0\} = K$, and let x be another point in X. For each $n \geq 1$, there exists, by Proposition 1, a unique solution u^n to (3) with $x = y$. Set $y_i = y_i^n = u_i^n - w_i$. Since A is accretive,

$$(y_{i+1} - 2y_i + y_{i-1}, Jy_i) \geq 0$$

for all $1 \le i \le n$. Hence

$$|y_i| \le \frac{1}{2}(|y_{i-1}| + |y_{i+1}|), \qquad |y_i| \le \max\{|y_0|, |y_{n+1}|\},$$

and

$$|u_i^n| \le |x| + 2K \tag{5}$$

for all $n \ge 1$ and $1 \le i \le n$. Now let $n_0 < n_1 < n_2$ and set $z_i = u_i^{n_1} - u_i^{n_2}$, $0 \le i \le n_1 + 1$. Since A is accretive and J is strongly monotone, we have

$$(z_{i+1} - 2z_i + z_{i-1}, Jz_i) \ge 0$$

and

$$(z_{i+1} - z_i, Jz_i) - (z_i - z_{i-1}, Jz_{i-1}) \ge M|z_i - z_{i-1}|^2$$

for all $1 \le i \le n_1$. Since

$$|z_k| = \sum_{i=1}^{k}(|z_i| - |z_{i-1}|) \le \sum_{i=1}^{k}|z_i - z_{i-1}|,$$

we see that

$$|z_k|^2 \le k \sum_{i=1}^{k} |z_i - z_{i-1}|^2 \le (k/M)(z_{k+1} - z_k, Jz_k) \le (k/2M)(|z_{k+1}|^2 - |z_k|^2)$$

for all $n_0 \le k \le n_1$. Therefore

$$\sum_{k=n_0}^{n_1} (M/k)|z_k|^2 \le |z_{n_1+1}|^2/2 \le 2(|x| + 2K)^2.$$

Since $|z_k| \le |z_{k+1}|$, we also have

$$M|z_i|^2 \left(\sum_{k=n_0}^{n_1} (1/k)\right) \le 2(|x| + 2K)^2$$

and

$$|u_i^{n_1} - u_i^{n_2}|^2 \le (2/M)(|x| + 2K)^2 \left(\sum_{k=n_0}^{n_1} 1/k\right)^{-1}$$

for all $0 \le i \le n_0$. Hence $u_i = \lim_{n\to\infty} u_i^n$ exists for each $i = 1, 2, \ldots$. The sequence $\{u = u_i : i = 0, 1, 2, \ldots\}$ is bounded by (5) and solves (1) because A is closed. If $v = \{v_i : i = 0, 1, 2, \ldots\}$ is another solution of (1), then

$$|u_i - v_i| \le \frac{1}{2}\left(|u_{i+1} - v_{i+1}| + |u_{i-1} - v_{i-1}|\right) \quad \text{for all } i \ge 1.$$

Since $\{|u_i - v_i|\}$ is also bounded, it must be non-increasing. Since $u_0 = v_0 = x$, $u_i = v_i$ for all i and the proof is complete.

Theorem 2 is the discrete analog of [9, Theorem 8]. The following result may be considered an analog of [9, Proposition 9].

Proposition 3. If u and v are two solutions of (1) with $u_0 = x$ and $v_0 = y$, and $z = u - v$, then

$$\sum_{i=1}^{\infty} |z_i - z_{i-1}|^2 \leq 2|x-y|^2/M. \tag{6}$$

Proof. Since A is accretive and J is strongly monotone, we have

$$(z_{i+1} - 2z_i + z_{i-1}, Jz_i) \geq 0$$

and

$$(z_{i+1} - z_i, Jz_i) - (z_i - z_{i-1}, Jz_{i-1}) \geq M|z_i - z_{i-1}|^2 \quad \text{for all } i \geq 1.$$

We also know that $\{|z_i|\}$ is non-increasing. Therefore

$$M\sum_{i=1}^{n} |z_i - z_{i-1}|^2 \leq (z_{n+1} - z_n, Jz_n) - (z_1 - z_0, Jz_0) \leq \frac{1}{2}(|z_{n+1}|^2 - |z_n|^2) + \frac{1}{2}|z_1|^2 + \frac{3}{2}|z_0|^2,$$

and the result follows.

Let A be an m-accretive operator with $0 \in R(A)$ and let z belong to $A^{-1}(0)$. Define $b : [0,\infty) \to [0,\infty)$ by $b(0) = 0$ and

$$b(t) = \inf\{(y, J(x-z))/|x-z| : [x,y] \in A \quad \text{and } |x-z| \geq t\}$$

for positive t. We shall say that A is coercive if there is a point $z \in A^{-1}(0)$ such that $\lim_{t\to\infty} b(t) = \infty$. We continue with an existence result for such operators.

Proposition 4. Let X be a Banach space with a strongly monotone duality map and let $A \subset X \times X$ be an m-accretive operator with $0 \in R(A)$. If A is coercive, $\{c_i\}$ is bounded away from zero, and $\{f_i\}$ is bounded, then problem (1) has a unique solution for all x in X.

Proof. Since A is coercive, there is a point $z \in A^{-1}(0)$ and a non-decreasing function $b : [0,\infty) \to [0,\infty)$ such that $b(0) = 0$, $\lim_{t\to\infty} b(t) = \infty$ and

$$(y, J(x-z)) \geq b(|x-z|)|x-z| \tag{7}$$

for all $y \in Ax$. For each $n \geq 1$ there exists, by Proposition 1, a unique solution $u = u^n$ to (3) with $x = y = z$. Denoting $\sup\{|f_i|/c_i : i = 1, 2, \ldots\}$ by K, and $u_i - z$ by v_i, we see that

$$(v_{i+1} - 2v_i + v_{i-1}, Jv_i) \geq c_i(b(|v_i|) - K)|v_i|$$

for all $1 \leq i \leq n$. We claim that $b(|v_i|) \leq K$ for all these i. If this were not true, there would be indices $1 \leq j \leq k \leq n$ such that $b(|v_{j-1}|) \leq K$, $b(|v_i|) > K$ for $j \leq i \leq k$, and $b(|v_{k+1}|) \leq K$. Since $(v_{i+1} - 2v_i + v_{i-1}, Jv_i) \geq 0$ for $j \leq i \leq k$, it follows that $|v_i| \leq (|v_{i+1}| + |v_{i-1}|)/2$ and $|v_{i-1}| \leq \max(|v_{j-1}|, |v_{k+1}|)$. Since b is non-decreasing, we have reached a contradiction. We conclude that $\{u^n\}$ is bounded by a bound which is independent of n. The proof of Theorem 2 can now be used to show that $u_i = \lim_{n\to\infty} u_i^n$ exists for each

$i = 1, 2, \ldots$, and that $u = \{u_i : i = 0, 1, 2, \ldots\}$ is a solution of problem (1) with $u_0 = z$. The existence of solutions for all initial points is now seen to be a consequence of Theorem 2 itself.

Proposition 4 improves upon a result mentioned on p. 448 of [8], where X is a Hilbert space, A is strongly accretive, and $f \in \ell^2(X)$.

It is of interest to note that the solutions of the continuous problem (2) can be approximated by the solutions of the discrete problem (1). To see this, we first note that the proofs of Theorems 4 and 8 of [9] show that the solutions of (2) can be approximated by the solutions to the problem

$$(8) \qquad \begin{cases} u''(t) = A_r u(t) + f(t), \ 0 < t < T \\ u(0) = u(T) = x, \end{cases}$$

where A_r is the Yosida approximation of A and $f \in W^{1,2}_{loc}(0, \infty; X)$. Therefore it suffices to show that the solutions to (3) with $A = A_r$ and $x = y$ approximate the solutions to (8). To this end, fix T and n, and let $h = T/(n+1)$, $c_i = h^2$, and $f_i = h^2 f(ih)$ for $1 \le i \le n$. Let u_T be the solution to (8) and u^n the solution to (3) with $A = A_r$ and $x = y$. Denoting $u_T(ih)$ by v_i, we see that

$$v_{i+1} - 2v_i + v_{i-1} = h^2 A_r v_i + f_i + g_i(h),$$

where $|g_i(h)|/h^3$ is bounded. Now let $w = u^n - v$. Since A is accretive, we have

$$(w_{i+1} - 2w_i + w_{i-1} + g_i(h), Jw_i) \ge 0.$$

Therefore

$$|w_i| \le \frac{1}{2}(|w_{i+1}| + |w_{i-1}|) + Kh^3$$

for some constant K. Hence

$$|w_i| \le |w_{i+1}| + 2iKh^3 \ \text{ and}$$
$$|w_i| \le n(n+1)Kh^3 \le KT^2 h$$

for all $1 \le i \le n$. In other words, for large n the solutions to (3) do indeed provide us with a good approximation to the solutions of (8).

We now turn our attention to the case when both $\{f_i\}$ and $\{c_i\}$ are periodic. We can assume, of course, that both sequences have the same period.

Theorem 5. Let X be a Banach space with a strongly monotone duality map and let $A \subset X \times X$ be an m-accretive operator. If problem (1) has a solution, and $\{f_i\}$ and $\{c_i\}$ are periodic of period N, then there is a solution of (1) which is also N-periodic.

Proof. Given an point x in X and an integer $m \ge 0$, there is by Theorem 2 a unique solution $v = \{v_i : i = m, m+1, \ldots\}$ of the problem

$$(9) \qquad \begin{cases} v_{i+1} - 2v_i + v_{i-1} \in c_i A v_i + f_i, i = m+1, m+2, \ldots \\ v_m = x \\ \sup\{|v_i| : i \ge m\} < \infty. \end{cases}$$

Therefore we can define, for $n \ge m$, operators $E(n, m) : X \to X$ by $E(n, m)x = v_n$. The proof of Theorem 2 shows that

$$|E(n, m)x - E(n, m)y| \le |x - y|.$$

We also have

$$E(n,m)E(m,k) = E(n,k)$$

for $n \geq m \geq k$ (by uniqueness) and

$$E(n+N, m+N) = E(n,m)$$

for $n \geq m$ (by periodicity). It follows that

$$E(m+N,m)^n = E(m+nN,m).$$

Consequently, the iterates of the nonexpansive map $E(m+N,m) : X \to X$ are bounded. Since X is uniformly convex by [10, Proposition 2.11], this map must have a fixed point. The result now follows by taking $m = 0$.

Corollary 6. If $\{f_i\}$ and $\{c_i\}$ are periodic sequences of period N, and u is a solution of problem (1), then the strong $\lim_{n\to\infty}(u_{m+nN} - u_{m+(n+1)N}) = 0$ for each $m \geq 0$.

Proof. Let v be the periodic solution of (1) the existence of which is guaranteed by Theorem 5. Since

$$u_{m+nN} - u_{m+(n+1)N} = \sum_{i=nN}^{(n+1)N} (u_{m+i} - v_{m+i} - (u_{m+i+1} - v_{m+i+1})),$$

the result follows from Proposition 3.

Theorem 5 can be improved when X is a Hilbert space.

Theorem 7. Let H be a Hilbert space and let $A \subset H \times H$ be a maximal monotone operator. If problem (1) has a solution u, and $\{f_i\}$ and $\{c_i\}$ are periodic of period N, then there is an N-periodic solution w of (1) such that the weak $\lim_{i\to\infty}(u_i - w_i) = 0$.

Proof. Using the notation of the proof of Theorem 5, we see that for each $m \geq 0$, $u_{m+nN} = E(m+N,m)^n u_m$ are the iterates of the nonexpansive map $E(m+N,m) : H \to H$. Since $\lim_{n\to\infty}(E(m+N,m)^n u_m - E(m+N,m)^{n+1} u_m) = 0$ by Corollary 6, we can conclude that the weak $\lim_{n\to\infty} u_{m+nN}$ exists for each $m \geq 0$. Denoting this limit by w_m, we clearly obtain an N-periodic sequence w. We already know, by Theorem 5, that problem (1) admits an N-periodic solution. Let v be such a solution. Since Proposition 3 implies that the strong $\lim_{n\to\infty}(u_{m+nN} - u_{m-1+nM} - (v_m - v_{m-1})) = 0$, we can also conclude that the strong $\lim_{n\to\infty}(u_{m+nN} - u_{m-1+nN}) = w_m - w_{m-1}$. Consequently,

$$\lim_{n\to\infty}(1/c_m)(u_{m+1+nN} - 2u_{m+nN} + u_{m-1+nN} - f_m, u_{m+nN}) = (1/c_m)(w_{m+1} - 2w_m + w_{m-1} - f_m, w_m).$$

Proposition 2.5 of [3] now implies that $w_m \in D(A)$ and that $w_{m+1} - 2w_m + w_{m-1} \in c_m A w_m + f_m$ for each $m \geq 0$. In other words, w us an N-periodic solution of (1). Finally, we observe that for $nN < m \leq (n+1)N$,

$$u_m - w_m = u_{nN} - w_0 + \sum_{i=nN+1}^{m} (u_i - u_{i-1} - (w_i - w_{i-1}))$$

and

$$\sum_{i=nN+1}^{m} |u_i - u_{i-1} - (w_i - w_{i-1})| \le \sqrt{N} \left(\sum_{i=nN+1}^{m} |u_i - u_{i-1} - (w_i - w_{i-1})|^2 \right)^{1/2} .$$

Applying Proposition 3 and recalling that the sequence $\{u_{nN}\}$ converges weakly to w_0, we see that the weak $\lim_{n\to\infty} (u_m - w_m) = 0$. The proof is complete.

Although we do not know if Theorem 7 is valid in the setting of Theorem 5, it does hold outside Hilbert space under more restrictive conditions.

Proposition 8. Let X be a Banach space with a Fréchet differentiable norm and a strongly monotone duality map, and let $A \subset X \times X$ be an m-accretive operator. If problem (1) has a solution u, $\{f_i\}$ and $\{c_i\}$ are periodic of period N, and $cl(D(A))$ is boundedly compact, then there is an N-periodic solution w of (1) such that the strong $\lim_{i\to\infty} (u_i - w_i) = 0$.

Proof. Using the argument leading to Theorem 7, we note that in the present case the weak $\lim_{n\to\infty} u_{m+nN} = w_m$ exists for each $m \ge 0$ by a corollary of the nonlinear mean ergodic theorem. Since $cl(D(A))$ is assumed to be boundedly compact, this limit is actually strong. The sequence w is now seen to be a solution of (1) because A is closed, and the result follows.

We present now another result on the asymptotic behavior of the solutions to (1). It is valid in all Banach spaces.

Proposition 9. Let X be a Banach space, $A \subset X \times X$ an m-accretive operator which is also strongly accretive, and u a solution of (1). If $\lim_{i\to\infty} f_i = f_\infty$ and $\lim_{i\to\infty} c_i = c_\infty > 0$, then $\lim_{i\to\infty} u_i = u_\infty$, where u_∞ is the unique solution to the inclusion $0 \in c_\infty A u_\infty + f_\infty$.

Proof. Denote $u_i - u_\infty$ by w_i and set $p = \lim_{i\to\infty} \sup |w_i|$. Since A is strongly accretive, there is a positive constant b such that

$$((w_{i+1} - 2w_i + w_{i-1} - f_i)/c_i + f_0/c_\infty, Jw_i) \ge b|w_i|^2$$

for all $i \ge 1$. Therefore

$$(b + 2/c_i)|w_i| \le (|w_{i+1}| + |w_{i-1}|)/c_i + |f_0/c_\infty - f_i/c_i|,$$

so that $(b + 2/c_\infty)p \le 2p/c_\infty$. Hence the result.

This result is the discrete analog of [9, Proposition 12], which provides an answer to a question raised on p. 219 of [6]. Note also that Theorem 4.4 in [10] contains a recent Hilbert space convergence result [7, Theorem 3.1].

Most of the results of this paper are proved under the assumption that the Banach space X is smooth and that its duality map is strongly monotone. The smoothness assumption can be dropped provided the strong monotonicity assumption (4) is rephrased as follows:

$$(x - y, x^* - y^*) \ge M|x - y|^2 \tag{10}$$

for all $x \in X$, $y \in X$, $x^* \in Jx$ and $y^* \in Jy$.

We do not know, however, if our results remain true when (10) is not assumed to hold. At any rate, we conclude this note with a new characterization of those Banach spaces which have a strongly monotone duality map.

Proposition 10. The duality map J of a Banach space X is strongly monotone if and only if its inverse $J^{-1} : X^* \to X$ is Lipschitzian.

Proof. Assume first that the duality map J of X is strongly monotone. Let $\{x_n^*\}$ be a sequence in the range of J which converges to x^*. Then $x_n^* \in Jx_n$ for some sequence $\{x_n\}$ in X and

$$M|x_n - x_m|^2 \leq (x_n - x_m, x_n^* - x_m^*) \leq |x_n - x_m||x_n^* - x_m^*|.$$

Therefore the sequence $\{x_n\}$ is Cauchy and converges to a point x in X. Since x^* must belong to Jx, we see that the range of J is closed. Since it is always dense, it must coincide with X^*. Hence X is reflexive, J^{-1} is defined on all of X and coincides with the duality map of X^*. If $x \in J^{-1}x^*$ and $y \in J^{-1}y^*$, then $x^* \in Jx$ and $y^* \in Jy$, so that

$$M|x - y|^2 \leq (x - y, x^* - y^*) \leq |x - y||x^* - y^*|.$$

Consequently, J^{-1} is single-valued and Lipschitzian.

Conversely, assume that the inverse J^{-1} of J is single-valued, defined on all of X and Lipschitzian with a Lipschitz constant L. Then X is reflexive and so is its dual X^*. Therefore J^{-1} coincides with the duality map of X^*. Denoting X^* by Y and J^{-1} by K, we observe that

$$(x, Ky) + (y, Kz) + (z, Kx) \leq \frac{1}{2}(|x|^2 + |y|^2) + \frac{1}{2}(|y|^2 + |z|^2) + \frac{1}{2}(|z|^2 + |x|^2) = |x|^2 + |y|^2 + |z|^2$$

forall x, y and z in Y. Therefore

$$(z - y, Kx - Kz) \leq (x - y, Kx - Ky)$$

and

$$(z - y, Kx - Ky) = (z - y, Kx - Kz) + (z - y, Kz - Ky) \leq (x - y, Kx - Ky) + (z - y, Kz - Ky).$$

Letting z belong to $z + (\frac{1}{2}\ L)J(Kx - Ky)$, we see that

$$(\frac{1}{2}\ L)|Kx - Ky|^2 \leq (x - y, Kx - Ky) + L \cdot (\frac{1}{2}\ L)^2 |Kx - Ky|^2$$

and $(x - y, Kx - Ky) \geq (\frac{1}{4}\ L)|Kx - Ky|^2$.

Now let $u, v \in X$, $u^* \in Ju$ and $v^* \in Jv$. Then $u = Ku^*$ and $v = Kv^*$. Hence $(u - v, u^* - v^*) \geq (\frac{1}{4}\ L)|u - v|^2$, and the proof is complete.

This result improves upon [12, Theorem 4]. We take this opportunity to remark that Theorem 2 of [12] (on the existence of fixed points of nonlinear semigroups) is indeed valid for all hyperconvex spaces.

We can also prove the sufficiency part of Proposition 10 by first showing that if J^{-1} is Lipschitzian, then the modulus of smoothness of X^* is of power type 2, and then using the first part of the proof of [10, Proposition 2.11]. We are thus led to the following result.

Theorem 11. For a Banach space X, the following are equivalent:

(A) X is uniformly convex with a modulus of convexity of power type 2;

(B) The duality map J of X is strongly monotone;

(C) The inverse $J^{-1} : X^* \to X$ of the duality map is Lipschitzian.

Spaces for which both J and J^{-1} are strongly monotone (and therefore Lipschitzian) are of interest in connection with the continuous problem (2) [9, p. 391], as well as in approximation theory [5]. More information on such spaces can be found on p. 549 of [5]

As a matter of fact, it can be shown that the following statements are also equivalent for a Banach space X (cf. [11, p. 337]):

(D) X is uniformly convex with modulus of convexity of power type $p \geq 2$;

(E) The duality map J of X satisfies

$$(x - y, x^* - y^*) \geq M|x - y|^p$$

on bounded sets, where M is positive and $x^* \in Jx$, $y^* \in Jy$;

(F) The inverse $J^{-1} : X^* \to X$ of the duality map is Hölder continuous with exponent $1/(p-1)$ on bounded sets.

Unfortunately, the proof of Theorem 2 breaks down when (10) is replaced by (E).

The first author was partially supported by the Byron K. Trippet Research Stipend at Wabash College and the second author was partially supported by the Fund for the Promotion of Research at the Technion and by the Technion VPR Fund.

REFERENCES

[1] PH. BÉNILAN, *Personal communication.*

[2] J. L. BONA AND S. GROSSMAN, *Price and interest rate dynamics in a transactions based model of money demand*, preprint.

[3] H. BREZIS, "Opérateurs Maximaux Monotones," North-Holland, Amsterdam, 1973.

[4] A. DROZDOWICZ AND J. POPENDA, *Asymptotic behavior of the solutions of the second order difference equation*, Proc. Amer. Math. Soc. **99** (1987), 135–140.

[5] C. FRANCHETTI AND W. LIGHT, *The alternating algorithm in uniformly convex spaces*, J. London Math. Soc. **29** (1984), 545–555.

[6] E. MITIDIERI, *Some remarks on the asymptotic behavior of the solutions of second order evolution equations*, J. Math. Anal. Appl. **107** (1985), 211–221.

[7] E.MITIDIERI AND G. MOROSANU, *Asymptotic behavior of the solutions of second order difference equations associated to monotone operators*, Numer. Funct. Anal. Optim. **8** (1986), 419–434.

[8] G. MOROSANU, *Second order difference equations of monotone type*, Numer. Funct. Anal. Optim. **1** (1979), 441–450.

[9] E. I. POFFALD AND S. REICH, *A quasi-autonomous second-order differential inclusion, in "Non-Linear Analysis"*, North-Holland, Amsterdam (1985), 387–392.

[10] E. I. POFFALD AND S. REICH, *An incomplete Cauchy problem*, J. Math. Anal. Appl. **113** (1986), 514–543.

[11] S. REICH, *Constructive techniques for accretive and monotone operators, in "Applied Nonlinear Analysis"*, Academic Press, New York (1979), 335–345.

[12] S. REICH, *Integral equations, hyperconvex spaces and the Hilbert ball, in "Nonlinear Analysis and Applications"*, Marcel Dekker, New York (1987), 517–525.

SPECTRUM ESTIMATIONS FOR THE GENERALIZED QUANTUM HENON-HEILES SYSTEM (*)

María J. Rodríguez and Luis Vázquez
Departamento de Física Teórica, Facultad de Ciencias Físicas
Universidad Complutense, 28040-Madrid (Spain)

ABSTRACT: We propose an explicit unitary discretization of the Heisenberg equations associated to a general quantum system with two degrees of freedom. In the framework of this approximation we extract information related to the energy spectrum of the generalized quantum Henon-Heiles system.

1. Introduction

Traditionally the quantum spectrum estimations have been obtained in the framework of the Schrödinger equation. Recently [1-5], the computation of the energy spectrum has been investigated by using a consistent unitary discretization of the Heisenberg equations of motion. Up to now, such estimations have been carried out for systems with either one or an infinite number of degrees of freedom. In this contribution we give a new explicit unitary discretization for the Heisenberg equations associated to a general quantum system with two degrees of freedom. On the other hand, by using this scheme we obtain partial information about the energy spectrum of the generalized quantum Henon-Heiles system.

2. The scheme and the computation of the energy spectrum

Let us consider the two-dimensional quantum system

$$H=\frac{1}{2}(p^2+P^2)+V(q,Q) \quad . \tag{1}$$

The Heisenberg equations of motion are

$$dq/dt = p \;, \quad dQ/dt = P$$
$$dp/dt = -\frac{\partial V}{\partial q} \qquad dP/dt = -\frac{\partial V}{\partial Q} \tag{2}$$

and the operators q(t), p(t), Q(t) and P(t) must satisfy the commutation relations

$$[q(t),p(t)] = i \qquad [Q(t),P(t)] = i \;. \tag{3}$$

A consistent unitary discretization of (2) is the following explicit scheme

$$(q_{K+1}-q_K)/\tau = p_{K+1} \quad , \quad (Q_{K+1}-Q_K)/\tau = P_{K+1}$$

$$(p_{K+1}-p_K)/\tau = -\frac{\partial V}{\partial q}(q_K,Q_K) \, , \quad (P_{K+1}-P_K)/\tau = -\frac{\partial V}{\partial Q}(q_K,Q_K) \tag{4}$$

where q_K, p_K, Q_K, P_K are the operators q,Q,p and P at time $t=k\tau$. This scheme is a generalization of the one studied in Ref.5 for a system with one degree of freedom. It is easy to verify that the scheme preserves the equal-time commutation relations (3) as follows

$$[q\ ,p\ ] = i \quad , \quad [Q\ ,P\ ] = i\ . \tag{5}$$

The simplest way to obtain energy eigenvalues is by studying the time-dependence of the operators q(t),Q(t),p(t) and P(t) for short times (one time step with the help of an unitary scheme). To this purpose, we introduce the set of Fock states |n,N> generated by the operators a,a^+,A and A^+ defined in terms of q_0 ,p_0 ,Q_0 and P_0 as follows

$$q_0 = \alpha(a+a^+)/\sqrt{2} \quad , \quad p_0 = (a-a^+)/i\alpha\sqrt{2}$$
$$Q_0 = \beta(A+A^+)/\sqrt{2} \quad , \quad P_0 = (A-A\ )/i\beta\sqrt{2}\ , \tag{6}$$

where α and β are variational parameters to be fixed by a simple self-consistency requirement. The operators a and A satisfy the commutation relations

$$[a,a^+] = 1 \qquad [A,A^+] = 1\ .$$

According to eqs. (4) we take the matrix elements

$$\langle\chi|\ q_1|\psi\rangle \quad , \quad \langle\chi|\ p_1|\psi\rangle$$
$$\langle\chi|\ Q_1|\psi\rangle \quad , \quad \langle\chi|\ P_1|\psi\rangle \tag{7}$$

where $|\psi\rangle = |n,N\rangle$, $\langle\chi| = \langle n+1,N+1| + \langle n+1,N| + \langle n,N+1| + \langle n,N|$. In the expressions of the matrix elements we only consider terms up to order τ , since the scheme is accurate to this order. On the other hand the matrix element of any operator which does not commute with the Hamiltonian (1) is of the form $\sum c\exp(i\omega t)$, ω denoting all possible energy differences. In this context we choose the parameters α and β such that to order τ only a single frequency ω appears in the matrix elements of the operators q_1 , p_1 , Q_1 and P_1:

$$\frac{\langle\chi|\ q_1|\psi\rangle}{\langle\chi|\ q_0|\psi\rangle} \simeq 1+i\omega_n\tau \qquad \frac{\langle\chi|\ p_1|\psi\rangle}{\langle\chi|\ p_0|\psi\rangle} \simeq 1+i\omega_n\tau$$
$$\frac{\langle\chi|\ Q_1|\psi\rangle}{\langle\chi|\ Q_0|\psi\rangle} \simeq 1+i\omega_N\tau \qquad \frac{\langle\chi|\ P_1|\psi\rangle}{\langle\chi|\ P_0|\psi\rangle} \simeq 1+i\omega_N\tau\ . \tag{8}$$

3. The generalized quantum Henon-Heiles system

By using Painlevé analysis it has been proved [6 , 7] that the classical generalized Henon-Heiles system given by the Hamiltonian:

$$H=\frac{1}{2}(p^2+P^2+A\,q^2+B\,Q^2)+D\,q^2\,Q-\frac{C}{3}Q^3, \qquad A\,,\,B>0\,, \tag{9}$$

is integrable for the values $\lambda=0$, $-1/6$, -1, where $\lambda=D/C$. For $\lambda=1$ the standard Henon-Heiles Hamiltonian is not integrable. On the other hand, numerical computations [6] have shown that the case $\lambda=-1/16$, $A=1$, $B=16$ is integrable while the case $\lambda=-5/16$, $A=5$, $B=16$ appears as nonintegrable. These facts suggest to us the importance of finding out what properties characterize the eigenvalues associated to the quantum Henon-Heiles system. This problem has been partially studied for certain values of the parameters using semiclassical methods [8].

In this contribution we present some estimations about the energy differences for the spectrum of the quantum system (9). Such results are obtained by using the method described in the above section. More precisely we get

$$\begin{aligned}
\langle\chi|\,q_1\,|\psi\rangle\,/\,\langle\chi|\,q_0\,|\psi\rangle &\simeq 1+iz/\alpha^2\\
\langle\chi|\,p_1\,|\psi\rangle\,/\,\langle\chi|\,p_0\,|\psi\rangle &\simeq 1+iz\Big(A\alpha^2+D\alpha^2\beta\sqrt{\tfrac{2(N+1)}{n+1}}\Big)\\
\langle\chi|\,Q_1\,|\psi\rangle\,/\,\langle\chi|\,Q_0\,|\psi\rangle &\simeq 1+iz/\beta^2\\
\langle\chi|\,P_1\,|\psi\rangle\,/\,\langle\chi|\,P_0\,|\psi\rangle &\simeq 1+iz\Big(B\beta^2+D\beta\alpha^2\frac{2n+1}{\sqrt{2(N+1)}}-C\beta^3\frac{2N+1}{\sqrt{2(N+1)}}\Big).
\end{aligned} \tag{10}$$

We choose α and β such that only a single energy difference appears in the matrix elements of the operators. Thus we have

$$\begin{aligned}
\omega_n &= 1/\alpha^2 \quad ; \quad \omega_n = A\alpha^2+D\alpha^2\beta\sqrt{\frac{2(N+1)}{n+1}}\\
\omega_N &= 1/\beta^2 \quad ; \quad \omega_N = B\beta^2+D\beta\alpha^2\frac{2n+1}{\sqrt{2(N+1)}}-C\beta^3\frac{2N+1}{\sqrt{2(N+1)}}\,.
\end{aligned} \tag{11}$$

The corresponding computed energy difference is given by

$$\Omega_{n,N}=\omega_n+\omega_N\,. \tag{12}$$

For the free case (D=C=0) we obtain

$$\Omega_{n,N}=E_{n+1,N+1}-E_{n,N}=\sqrt{A}+\sqrt{B}\,.$$

Two important remarks must be considered:

1. N and n are quantum numbers associated to the system only when D=C=0, but we also use them to label the energy differences obtained with our approximation in the general case.
2. In our approximation, (which is equivalent to the one-pole approximation for a Green's function [2]), we obtain a representative sector of the discrete spectrum associated to the generalized quantum Henon-Heiles system.

From (10) we obtain the equation

$$B\beta^4 + D\beta^3 \frac{2n+1}{\left(2(N+1)\left(A + D\beta\sqrt{2(N+1)}/(n+1)\right)\right)^{1/2}} - C\beta^5 \frac{2N+1}{\sqrt{2(N+1)}} - 1 = 0 . \quad (13)$$

It is easy to check that when λ = D/C<0 equation (13) has a unique positive root. For λ >0 there are either two or no roots according to the values of n and N. This result has been checked numerically for different cases. In table I we give the numerical results for two representative examples.

Our results show that the structure of the discrete spectrum, obtained in the framework of our approximation, depends only on the sign of λ . On the other hand, the classical integrability of the system depends on the sign of λ and the values of the frequencies A and B. This result fits with the results of Casati et al [9], in the sense that the classical phenomenology of the system is richer than the quantum one.

TABLE I: Separation of the energy levels for two sets of parameters of the quantum Henon-Heiles system.

		Classically integrable $\lambda=-1/6$, $A=B=1$, $C=-6$, $D=1$	Classically non-integrable $\lambda=1$, $A=B=C=D=1$
n	N	$\Omega_{n,N}$	$\Omega_{n,N}$
0	0	3.55366873	2.20754272 ; 2.37093551
1	0	3.78896664	1.99024493 ; 2.89126969
1	1	4.21723347	-----
0	1	4.15047002	-----
2	0	4.12210349	1.90563096 ; 3.41321513
2	1	4.37506960	2.20536136 ; 2.72144757
2	2	4.58392413	-----
0	2	4.49158783	-----
1	2	4.49189186	-----
3	0	4.49146826	1.85652745 ; 3.91588660
3	1	4.57067960	2.02220872 ; 3.20344004
3	2	4.71272642	-----
3	3	4.84817007	-----
0	3	4.74014481	-----
1	3	4.70124095	-----
2	3	4.75636376	-----

References

(*) Partially supported by U.S.-Spanish Joint Committee for Scientific and Technological Cooperation under grant Nº CCB-8509/001.

1. C.M.Bender, K.A.Milton, D.H.Sharp and M.Simmons, Jr., Discrete time quantun mechanics, Phys. Rev. D32 (1985), 1476-1485.

2. C.M.Bender and K.A.Milton, Approximate determination of the mass gap in quantum field theory using the method of finite elements, Phys. Rev. D34 (1986), 3149-3155.

3. L.Vázquez, The two dimensional quantum field theory $\Box\phi+6\phi+2\phi^3=0$ on a Minkowski lattice, Phys. Rev. D35 (1987), 1409-1411.

4. L.Vázquez, Particle spectrum estimations for the quantum field theory $\Box\phi+(m^3/\sqrt{2})\sin(\sqrt{2}\phi/m)=0$ on a Minkowski lattice, Phys. Rev. D35 (1987), 3274-3276.

5. L. Vázquez, On the discretization of certain operator field equations, Z. Naturforsch. 41a (1986), 788-790.

6. Y.F.Chang, M.Tabor and J.Weiss, Analytic structure of the Henon-Heiles Hamiltonian in integrable and non-integrable systems, J. Math. Phys. 23 (1982), 531-538.

7. B.Grammaticos, B.Dorizzi and A.Ramani, Integrability of Hamiltonian with third- and fourth-degree polynomial potentials, J. Math. Phys. 24 (1983), 2289-2295.

8. G.Casati, B.V.Chirikov, I.Guarnieri and D.L.Shepelyansky, Dynamical stability of quantum "chaotic" motion in a hydrogen atom, Phys. Rev. Lett. 56 (1986), 2437-2440.

A SURVEY OF LOCAL EXISTENCE THEORIES FOR ABSTRACT NONLINEAR INITIAL VALUE PROBLEMS

Eric Schechter, Mathematics Department
Vanderbilt University, Box 21, Station B
Nashville, Tennessee 37235, U.S.A.

Abstract. This paper surveys the abstract theories concerning local-in-time existence of solutions to differential inclusions, $u'(t) \in F(t, u(t))$, in a Banach space. Three main approaches assume generalized compactness, isotonicity in an ordered Banach space, or dissipativeness. We consider different notions of "solution," and also the importance of assuming or not assuming that $F(t, x)$ is continuous in x. Other topics include Carathéodory conditions, uniqueness, semigroups, semicontinuity, subtangential conditions, limit solutions, continuous dependence of u on F, and bijections between u and F.

1. Introduction. In this paper we consider differential inclusions of the form

$$u'(t) \in F(t, u(t)) \qquad (0 \le t \le T). \tag{1.1}$$

Here $u(t)$ takes values in a Banach space $(X, \|\ \ \|)$, and F is a mapping from some subset of $[0, T] \times X$ into the set of all subsets of X. We ask what hypotheses on X, F and a given initial value $u(0)$ guarantee that a solution of (1.1) exists for some $T > 0$. We permit T to be small, and to depend on $u(0)$ — we are concerned primarily with *local* existence, as explained further in §2. Intuitively, t represents time, and $u(t)$ represents the state of some system which is evolving as time passes, according to a rule or environment described by F. Thus u is sometimes called an *evolution*, and (1.1) an *evolution equation*. The operator F is said to be the *generator* of the evolution. An important special case to which we shall devote some additional attention is the *autonomous* problem

$$u'(t) \in G(u(t)) \qquad (0 \le t \le T) \tag{1.2}$$

(i.e., where $F(t, x) = G(x)$ is independent of t).

Note that F may be point-valued, or interval-valued; thus (1.1) includes differential equations and differential inequalities as special cases. On the other hand, F may be set-valued; thus implicit differential equations of the form $H(t, u(t), u'(t)) = 0$ are also included as special cases, even if $H(t, x, \cdot)$ is not invertible for some t and x. Also F may be discontinuous and X may be infinite-dimensional; thus (1.1) may represent a partial differential equation. Problems of the form (1.1) also include problems from control theory, integral equations, functional differential equations, population models, and numerous other applications; but we are concerned here with the abstract theory, not the applications.

We are far from a complete understanding of local existence of solutions to (1.1), or even (1.2). Some of the known sufficient conditions for existence have partial converses, but we are still quite far from knowing full necessary and sufficient conditions. A handful of examples are known in which (1.1) has no solution, but most of these examples are variants of a single example of Dieudonné (discussed in §2). They are not diverse enough to adequately motivate the known sufficient conditions.

We are even far from a *unified* understanding of (1.1) or (1.2). The literature includes several essentially separate schools of thought, which use definitions and hypotheses so different in nature that it is difficult to make comparisons between them. Three main approaches assume F satisfies a condition of generalized compactness, isotonicity, or dissipativeness (introduced in §7-10); the dissipativeness school of thought also has several separate subschools. Various papers also differ substantially in their definitions of "solution" (see §3, 7, 13, and 14), as well as in their other assumptions about F — e.g., whether F is assumed continuous, semicontinuous, or not continuous at all (see especially §5 and 12).

Wide gaps exist between the theories. However, this survey will show certain similarities and analogies between the several theories. These suggest that deeper connections may be uncovered in the future; that is a goal of this author's future research. This paper is intended as preparation for that research.

This survey contains no new results; it is merely a review of the literature. However, it is our hope that this survey may convey to some readers a broad perspective including some new insights. This survey is ordered pedagogically — i.e., simpler topics first — with the intention that it can be read by newcomers to the subject. Mathematicians who are already familiar with one or more of the theories of existence may wish to skip ahead to §11 and 12, which unite many of the simpler ideas of §5-10 into a single framework.

The literature concerning (1.1) is enormous, and so this survey is necessarily incomplete. We omit much historical background, and concentrate on the most recent results. We make no attempt to deal with applications. Aside from an occasional indication of some of the simpler ideas involved, we omit all proofs. For brevity, we give only the simplest and/or most general versions of certain ideas or theorems, omitting many other versions. We sometimes mention technicalities (boundedness, measurability, etc.) without giving their details. We omit some important and widely-studied topics — e.g., additional existence results which use linear operators on Hilbert spaces or the dual space X^* — because these topics are too specialized for the purposes of this survey. Decisions of what to include and what to omit were guided by the goal of a broad perspective, mentioned above, but were inevitably biased by the present author's own interests and ignorances. If we have omitted the reader's favorite result, or some cherished technical details from that result, we apologize. The present author would be grateful for communications about such omissions and about more recent results; some of these might be reported in an addendum to this survey a year or two hence.

A small portion of this paper was presented at the conference at Howard University in August 1987. This paper owes much to related surveys of Crandall [26], Hájek [48], Lakshmikantham and Leela [73], and Volkmann [124]. The author is grateful to A. Bressan, M. Freedman, M. Parrott, T. Seidman, P. Takac, P. Volkmann, and others for preprints and reprints and for their helpful comments on earlier versions of this paper.

2. Boundedness, local existence, and nonexistence. We begin by recalling two classical existence theorems, with G single-valued and continuous:

2.1. THEOREM. *Given any* $u(0) \in X$, *the initial value problem* (1.2) *has a unique solution for all* $t \geq 0$, *if* G *satisfies the Lipschitz condition*

$$\|G(x) - G(y)\| \leq \omega\|x - y\| \qquad (x, y \in X) \tag{2.2}$$

for some constant ω.

Such a function G is said to be *Lipschitzian*, and ω is called the *Lipschitz constant* of G.

2.3. THEOREM (Peano, 1885). *If* X *is a finite-dimensional Banach space and* $G : X \to X$ *is continuous, then for each* $u(0) \in X$ *the initial value problem* (1.2) *has at least one solution for some* $T > 0$.

Both of these results have many different proofs which can be found in many books; see for instance [32]. We sketch one proof: If F is single-valued, continuous, and defined everywhere on $[0, T] \times X$, then (1.1) is equivalent to the integral equation

$$u(t) = u(0) + \int_0^t F(s, u(s))\, ds \qquad (0 \leq t \leq T). \tag{2.4}$$

A solution of (2.4) is the same as a fixed point of the operator Φ defined by

$$(\Phi u)(t) = u(0) + \int_0^t F(s, u(s))\, ds \qquad (0 \leq t \leq T). \tag{2.5}$$

Under the hypotheses of Theorems 2.1 or 2.3, for sufficiently small T, Φ has a fixed point, by Banach's fixed point theorem for contraction mappings or by Schauder's fixed point theorem for compact mappings, respectively. (See [116] for an introduction to fixed point theory.) Under the hypotheses of Theorem 2.1, or under those of 2.3 with G bounded, we can repeat this argument on the intervals $[0,T]$, $[T,2T]$, $[2T,3T]$, etc., and thus obtain a solution for all positive t.

We shall use (2.4) and (2.5) again occasionally in this paper, but we shall not survey the theories of integral equations or fixed points. Those theories give an elegant approach to (1.1), but it is not the only approach, and it is generally less successful than other approaches which focus more directly on initial value problems. When F is permitted to be discontinuous and set-valued, and when the domain of $F(t,\cdot)$ is permitted to vary with t, then the conditions corresponding to (2.4) and (2.5) become very complicated.

The hypotheses of Peano's Theorem (2.3) do not guarantee global existence. For instance, the equation $u'(t) = u(t)^2$ with initial value $u(0) = 1$ has unique solution $u(t) = 1/(1-t)$, which blows up as t increases to 1. This behavior is not pathological, but actually typical of nonlinear differential equations, and so we concern ourselves with *local* existence of solutions. Global continuability versus finite-time blowup is a separate question which has received much attention in the literature but will not be studied in depth here. See [6] and references cited therein, for an introduction to this subject. We remark that for continuability of solutions, the crucial question is not whether $\|u(t)\|$ blows up as $t \uparrow T$, but whether $\{u(t) : 0 \leq t < T\}$ is a relatively compact set. Dieudonné [33] gives a simple example of noncontinuability without blowup.

Most abstract existence theorems can be stated in either a local or global form. For instance, the Lipschitz hypothesis (2.2) could be weakened to a *local* Lipschitz condition: we could replace the constant ω with a continuous function of x and y; then the conclusion would be just local existence. On the other hand, if we add to Peano's theorem the assumption that G is uniformly bounded, then we gain the conclusion that existence is guaranteed for all positive t. (Locally this is no real change in the theorem, since any continuous G must be bounded at least on some neighborhood of $u(0)$.) Most abstract existence results in the literature can be modified similarly. In effect, most global existence results can be decomposed into a local existence result plus a global continuability result, the latter being obtained from a global estimate.

The verification of such global estimates in specific problems is a basic part of applied mathematics. The methods needed for such verification vary greatly from one differential equation to another; they do not seem to fit into just a few abstract theories. In fact, the main question addressed in this paper — what conditions on F are sufficient for local existence — may seem alien or even trivial to the applied mathematician, who generally begins with an observed physical phenomenon for which existence is already known.

On the other hand, local existence does fit into just a few abstract theories. A pure mathematician pursuing abstract existence theory may choose to simply assume as a hypothesis that F is uniformly bounded, and focus his or her attention on other difficulties not substantially affected by that assumption. Of course, some researchers choose to investigate the subtleties of weaker, more complicated, and more appropriate boundedness hypotheses on F. For instance, Himmelberg and Van Vleck [57] say that F is "locally weakly integrably bounded" if for each $\rho > 0$, there exists a function $m_\rho \in L^1[0,T]$ such that $\inf\{\|y\| : y \in F(t,x)\} \le m_\rho(t)$ for all x, t with $\|x\| \le \rho$. For another example, Dollard and Friedman [34] work with (1.1) where $F(t,\cdot)$ is continuous and linear for each t; their boundedness assumption is that the upper integral

$$\overline{\int_0^T} \|F(s,\cdot)\|\, ds \equiv \inf\{\textstyle\int_0^T \psi(s)\, ds : \psi \in L^1[0,T]; \quad \|F(s,\cdot)\| \le \psi(s) \text{ for all } s\}$$

be finite. The upper integral coincides with the ordinary (Lebesgue) integral if $\|F(s,\cdot)\|$ is a measurable function of s. A similar estimate is applied to nonlinear problems in [108].

With the abstract viewpoint indicated above, we see local and global existence theorems as superficially different presentations of the same basic ideas. The global formulation is usually simpler in notation than the local formulation, and so it is usually the version used in the literature. In this respect 2.1 is typical, while 2.3 is not. This paper is concerned with existence results (and closely related results, such as uniqueness and continuous dependence) that are local *in essence*, even if they are sometimes formulated in global terms for simplicity of notation; we shall not pursue the verification of the global estimates. For our purposes, nonexistence of a solution to (1.1) means that no solution exists *no matter how small we choose* T.

Our questions of local existence can be studied in a very abstract setting, and so they may be simple in appearance, but they are by no means trivial. For instance, Peano's existence result 2.3 *fails* in infinite-dimensional Banach spaces. Dieudonné [33] gave a simple counterexample in the space c_0 of sequences converging to 0: Let $G(\{x_n\}) = \{(1/n) + \sqrt{|x_n|}\}$ and $u(0) = 0$. Then G is continuous from c_0 into c_0, but (1.2) has *no* solution in c_0, no matter how small we choose T. Several other authors subsequently gave examples in other spaces, by modifying Dieudonné's example, and finally Godunov [45] showed that Peano's conclusion fails in *every* infinite-dimensional Banach space.

Numerous different notions of "solution" can be found in the literature; these will be discussed in §3, 7, 13, and 14. But the particular choice of the definition of "solution" does not matter in the examples of Dieudonné *et al.* A "solution" $u(t)$ for Dieudonné's initial value problem can be explicitly constructed componentwise in the space ℓ^∞ of bounded sequences. This function $u(t)$ is uniquely determined, and is the only possible candidate for a "solution" in any reasonable sense. But this function $u(t)$ takes its values outside of the chosen Banach space c_0, and so is disqualified as a solution of (1.2). Similar remarks apply to the examples of Godunov *et al.*, which are variants of Dieudonné's example.

3. Carathéodory conditions, Carathéodory solutions, and other differentiable solutions. Numerous different notions of "solution" are used in the literature. In this section we discuss those solutions $u(t)$ which are differentiable, i.e. which actually satisfy the differential inclusion (1.1) or (1.2) in some fairly direct sense. Other notions of "solution," more general and useful but less directly appealing to our intuition, will be discussed in §7, 13, and 14.

If F or G is continuous, then a "solution" of (1.1) or (1.2) usually means a continuously differentiable function $u(t)$ which satisfies the differential equation. Such solutions are sometimes called *Newton solutions*, or "strong" or "classical" solutions; however, those last two terms also have other meanings in the literature. If F or G is not continuous, then a more general "solution" may be needed. For motivation we first consider some classical notions of Carathéodory:

We say that a single-valued function $F(t,x)$ satisfies *Carathéodory conditions* if it is continuous in x and if $\sup\{\|F(t,x)\| : x \in S\}$ is majorized by some locally integrable function $m_S(t)$ for each bounded set S or each relatively compact set S. Both definitions — with S bounded or with S relatively compact — are used in the literature, and of course they coincide when X is finite-dimensional. For an arbitrary Banach space, the two definitions still agree locally, i.e. on a sufficiently small neighborhood of $u(0)$; see Theorem 4.7 in [108]. The condition involving compact sets generalizes joint continuity of F.

3.1. THEOREM (Carathéodory, 1927). *Assume $F : [0,\infty) \times X \to X$ satisfies Carathéodory conditions, as defined above. Let $u(0) \in X$. If X is finite-dimensional, then* (2.4) *has a solution u on $[0,T]$ for some $T > 0$.*

This theorem can be found in [24], for instance. Of course, the conclusion fails when X is infinite-dimensional, but variants of this theorem are valid in arbitrary Banach spaces, as we shall see later.

The "solution" u whose existence is asserted above need not be continuously differentiable, since F may be discontinuous. In an arbitrary Banach space, for single-valued F, we say u is a *Carathéodory solution* of the initial value problem (1.1) if u is the solution of the corresponding integral equation (2.4). The integral is understood in the sense of Bochner integrals, i.e., Banach-space-valued Lebesgue integrals. (For an introduction to such integrals, see [35], [74], [129].) Of course, if F is jointly continuous, then a Carathéodory solution is the same thing as a Newton solution.

More generally, whether F is single-valued or not, we say that a *Carathéodory solution* of the differential inclusion (1.1) is a function $u(t)$ which is absolutely continuous on $[0,T]$, which is differentiable almost everywhere on $[0,T]$, which satisfies the differential equation almost everywhere on $[0,T]$, and which satisfies the initial condition if one is given. (For single-valued F, this is equivalent to (2.4). In finite-dimensional spaces, absolute continuity of u implies existence of $u'(t)$ for almost all t, but that implication is not valid in an arbitrary Banach space.) This notion of "solution" is used widely, and in §4-9 of this paper a "solution" will mean a Carathéodory solution unless specified otherwise. Carathéodory solutions are also sometimes referred to in the literature as "strong" or "classical" solutions, but those terms also sometimes refer to Newton solutions, or have still other meanings in the literature.

Because integrals occur naturally in (2.4), Carathéodory solutions are in some sense more natural than Newton solutions, and Carathéodory solutions have been studied extensively in the literature. An interesting result is that of Binding [12], who considers the autonomous differential equation $u'(t) = G(u(t))$ for a single-valued function $G : \mathbf{R} \to \mathbf{R}$ in one dimension. Among other results, Binding gives *necessary and sufficient* conditions on G for existence of solutions, given an initial value. These conditions are as follows. If it is not constant, the solution for such a differential equation cannot cross itself, hence it must be monotone; hence G must be nonnegative (respectively, nonpositive) almost everywhere on an interval to the right (respectively, left) of the initial value. Also, the set where G is infinite or undefined must have measure zero. Finally, if G is not zero almost everywhere, then $1/G$ must be integrable on that interval on one side of the initial value.

Though the notion of Carathéodory solutions is fairly simple and intuitively appealing, it is not adequate for discontinuous G, or for $F(t,x)$ discontinuous in x. For example, the differential equation

$$u'(t) = G(u(t)) = \begin{cases} -1 & \text{if } u(t) \geq 0, \\ 1 & \text{if } u(t) < 0, \end{cases}$$

with initial condition $u(0) = 0$, has no Carathéodory solutions. This problem exhibits what Binding [12] calls *jamming*. Intuitively, we might feel that $u(t) \equiv 0$ *should* be a solution, and indeed it is in the sense of Krasovskij or Filippov:

A *Krasovskij solution*, respectively a *Filippov solution* of (1.1) is a Carathéodory solution of $u'(t) \in \mathbf{K}F(t,u(t))$, respectively $u'(t) \in \mathbf{F}F(t,u(t))$, where

$$\mathbf{K}F(t,x) = \bigcap_{\varepsilon>0} \overline{\text{conv}}F(t, x + \varepsilon B),$$

$$\mathbf{F}F(t,x) = \bigcap_{\varepsilon>0} \bigcap_{\text{null } Z} \overline{\text{conv}}F(t, (x + \varepsilon B)\setminus Z).$$

Here B is the open unit ball, and $\overline{\text{conv}}$ means convex closure. In both of these definitions, "bad" points which are in some sense isolated and atypical of the behavior of F or u are discarded. In the definition of $\mathbf{F}F$, sets $Z \subset X$ having Lebesgue measure zero are discarded. This definition is only meaningful when the Banach space X is finite-dimensional, since Lebesgue measure has no natural analogue on infinite-dimensional spaces. In finite dimensions, $\mathbf{F}F(t,x) \subseteq \mathbf{K}F(t,x)$, so any Filippov solution is also a Krasovskij solution. Krasovskij and Filippov solutions, as well as Carathéodory solutions and Hermes solutions (introduced in §11 of this paper) are surveyed by Hájek [48], at least for the finite-dimensional case.

Still other "solutions" weaken the notion of derivative. The *contingent derivative* of a function $u(t)$ at a point t_0 is the set of all limits (or, in some papers, all weak limits) of sequences of the form $\{[u(t_0+h_n)-u(t_0)]/h_n\}_{n=1}^{\infty}$, where $h_n \to 0$. That set is denoted by $Du(t_0)$. A solution of the *contingent differential equation* $Du(t) \subset F(t,u(t))$ is a function u which satisfies that relation for almost all t (or, in some papers, for all but at most denumerably many t). Some results on contingent differential equations in Banach spaces are given by Chow and Schuur [23].

If $F(t,\cdot)$ is linear for each t, and $F(t,x)$ is written $F(t)x$, then a *weak solution*, or *$*$-solution*, of (1.1) is a function u satisfying

$$< u(t),y > \quad = \quad < u(0),y > + \int_0^t < u(s),F(s)^*y > ds \qquad (0 \le t \le T),$$

for every y in the dual space X^* or in some dense subset of X^*. (The term "weak solution" also has other meanings.) For some recent results concerning $*$-solutions, see Dawson and Gorostiza [31].

A $*$-solution need not be differentiable in the topology of the norm of X. In §11 we shall consider some "solutions" u which need not be differentiable in any sense at all. Thus, the solution $u(t)$ of (1.1) need not actually satisfy (1.1) in any direct sense. Equation (1.1) is only used as an abbreviation for a much longer and more complicated definition of "solution" which does involve u and F. Although we shall not discuss such solutions in any detail until §11, they should be kept in mind in the discussion of evolution operators at the end of §4.

4. Uniqueness, Kamke functions, and semigroups. The hypotheses of Peano's Theorem (2.3) do not guarantee uniqueness. For instance, the equation $u'(t) = 2\sqrt{|u(t)|}$ with initial value $u(0) = 0$ has solution $u(t) = (\max\{0, t-b\})^2$ for any number $b \ge 0$. Among the three major hypotheses of generalized compactness, isotonicity, or dissipativeness, introduced in §7-10 below, only dissipativeness guarantees uniqueness of solutions — and even that uniqueness is lost when we consider some generalizations in §15. However, even without uniqueness, the theory associated with (1.1) is rich and interesting; see for instance inequality (7.5) and the remarks about continuous and semicontinuous dependence in §7 and 13. Thus, uniqueness is not essential to the theory of existence of solutions. Still, some of the concepts of uniqueness theory will be useful in our study of existence, and so we briefly introduce them here. For a more detailed introduction to uniqueness, see [51].

A function $\omega : [0,T] \times [0,+\infty) \to [0,+\infty)$ is a *Kamke function* (or *uniqueness function*) if ω satisfies Carathéodory conditions, $\omega(t,0) = 0$ for all t, and ω has the property that the only Carathéodory solution of $p'(t) \le \omega(t,p(t))$ on $[0,T]$ with $p(0) = 0$ is the trivial solution $p \equiv 0$. Examples of Kamke functions are $\omega(t,r) = kr$ or $\omega(t,r) = kr\ln(1+r)/\sqrt{t}$ (k = constant) or $\omega(t,r) = r/t$, but *not* $\omega(t,r) = 2r/t$. (Some papers on uniqueness use slightly different definitions; [9] gives a comparison of some of the different classes of Kamke functions.) One variant (given by [24]) of Kamke's classical uniqueness result is as follows: if ω is a Kamke function, F satisfies Carathéodory conditions, and

$$\|F(t,x) - F(t,y)\| \le \omega(t, \|x-y\|) \tag{4.1}$$

for all t, x, y, then (1.1) has at most one Carathéodory solution for each initial value $u(0)$. (The main idea of the proof is that if u_1 and u_2 are solutions of (1.1), then we may apply the definition of the Kamke function with $p(t) = u_1(t) - u_2(t)$.)

Hypothesis (4.1) can be generalized substantially. For instance, let ω be a Kamke function, but instead of (4.1) assume that F satisfies

$$\liminf_{h\downarrow 0} \frac{V(t,x,y) - V(t-h, x-hF(t,x), y-hF(t,y))}{h} \le \omega(t, V(t,x,y)) \tag{4.2}$$

where V is locally Lipschitzian in x and y, V is nonnegative, and $V(t,x,y) = 0$ if and only if $x = y$. Then (1.1) has at most one solution u for each initial value $u(0)$. Roughly, the idea of the proof is that $D_t V(t, u_1(t), u_2(t)) \le \omega(t, V(t, u_1(t), u_2(t)))$ for any solutions u_1, u_2. The relation between (4.1) and (4.2) will be discussed further at the end of §9. Many other uniqueness results, more general and more complicated, can be found in the literature. One particularly general and recent result is [104]; see its bibliography for earlier results.

In initial value problems where uniqueness is known, we can use the notation of semigroups and evolution operators. An *evolution* (or *evolution operator*) on a set Ω is a two-parameter family of self-mappings of Ω:

$$U(t,s) : \Omega \to \Omega \qquad (-\infty < s \le t \le +\infty), \tag{4.3}$$

such that

$$U(t,t) = \text{identity} \qquad \text{and} \qquad U(t,s) \circ U(s,r) = U(t,r) \qquad (r \le s \le t). \tag{4.4}$$

The evolution is *linear* if Ω is a linear subspace of X and $U(t,s)$ is linear for each t and s.

An important special case is that in which $U(t,s)$ depends on t, s only through the value $t-s$. Then we can write $U(t,s) = S(t-s)$, where S is a *semigroup* on Ω, i.e., a one-parameter family of mappings

$$S(t) : \Omega \to \Omega \qquad (t \ge 0) \tag{4.5}$$

with the properties

$$S(0) = \text{identity}, \qquad S(t+s) = S(t) \circ S(s). \tag{4.6}$$

If Ω is a subset of a Banach space, then the semigroup S on Ω is *of type* ω if $(t,x) \mapsto S(t)x$ is a jointly continuous function on $[0,+\infty) \times \Omega$ and

$$\|S(t)x - S(t)y\| \le \|x-y\| \exp(t\omega) \tag{4.7}$$

for all t, x, y. If this is true for $\omega = 0$, we say S is *nonexpansive* (or, in some papers, *contractive*). If G satisfies (2.2), then the semigroup generated by G (in the sense of (4.8), below) is of type ω.

Joint continuity is more natural in semigroups $S(t)$ than in temporally inhomogeneous evolutions $U(t,s)$. Indeed, Ball [5] has shown that if S is a semigroup on a metric space Ω, and $S(t)x$ is measurable in t and continuous in x, then $(t,x) \mapsto S(t)x$ is jointly continuous on $(0,+\infty) \times \Omega$. (Joint continuity at $t=0$ does not necessarily follow, even if S is separately continuous at $t=0$; see [22].) Measurability does not imply continuity for temporally inhomogeneous evolutions, even bounded linear ones. For instance, take $U(t,s)x = \exp[ig(t) - ig(s)]x$, where $g : \mathbf{R} \to \mathbf{R}$ is measurable but not continuous. Thus, temporally homogeneous and inhomogeneous evolutions differ substantially: a temporally inhomogeneous evolution may have jumps, but a semigroup (if measurable) may not.

Evolution operators arise in the study of (1.1) as follows: Let F be some operator — possibly discontinuous and set-valued — in a Banach space X. Assume some notion of "solution" has been specified for the differential inclusion $u'(t) \in F(t,u(t))$ — e.g., one of the notions discussed in any of §3, 11, 13, 14. Assume that the notion of "solution" is such that if u is a solution on an interval I_1, and I_2 is an interval contained in I_1, then the restriction of u to I_2 is a solution on I_2. For simplicity, assume global existence on some set $\Omega \subseteq X$ — i.e., assume that for each initial time $a \in \mathbf{R}$ and each initial value $x \in \Omega$, there exists a solution $u(t) = u(t;a,x)$ for the initial value problem

$$\begin{cases} u'(t) \in F(t,u(t)) & (a \le t < +\infty), \\ u(a) = x, \end{cases}$$

with $u(t)$ taking values in Ω. Also assume forward uniqueness — i.e., assume that whenever $u_1(t)$ and $u_2(t)$ are solutions of $u'(t) \in F(t,u(t))$ on some interval $[a,b]$, and $u_1(a) = u_2(a)$, then $u_1(t) = u_2(t)$ for all $t \in [a,b]$. Under these conditions, it follows that $U(t,s)x \equiv u(t;s,x)$ defines an evolution operator on Ω. We say that $U(t,s)$ (or $u(t)$) is the evolution *generated* by F, and that F is the *generator* of U.

An important special case is that in which $F(t,x) \equiv G(x)$ does not depend on t. For that case, it follows that $U(t,s) = S(t-s)$ defines a semigroup S, called the semigroup *generated* by G, and G is called the *generator* of that semigroup. That semigroup is often denoted $S(t) = \exp(tG) = e^{tG}$, because for many G's we have the *exponential formula*

$$\exp(tG)x = \lim_{n\to\infty} \left(I - \frac{t}{n}G\right)^{-n} x, \tag{4.8}$$

generalizing a familiar formula from undergraduate calculus. When G is continuous and linear, then (4.8) and the other familiar formulas $\exp(tG) = \lim_{n\to\infty}(I + (t/n)G)^n$ and $\exp(tG) = \sum_{n=0}^{\infty} t^n G^n/n!$ are valid; but (4.8) is also valid for many G's which are discontinuous and/or nonlinear — see Theorems 10.1 and 14.1.

For the benefit of newcomers who are unfamiliar with semigroup notation, here are two very elementary examples for study: If $X = \mathbf{R}$ and $G(x) = -(x+5)^3$, then $\exp(tG)x = [2t+(x+5)^{-2}]^{-1/2} - 5$. Continuity of G is not required: If X is a space of functions $x(\theta)$ from $\mathbf{R}$ into $\mathbf{R}$, and $G(x) = 2x+6+5dx/d\theta$ (with domain equal to some suitable subspace of X), then $[\exp(tG)x](\theta) = \exp(2t)x(\theta + 5t) + 3\exp(2t) - 3$. These examples are atypical in that we are able to give explicit formulas for $\exp(tG)$. In many applications, the most we can give is an approximating formula such as (4.8), and perhaps also some other properties of the limit.

Notations like those above are also sometimes used when existence of solutions to (1.1) or (1.2) is only known locally, not globally; but then the notation must be modified slightly and it becomes somewhat more complicated. Even for locally defined solutions, however, the exponential notation remains the simplest way to express certain ideas, e.g. the Trotter-Lie-Kato product formula (13.3); see for instance [111]. The exponential notation is used less often in the study of temporally inhomogeneous evolutions (i.e., (4.3)-(4.4)), but it can be extended to that context too; see for instance [34].

5. Generalizations of continuity. The literature concerning (1.1) varies considerably in its assumptions about continuity of F. Some papers using dissipativeness conditions (discussed in §10 below) make no assumption of continuity at all; these results have direct applications to partial differential equations. Most other existence results for (1.1) assume $F(t, x)$ is continuous in x, or satisfies some condition generalizing continuity, as discussed below. Still, the continuous theory might well be of interest even to researchers concerned primarily with discontinuous problems, for at least a couple of reasons. First, many of the difficulties in the discontinuous theory are still present, albeit in simpler form, when we assume continuity. Second, it may be possible to weaken or remove the continuity hypotheses from some of the continuous theory, particularly by methods discussed in this section or in §12.

Numerous papers deal with (1.1) or (1.2) using assumptions of semicontinuity. A set-valued function G is *upper semicontinuous* (respectively, *lower semicontinuous*) at a point x_0 if for each open set V which contains (respectively, meets) $G(x_0)$, the set $\{x : V \text{ meets } G(x)\}$ (respectively, $\{x : V \text{ contains } G(x)\}$) is a neighborhood of x_0.

Some consequences: If the values of G are compact sets, then G is both upper and lower semicontinuous if and only if G is continuous with respect to the Hausdorff metric on closed bounded sets. If the values of G are closed subsets of a single compact set, then upper semicontinuity is equivalent to closed graph. If the values of G are points, then either upper or lower semicontinuity in the sense above is equivalent to continuity in the usual sense of functions.

Most of the results for semicontinuous set-valued functions assume the values of G are closed nonempty subsets of $\mathbf{R}^n$. An introduction to this theory can be found in the book [4]. The main existence results are as follows: Let F be a mapping from $[0,T] \times \mathbf{R}^n$ into the set of all nonempty closed subsets of $\mathbf{R}^n$. Also assume F satisfies some conditions of measurability and boundedness (which vary from one paper to another; see §2). Then (1.1) has a Carathéodory solution if either

(5.1) for each t, $F(t,\cdot)$ is lower semicontinuous, or

(5.2) for each t, $F(t,\cdot)$ is upper semicontinuous and has convex sets for its values.

Condition (5.1) is used in [15], [77]; condition (5.2) is surveyed in [4]. Convexity cannot be omitted from hypothesis (5.2), as the following simple example from [87] shows: With $X = \mathbf{R}$, let

$$G(x) = \begin{cases} \{1\} & \text{when } x < 0, \\ \{1,-1\} & \text{when } x = 0, \text{ and} \\ \{-1\} & \text{when } x > 0. \end{cases}$$

Then G is upper semicontinuous and has closed values, but (1.2) has no Carathéodory solution for $u(0) = 0$. (However, (1.2) has a Krasovskij solution; see §3.)

The resemblance between upper and lower semicontinuity is only superficial. Hypotheses (5.1) and (5.2) lead to two separate theories, whose solutions have fundamentally different properties. For instance, under assumption (5.2), the solution set is closed and connected, and depends on the initial value in an upper semicontinuous fashion; see [4]. Analogous conclusions are not valid under assumption (5.1).

Some attempts have been made to unify these two approaches, but so far the "unifications," though more general, have also been more complicated. Łojasiewicz [78] assumed, in addition to measurability and boundedness conditions, that

> for almost every t, for each x, either (a) $F(t,x)$ is convex and $F(t,\cdot)$ has closed graph at x, or (b) $F(t,\cdot)$ is lower semicontinuous on some neighborhood of x.

Similarly, Himmelberg and Van Vleck [57] assumed, in addition to measurability and boundedness conditions, that

> for each t, $F(t,\cdot)$ has closed graph and, at each x, either (a) $F(t,x)$ is convex, or (b) $F(t,\cdot)$ is lower semicontinuous at x.

It is not yet known whether a single sufficient condition can be given which is weaker than both (5.1) and (5.2) and which is also *simpler.* A recent step in that direction was taken in [17]:

5.3. THEOREM (Bressan, 1987). *Let $F : [0,T] \times \mathbf{R}^n \to \{$compact nonempty subsets of $\mathbf{R}^n\}$ be bounded and lower semicontinuous. Then there exists an upper semicontinuous function H with compact convex nonempty values such that every Carathéodory solution of $x'(t) \in H(t, x(t))$ is also a Carathéodory solution of $x'(t) \in F(t, x(t))$.*

Bressan's proof involves "directional continuity," which is also of interest for its own sake. Directional continuity is a weakened version of continuity, for single-valued functions. It was first used by [19]. One simple application of it is as follows:

5.4. THEOREM (Pucci, 1971). *Let δ be a positive constant. Let $G : \mathbf{R}^n \to \mathbf{R}^n$ be a bounded mapping which is continuous at each x where $G(x) = 0$. At each x where $G(x) \neq 0$, assume the following "directional continuity" condition:*

$$x_k \to x, \quad x_k \neq x, \quad \left\| \frac{x_k - x}{\|x_k - x\|} - \frac{G(x)}{\|G(x)\|} \right\| < \delta \quad \Rightarrow \quad G(x_k) \to G(x).$$

Let $u(0) \in \mathbf{R}^n$ be given. Then (1.2) has a Carathéodory solution for some $T > 0$.

Finally, we mention one other way of generalizing continuity. Discontinuous operators become continuous if we restrict them to smaller domains; in some cases this restriction does not entirely destroy their usefulness. For instance, the differential operator d/dx is uniformly continuous from the metric of $L^2(\mathbf{R})$ to the metric of $L^2(\mathbf{R})$, when restricted to a bounded subset of the Sobolev space $H^2(\mathbf{R})$; see [109]. Consequently, some techniques of ordinary differential equations can be applied to some partial differential equations; see [109], [110], [111].

6. Subtangential conditions. In some of the literature concerning (1.2), it is assumed that G is defined on all of X, or at least on some neighborhood of the given initial value $u(0)$. That assumption does simplify some of the resulting local existence theory, but it precludes many important applications, especially to partial differential equations. If the domain of G does not include a neighborhood of $u(0)$, then generally we must assume some condition at the boundary of the domain of G. Such conditions — known variously as *subtangential conditions*, *inwardness conditions*, *range conditions*, and *Nagumo conditions* — are satisfied trivially if G is defined everywhere, and so such conditions are not even mentioned in most papers concerned with everywhere-defined functions.

We now introduce the main subtangential conditions, although some of them will not be used until §10 and 12. In the conditions below, "Ran" stands for range,

"Dom" for domain, "cl" for closure, and "dist" for distance. Some subtangential conditions, in order of increasing generality, are:

(6.1) $\mathrm{Ran}(I - hG) = X$ for all $h > 0$ sufficiently small;

(6.2) $\mathrm{Ran}(I - hG) \supseteq \mathrm{cl}(\mathrm{Dom}(G))$ for all $h > 0$ sufficiently small; and

(6.3) $\liminf_{h\downarrow 0} \frac{1}{h}\mathrm{dist}(x, \mathrm{Ran}(I - hG)) = 0$ for all $x \in \mathrm{cl}(\mathrm{Dom}(G))$.

That last condition is the case $N = 1$ of the following condition:

(6.4) for each $\varepsilon > 0$ and each $x_0 \in \mathrm{cl}(\mathrm{Dom}(G))$, there exist a $\delta \in (0, \varepsilon]$, an integer N, and numbers $h_i > 0$ and points $y_i \in G(x_i)$ such that $\sum_{i=1}^{N} h_i = \delta$ and $\sum_{i=1}^{N} \|x_i - x_{i-1} - h_i y_i\| < \varepsilon\delta$.

If G is single-valued, continuous, and bounded, then (6.3) is equivalent to this classical subtangential condition:

$$\liminf_{h\downarrow 0} \frac{\mathrm{dist}(x + hG(x), D)}{h} = 0, \tag{6.5}$$

where D is the domain of G. This condition is appealing because of its obvious geometrical interpretation: it says, roughly, that the vector starting at x and pointing in the direction of $G(x)$ points "into" — or at least not out of — the domain. Condition (6.5), or one equivalent to it, is used in many different papers which assume G is single-valued and continuous. In some of these papers, D is a closed set. A few papers assume, more generally, that D is a *locally closed set*, i.e., the intersection of an open set and a closed set — equivalently, that each point $x \in D$ has a neighborhood N such that $D \bigcap N$ is closed.

It is easy to see that if (1.2) has a continuously differentiable solution u which begins at $u(0) = x$ and which remains in D, then (6.5) must be satisfied at x. Indeed, we have then

$$\frac{\mathrm{dist}(x + hG(x), D)}{h} \le \frac{\|[x + hG(x)] - u(h)\|}{h} = \frac{\|[u(0) + hu'(0)] - u(h)\|}{h}$$

which vanishes as $h \downarrow 0$. That is, the tangent to the trajectory points into the domain. Thus, conditions like (6.5) are *necessary* for the existence of solutions to (1.2). In the presence of certain other hypotheses, such as dissipativeness or compactness (discussed in §7, 9), a condition like (6.5) is both necessary and sufficient for the existence of solutions to (1.1). In finite dimensions, the earliest such result apparently was due to Nagumo (1942); see [63] for further discussion and references.

Variants of (6.5) also apply to set-valued operators. For instance:

6.6. THEOREM (Bressan, 1983). *Let D be a compact subset of a Banach space X, and let B be a bounded subset of X. Let $G : D \to \{$nonempty closed subsets of $B\}$ be lower semicontinuous. Also suppose that*

$$\text{(6.7)} \qquad \textit{for all } x \in D \textit{ and all } y \in G(x), \qquad \lim_{h \downarrow 0} \frac{\operatorname{dist}(x + hy, D)}{h} = 0.$$

Let $u(0) \in D$ be given. Then $u'(t) \in G(u(t))$ has a solution for all $t \geq 0$.

Martin [82] shows the importance of (6.5) without any assumptions of compactness or dissipativeness. Let D be a locally closed subset of a Banach space X, and let $G : D \to X$ be single-valued, bounded and uniformly continuous. Then (6.5) is necessary and sufficient for the existence of a class of *approximate solutions* for (1.2). Approximate solutions will be discussed further in §11.

Temporally inhomogeneous conditions corresponding to (6.5) and (6.3) are, respectively,

$$\text{(6.8)} \quad \liminf_{h \downarrow 0} \frac{\operatorname{dist}(x + hF(t,x), D_{t+h})}{h} = 0 \qquad \text{for all } (t,x) \in \operatorname{Dom}(F(t,\cdot));$$

(6.9)

$$\liminf_{h \downarrow 0} \frac{\operatorname{dist}(x, \operatorname{Ran}(I - hF(t+h,\cdot)))}{h} = 0 \qquad \text{for all } x \in \operatorname{cl}(\operatorname{Dom}(F(t,\cdot))).$$

These will be used in §9 and 10, respectively.

Closely related to subtangential conditions is the notion of invariant sets. Invariance has at least two slightly different meanings. Let S be a subset of X, not necessarily the domain of G. Generally, we say S is *invariant* for (1.2) if either

(6.10) every solution of (1.2) which is in S at some time t must remain in S at all later times, or

(6.11) for each initial value $u(0) \in S$ there exists at least one solution u of (1.2) remaining in S for some $T > 0$.

Clearly, if uniqueness of solutions is known, then (6.11) $\Rightarrow$ (6.10); if existence of solutions is known, then (6.10) $\Rightarrow$ (6.11). In most of the literature on invariance, both uniqueness and existence are known, so the two notions of "invariance" coincide. See [50] for a theorem concerning (6.11) without uniqueness. See [99] for results concerning (6.10) with a discussion of different kinds of uniqueness functions.

For solutions which remain in S, the behavior of G outside S is irrelevant, so G might as well be undefined outside S. Thus, for most definitions of "solution," (6.11) is equivalent to the existence of solutions of (1.2) if we replace G with its restriction to the set $\mathrm{Dom}(G) \bigcap S$. Invariance generally is guaranteed by conditions similar to the subtangential conditions described earlier in this section. An important problem is to identify invariant sets by some of their other properties. For further discussion of invariant sets see [21] [50] [63] [95] [99] [100] [120] and other papers cited therein.

7. Generalized compactness. As Dieudonné's example (§2) showed, Peano's local existence theorem 2.3 fails in infinite-dimensional Banach spaces. However, that theorem can be extended to arbitrary Banach spaces in a natural way, if we restate the the finite-dimensional results a bit differently. Let us add to Peano's theorem the hypothesis that G is a *compact mapping*; i.e., G maps bounded sets to relatively compact sets. (More generally, we could assume G maps some neighborhood of each point to a relatively compact set.) Then Peano's theorem remains unchanged in finite-dimensional spaces, since bounded sets in such spaces are relatively compact; but Peano's conclusion becomes valid in infinite-dimensional spaces as well. Peano's theorem can be generalized further:

For bounded sets S in a Banach space X, we define *Kuratowski's measure of noncompactness*,

$$\alpha(S) = \inf\{r \ : \ S \text{ can be covered by finitely many sets with diameter} \leq r\}.$$

Slightly less often used is the *Hausdorff* (or *ball*) *measure of noncompactness*,

$$\beta(S) = \inf\{r \ : \ S \text{ can be covered by finitely many sets with radius} \leq r\}.$$

These functions measure how far S is from being compact; they vanish precisely when S is relatively compact. In fact, $\beta(S)$ is the distance from $\mathrm{cl}(S)$ to the nearest compact set, in the Hausdorff metric (a metric on the space of closed bounded sets). These two measures are equivalent in the sense that $\beta(S) \leq \alpha(S) \leq 2\beta(S)$. Other measures of noncompactness, and more general notions of such measures, can be found in the book [8].

Now let ω be a Kamke function (see §4). Assume that $F : [0,T] \times X \to X$ is bounded and satisfies Carathéodory conditions, and that

$$\alpha(F(t,S)) \leq \omega(t, \alpha(S)) \tag{7.1}$$

for all bounded sets S. Under various mild additional assumptions (discussed below), it follows that (1.1) has a Carathéodory solution. The proofs make use of the Kamke function ω in different ways. A typical argument (from [82]) is roughly as follows: A sequence of approximate solutions $\{u_n\}$ is carefully constructed, with $u_n(0) \to u(0)$. Then (7.1) is used to show that the function $p(t) = \alpha(\{u_1(t), u_2(t), u_3(t), \ldots\})$ satisfies $p'(t) \leq \omega(t, p(t))$. Since $p(0) = 0$, it follows that $p(t) = 0$ for all t. Thus the sequence $\{u_n\}$ is relatively compact, and some subsequence converges to a limit, which is then shown to satisfy (1.1).

Some of the earlier results in this direction were by Ambrosetti, Goebel and Rzymowski; we omit the details. Among more recent and more general theorems, some of the most interesting results are as follows: Szufla [119] showed that a solution exists if $F(t,x)$ is uniformly continuous as a function of (t,x) and (7.1) holds. Pianigiani [91] showed that a solution exists if F satisfies Carathéodory conditions and (7.1) is replaced by this slightly stronger assumption:

$$\lim_{\delta \downarrow 0} \alpha(F([t-\delta, t+\delta] \times S)) \leq \omega(t, \alpha(S)).$$

Li [76] observed that (7.1) can be weakened slightly, to

$$\text{(7.2)} \qquad \frac{\alpha(S) - \alpha([I - hF(t,\cdot)](S))}{h} \leq \omega(t, \alpha(S)) \qquad (h > 0),$$

if F is uniformly continuous and bounded. Martin [82] used a similar condition for the autonomous problem (1.2): Assume $D = \mathrm{Dom}(G)$ is closed and bounded, and G is bounded and uniformly continuous and satisfies

$$\text{(7.3)} \qquad \frac{\alpha(S) - \alpha([I - hG](S))}{h} \leq \omega\alpha(S) \qquad (h > 0)$$

with ω constant; then (1.2) has a solution. Martin also obtained some other interesting consequences of (7.3): Define

$$\text{(7.4)} \qquad W(t)z = \{u(t) : u \text{ satisfies (1.2) with initial value } z\}.$$

Then W is a semigroup (in the sense of (4.5), (4.6)) on $\Omega = \{\text{subsets of } D\}$, and

$$\text{(7.5)} \qquad \alpha\left(\bigcup_{z \in S} W(t)z\right) \leq \alpha(S)\exp(t\omega) \qquad (t \geq 0),$$

generalizing (4.7). Also Martin showed that $W(t)z$ depends on z in an upper semicontinuous fashion.

Mönch and von Harten [83] showed that (7.1) suffices for existence if $F(t,x)$ is jointly continuous as a function of (t,x) and ω is 1/2 times a Kamke function. In general, it is not known whether the factor of 1/2 can be omitted. This depends on certain questions of measurability which are pursued further by Heinz [52] but are still not fully resolved.

As we noted in §4, the literature varies slightly on the precise definition of a "Kamke function." For uniformly continuous F's, Banaś [7] discusses the requirements on the Kamke function ω . He observes that

$$\omega(t,r) = \sup\{\alpha(F(t,S)) : \alpha(S) = r\}$$

is the smallest function which satisfies (7.1). He shows that this function satisfies certain regularity conditions; hence, assuming fewer regularity hypotheses about ω does not permit wider choices of F. See [7] for the technical details.

We have attempted to indicate the main ideas in this theory, but the literature abounds with generalizations, mostly more complicated. We give two indications of the literature's diversity: Tolstogonov [121] uses measure of noncompactness with set-valued functions F, using some of the concepts discussed in §5. Rzepecki [103] introduces Banach-space-valued measures of noncompactness, and then uses them to solve some systems of ordinary differential equations in Banach spaces.

8. Isotonicity. Let K be a cone (closed, convex, and invariant under multiplication by positive scalars) in a Banach space X. Then K is the *positive cone* for a partial ordering on X defined by: $x \leq y$ if and only if $y - x \in K$. We say K is *regular* (in the sense of Kransnosel'skiĭ) if every monotone, order-bounded sequence is convergent in norm to an element of X. A mapping $G : X \to X$ is *isotone* if $x \leq y \Rightarrow G(x) \leq G(y)$. (Such mappings are also called *order-preserving*, or *monotone*; but the latter term also has another meaning indicated in §9 below.)

Let Y be an ordered Banach space. (Below, we shall take Y to be a space of functions from $[0,T]$ into X.) It is easy to show that if $\Phi : Y \to Y$ is continuous (or, more generally, continuous from above or from below) and isotone, and Φ leaves invariant some order-interval $J = \{y : a \leq y \leq b\}$, then Φ has at least one fixed point in that order-interval. Indeed, we have $\Phi(a), \Phi(b) \in J$, and $\Phi(a) \leq \Phi(b)$; hence $a \leq \Phi(a) \leq \Phi(b) \leq b$. Continuing in this fashion, we obtain sequences

$$a \leq \Phi(a) \leq \Phi^2(a) \leq \ldots \leq \Phi^n(a) \leq \Phi^n(b) \leq \ldots. \leq \Phi^2(b) \leq \Phi(b) \leq b.$$

Since the cone is regular, both the sequences $\{\Phi^n(a)\}$ and $\{\Phi^n(b)\}$ must converge to limits. Continuity (or one-sided continuity) of Φ implies that both of those limits (or one of those limits) are fixed points of Φ.

This argument and some variants can be found in [69]. Also, some variants of this argument have been used in solving a number of partial differential equations, especially elliptic ones, using the maximal principle; see for instance [1] or [106].

In the paragraphs below, however, we shall only consider the abstract initial value problem (1.1). A solution of (1.1) is the same thing as a fixed point for Φ, if Φ is the integral operator given by (2.5). For Φ to be isotone, it suffices that $F(t, \cdot)$ be isotone for each t. The remaining problem is to guarantee that Φ leaves invariant some order-interval.

If F is bounded, then Φ maps into a bounded set. If the positive cone K has nonempty interior, then every bounded set is contained in an order-interval. Thus we have sketched a proof of:

8.1. THEOREM (Stecenko, 1961). *Let X be a Banach space partially ordered by a regular cone with nonempty interior. Suppose $F : [0,T] \times X \to X$ is single-valued, continuous, and bounded, and*

$$(8.2) \qquad F(t,x) \le F(t,y) \quad \textit{whenever} \quad x \le y.$$

Then (1.1) *has at least one solution.*

Stecenko's hypotheses on K are really very strong. They are not satisfied by the usual positive cones in any of the most familiar Banach spaces of real-valued functions. Indeed, in $L^\infty(a,b)$ and in $C[a,b]$, the usual positive cone does have nonempty interior, but it is not regular. The same is true for $B(S)$, the space of bounded functions on a set S, if that set S is infinite. In c_0 and in $L^p(a,b)$ $(1 \le p < \infty)$, the usual positive cone is regular but has empty interior.

Moreover, it is not possible to include those Banach spaces by proving a better theorem. Volkmann [123], [124] gives examples showing that Stecenko's conclusion fails when K is the usual positive cone in c_0 or in $C[-1,1]$ — i.e., Volkmann gives examples of bounded, continuous, isotone F for which (1.1) has no solution. Volkmann's examples are variants of Dieudonné's example, mentioned in §2. Like Dieudonné, Volkmann determines solutions componentwise and then shows that they do not lie in an appropriate space; hence the particular choice of a notion of "solution" (discussed in §3, 11, 13, 14) is not at issue here.

What about using some cone K other than the usual cone of nonnegative-valued functions? Trivial, degenerate choices will not satisfy Stecenko's hypotheses: a cone which is too small has no interior, and a cone which is too big is not regular. But Volkmann [124] notes that Stecenko's conditions are satisfied in *any* Banach space X, using a cone K constructed as follows: Let q be any nonzero element of X, and let ρ be any constant in $(0, \|q\|)$. Let S be the closed ball centered at q with radius ρ, and let $K = \bigcup_{\lambda \ge 0} \lambda S$. Then it can be shown that Stecenko's hypotheses are satisfied.

Volkmann [123] also notes that Stecenko's regularity hypothesis can be omitted at least in one important case: Let S be an arbitrary set, and let X be the space $B(S)$ of bounded, real-valued functions from S, equipped with the supremum norm. Let K be the usual positive cone, i.e., those functions which are nonnegative everywhere on S. Then K is regular precisely when S is finite; but (1.1) has a solution even if S is infinite.

Volkmann [123] also observes that condition (8.2) can be weakened to the assumption that

$$(8.3) \qquad F(t,y) - F(t,x) \ge -\omega(y-x) \quad \text{whenever} \quad x \le y$$

for some real constant ω, since then we can apply the preceding theorem to the function

$$H(t,x) = e^{\omega t} F(t, e^{-\omega t} x) + \omega x$$

and obtain existence of the function $v(t) = e^{\omega t} u(t)$.

Because their hypotheses are so strong, Stecenko's theorem and related results have few known applications. We have included these results simply because isotonicity is so very different from generalized compactness or dissipativeness (§7 and 9). Two other related results, in finite dimensions, are also simple enough to deserve mention:

Biles [11] has shown that if $F : [0,T] \times \mathbf{R} \to \mathbf{R}$ is bounded and measurable, and $F(t,x)$ is both right-continuous and upper semicontinuous in x for each fixed t, then (1.1) has a solution. Biles' two assumptions of one-sided continuity may be restated as follows:

$$\limsup_{y \to x-} F(t,y) \le F(t,x) = \lim_{y \to x+} F(t,y).$$

When $X = \mathbf{R}$, Biles' result extends Stecenko's. We wonder if Biles' result or one like it might be extended to spaces of higher dimension.

Wend [127] considers $\mathbf{R}^n$ with its usual positive cone. He shows that if $F : [0,\infty) \times \mathbf{R}^n \to \mathbf{R}^n$ takes values in the positive cone, and $F(t,x)$ is a nondecreasing function of both t and x, then (1.1) has a Carathéodory solution for some $T > 0$. No continuity or semicontinuity of F is assumed explicitly, but Wend makes use of the fact that any isotone function from $\mathbf{R}$ into $\mathbf{R}^n$ is continuous almost everywhere.

9. Dissipativeness. Some introductions to dissipativeness are [26] and [73]. For completeness we give a brief introduction here, with different emphases than those two sources.

Let G be any map from the Banach space X into the set of all subsets (not necessarily nonempty) of X. We say G is *single-valued*, or *point-valued*, if $G(x)$ contains at most one point for each x. The *effective domain*, the *range*, and the *resolvent* of G are defined by

$$\mathrm{Dom}(G) = \{x \in X : G(x) \text{ is nonempty}\},$$

$$\mathrm{Ran}(G) = \bigcup_{x \in X} G(x) = \bigcup_{x \in \mathrm{Dom}(G)} G(x),$$

$$J_\lambda(x) = (I - \lambda G)^{-1} x = \{y : x \in y - \lambda G(y)\} \quad (\lambda > 0).$$

The operator G is *dissipative* if

> for each $\lambda > 0$, the resolvent J_λ is single-valued on $\mathrm{Dom}(J_\lambda) = \mathrm{Ran}(I - \lambda G)$ and is Lipschitzian with Lipschitz constant ≤ 1.

Equivalently, we say $-G$ is *accretive* — or, in Hilbert spaces, $-G$ is *monotone*, but that word also has another meaning (see §8). Much of the related literature is written in terms of accretive operators, rather than dissipative operators, because the resulting notation is more consistent with the traditional notation of hyperbolic partial differential equations — one of the most important applications of accretive/dissipative operators. But we shall use the dissipative notation, since it is more consistent with the other theories surveyed in this paper.

Locally Lipschitz mappings satisfy a local dissipativeness condition; hence so do continuously differentiable mappings. Thus, in most studies of ordinary differential equations, both compactness and dissipativeness are available for existence proofs. Many nonlinear partial differential operators are dissipative, or satisfy a similar condition, in suitably normed Banach spaces; see [25], [37], [38], [109]. For motivation, consider that differential operators tend to be discontinuous under most natural topologies, but in many cases their inverses — or their resolvents, which are inverses of perturbations of those differential operators — are integral operators, which tend to be continuous under most natural topologies. For this reason, many of the inequalities and other conditions in the dissipative theory are formulated in terms of the resolvent J_λ, rather than directly in terms of the generator G.

Dissipativeness can be restated in terms of the directional derivative of the norm:

$$(9.1) \qquad [p,q]_- \equiv \sup_{h<0} \frac{\|p+hq\| - \|p\|}{h} = \lim_{h\uparrow 0} \frac{\|p+hq\| - \|p\|}{h}.$$

The last equality follows from the fact that $\|p+hq\|$ is a convex function of h, and hence $(\|p+hq\| - \|p\|)/h$ is a nondecreasing function of h. Slight variants of (9.1) are used in various papers; here we follow the notation of [73]. A consequence of (9.1) is

$$(9.2) \qquad |\,[p,q]_-\,| \leq \|q\|.$$

For motivation note that if X is a Hilbert space with inner product $(\ ,\ )$, then $[p,q]_- = \mathrm{Re}(p,q)/\|p\|$.

For any $\omega \in \mathbf{R}$, we say G is ***ω-dissipative*** if $G - \omega I$ is dissipative. That condition holds if and only if

$$(9.3) \qquad [x_1 - x_2, y_1 - y_2]_- \leq \omega\|x_1 - x_2\|$$

for all $x_1, x_2 \in \mathrm{Dom}(G)$ with $x_1 \neq x_2$, and all $y_1 \in G(x_1)$, $y_2 \in G(x_2)$. A few papers call this "quasi-dissipative," but that term usually has another meaning: We say G is ***ω-quasi-dissipative*** if

$$(9.4) \qquad [x_1 - x_2, y_1]_- + [x_2 - x_1, y_2]_- \leq \omega\|x_1 - x_2\|$$

for all $x_1, x_2 \in \mathrm{Dom}(G)$ with $x_1 \neq x_2$, and all $y_1 \in G(x_1)$, $y_2 \in G(x_2)$. This is generally weaker than ω-dissipativeness (but not if X is a Hilbert space). A stronger condition than ω-dissipativeness, used in a few papers, is *strict ω-dissipativeness*; this condition is obtained by using $t \downarrow 0$ instead of $t \uparrow 0$ in (9.1).

Dissipativeness is sometimes referred to as a "one-sided Lipschitz condition," because if (2.2) holds then (by (9.2)) both G and $-G$ are ω-dissipative. However, the converse is not valid. For instance, if A is a linear, self-adjoint operator in a complex Hilbert space, then both iA and $-iA$ are dissipative, but A need not be Lipschitz or even continuous. Dissipativeness does generalize another property of Lipschitz operators mentioned in §4: If G is ω-dissipative for some constant ω, then each solution of (1.2) is uniquely determined by its initial value, and the semigroup so determined is of type ω — i.e., satisfies (4.7). More generally, if $F(t,\cdot)$ is $\omega(t)$-dissipative for each t, and u and v are solutions of (1.1), then

$$\|u(t)-v(t)\| \leq \|u(0)-v(0)\| \exp[\textstyle\int_0^t \omega(s)\,ds]. \tag{9.5}$$

For continuous F or G, these inequalities actually characterize dissipativeness.

The existence theory takes its simplest form when F is continuous:

9.6. THEOREM. *Let D be a locally closed subset of a Banach space X. Let $F : \mathbf{R} \times D \to X$ be single-valued and jointly continuous. For each t, suppose $F(t,\cdot)$ satisfies* (6.5) *and is $\omega(t)$-dissipative, where $\omega(\cdot)$ is a continuous function. Then for each $u(0) \in D$,* (1.1) *has a solution at least for some $T > 0$. If D is closed, then* (1.1) *has a solution $u(t)$ for all $t \geq 0$.*

The local existence result is essentially due to Martin [81], although he assumed strict ω-dissipativeness. That additional assumption was dropped in later papers which extended Martin's result in other ways as well. Most notably: [108] weakens the assumption of joint continuity to an assumption that F satisfies Carathéodory conditions. In that paper, the domain of F is still "cylindrical," i.e., of the form $\mathbf{R} \times D$. Several other papers assume joint continuity, but permit F to have a noncylindrical domain. It is assumed, roughly, that the set $D_t = \mathrm{Dom}(F(t,\cdot))$ is upper semicontinuous from the left, as a function of t. Also, the subtangential condition (6.5) is replaced with (6.8). Kenmochi and Takahashi [67] observe that, in the presence of the other hypotheses, this subtangential condition is not only sufficient but also necessary for the existence of solutions. Iwamiya [59] weakens the dissipativeness hypothesis, assuming instead only that

$$[x-y, F(t,x)-F(t,y)]_- \leq \omega(t, \|x-y\|) \tag{9.7}$$

for some Kamke function ω.

Note that (9.7) is a special case of (4.2), with $V(t,x,y) = \|x-y\|$. Also (9.7) $\Rightarrow$ (4.1), in view of (9.2).

10. Dissipativeness without continuity. We now turn to the existence results in which $F(t,\cdot)$ is dissipative but not necessarily single-valued or continuous. In most of these results, the solution obtained is not a Carathéodory solution (defined in §3), but a still weaker "limit solution." It may be rather ill-behaved, but that is not necessarily a disadvantage to this theory, for it means that the theory can even be applied to differential equations whose solutions are known to be ill-behaved — for instance, hyperbolic partial differential equations with shocks. Limit solutions will be discussed in §11.

10.1. THEOREM (Crandall and Liggett, 1970). *Suppose G is an ω-dissipative operator in a Banach space X, satisfying* (6.2). *Then G generates a strongly continuous semigroup $S(t) = \exp(tG)$, defined as in* (4.8), *on* $\mathrm{cl}(\mathrm{Dom}(G))$*; and that semigroup is of type ω (i.e., satisfies* (4.7)*).*

The Crandall-Liggett theorem has been extended in a few ways:

Kobayashi [68] replaces hypothesis (6.2) with the weaker subtangential condition (6.3). This requires that we replace the simple product of resolvents in (4.8) with a more complicated product of approximate resolvents, such as that in (11.4) below. More recently, Kobayashi (unpublished result; mentioned in [26]) has shown that (6.3) can be weakened further to (6.4). In fact, for G dissipative, (6.4) is *necessary and sufficient* for existence of a solution to (1.2).

The operator G need not be ω-dissipative; it suffices for G to be ω-quasi-dissipative, as in (9.4). This weaker hypothesis was used by Kobayashi [68]. The extension from dissipative to quasi-dissipative is primarily of theoretical, not applied, interest. Kobayashi gives an example of an ω-quasi-dissipative operator which is not ω-dissipative, but the example is quite artificial. Apparently no examples are yet known which have real application.

A different direction of generalization is taken by Picard [92]. Let ϕ be a convex mapping from a topological vector space X into $\mathbf{R}$. The vector space X is *ϕ-complete* if $\phi(x_m - x_n) \to 0$ implies that $\{x_n\}$ converges to a limit in X. An operator U in X is *ϕ-nonexpansive* if $\phi(Ux-Uy) \le \phi(x-y)$ for all $x, y \in \mathrm{Dom}(U)$. An operator G is *ϕ-dissipative* if the resolvent $J_\lambda = (I - \lambda G)^{-1}$ is ϕ-nonexpansive for all $\lambda > 0$. (Actually, Picard calls such U "ϕ-contractive," and says that $-G$ is "ϕ-accretive.") Picard finds that if X is a normed vector space, $\phi : X \to \mathbf{R}$ is convex and Lipschitzian, X is ϕ-complete, G is ϕ-dissipative, and G satisfies (6.2), then the limit $S(t)$ defined in (4.8) exists on $\mathrm{cl}(\mathrm{Dom}(G))$, and $S(t)$ is ϕ-nonexpansive for each t.

The Crandall-Liggett Theorem 10.1 and most of its extensions assert global existence, i.e., existence for all $t \ge 0$. This follows from the fact that, for simplicity, we have taken ω to be constant, and also from the global nature of our "range condition" (6.2) or (6.3). Weakening these conditions to local ones yields local existence results. However, for simplicity of notation, most of the abstract theory of dissipative operators has been developed with global estimates. In fact, much of the literature only considers the case of $\omega = 0$. These restrictions are without substantial loss of generality, since most of the statements and proofs can be extended to more general choices of ω without great difficulty. Of course, for some applications, ω must be nonzero or even nonconstant; see for instance [109]. In most of the remainder of this survey, we shall only discuss dissipativeness for the case of $\omega = 0$.

We shall have more to say about the autonomous problem (1.2) in later sections, but now let us turn to the nonautonomous problem (1.1). The theory for this problem, with $F(t,x)$ dissipative in x, is not unified; there are at least three substantially separate approaches to it. One approach, mentioned earlier, is descended from Martin's result, Theorem 9.6. A second approach is discussed below; a third approach will be discussed in §14.

The Crandall-Liggett Theorem 10.1 has been generalized to temporally inhomogeneous problems (1.1) — i.e., with $F(t,\cdot)$ dissipative and discontinuous — in a number of different ways, none of them wholly satisfactory. All such generalizations involve regularity assumptions about the dependence of $F(t,x)$ on t which, though fairly weak, seem somewhat unnatural and hard to motivate. Research in this area continues. A good introduction to this area is given by the survey paper [86] and the book [89].

The earliest result in this direction is that of Crandall and Pazy [28]. They make regularity assumptions of the form

$$\|J_\lambda(t)x - J_\lambda(s)x\| \leq \lambda|g(t) - g(s)|B(s,x), \tag{10.2}$$

where $J_\lambda(t) = (I - \lambda F(t,\cdot))^{-1}$. They give two versions. In one version (condition (C.1) in [28]), they assume g is continuous and $B(s,x) = L(\|x\|)$ for some increasing function L independent of s. In the other version (condition (C.2)), they assume g is continuous and of bounded variation, and $B(s,x)$ depends on $\|x\|$ and on the behavior of $J_\lambda(s)$ near x as $\lambda \downarrow 0$; the details are complicated and are omitted here. Evans [36] keeps the form of (10.2) but replaces continuity of g with integrability. The hypotheses of Crandall and Pazy and of Evans imply that $\mathrm{cl}(\mathrm{Dom}(F(t,\cdot)))$ is independent of t, but that restriction is weakened in later papers mentioned below.

Some subsequent papers have departed in form from (10.2). Hypotheses used are of roughly the following form:

$$\|J_\lambda(t_1)x_1 - J_\lambda(t_2)x_2\| \leq \|x_1 - x_2\| + f(t_1,t_2)L(\|x_1\|) \tag{10.3}$$

or, alternatively,

(10.4)

$$[x_1 - x_2, y_1 - y_2]_- \leq f(t_1,t_2)L(\|x_1\|) \qquad \text{whenever} \quad y_i \in F(t_i,x_i) \quad (i = 1,2),$$

in both cases for $t_1 \leq t_2$. Here f is assumed integrable on $[0,T]\times[0,T]$, and f is also assumed to satisfy some other conditions. Condition (10.2) is included by taking

$f(t,s) = g(t) - g(s)$. Iwamiya, Oharu, and Takahashi [62] have recently proven existence assuming f is integrable on $[0,T] \times [0,T]$, $f(t,t) = 0$, and f is continuous on the diagonal. Pavel [88] uses a quasi-dissipativeness condition, replacing the left side of (10.4) with $[x_1 - x_2, y_1]_- + [x_2 - x_1, y_2]_-$. The papers [61], [93], [94] assume regularity conditions similar to (10.3), but applied to the "tangential" semigroups $\exp(hF(t,\cdot))$ rather than to the resolvents $J_\lambda(t) = (I - \lambda F(t,\cdot))^{-1}$.

In some papers, the factor $L(\|x_1\|)$ is omitted. That omission might seem to be a severe restriction, in that it requires the regularity of $F(t,x)$ in t to be uniform for all x in $\mathrm{Dom}(F(t,\cdot))$. But the factor $L(\|x_1\|)$ is unnecessary, as explained in [60]: If Crandall and Pazy's condition (10.2) is satisfied, then — once one has selected the initial value $u(0)$ — one can restrict F to a smaller set, still large enough to contain the solution and all needed approximate solutions, but small enough so that — with a suitable choice of f — the quantity $[x_1 - x_2, y_1 - y_2]_- / f(t_1, t_2)$ (or other relevant quantity) remains bounded.

Even with Crandall and Pazy's condition (10.2), the regularity in t (as $t \to s$, with s fixed) is uniform for x in sets where $B(s,x)$ is bounded, and those may be rather large sets. In contrast, if F is jointly continuous (or more generally, satisfies Carathéodory conditions), then F behaves regularly in t uniformly on compact sets of x — but compact sets are not "large," in infinite-dimensional spaces. It appears that Theorem 9.6 is not a special case of the theories discussed in the last few paragraphs. Indeed, a simple example is given in the introduction of [108], in which $F : \mathbf{R} \times X \to X$ satisfies the hypotheses of Theorem 9.6 (and moreover $F(t,\cdot)$ is a bounded linear operator for each t), but F does not satisfy $\|F(t,x) - F(s,x)\| \leq |g(t) - g(s)| B(s,x)$ for any choice of B and integrable g. An easy modification of the computations in [108] shows that this same F also does not satisfy (10.2). One wonders whether this F can be made to fit (10.2) by a suitable restriction of domain.

Conditions (10.2)-(10.4) are sufficient to make various existence proofs work, but motivation for those hypotheses is not entirely clear. In light of the example just noted, those hypotheses seem somewhat unnatural. It is not yet clear what hypotheses would be more natural. A first guess would be to impose Carathéodory conditions on the resolvents $J_\lambda(t) = [I - \lambda F(t,\cdot)]^{-1}$, rather than on F itself. However, an example given by Freedman [41] shows that that does not work. In Freedman's example, $F(t,\cdot)$ is a bounded linear operator for each t, and the resolvent $J_\lambda(t)x$ is a jointly continuous function of λ, t, and x. Yet (10.2) is not satisfied and (1.1) has no solution. Freedman's example is also noteworthy in that he proves nonexistence using a technique substantially different from that of Dieudoné [33] (sketched in §2). Most of the examples of nonexistence to be found in the literature are slight variants of Dieudonné's example.

11. Limit solutions. There are a number of initial value problems, arising in control theory or in fluid mechanics, for which a "solution" is known to exist by physical considerations, but for which that "solution" is not differentiable [25],

[37], [38]. An abstract theory to support these applications is desirable, even if it requires somewhat complicated notions of "solution." As we remarked at the end of §3, such a solution need not satisfy (1.1) in any direct sense; the relation (1.1) is kept merely as an abbreviation for the more complicated notion of "solution" to be used.

We say $u(t)$ is a *limit solution* of (1.1) if there exist a sequence of numbers $\varepsilon_n \downarrow 0$ and a sequence of functions $v_n(t) \to u(t)$, such that v_n is an ε_n-approximate solution of (1.1). A function $v(t)$ is an *ε-approximate solution* of (1.1) if v comes within ε of satisfying (1.1) in some suitable sense. The precise definition of "ε-approximate solution" varies from one paper to another; some typical versions are indicated below. We shall not attempt to give all the variants in technical detail. (In fact, many of the papers surveyed do not use the terminology given here, but their results can easily be reformulated in this terminology.) Usually the definitions require that $u(t)$ be continuous and that the convergence $v_n(t) \to u(t)$ be uniform over all $t \in [0, T]$, but these requirements are weakened in some papers discussed below. The existence theorems usually do not assert that every sequence of ε-approximate solutions converges to u as $\varepsilon \downarrow 0$, but only that some such convergent sequence exists. In some results with compactness, the limit solution u is not unique, and different approximating sequences $\{v_n\}$ may converge to different limit solutions u.

Even when a Carathéodory solution exists, one common method of constructing that solution is by taking the limit of a sequence of approximate solutions. Generally, if F is continuous in its second argument, then every limit solution is a Carathéodory solution. If F is not continuous in its second argument, then a limit solution $u(t)$ need not be a Carathéodory solution, and in fact $u(t)$ need not be differentiable for any t. A simple example of a nowhere differentiable solution is given in [27]. Moreover, $u(t)$ need not take values in $\mathrm{Dom}(F(t,\cdot))$ — but under most definitions of "limit solution," $u(t)$ does take values in $\mathrm{cl}(\mathrm{Dom}(F(t,\cdot))$.

The simplest type of approximate solution is an "outer perturbation": v is an ε-approximate solution of (1.1) in this sense if v is absolutely continuous, differentiable almost everywhere, and satisfies

$$\|v'(t) - F(t, v(t))\| \le \varepsilon \qquad \text{for} \qquad 0 \le t \le T.$$

This condition generalizes to multivalued F as well: use the distance from $v'(t)$ to $F(t, v(t))$ instead of the norm of the difference. Equivalently, v is an ε-approximate solution of (1.1) if v is a Carathéodory solution of

$$v'(t) \in F(t, v(t)) + \varepsilon B \tag{11.1}$$

where B is the closed unit ball.

A more general approach involves "inner perturbations;" we replace (11.1) with

$$v'(t) \in F(t + [-\varepsilon, \varepsilon], v(t) + \varepsilon B) \qquad (0 \le t \le T). \tag{11.2}$$

It is easy to see that any limit of outer perturbations is a limit of inner perturbations. The converse holds if F is single-valued and continuous, but not in general. In the terminology of Hájek [48] (modified slightly), u is a *Hermes-∗ solution* of (1.1) if u is a uniform limit of ε-approximate solutions in the sense of (11.2).

In many of the constructions found in the literature, the interval $[0, T]$ is (for fixed ε) partitioned into subintervals:

$$0 = t_0 < t_1 < t_2 < \ldots < t_m = T,$$

where $h_j = t_j - t_{j-1} < \varepsilon$ for all j. Then the approximate solution $v(t_j)$ is defined at each partition point t_j, typically by some recursion on j, using some subtangential condition such as (6.8) or (6.9). The approximate solution $v(t)$ is then defined either as a step-function which is constant on each open subinterval (t_{j-1}, t_j), or as a piecewise-linear function which is affine on each closed subinterval $[t_{j-1}, t_j]$. These two methods for defining $v(t)$ differ only superficially, since both methods must yield the same uniform limit $u(t)$ if that limiting function is continuous. The method of step-functions — in particular, the backward difference method, discussed further below — has been especially popular among papers with discontinuous $F(t, \cdot)$ (§10), perhaps because this brings the definition of "solution" closer to the method of proof. On the other hand, the piecewise-linear approach has an advantage in that it makes the definition of "approximate solution" simpler and more intuitive: Each approximate solution is itself the exact (Carathéodory) solution of an approximating differential equation, such as (11.2).

Generally, $v(t_j)$ is chosen recursively from $v(t_{j-1})$ so that $[v(t_j) - v(t_{j-1}]/h_j$ is exactly or approximately equal to an element of one of the following sets:

> $F(t_{j-1}, v(t_{j-1}))$, used in *forward difference schemes*, also known as *Euler polygonal approximations*;
>
> $F(t_j, v(t_j))$, used in *backward difference schemes*;
>
> $(t_j - t_j - 1)^{-1} \int_{t_{j-1}}^{t_j} F(s, v(t_{j-1}))\, ds$, used in *Euler-Lebesgue approximation*;

or some other function involving $F(\sigma, v(\tau))$ for one or more values of σ and τ in the interval $[t_{j-1}, t_j]$.

The backward difference method uses a product of resolvents, as in the Crandall-Liggett Theorem 10.1, or an approximate product of resolvents, as in the example below. In the papers discussed in §10, the definition of "solution" is very complicated, and must be bewildering to newcomers to this subject. A typical definition is as follows: v is an ε-approximate solution to (1.1) (with initial value $u(0)$ given) if there exists a partition

$$0 = t_0 < t_1 < \ldots < t_{N-1} < t_N = T \tag{11.3}$$

and points $x_k \in \mathrm{Dom}(F(t_k, \cdot))$ and $p_k \in X$, such that

$$(11.4)\quad \begin{cases} \|x_0 - u(0)\| < \varepsilon; \\ t_k - t_{k-1} < \varepsilon \qquad (k = 1, 2, \ldots, N); \\ v(0) = x_0, \text{ and } v(t) = x_k \text{ for } t \in (t_{k-1}, t_k] \quad (k = 1, 2, \ldots, N); \\ (x_k - x_{k-1})/(t_k - t_{k-1}) - p_k \in F(t_k, x_k) \qquad (k = 1, 2, \ldots, N); \text{ and} \\ \sum_{k=1}^{N} (t_k - t_{k-1}) \|p_k\| < \varepsilon. \end{cases}$$

It may be helpful to observe that

$$x_k \in \{I - (t_k - t_{k-1})[p_k + F(t_k, \cdot)]\}^{-1}(x_{k-1}) \qquad (k = 1, 2, \ldots, N),$$

and thus in a sense $x_k \approx J_\lambda(t_k)x_{k-1}$ with $\lambda = t_k - t_{k-1}$.

A "solution" in these papers is a continuous function $u(t)$ which is a uniform limit of such approximations $v(t)$, as $\varepsilon \downarrow 0$. The existence of such approximate solutions follows easily from a subtangential condition such as (6.9). The hard part of this theory is proving their convergence to a limit (which proof we shall *not* attempt to sketch here).

Any such limit solution is also a Hermes-$*$ solution, as we shall now show. For each ε-approximate solution v and error terms p_k as in (11.4), let w be the piecewise-linear function which is affine on each interval $[t_{k-1}, t_k]$ and agrees with v at the endpoints of that interval; then $w'(t) \in p_k + F(t_k, x_k)$ on that interval. Let $p(t)$ be the step-function which satisfies $p(t) = p_k$ on the interval (t_{k-1}, t_k), and let $z(t) = w(t) - \int_0^t p(s)\, ds$. Then $\int_0^t \|p(s)\|\, ds \le \varepsilon$, hence $\|z(t) - w(t)\| \le \varepsilon$ for all t, and $z'(t) \in F(t_k, x_k) \subseteq F([t - \varepsilon, t + \varepsilon], x_k)$. As $\varepsilon \downarrow 0$, the v's converge uniformly to u; hence so do the w's; hence so do the z's; hence $\|v - z\|_{\sup} \downarrow 0$. Since u is continuous, $\sup\{\|z(t) - x_k\| : t \in [x_{k-1}, x_k]\} \downarrow 0$ also. Thus $z'(t) \in F([t - \delta, t + \delta], z(t) + \delta B)$ for some δ which goes to 0 when $\varepsilon \downarrow 0$.

Thus, any backward difference scheme limit solution is also a Hermes-$*$ solution; perhaps this observation will be helpful to beginners trying to come to grips with (11.4). It is not known (at least, to this author) whether the converse holds — i.e., whether every Hermes-$*$ solution is also a backward difference scheme limit solution. Such a converse would follow if dissipativeness of $F(t, \cdot)$ is enough to guarantee uniqueness of Hermes-$*$ solutions of (1.1).

With most definitions of "limit solution" — including all of those discussed above — the limit $u(t)$ is required to be a continuous function of t. In §13 and 14 we shall discuss some other approaches to (1.1) which permit discontinuities in u for various theoretical reasons. Moreau [84] permits discontinuities for more practical reasons: For some differential equations with applications to elastoplastic mechanical systems, the physically "natural" solution $u(t)$ may have jump discontinuities. Moreau [85] observes that if $u(t)$ is discontinuous, then it is not natural to require approximate solutions $v_n(t)$ to converge uniformly in t. Indeed, if $u(t)$ has a jump at t_0 of size r or larger, and if $\|u - v_n\|_{\sup} < r$, then v_n must also have a jump (possibly of a different size) at the exact same location t_0. It is more natural to require v_n to have a jump *near* t_0. This effect is achieved by requiring the sequence $\{v_n\}$ to converge to u *in graph*, rather than uniformly.

All the notions of "solution" discussed so far depend only on the topological vector space structure of the Banach space X. If we replace the norm on X with an equivalent norm, then a solution of (1.1) remains a solution. Intuitively, we expect the same to be true for our methods of proving existence of solutions. But our methods of proof are very norm-dependent. To apply existence theorems such as those discussed in §7-10, we must choose precisely the right norm on X; an equivalent norm will not necessarily work. For instance, if an operator G is dissipative with respect to one norm, the same operator G is not necessarily dissipative (or ω-dissipative, or even ω-quasi-dissipative) with respect to any other, equivalent norm.

Some interesting approaches have been used to get around this difficulty. For instance, in studying nonautonomous problems (1.1), [64] and [126] use a whole family of norms $\| \quad \|_t$, parametrized by the time variable t; each instantaneous operator $F(t,\cdot)$ is shown to satisfy dissipativeness and other conditions with respect to $\| \quad \|_t$. Ideally, we would prefer to find some topological vector space condition, not dependent on particular norms, which generalizes dissipativeness and which is satisfied by $F(t,\cdot)$ for all t. The notion of "dissipative," though very useful for applications, is not completely satisfactory from a purely theoretical point of view.

In the *linear* theory of dissipative operators, this is not a problem. The norm on the Banach space X can always be replaced by an equivalent norm, so that a semigroup S satisfying $\|S(t)\| \leq Me^{\omega t}$ for some constant M becomes a semigroup of type ω. Thus, M gets replaced by 1 (see [46]). An analogous trick does not work for nonlinear semigroups. The best we can do is replace the norm with an equivalent *metric*, making the Banach space X into a Banach manifold. That approach has been investigated by Marsden [80], but it requires more smoothness than is customary in the theory of nonlinear semigroups. The smoothness requirements have recently been weakened somewhat by Pimbley [97].

12. Identification of the limit. Usually, a proof of existence of a solution to (1.1) can be divided into three main steps — approximability, convergence, and identification — discussed in greater detail below. This outline is indicated in [73], for instance. (A variant of this outline is needed for proofs which use fixed point theorems; see the discussion after (13.4).) Of course, even in papers where these three steps are followed, they may not be mentioned explicitly; most papers in existence theory combine the steps. But we may understand these three steps more clearly if we study them separately; they involve different techniques and different hypotheses. The first two steps can be described briefly, since they involve ideas which we have already discussed at some length:

(1) Existence of approximate solutions. We must show that ε-approximate solutions v exist, for arbitrarily small values of ε. Generally the approximate solutions must be constructed with some care, so that the next two steps will be possible. Existence of approximate solutions follows from a subtangential condition such as those discussed in §6.

(2) Convergence to a limit. We must show that some sequence of approximate solutions v_n (with $\varepsilon_n \downarrow 0$) converges to some limit u. This usually follows from a hypothesis of generalized compactness, isotonicity, or dissipativeness — discussed in §7-10. Thus, those three substantially different hypotheses play analogous roles. Hence there is some hope of unifying their three separate theories, as discussed in §15.

The third step will require a lengthier discussion:

(3) Identification of the limit. Roughly, the goal of this step is to show that the limit $u(t)$ obtained in Step 2 is really a "solution" in some sense, i.e., to establish some connection between $u(t)$ and the differential inclusion (1.1). Step 3 is optional: Some connection is implicit in the definition of ε-approximate solution used in Step 1, and so a further connection is not absolutely necessary. However, that implicit connection is rather indirect, and not fully satisfactory. If one's notion of "solution" is too general, then *everything* becomes a solution, and existence theory becomes meaningless. It is desirable to show that the solution $u(t)$ obtained in Step 2 is uniquely determined by (1.1), or at least that the set of solutions is narrowly restricted by some properties connecting it more closely to the differential equation (1.1).

The simplest case is that in which $F(t, x)$ is single-valued, bounded, and continuous in its second argument. Then, generally, the limit solution is also a Carathéodory solution. A typical proof runs thus: We observe that the approximate solutions satisfy an approximate version of the integral equation (2.4). Applying Lebesgue's Dominated Convergence Theorem, we find that the limit $u(t)$ satisfies (2.4). Variants of this argument sometimes work when F is set-valued and semicontinuous, as discussed in §5.

Thus, one of the chief uses of a hypothesis of continuity of F (in conjunction with hypotheses of compactness or isotonicity) is for identification of the limit. If we are satisfied with the identification property already implicit in our definition of ε-approximate solutions, then continuity of F becomes less important. For instance, Hájek [48] observes that if $F : [0, T] \times \mathbf{R}^n \to \mathbf{R}^n$ is bounded, then there exists a convergent sequence of Euler-polygonal approximations for (1.1). We need not assume F is continuous — or even measurable! These results extend easily to multivalued F in infinite dimensional spaces, if we assume $\mathrm{Ran}(F)$ is compact. We speculate that, analogously, the continuity hypotheses in some other known existence theorems might be weakened or removed if we weaken the method of identification of the limit (as suggested in the paragraphs below). The present author hopes to research this idea further in the near future.

Even if the solution $u(t)$ is not differentiable, other methods of identifying the limit may be available. Such methods have been developed especially for $F(t, \cdot)$ dissipative and discontinuous.

The concept of "envelope solutions" was introduced by Pierre [93], [94] and developed further by Iwamiya, Oharu, and Takahashi [61]. A function u is an *envelope solution* of (1.1) if

$$\lim_{h\downarrow 0}\frac{1}{h}\int_0^{T-h}\|u(t+h)-\exp(hF(t,\cdot))u(t)\|\,dt=0.$$

Roughly, the idea is that at each instant t the trajectory $u(t)$ is "tangent" to the semigroup $\exp(hF(t,\cdot))$, and thus behaves like that semigroup, at least momentarily. The solution $u(t)$ need not be differentiable, but the tangential semigroups need not be differentiable either.

Another approach is that of integral inequalities. An evolution operator $U(t,s)$ is the *integral solution* of (1.1) if it satisfies

$$(12.1)\quad \|U(t,r)x_1-x_2\|\leq\|x_1-x_2\|-\int_r^t\{[U(s,r)x_1-x_2,-w]_-+f(s,r)\}\,ds$$

for all $[r,t]\subseteq[0,T]$, all $x_1\in\mathrm{Dom}(F(r,\cdot))$, all $q\in[0,T]$, and all $w\in F(q,z)$. Here f is the function which appears in (10.3) or (10.4). Under appropriate hypotheses on F, (12.1) determines $U(t,r)$ uniquely. This inequality is given in [86], for instance. It generalizes an inequality which was developed by Bénilan for the quasiautonomous problem, i.e., the problem in which F can be written in the form $F(t,x)=G(x)+g(t)$.

Another approach to identification involves continuous dependence (discussed further in the next section). For simplicity, suppose that for each $(u(0),F)$ in some problem space $\mathcal{P}$, the construction described above as Steps 1 and 2 yields a unique solution u which is a point in some solution space $\mathcal{S}$. Define a mapping $\Gamma:\mathcal{P}\to\mathcal{S}$ by $u=\Gamma(u(0),F)$. Suppose that some topologies on $\mathcal{P}$ and $\mathcal{S}$ can be described simply and naturally, and that they make the mapping Γ continuous. Moreover, suppose that u is a Carathéodory solution (or other easily-motivated solution) of (1.1), for all $(u(0),F)$ in some dense set $\mathcal{P}_0\subset\mathcal{P}$. Then it is reasonable to call $u=\Gamma(u(0),F)$ a "solution" of (1.1) in a generalized sense, even for $(u(0),F)$ outside that dense set. If Γ is *uniformly* continuous, then we can replace $\mathcal{P}$ with its completion, thus motivating solutions for a possibly larger class of initial value problems. Even if the solution u is not uniquely determined by $u(0)$ and F, the preceding argument is applicable if we can show that the *set of solutions* depends continuously on the data, as indicated in the next section.

13. Continuous dependence. Assume (for the moment) that for each F in some suitable class of operators, and each $u(0)$ in some suitable subset of X, the initial value problem (1.1) has a solution u. Does u depend continuously on the initial value $u(0)$ and the generator F? This question is important for many reasons: (i) As we noted at the end of the previous section, continuous dependence results can be viewed as a method of identification of the limit. (ii) Continuity of a map such as $(u(0),F)\mapsto u$ is one of the principal ingredients in applications of the Schauder Fixed Point Theorem — see for instance the discussion after (13.4), below. (iii) The data $u(0)$ and F may be based on a physical experiment, and so the specification of $u(0)$ and F may involve inexact measurements; we would like to know that small errors in the data cause only small errors in u. (iv) If F is discontinuous or otherwise badly behaved — or if $u(0)$, which itself may lie in some function space, is badly behaved — we may wish to replace F or $u(0)$ with a "nearby" choice which is better behaved, to facilitate computations or proofs; we would like to know that this will not change u greatly either. (v) The convergence of approximate solutions to a limit can be viewed as a special case of continuous dependence results, since — as we noted in (11.2) — approximate solutions to an initial value problem are themselves excact solutions to approximating problems. (vi) Continuous dependence results can be applied in existence proofs in other ways, too. For instance, in [108], a continuous dependence result is used to show that a certain "nice" class of initial value problems has solutions taking values in a separable subset of X. That implies that a certain "bad" set of values of t has Lebesgue measure 0. That fact, in turn, is used to prove convergence of some of those "nice" solutions, and hence to prove existence of limit solutions to some not-so-nice initial value problems.

Let us survey first the continuous dependence of u on $u(0)$. We have already seen that if $F(t,\cdot)$ satisfies a dissipativeness condition for each t, then $u(t)$ depends on $u(0)$ in a Lipschitz fashion (9.5). Without dissipativeness, other hypotheses may be needed to guarantee uniqueness, as noted in §4. In finite dimensions, for F bounded but possibly discontinuous, it is known that uniqueness of solutions implies continuous dependence on initial values; see [49]. In infinite dimensions, however, this result fails, even for continuous F; see [42].

Continuous dependence is meaningful even without uniqueness of solutions. If F is a bounded, upper semicontinuous map from $[0,T] \times \mathbf{R}^n$ into the nonempty compact convex subsets of $\mathbf{R}^n$, then the set of solutions of (1.1) is a compact set which depends in an upper semicontinuous fashion on the initial data; see [4]. Similar results apply to at least some initial value problems in infinite-dimensional spaces; for instance, see the remark after (7.5).

Next we shall survey the dependence of u on F. We shall not give complete details here, but shall try to indicate some of the most interesting ideas. The theory of dependence on F is unavoidably more complicated than that of the dependence on $u(0)$, for $u(0)$ is just a point in X, while F is a function of two arguments, possibly discontinuous and set-valued. Moreover, as we vary the choice of F, even the set $\mathrm{Dom}(F) \subseteq [0,T] \times X$ may vary.

For simplicity of notation, let $F = F(t,x,\theta)$, where θ takes values in some parameter space. For each fixed θ, assume $F(\cdot,\cdot,\theta)$ is a generator, i.e., its initial value problem is "solvable" in some sense. For motivation, at first we shall assume that $F(\cdot,\cdot,\theta)$ is single-valued and defined everywhere on $[0,T] \times X$, and that the solution to the initial value problem is unique. For part of the discussion below, we shall need to vary the initial time, so let us rewrite the initial value problem (1.1) as

$$\begin{cases} u'(t) \in F(t,u(t),\theta) & (a \le t \le b) \\ u(a) = z. \end{cases}$$

Let the unique solution be denoted $u(t) = u(t; a, z, \theta)$ to display its dependence on the various arguments. The continuous dependence problem now is to show that if F depends continuously on θ in some appropriate sense, then so does u.

It is not hard to choose a topology for u. In most of the literature, u is a continuous function of t, and so we consider $u(\cdot; a, z, \theta)$ as an element of the Banach space $C([a, b]; X)$ with the usual supremum norm. We want $\theta \mapsto u(\cdot; a, z, \theta)$ to be continuous from the parameter space into $C([a, b]; X)$ for each choice of a and z.

The choice of a topology for F is not so simple. Numerous different topologies have been used in the literature (and the choice will become more complicated when we permit F to be set-valued). The simplest choice is *pointwise continuity* — i.e., $F(t, x, \theta)$ is a continuous function of θ, separately for each t. This assumption can be made separately for each x, or uniformly on certain sets of x's — e.g., for bounded x, or for x lying in a compact set, etc. This kind of hypothesis is found in some textbooks [51]. It is sufficient to guarantee continuous dependence of u on θ, in the presence of various additional technical assumptions about $F(\cdot, \cdot, \theta)$. In fact, variants of this pointwise continuity assumption are sufficient even if F is discontinuous in t and x, set-valued and not everywhere defined. For instance, Crandall and Pazy [28] assume that $F(t, x, \theta)$ is dissipative in x for each t and θ, and that the resolvent $J_\lambda(t, \theta)x = (I - \lambda F(t, \cdot, \theta))^{-1}x$ depends continuously on θ for each fixed λ, t, and x. Under these assumptions, plus some mild technical hypotheses, it follows that u depends continuously on θ.

For another topology on F, consider the mapping $t \mapsto F(t, x, \theta)$ as an element in $L^1([a, b]; X)$ for each fixed x and θ. We shall say F depends *L^1-continuously* on θ if the mapping $\theta \mapsto F(\cdot, x, \theta)$ is continuous from the parameter space $\{\theta\}$ into $L^1([a, b]; X)$, for each fixed x (or uniformly on certain sets of x's). This hypothesis does not differ greatly from pointwise continuity: any L^1-convergent sequence of functions on $[a, b]$ has a pointwise-convergent subsequence, while any bounded, pointwise convergent sequence of functions on $[a, b]$ converges also in L^1.

One example of L^1-continuity has been studied extensively: An operator G is *m-dissipative* if it is dissipative and satisfies (6.1). Let G be an m-dissipative operator in a Banach space X. Then for each $z \in \mathrm{cl}(\mathrm{Dom}(G))$ and each $g \in L^1([0, T]; X)$, the *quasiautonomous problem*

$$\begin{cases} u'(t) \in G(u(t)) + g(t) & (0 \le t \le T), \\ u(0) = z \end{cases} \tag{13.1}$$

has a unique solution $u(t) = u(t; z, g)$. Using (12.1), it can be shown that

$$\|u(t; z_1, g_1) - u(t; z_2, g_2)\| \le \|z_1 - z_2\| + \int_0^t \|g_1(s) - g_2(s)\|\, ds,$$

and hence $u(t; z, \cdot)$ is continuous on $L^1([0, T]; X)$. (We remark that this quasi-autonomous problem in X can be reduced to an autonomous problem in $X \times L^1([0, +\infty); X)$; see [29].)

For a weaker hypothesis than pointwise- or L^1-continuity, we shall say F depends *integral-continuously* on θ if

$$\Phi(t,x,\theta) = \int_0^t F(s,x,\theta)\,ds \tag{13.2}$$

is a continuous function of θ. For many classes of F's (discussed below), it can be shown that u depends continuously on θ for all choices of t, a, and z *if and only if* F depends integral-continuously on θ for all choices of t and x. (It is necessary to permit the initial time a to vary, since the behavior of $F(s,x,\theta)$ for $s < a$ is unrelated to the behavior of $u(t;a,z,\theta)$.) Of course, the "if" part of this statement is the practical part for most applications, since we generally can verify conditions on F more directly and more easily than conditions on u. The "only if" part is of interest because it tells us our theory is headed in the right direction: in some sense, the practical part can't be improved. Integral-continuity yields a topology on the F's weak enough so that many sets are compact; this compactness has some applications mentioned below. The fact that the topology can't be improved — i.e., weakened further — follows from the theorem in point-set topology that no Hausdorff topology on a set can be strictly weaker than a compact Hausdorff topology.

Obviously, L^1-continuity implies integral-continuity. Integral-continuity is in fact a weaker hypothesis, especially in cases where F oscillates rapidly as t varies. For instance, suppose $H(s,x)$ is periodic in s with period 1 (e.g., let $H(s,x) = \sin(s/2\pi)$). Let

$$F(t,x,\theta) = \begin{cases} H(t/\theta,x) & \text{when } \theta \neq 0, \\ G(x) \equiv \int_0^1 H(s,x)\,ds & \text{when } \theta = 0. \end{cases}$$

Then $F(t,x,\theta)$ generally is discontinuous at $\theta = 0$ in both the pointwise- and L^1-senses, but $\Phi(t,x,\theta)$ is continuous there. In cases where integral-continuity is sufficient for continuous dependence of u, this tells us that the solution of $u'(t) \in H(t/\theta, u(t))$, with $u(0)$ given, converges as $\theta \to 0$ to the solution of $u'(t) \in G(u(t))$.

An interesting special case is that in which $H(s,x) = B(x)$ for $0 < s < 1/2$ and $H(s,x) = A(x)$ for $1/2 < s < 1$, where A and B are two m-dissipative operators. Then the convergence result just described reduces to the *Trotter-Lie-Kato product formula*:

$$e^{t(A+B)}x = \lim_{n\to\infty} \left[\exp\left(\frac{t}{n}A\right)\exp\left(\frac{t}{n}B\right)\right]^n x. \tag{13.3}$$

This formula is valid for many, but not all, choices of A and B. See [65] [66] [70] [75] [96] for some recent discussions of this formula. A particularly simple example in which the formula fails is given by [112]: Let X be the complex Banach space of bounded sequences of complex numbers, with the supremum norm; let

$A(\{x_k\}) = \{ikx_k + 1\}$ and $B(\{x_k\}) = \{ikx_k - 1\}$. Then (13.3) can be shown to fail at $x = 0$.

The assumption of integral-continuity apparently was first used by Gihman [44], to prove continuous dependence for a very simple class of ordinary differential equations in finite dimensions. The converse ("only if") part apparently was first noted by Artstein [3]. Gihman's continuous dependence principle has been extended to many other classes of initial value problems; the rather strong assumptions about F suggested above can be weakened considerably in numerous different directions. We note a few of those directions below, but we omit the many technical details, which vary from one paper to another. Although the several existence theories for (1.1) are far from unified, each has a version of Gihman's principle which is true for at least some F's. These different versions are proved separately by different methods, but their similarity suggests that there may be a single theory underlying them all. The present author hopes to research this idea further in the near future.

The paper [108] proves a version of Gihman's continuous dependence result in infinite dimensions, assuming $F(t, x, \theta)$ continuous and dissipative in x. The paper [110] extends Gihman's result to a case where $F(t, \cdot, \theta)$ is actually discontinuous, but is uniformly continuous and dissipative when restricted to suitable subsets of the Banach space X. These hypotheses are weak enough to apply to some partial differential equations with smooth coefficients.

A variant of Gihman's principle applies to the quasiautonomous problem (13.1), at least for some choices of G. Following the terminology of [112], we say an m-dissipative operator G has *Gihman's property* if the solution $u(t; z, g)$ of (13.1) depends continuously on the indefinite integral of g, as g varies over a weakly compact set of integrable functions. Many, but not all, m-dissipative operators have this property; see [112]. This continuous dependence property has consequences for existence theory, as follows: Let G be an m-dissipative operator (not necessarily satisfying any compactness condition), and let $H : X \to X$ be a continuous mapping with relatively compact range (not necessarily satisfying any dissipativeness condition). Let $z \in \mathrm{cl}(\mathrm{Dom}(G))$. Does the "dissipative plus compact" problem

$$(13.4) \qquad \begin{cases} v'(t) \in (G + H)(v(t)) & (0 \le t \le T), \\ v(0) = z \end{cases}$$

necessarily have any solutions? It does if we impose some additional mild hypotheses on G or H or X, but in general the answer is not yet known. The best results in this direction use the following observation: A solution of (13.4) is the same thing as a fixed point of the map $v \mapsto u(\cdot; z, H \circ v)$, where $u(t; z, g)$ is the solution of (13.1). Such a fixed point exists by Schauder's Fixed Point Theorem (or a variant thereof) if G has Gihman's property (or some variant thereof). Here we make use of the many compact sets made available by the topology of integral-convergence, mentioned earlier. The "dissipative plus compact" problem will be discussed further in §15.

For applications of Gihman's principle, the solution u of (1.1) need not be uniquely determined by F. Nonuniqueness can be dealt with in at least a couple of ways. In [30] [117], the functions F and Φ are set-valued. Continuity is defined for Φ and for the set of solutions $\{u\}$ by using the Hausdorff metric to measure the distance between two sets. Again, a variant of Gihman's integral convergence is shown sufficient for continuous dependence.

Another approach to nonuniqueness is taken by [11] [107]. In those papers, the Banach space X considered is the real line. The generator F considered — discontinuous in both papers — does not uniquely determine the solution u, but among the solutions is a *maximal* (i.e., largest) solution. One-sided continuous dependence results (i.e., involving lim sup's and inequalities, instead of limits and equations), analogous to Gihman's result, are proved for the maximal solution.

Variants of Gihman's hypothesis also suffices for continuous dependence for some stochastic differential equations; see [43] [122].

Continuous dependence results such as those described above can be taken as a basis for extending the notion of "differential equation." The following discussion is based on Kurzweil [71]. (We now drop the θ from our notation.) If we substitute (13.2), then the integral equation (2.4) can be restated as

$$(13.5)\quad u(t)-u(0) = \int_0^t d\,\Phi(s,u(s)) \equiv \lim \sum_{k=1}^{N} [\Phi(t_k,u(\tau_k)) - \Phi(t_{k-1},u(\tau_k))]$$

where $\tau_k \in (t_{k-1}, t_k)$ and the limit is taken over partitions (11.3) partially ordered by refinement. Many of the main equations, inequalities, and theorems about existence and continuous dependence — including a version of Gihman's principle — can then be restated in terms of Φ and Kurzweil's integral (13.5), without ever mentioning the original function F, and much of the theory then takes a simpler form. Now define u to be a *solution of the generalized differential equation*

$$u'(t) = D_t\Phi(t,u(t))$$

if u satisfies (13.5). Then the theory of existence and continuous dependence of such solutions can be developed entirely in terms of Φ. We have mentioned F only for motivation; we do not need to assume that Φ arises as in (13.2). Some assumptions must be made about Φ, of course, but the natural hypotheses on Φ are in many cases simpler and weaker than those implicit in (13.2). In particular, neither $\Phi(t,x)$ nor $u(t)$ need be a continuous function of t. Schwabik [114] applies these ideas to differential equations with impulses.

Kurzweil's integral is a variant of the Stieltjes integral. Schwabik [115] compares Kurzweil's integral with those of Perron and Young. For equations using the usual Stieltjes integral, Binding [13] [14] develops a theory including some results on existence (generalizing Caratheódory's Theorem 3.1), uniqueness, continuous dependence, and other results. (However, Binding does not give an analogue of Gihman's convergence principle.)

Finally, we note that integral-continuity plays an important role in the study of nonautonomous differential equations as dynamical systems (invariance principles, stability, etc.): Let Ω be a subset of a Banach space X, and let $\mathcal{F}$ be some family of mappings from $\mathbf{R} \times \Omega$ into X. Assume that each $F \in \mathcal{F}$ generates an evolution on Ω, in the sense of (4.3)-(4.4); denote that evolution by $U(t, s; F)$. Also assume that $F_h \in \mathcal{F}$ whenever $h > 0$ and $F \in \mathcal{F}$, where $F_h(t, x) \equiv F(t + h, x)$. Then it follows that $U(t, s; F_h) = U(t + h, s + h; F)$. Hence

$$\mathcal{S}(h)\begin{pmatrix} x \\ F \end{pmatrix} = \begin{pmatrix} U(h, 0; F)x \\ F_h \end{pmatrix}$$

defines a semigroup (in the sense of (4.5)-(4.6)) on $\Omega \times \mathcal{F}$. For some choices of $\mathcal{F}$, the topology given by integral-convergence is compact. Now stability results, invariance results, etc. can be applied to the semigroup $\mathcal{S}$; they yield corresponding results for the evolution U. The semigroup $\mathcal{S}$ is called the *skew product semidynamical system*. It was first studied by Miller and Sell, and later by Wakeman, Artstein, and others; an introduction to this subject is given by [105].

14. Bijections. In the continuous dependences $\Gamma : F \mapsto u$ described in the preceding two sections, we know the domain $\mathcal{P}$ (i.e., the set of F's) and the codomain $\mathcal{S}$ (i.e., a space *containing* the set of u's), but generally we do not have a characterisation of the *range* $\Gamma(\mathcal{P})$. Under stronger h ypotheses than those of the previous section, we are sometimes able to give bijections $F \leftrightarrow u$, where both domain and range are known. All such results are motivated by the following classical result:

14.1. THEOREM (Hille and Yosida, 1948). *Formula* (4.8) *gives a one-to-one correspondence between linear, strongly continuous semigroups* $S(t) = \exp(tG)$ *of type* ω *(i.e., satisfying* (4.7)*) on a Banach space* X*, and linear, densely defined operators* G *on* X *such that* $G - \omega I$ *is m-dissipative. Moreover,* G *can be recovered from* S *by the formula*

$$G(x) = \lim_{t \downarrow 0} \frac{S(t)x - x}{t}$$

with $\mathrm{Dom}(G)$ *consisting of those* x *for which the limit exists.*

The theory of linear semigroups is extensive; see [46] for a thorough introduction. Part of the Hille-Yosida Theorem extends to nonlinear operators in an arbitrary Banach space, in Theorem 10.1. However, the bijection between $S(t)$ and G does not extend to the setting of that theorem. Crandall and Liggett [27] give an example with $X = \mathbf{R}^2$, normed by $\|(x_1, x_2)\| = \max\{|x_1|, |x_2|\}$, in which many different generators G yield the same semigroup S.

All the results stated in 14.1 do extend to nonlinear semigroups in a sufficiently nice Banach space — e.g., a Hilbert space, or more generally a uniformly smooth Banach space, or still more generally a reflexive Banach space with uniformly

Gateaux differentiable norm. The nonlinear semigroup is defined on a closed convex nonexpansive retract C of the Banach space, and the limit defined in (4.8) exists for all x in a dense subset of C; see Reich [101] [102] for further details.

We turn now to the temporally inhomogenous theory. A number of bijections have been established between classes of evolution operators U and their generators F. In most of these results, $F(t,\cdot)$ and $U(t,s)$ take their values in the space $B(X)$ of bounded linear operators on the Banach space X. However, $U(t,s)x$ need not be a continuous function of t. In the temporally inhomogeneous case, discontinuity in t is quite natural, as we noted in §4 and 13.

Many of the bijection results use the following notation, or a close variant: Let W be a mapping, not necessarily multiplicative, from $\{(t,s) : -\infty < s \leq t < \infty\}$ into $B(X)$. We define the "sum integral" and "product integral," respectively, as

$$\sum_a^b W = \lim\left[W(t_n, t_{n-1}) + \ldots + W(t_2, t_1) + W(t_1, t_0)\right]$$

$$\prod_a^b W = \lim\left[W(t_n, t_{n-1}) \circ \cdots \circ W(t_2, t_1) \circ W(t_1, t_0)\right]$$

when these limits exist; the limits are taken over of partitions of $[a,b]$:

$$a = t_0 < t_1 < t_2 < \ldots < t_n = b \tag{14.2}$$

with those partitions partially ordered by refinement. (In some papers, the symbol $\int$ is used in place of Σ.) Note that if $V(t,s) = \sum_s^t W$ exists for all $[s,t] \subseteq [a,b]$, then it is additive — i.e., $V(t,s) + V(s,r) = V(t,r)$. Similarly, if $U(t,s) = \prod_s^t W$ exists for all $[s,t] \subseteq [a,b]$, then it is multiplicative (i.e., an evolution, in the sense of (4.3), (4.4)). We say W has *bounded variation* on an interval $[a,b]$ if

$$\sup\{\|W(t_n, t_{n-1})\| + \ldots + \|W(t_2, t_1)\| + \|W(t_1, t_0)\|\} < \infty,$$

where the supremum is taken over all refinements (14.2) of some partition of $[a,b]$. (Clearly, if W is an additive function given by $W(t,s) = w(t) - w(s)$, then W has bounded variation in the sense above if and only if w has bounded variation in the usual sense of real analysis.) Helton [53] has proven that if W has bounded variation on $[a,b]$, then $\sum_a^b W$ exists if and only if $\prod_s^t (I + W)$ exists for all $[s,t] \subseteq [a,b]$.

To relate these results to differential equations, take (13.5) as a starting point, but with $\Phi(s,\cdot)$ linear. Then we write $\Phi(s, u(s)) = \Phi(s)u(s)$, and the condition we wish to satisfy is the Stieltjes integral equation

$$u(t) - u(r) = \int_r^t d\Phi \cdot u \tag{14.3}$$

for all r, t in the time-interval being considered. The solution is given by $u(t) = U(t,r)u(r)$, where $U(t,r)$ is a multiplicative function (i.e., an evolution). Thus (14.3) can be restated in terms of U:

$$U(t,r) = I + \int_r^t d\Phi \cdot U(\cdot, r).$$

Define an additive function $V(t,r) = \Phi(t) - \Phi(r)$; then the function Φ or V can be retrieved from the evolution U by

$$V(t,r) = \Phi(t) - \Phi(r) = \int_r^t U(s,b)\, dU(b,s).$$

Here s is the variable of integration, and b is an arbitrary constant; the value of the Stieltjes integral can be shown independent of the choice of b. This formula is equivalent to

$$V(t,r) = \sum_r^t [U - I]. \tag{14.4}$$

Conversely, we can obtain U from V by the formula

$$U(t,r) = \prod_r^t [I + V]. \tag{14.5}$$

In fact, formulas (14.4) and (14.5) give a bijection between additive V's with bounded variation and multiplicative U's with $U - I$ having bounded variation. For an introduction to this subject and proof of this bijection see MacNerney [79].

A number of subsequent papers have extended MacNerney's bijection; we mention a few of the simplest and most interesting extensions. Operators are continuous and linear except where otherwise noted.

Herod [54] permitted the operators to be nonlinear. However, the other conditions involved in Herod's bijection are complicated.

Freedman [40] added an assumption of continuity, but weakened MacNerney's hypothesis of bounded variation to a hypothesis of bounded p-variation. The *p-variation* of a function W over an interval $[a,b]$ (in the sense of Wiener [128]) is the supremum of the quantities

$$\left[\|W(t_n,t_{n-1})\|^p + \ldots + \|W(t_2,t_1)\|^p + \|W(t_1,t_0)\|^p\right]^{1/p}$$

over all partitions (14.2) of the interval. For further references concerning the p-variation, see [40].

Herod and McKelvey [55] permitted their operators $V(t,s)x$ to be discontinuous in x, in a somewhat weak sense. They did this by using a scale of Banach spaces

$X_0 \subset X_1 \subset X_2 \subset \cdots \subset X_N$. Each generator $V(b,a)$ is assumed to be continuous from X_p to X_{p+1}, but possibly discontinuous from X_p to X_p. Their somewhat complicated bijection has the virtue of including the classical Hille-Yosida Theorem 14.1 as a special case. Freedman [39] extended this result further, so as to include evolutions $U(t,s)x$ which may be discontinuous in t.

The results of Hinton [58] take us a bit further from the simple initial value problem (1.1), but weaken MacNerney's hypotheses slightly and also strengthen the symmetry: the limit procedures in (14.4) and (14.5) are special cases of a single mapping. Let $\mathcal{F}$ be the class of all functions $F : [a,b] \times [a,b] \to B(X)$ with these properties: (i) $F(t,t) = I$ for all t; (ii) for each fixed r, $F(\cdot,r)$ has left- and right-hand limits at every point of (a,b), and one-sided limits at a and at b; and (iii) for each fixed t, $F(t,\cdot)$ has bounded variation. (Note that F is not assumed multiplicative or additive.) Hinton shows that for each $F \in \mathcal{F}$ there is a unique $M \in \mathcal{F}$ satisfying the Stieltjes-Volterra integral equation

$$M(t,r) = I + (L)\int_r^t dF(t,s) \cdot M(s,r)$$

for all $[r,t] \subseteq [a,b]$. (Here $(L)\int$ denotes a left Cauchy integral, with s being the variable of integration; see [58].) Thus $M = \mathcal{C}(F)$ defines a mapping from $\mathcal{F}$ into $\mathcal{F}$. Hinton shows that this mapping $\mathcal{C}$ is in fact a bijection from $\mathcal{F}$ onto $\mathcal{F}$, and moreover $\mathcal{C}\mathcal{C}(F) = F$ for each $F \in \mathcal{F}$. When U and V are an evolution and generator satisfying the hypotheses of MacNerney [79] and related by (14.4) and (14.5), then Hinton shows that $U, V \in \mathcal{F}$ and that

$$U = \mathcal{C}(I - V) \qquad \text{and} \qquad V = I - \mathcal{C}(U).$$

15. Unification. The ultimate problem in existence theory is to find necessary and sufficient conditions on $u(0)$ and F for the existence of a solution to (1.1). This problem is too hard for the near future, but it does suggest various subproblems which are worth pursuing. Foremost of these is the unification of the known sufficient conditions for existence of solutions.

As we have indicated in §12, the conditions of generalized compactness, isotonicity, and dissipativeness play analogous roles; each is a hypothesis used in proving convergence of approximate solutions. These three conditions are quite different, and have led to three largely separate theories. Still, those three theories have some analogous structures, as we have tried to indicate by the use of the same letter ω in inequalities (7.1), (8.3), (9.7). Does a single, weaker condition underlie the three hypotheses for convergence — or at least two of them?

Little has been done to find hypotheses weaker than that of isotonicity (§8). One approach worth noting is that of Calvert [18], who defines K-dissipativeness, analogous to dissipativeness, in terms of a cone K. Existence of solutions is proved under the assumption that the cone is normal, i.e., that the ordering satisfies

$0 \leq x \leq y \Rightarrow \|x\| \leq \|y\|$. However, Picard [92] shows that if G is K-dissipative and K is normal, then $\| \quad \|$ can be replaced by an equivalent norm which makes G dissipative. Of less interest for the goals of this survey, but still worth mentioning, are some results with *stronger* hypotheses, involving both isotonicity and dissipativeness [2] [10], or involving both isotonicity and compactness [72].

More has been done in unifying compactness and dissipativeness. Martin [82] and Li [76] observed that (7.3) is satisfied if G is ω-dissipative *or* if G satisfies a generalized compactness condition of type (7.1). A drawback to the approach of Martin and Li is that they require their operators to be uniformly continuous, thus excluding the many applications of the dissipative theory to partial differential equations. It would be interesting to weaken or remove that uniform continuity assumption.

The sum of two uniformly continuous operators satisfying (7.3) is another such operator (add the ω's — see [82]), and so the existence results with hypothesis (7.3) include, as a special case, some results on the existence of solutions to differential inclusions of the form (13.4), where G is m-dissipative and H satisfies some sort of generalized compactness condition such as (7.1). However, stronger results for (13.4) have been proved by other methods not involving (7.3).

Without further assumptions, does (13.4) necessarily have a solution? In general, the answer is not yet known. Various partial answers can be divided into two main classes:

(a) Assume G is continuous, and defined on all of X. Without further assumptions, it is not known whether (13.4) necessarily has a solution. But (13.4) is known to have a solution if G or H is uniformly continuous, or if $G + H$ is uniformly continuous, or if the Banach space X is uniformly smooth [113]. See the survey [124] for further references.

(b) Assume H has relatively compact range. Again, existence is known if any one of several additional assumptions holds, but not otherwise. For details see the surveys [47] [112], and other papers cited therein; for more recent results not mentioned in those surveys see also [56] [125]. A chief method here is that of fixed point theory, as discussed after (13.4).

Even if we can unify the three known hypotheses for convergence, we are still a long way from a full understanding of the autonomous problem (1.2). To see this, note that any nonautonomous problem (1.1) in a Banach space X can be transformed to an autonomous problem $v'(t) \in G(v(t))$ in $\mathbf{R} \times X$, via the transformation

$$v(t) = \begin{pmatrix} t \\ u(t) \end{pmatrix}, \qquad G\begin{pmatrix} t \\ x \end{pmatrix} = \begin{pmatrix} 1 \\ F(t,x) \end{pmatrix}. \tag{15.1}$$

However, when we apply this transformation to existence theorems for (1.1), it yields existence theorems for (1.2) which are not at all understood except via this transformation. For instance, some of the theorems discussed in §9-10 assume, roughly, that $F(t, x)$ is dissipative in x, and integrable in t in some sense.

For $t_1 \neq t_2$, the functions $F(t_1, \cdot)$ and $F(t_2, \cdot)$ may be almost entirely unrelated. Hence, although the operator G given by (15.1) must satisfy some conditions sufficient for existence of solutions, those conditions have very little resemblance to dissipativeness, compactness, or isotonicity.

16. References. For brevity, we omit some older papers which, though important, are cited by more recent papers listed here.

1. H. Amann, Fixed point equations and nonlinear eigenvalue problems in ordered Banach spaces, *SIAM Review* **18** (1976), 620-709.
2. W. Arendt, P. R. Chernoff, and T. Kato, A generalization of dissipativity and positive semigroups, *J. Oper. Th.* **8** (1982), 167-180.
3. Z. Artstein, Continuous dependence of parameters: on the best possible results, *J. Diff. Eqns.* **19** (1975), 214-225.
4. J. P. Aubin and A. Cellina, *Differential Inclusions*, Grundlehren der mathematischen Wissenschaften **264**, Springer-Verlag, Berlin, 1984.
5. J. M. Ball, Measurability and continuity conditions for nonlinear evolutionary processes, *Proc. Amer. Math. Soc.* **55** (1976), 353-358.
6. J. M. Ball, Finite time blow-up in nonlinear problems, pp.189-205 in: *Nonlinear Evolution Equations*, proc. of a symposium held in Madison, Wisconsin, October 1977), ed. M. G. Crandall; Academic Press, N.Y., 1978.
7. J. Banaś, On existence theorems for differential equations in Banach spaces, *Bull. Austral. Math. Soc.* **32** (1985), 73-82.
8. J. Banaś and K. Goebel, *Measures of noncompactness in Banach spaces,* Lecture Notes in Pure and Applied Math. **60**, Marcel Dekker, New York and Basel, 1980.
9. J. Banaś, A. Hajnosz, and S. Wędrychowicz, Relations among various criteria of uniqueness for ordinary differential equations, *Commentationes Math. Univers. Carolinae* **22** (1981), 59-70.
10. P. Bénilan and J. I. Diaz, Comparison of solutions of nonlinear evolution problems with different nonlinear terms, *Israel J. Math.* **42** (1982), 241-257.
11. D. Biles, *Existence and Continuous Dependence of Solutions of Discontinuous Differential Equations*, Ph.D. thesis, Vanderbilt University, Nashville, Tennessee, 1987.
12. P. Binding, The differential equation $x' = f \circ x$, *J. Diff. Eqns.* **31** (1979), 183-199.
13. P. Binding, Bounded variation evolution equations, *J. Math. Anal. Appl.* **48** (1974), 70-94.
14. P. Binding, Corrections to: "Bounded variation evolution equations," *J. Math. Anal. Appl.* **74** (1980), 635-636.
15. A. Bressan, On differential relations with lower continuous right-hand side: an existence theorem, *J. Diff. Eqn.* **37** (1980), 89-97.

16. A. Bressan, Solutions of lower semicontinuous differential inclusions on closed sets, *Rend. Sem. Mat. Univ. Padova* **69** (1983), 99-107.
17. A. Bressan, Upper and lower semicontinuous differential inclusions: a unified approach, to appear in the proceedings of the workshop "Controllability and Optimal Control" held at Rutgers Univ., May 1987; H. Sussman, editor; Marcel-Dekker, publisher.
18. B. D. Calvert, Semigroups in an ordered Banach space, *J. Math. Soc. Japan* **23** (1971), 311-319.
19. A. Cambini and S. Querci, Equazioni differenziali del primo ordine con secondo membro discontinuo rispetto all'incognita, *Rend. Ist. Mat. Univ. Trieste* **1** (1969), 89-97.
20. C. Carathéodory, *Vorlesungen uber reele Funktionen*, Leipzig, 1927; reprinted, New York, 1948, pp. 665-688.
21. J. Caristi, Fixed point theorems for mappings satisfying inwardness conditions, *Transac. Amer. Math. Soc.* **215** (1976), 241-251.
22. P. Chernoff, A note on continuity of semigroups of maps, *Proc. Amer. Math. Soc.* **53** (1975), 318-320.
23. S.-N. Chow and J. D. Schuur, Fundamental theory of contingent differential equations in Banach space, *Transac. Amer. Math. Soc.* **179** (1973), 133-144.
24. E. A. Coddington and N. Levinson, *Theory of Ordinary Differential Equations*, McGraw-Hill Book Co., N.Y., 1955.
25. M. G. Crandall, The semigroup approach to first order quasilinear equations in several space variables, *Israel J. Math.* **12** (1972), 108-132.
26. M. G. Crandall, Nonlinear semigroups and evolution governed by accretive operators, in *Nonlinear Functional Analysis and Its Applications* (proceedings of the 31st Summer Research Institute of the A.M.S., at the University of California at Berkeley, 1983), ed. by F. Browder, *Proc. Sympos. Pure Math.* **45**, part 1 (1986), 305-337.
27. M. G. Crandall and T. M. Liggett, Generation of semi-groups of nonlinear transformations on general Banach spaces, *Amer. J. Math.* **93** (1971), 265-298.
28. M. G. Crandall and A. Pazy, Nonlinear evolution equations in Banach spaces, *Israel J. Math.* **11** (1972), 57-94.
29. M. G. Crandall and A. Pazy, An approximation of integrable functions by step functions with an application, *Proc. Amer. Math. Soc.* **76** (1979), 74-80.
30. M. Dawidowski, On some generalization of Bogoliubov averaging theorem, *Functiones et Approximatio* **7** (1979), 55-70.
31. D. A. Dawson and L. G. Gorostiza, $*$-solutions of evolution equations in Hilbert space, *J. Diff. Eqns.* **68** (1987), 299-319.
32. K. Deimling, *Ordinary Differential Equations in Banach Spaces*, Springer Lecture Notes in Math. **596** (1977), Springer-Verlag, Berlin.
33. J. Dieudonné, Deux exemples singuliers d'équations différentielles, *Acta Sci. Math. (Szeged)* **12B** (1950), 38-40.

34. J. D. Dollard and C. N. Friedman, On strong product integration, *J. Func. Anal.* **28** (1978), 309-354.

35. N. Dunford and J. T. Schwartz, *Linear Operators I*, Wiley, N.Y., 1957.

36. L. C. Evans, Nonlinear evolutions in an arbitrary Banach space, *Israel J. Math.* **26** (1977), 1-42.

37. L. C. Evans, Applications of nonlinear semigroup theory to certain partial differential equations, pp.163-188 in: *Nonlinear Evolution Equations*, proceedings of a symposium held at the Math. Research Center, Univ. of Wisconsin, Madison, Wisconsin, 1977; ed. by M. G. Crandall. Publ. of the Math. Research Center, Univ. of Wisconsin, Madison, no. **40**. New York: Academic Press, 1978.

38. L. C. Evans, Nonlinear semigroup theory and viscosity solutions of Hamilton-Jacobi PDE, pp.63-77 in *Nonlinear Semigroups, Partial Differential Equations and Attractors (proc. conference at Howard University, Washington D.C., 1985)*, ed. by T. L. Gill and W. W. Zachary, *Springer Lecture Notes in Mathematics* **1248** (1987).

39. M. A. Freedman, Necessary and sufficient conditions for discontinuous evolutions with applications to Stieltjes integral equations, *J. Integral Eqns.* **5** (1983), 237-270.

40. M. A. Freedman, Operators of p-variation and the evolution representation problem, *Transac. Amer. Math. Soc.* **279** (1983), 95-112.

41. M. A. Freedman, Product integrals of continuous resolvents: existence and nonexistence, *Israel J. Math.* **46** (1983), 145-160.

42. B. M. Garay and J. J. Schäffer, More on uniqueness without continuous dependence in infinite dimension, *J. Diff. Eqns.* **64** (1986), 48-50.

43. S. Geman, A method of averaging for random differential equations with applications to stability and stochastic approximations, pp.49-85 in *Approximate Solution of Random Equations*, ed. by A. T. Bharucha-Reid, New York: North-Holland Series in Probability and Applied Math., 1979.

44. I. I. Gihman, Concerning a theorem of N. N. Bogolyubov, *Ukrain. Math. J.* **4** (1952), 215-219 (Russian). (For an English summary see *Math. Reviews* **17**, p.738.)

45. A. N. Godunov, Peano's theorem in Banach spaces, *Funktsional'nyi Analiz i Ego Prilozheniya* **9** (1975), 59-60 (Russian.) English translation in: *Func. Anal. and its Applic.* **9** (1975), 53-55.

46. J. A. Goldstein, *Semigroups of Operators and Applications*, Oxford Univ. Press, N.Y., and Clarendon Press, Oxford; 1985.

47. S. Gutman, Existence theorems for nonlinear evolution equations, *Nonlin. Analysis, Theory, Methods, Applicns.* **11** (1987), 1193-1206.

48. O. Hájek, Discontinuous differential equations I, *J. Diff. Eqns.* **32** (1979), 149-170.

49. O. Hájek, Discontinuous differential equations II, *J. Diff. Eqns.* **32** (1979), 171-185.

50. P. Hartman, On invariant sets and on a theorem of Ważewski, *Proc. Amer. Math. Soc.* **32** (1972), 511-520.

51. P. Hartman, *Ordinary Differential Equations*, second edition, 1973, reprinted 1982, Birkhäuser Boston.

52. H.-P. Heinz, On the behaviour of measures of noncompactness with respect to differentiation and integration of vector-valued functions, *Nonlinear Anal.* **7** (1983), 1351-1371.

53. J. C. Helton, Mutual existence of sum and product integrals, *Pacific J. of Math.* **56** (1975), 495-516.

54. J. V. Herod, A pairing of a class of evolution systems with a class of generators, *Transac. Amer. Math. Soc.* **157** (1971), 247-260.

55. J. V. Herod and R. W. McKelvey, A Hille-Yosida theory for evolutions, *Israel J. Math.* **36** (1980), 13-40.

56. A. Hertzog, Résolution du problème $u'(t) - (A + B)(u(t)) \in C(t, u(t))$, in *Analyse non linéaire, Années 1982 et 1983*, Exp. no.1, 32 pp., *Publ. Math. Fac. Sci. Basançon*, **7**, Univ. Franche-Comté, Besançon (1983).

57. C. J. Himmelberg and F. S. Van Vleck, Existence of solutions for generalized differential equations with unbounded right-hand side, *J. Diff. Eqns.* **61** (1986), 295-320.

58. D. B. Hinton, A Stieltjes-Volterra integral equation theory, *Canad. J. Math.* **18** (1966), 314-331.

59. T. Iwamiya, Global existence of solutions to nonautonomous differential equations in Banach spaces, *Hiroshima Math. J.* **13** (1983), 65-81.

60. T. Iwamiya, S. Oharu, and T. Takahashi, On the class of nonlinear evolution operators in Banach space, *Nonlin. Anal. Theory Methods Applicns.* **10** (1986), 315-337.

61. T. Iwamiya, S. Oharu, and T. Takahashi, Envelopes of nonlinear semigroups, preprint.

62. T. Iwamiya, S. Oharu, and T. Takahashi, manuscript in preparation.

63. J. L. Kaplan and J. A. Yorke, Toward a unification of ordinary differential equations with nonlinear semi-group theory, pages 424-433 in *International Conference on Differential Equations* (proceedings of a conference at U. Southern California, 1970), ed. by H. A. Antosiewicz; N.Y.: Academic Press, 1975.

64. T. Kato, Linear evolution equations of "hyperbolic" type, *J. Fac. Sci., Univ. Tokyo, Sec.1,* vol. **17** (1970), 241-258.

65. T. Kato, Trotter's product formula for an arbitrary pair of selfadjoint contraction semigroups, in: *Topics in Functional Analysis*, ed. by I. Gohberg and M. Kac; Academic Press, N.Y., 1978.

66. T. Kato and K. Masuda, Trotter's product formula for nonlinear semigroups generated by the subdifferentials of convex functionals, *J. Math. Soc. Japan* **30** (1978), 169-178.

67. N. Kenmochi and T. Takahashi, Nonautonomous differential equations in Banach spaces, *Nonlin. Anal.* **4** (1980), 1109-1121.
68. Y. Kobayashi, Difference approximation of Cauchy problems for quasi-dissipative operators and generation of nonlinear semigroups, *J. Math. Soc. Japan* **27** (1975), 640-665.
69. M. A. Krasnosel'skiĭ, *Positive Solutions of Operator Equations*, P. Noordhoff Ltd., The Netherlands, 1964.
70. T. G. Kurtz and M. Pierre, A counterexample for the Trotter product formula, *J. Diff. Eqns.* **52** (1984), 407-414.
71. J. Kurzweil, Problems which lead to a generalization of the concept of an ordinary nonlinear differential equation, *Differential Equations and Their Applications (Proceedings of a Conference held in Prague, September, 1962)*, pages 65-76, Academic Press, N.Y., 1963.
72. V. Lakshmikantham, The method of upper and lower solutions for differential equations in a Banach space, pp.387-391 in: *Nonlinear Analysis and Applications*, proceedings of a conference held at Memorial Univ. of Newfoundland in St. Johns, June 1981; ed. by S. P. Singh and J. H. Burry; New York: Marcel Dekker, Inc. 1982.
73. V. Lakshmikantham and S. Leela, *Nonlinear Differential Equations in Abstract Spaces*, Oxford/New York: Pergamon Press Internat. Series in Nonlin. Math.: Theory, Methods, and Applicns. **2**, 1981.
74. S. Lang, *Real Analysis*, Addison-Wesley, Reading, Massachusetts, 1969.
75. M. L. Lapidus, Generalization of the Trotter-Lie formula, *Integral Eqns. and Operator Theory* **4** (1981), 366-415.
76. T.-Y. Li, Existence of solutions for ordinary differential equations in Banach spaces, *J. Diff. Eqns.* **18** (1975), 29-40.
77. S. Łojasiewicz, Jr., The existence of solutions for lower semicontiniuous orientor fields, *Bull. Acad. Polon. Sér. Sci. Math.* **28** (1980), 483-487.
78. S. Łojasiewicz, Jr., Some theorems of Scorza-Dragoni type for multifunctions with applications to the problem of existence of solutions for differential multivalued equations, pp.625-643 in: *Mathematical Control Theory (16th Semester at the Stefan Banach Internat. Math. Center, 1980)*; ed. by C. Olech, B. Jakubczyk, and J. Zabczyk; Banach Center Publications **14**, Polish Academy of Sciences, Warszawa: Polish Scientific Publishers, 1985.
79. J. S. MacNerney, Integral equations and semigroups, *Illinois J. Math.* **7** (1963), 148-173.
80. J. Marsden, On product formulas for nonlinear semigroups, *J. Funct. Anal.* **13** (1973), 51-72.
81. R. H. Martin, Jr., Differential equations on closed subsets of a Banach space, *Transac. Amer. Math. Soc.* **179** (1973), 399-414.
82. R. H. Martin, Jr., Approximation and existence of solutions to ordinary differential equations in Banach spaces, *Funkcialaj Ekvacioj* **16** (1973), 195-211.

83. H. Mönch and G.-F. von Harten, On the Cauchy problem for ordinary differential equations in Banach spaces, *Arch. Math.* **39** (1982), 153-160.

84. J. J. Moreau, Evolution problem associated with a moving convex set in a Hilbert space, *J. Diff. Eqns.* **26** (1977), 347-374.

85. J. J. Moreau, Approximation en graphe d'une évolution discontinue, *R. A. I. R. O. Analyse numérique/ Numerical Analysis* **12** (1978), 75-84.

86. S. Oharu, A class of nonlinear evolution operators: basic properties and generation theory, in *Semigroups, theory and applications, vol. 1*, ed. by H. Brezis, M. G. Crandall, and F. Kappel, Pitman Research Notes in Mathematics **141** (1986), 186-196.

87. C. Olech, An existence theorem for solutions of orientor fields, pp.63-66 in *Dynamical Systems, an International Symposium, vol. 2*; ed. by L. Cesari, J. K. Hale, and J. P. LaSalle; New York: Academic Press (1976).

88. N. H. Pavel, Nonlinear evolution equations governed by f-quasi-dissipative operators, *Nonlin. Anal.* **5** (1981), 449-468.

89. N. H. Pavel, *Nonlinear evolution operators and semigroups; applications to partial differential equations*, Berlin-New York: Springer-Verlag Lecture Notes in Math. **1260** (1987).

90. G. Peano, Sull'integrabilita delle equazioni differenziali di primo ordine, *Accad. Sci. Torino* **21** (1885), 677-685.

91. G. Pianigiani, Existence of solutions for ordinary differential equations in Banach spaces, *Bull. Acad. Polon. Sci., Sér. sci. math. astron. phys.* **23** (1975), 853-857.

92. C. Picard, Opérateurs φ-accrétifs et génération de semi-groupes non linéaires, *C. R. Acad. Sci. Paris* **275A** (1972), 639-641.

93. M. Pierre, Enveloppe d'une famille de semi-groupes non linéaires et équations d'évolution, *Seminaire d'analyse non linéaire*, Univérsité Besançon, 1976-77.

94. M. Pierre, Enveloppe d'une famille de semi-groupes dans un espace de Banach, *C. R. Acad. Sci. Paris* **284A** (1977), 401-404.

95. M. Pierre, Invariant closed subsets for nonlinear semigroups, *Nonlin. Anal.* **2** (1978), 107-111.

96. M. Pierre and M. Rihani, About product formulas with variable step-size, University of Wisconsin at Madison: *Math. Research Ctr. Tech. Summary Report* **2783** (1984).

97. G. H. Pimbley, A semigroup for Lagrangian one-dimensional isentropic flow, preprint.

98. A. Pucci, Sistemi di equazioni differenziali con secondo membro discontinuo rispetto all'incognita, *Rend. Ist. Mat. Univ. Trieste* **3** (1971), 75-80.

99. R. M. Redheffer, The theorems of Bony and Brézis on flow-invariant sets, *Amer. Math. Monthly* **79** (1972), 740-747.

100. S. Reich, Approximate selections, best approximations, fixed points, and invariant sets, *J. Math. Anal. Applic.* **62** (1978), 104-113.

101. S. Reich, Product formulas, nonlinear semigroups, and accretive operators, *J. Functional Anal.* **36** (1980), 147-168.
102. S. Reich, Convergence and approximation of nonlinear semigroups, *J. Math. Anal. Applic.* **76** (1980), 77-83.
103. B. Rzepecki, On measures of noncompactness in topological vector spaces, *Comment. Math. Univ. Carolin.* **23** (1982), 105-116.
104. M. Samimi and V. Lakshmikantham, General uniqueness criteria for ordinary differential equations, *Appl. Math. and Comput.* **12** (1983), 77-88.
105. S. H. Saperstone, *Semidynamical Systems in Infinite Dimensional Spaces*, Applied Mathematical Sciences **37**, New York: Springer-Verlag, 1981.
106. D. H. Sattinger, Monotone methods in nonlinear elliptic and parabolic boundary value problems, *Indiana Univ. Math. J.* **21** (1972), 979-1000.
107. E. Schechter, One-sided continuous dependence of maximal solutions, *J. Diff. Eqns.* **39** (1981), 413-425.
108. E. Schechter, Existence and limits of Carathéodory-Martin evolutions, *Nonlin. Anal.* **5** (1981), 897-930.
109. E. Schechter, Interpolation of nonlinear partial differential operators and generation of differentiable evolutions, *J. Diff. Eqns.* **46** (1982), 78-102.
110. E. Schechter, Necessary and sufficient conditions for convergence of temporally irregular evolutions, *Nonlin. Analysis Theory Methods Applicns.* **8** (1984), 133-153.
111. E. Schechter, Sharp convergence rates for nonlinear product formulas, *Mathematics of Computation* **43** (1984), 135-155.
112. E. Schechter, Compact perturbations of linear m-dissipative operators which lack Gihman's property, in *Nonlinear Semigroups, Partial Differential Equations and Attractors (proc. conference at Howard University, Washington D.C., 1985)*, ed. by T. L. Gill and W. W. Zachary, *Springer Lecture Notes in Mathematics* **1248** (1987), 142-161.
113. S. Schmidt, Zwei Existenzsätze für gewöhnliche Differentialgleichungen, preprint.
114. S. Schwabik, Stetige Abhängigkeit von einem Parameter für ein Differentialgleichungssystem mit Impulsen, *Czech. Math. J.* **21** (1971), 198-212. (*Math. Reviews* **44** no.7019.)
115. S. Schwabik, On a modified sum integral of Stieltjes type, *Časopis Pěst. Mat.* **98** (1973), 274-277.
116. D. R. Smart, *Fixed points*, Cambridge Tracts in Math. **66** (1974), Cambridge U. Press, Cambridge.
117. G. I. Stassinopoulos and R. B. Vinter, Continuous dependence of solutions of a differential inclusion on the right hand side with applications to stability of optimal control problems, *SIAM J. on Control and Applicns.* **17** (1979), 432-449.
118. V. Ya. Stecenko, K-regular cones, *Dokl. Akad. Nauk SSSR* **136** (1961), 1038-1040 (Russian); translated as *Soviet Math. Dokl.* **2** (1961), 170-172.

119. S. Szufla, Measure of non-compactness and ordinary differential equations in Banach spaces, *Bull. Acad. Polon. Sci., Sér. sci. math. astrono. phys.* **19** (1971), 831-835.
120. T. Takahashi, Semigroups of nonlinear operators and invariant sets, *Hiroshima Math. J.* **10** (1980), 55-67.
121. A. A. Tolstonogov, On differential inclusions in Banach spaces and continuous selections, *Dokl. Akad. Nauk. CCCP* **244** (1979), 1088-1092. (Russian). Translated to English in *Soviet Math. Dokl.* **20** (1979), 186-190.
122. C. Vârsan, Continuous dependence and time change for Ito equations, *J. Diff. Eqns.* **58** (1985), 295-306.
123. P. Volkmann, Équations différentielles ordinaires dans les espaces des fonctions bornées, *Czech. Math. J.* **35** (1985), 201-211.
124. P. Volkmann, Existenzsätze für gewöhnliche Differentialgleichungen in Banachräumen, pages 271-287 in *Mathematica ad diem natalem septuagesium quintum data: Festschrift Ernst Mohr zum 75. Geburtstag am 20. April, 1985*, ed. by K.-H. Förster. Universitätsbibliothek, Berlin, 1985.
125. I. I. Vrabie, Compact perturbations of weakly equicontinuous semigroups, pages 267-277 in *Differential Equations in Banach Spaces (Proceedings of a Conference Held in Bologna, 1985)*, ed. by A. Favini and E. Obrecht; Springer Lecture Notes in Mathematics **1223** (1986).
126. G. F. Webb and M. Badii, Nonlinear nonautonomous functional differential equations in L^p spaces, *Nonlin. Anal.* **5** (1981), 203-223.
127. D. V. V. Wend, Existence and uniqueness of solutions of ordinary differential equations, *Proc. Amer. Math. Soc.* **23** (1969), 27-33.
128. N. Wiener, The quadratic variation of a function and its Fourier coefficients, *J. Math. Phys. Sci.* **3** (1924), 73-94.
129. K. Yosida, *Functional Analysis*, Springer-Verlag, Berlin, first and later editions, 1964 and later.

THE TRANSIENT SEMICONDUCTOR PROBLEM WITH GENERATION TERMS, II[1]

Thomas I. Seidman

Department of Mathematics and Statistics
University of Maryland Baltimore County
Baltimore, MD 21228

ABSTRACT

We consider a modified time-dependent van Roosbroeck system with saturating velocities and a source term subject to assumptions designed to permit a standard generation term corresponding to impact ionization. This is then a nonlinear parabolic system with a right hand side of the form $S = S(., \mathbf{u}, \nabla\mathbf{u}, \mathcal{E})$, coupled with the map $\mathbf{L} : \mathbf{u} \mapsto \mathcal{E}$ defined by the Poisson equation at each t. The analysis leads to existence of solutions for all time as well as uniqueness (and continuous dependence on data) under stronger conditions.

KEYWORDS: Semiconductors, device modeling, transient, van Roosbroeck, impact ionization, nonlinear, partial differential equations, system, parabolic, existence, uniqueness.

AMS SUBJECT CLASSIFICATIONS: **35K60, 35Q20, 78A35.**

1 Introduction

We wish here to consider basic issues of existence and uniqueness for a standard model of semiconductor device physics. The 'drift/diffusion' model treated is a version of the van Roosbroeck model [15], [13] permitting the consideration of generation terms corresponding to impact ionization. From a mathematical viewpoint, the essential feature of this consideration is that it precludes the use of maximum principle arguments which formed an essential part of previous work (compare, e.g., the approaches of [6], [8], [12], [5], [3], involving more restricted models without generation terms). While specifically motivated by the semiconductor setting, the model is general enough to be applicable to a wide variety of physical situations involving reaction/diffusion systems of charged particles for which convection is determined by the electrostatic field (e.g., electrophoresis).

The results presented here represent an improved version of [10] in that we now have an existence proof without the geometric restriction imposed there and with the possibility of field-dependent diffusion. Additionally, under the same

[1]This research was partially supported by the Air Force Office of Scientific Research under grants #AFOSR-82-0271 and #AFOSR-87-0190.

hypotheses as were used in [10] we can now show uniqueness by a somewhat more constructive version of the argument in [10].

The relevant variables are $\mathbf{u} = [u_1, \ldots, u_K]$ (*concentrations* for K species of charged particles) and $\mathcal{E} = \mathcal{E}_{tot} - \mathcal{E}_0$ (interactive contribution to the *electrostatic field*). The *drift/diffusion model* for semiconductor device physics consists of two parts: (1) the Poisson equation for the electrostatic potential ψ_{tot} (giving the field $\mathcal{E}_{tot}$ as $-\nabla\psi_{tot}$; we actually work with the *linear* map $\mathbf{L} : \mathbf{u} \mapsto \mathcal{E} = \mathcal{E}_{tot} - \mathcal{E}_0$ defined by the Poisson equation after subtracting the known inhomogeneities) and (2) the conservation laws governing the set $\mathbf{u}$ of concentrations.

The next section presents the basic formulation to be used and two sets of hypotheses **(H-1)**, **(H-2)**. In Section 3 we obtain existence of solutions for this system under the first (quite mild) set of hypotheses **(H-1)**, using an approach based on the Schauder Fixpoint Theorem. Then, in Section 4, it is shown under the somewhat stronger set of hypotheses **(H-2)** that a related Picard map is contractive with respect to a suitably constructed metric, giving uniqueness as well by the Banach Contractive Mapping Principle. The final section discusses briefly the relation of the present results to the physical setting and hopes for the future.

Notation: Spatially, we will be working with $\mathbf{u}$ on a fixed bounded region Ω in $\mathbb{R}^d$ whose boundary $\partial\Omega$ is partitioned into $\Gamma = \Gamma_D$ (at which we prescribe Dirichlet conditions) and $\Gamma_N = \partial\Omega\backslash\Gamma$ (at which we impose flux conditions). The evolution problem will be considered on a fixed bounded interval $[0,T]$ — but with $T > 0$ arbitrarily large so our results are actually global; we set $\mathcal{Q} = \mathcal{Q}_T := [0,T] \times \Omega$ and $\Sigma := [0,T] \times \partial\Omega$. We use $\|\cdot\|_p$ for the $L^p(\Omega)$-norm ($p = 2$ if the subscript is omitted). We abbreviate $L^\infty([0,T] \to L^p(\Omega))$, for example, by $L^\infty(\to L^p)$ and do not distinguish notationally between this as a space of $\mathbb{R}$-valued functions and as a space of $\mathbb{R}^K$-valued functions. We also permit the useful confusion between $\mathbf{L} : \mathbf{u} \mapsto \mathcal{E}$ as a map between functions on Ω (with t appearing only parametrically) and the induced map between functions on $\mathcal{Q}$. We assume that Ω is connected and Γ_D is a suitably large part of $\partial\Omega$ so we can simply take $\|\nabla u\|$ as the norm $\|u\|_{(1)}$ on the space $\mathcal{H}^1_\Gamma := \{v \in H^1(\Omega) : v|_\Gamma = 0\}$ (i.e., the $H^1(\Omega)$-closure of $\{v \in C^\infty(\bar{\Omega}) : \operatorname{supp}\{v\} \cap \Gamma = \emptyset\}$). Finally, we introduce the space $\mathcal{X}_2 := L^\infty(\to L^2) \cap L^2(\to H^1)$ with the norm

$$\|v\|_{\mathcal{X}} := (\sup_{[0,T]}\{e^{-rt}\int_\Omega |v(t)|^2\} + \int_{\mathcal{Q}} e^{-rt}|\nabla v|^2)^{1/2} \tag{1.1}$$

where r is to be specified later (large enough).

Acknowledgments: Apart from thanks to T. Gill and W. Zachary as organizers of this conference and to the AFOSR for its support, a number of other acknowledgments are in order: to P. Markowich for the original stimulation to this work during a visit to T. U. Wien, to the Institute for Mathematics and its Applications for its hospitality and stimulating atmosphere during the Workshop on Computational Aspects of VLSI Design and Semiconductor Device Simulation (April-May, 1987), to Jan Klika for his aid in the writing of this report, to Marjorie for an undisclosed variety of contributions.

2 Formulation and Hypotheses

In this section we present the formalism within which we will be working and state two sets of hypotheses: a weaker set (**H-1**) which will suffice for the existence result in Section 3 and a stronger set (**H-2**) under which we can prove uniqueness in Section 4.

The electrostatic field $\mathcal{E}_{tot}$ is determined as $\mathcal{E}_{tot} = -\nabla\psi_{tot}$ where the potential ψ_{tot} satisfies the Poisson equation

$$-\nabla\cdot\underline{\epsilon}\nabla\psi_{tot} = N + \mathbf{q}\cdot\mathbf{u} = N + \sum q_k u_k \tag{2.1}$$

and the associated boundary conditions; here, N is the net charge distribution due to the 'doping' which defines the device. Note that it may be appropriate to consider (2.1) as holding for a spatial region Ω' other than the Ω considered for the concentrations. This region Ω' should then contain Ω and we assume that $\mathbf{u}$, as it appears on the right of (2.1), is given *a priori* on $\Omega'\setminus\Omega$. The boundary conditions for ψ are, of course, to be imposed at $\partial\Omega'$ and might include 'conditions at infinity'.

If we let ψ_0 be the solution of (2.1) corresponding to $\mathbf{u} = 0$ and set $\mathcal{E}_0 = -\nabla\psi_0$ then $\mathcal{E}_0$ contains all the relevant information given by N and the inhomogeneous boundary data. Setting $\mathcal{E} = \mathcal{E}_{tot} - \mathcal{E}_0$, we now have $\mathcal{E} = -\nabla\psi$ (restricted to Ω, if necessary) where ψ is the solution of the Poisson equation

$$-\nabla\cdot\underline{\epsilon}\nabla\psi = \mathbf{q}\cdot\mathbf{u} \tag{2.2}$$

with *homogeneous* boundary conditions of the same form as those used for (2.1). Clearly the map $\mathbf{L}: \mathbf{u} \mapsto \mathcal{E}$ defined in this way is *linear*.

For the concentration vector $\mathbf{u} = [u_1, \ldots, u_K]$, we assume a coupled system of conservation laws of the usual general form:

$$\dot{u}_k - \nabla\cdot[D_k\nabla u_k + u_k\mathcal{V}_k] = S_k \tag{2.3}$$

for $k = 1, \ldots, K$. Here, suppressing the index k as we will do in general, $J := D\nabla u + u\mathcal{V}$ is the particle flux, S is the (net) *source*, D is the *diffusion coefficient*, and $\mathcal{V}$ is the *drift velocity*.

We also have *constitutive relations*

$$D = D_k(\cdot, \mathcal{E}), \qquad \mathcal{V} = \mathcal{V}_k(\cdot, \mathcal{E}), \qquad S = S_k(\cdot, \mathbf{u}, \nabla\mathbf{u}, \mathcal{E}). \tag{2.4}$$

We will not distinguish notationally between the *given* constitutive relation, $S = S_k$ as a function: $\mathcal{Q} \times \mathbb{R}^K \times \mathbb{R}^{Kd} \times \mathbb{R}^d \to \mathbb{R}$ and the derived 'substituted function' — $S(\cdot, \mathbf{u}(\cdot), \nabla \mathbf{u}(\cdot), \mathcal{E}(\cdot)) : \mathcal{Q} \to \mathbb{R}$. Note also that in writing the dependence of D_k in terms of $\mathcal{E}$ (rather than $\mathcal{E}_{tot}$), etc., we are taking advantage of the fact that $\mathcal{E}_0$ is known *a priori*. The relevant hypotheses about N, Ω' (in relation to Ω), the boundary conditions and data for (2.1) are subsumed in our hypotheses on $\mathbf{L}$ and the functions of (2.4) (Recall that we omitted temperature coupling for expository simplicity — else $D_k, \mathcal{V}_k, S_k$ would also depend on the temperature).

We next state the set of hypotheses **(H-1)**, under which we will show existence in the next section. The generic constant γ need not be the same at each appearance.

(H-1.i) We have:
- Ω is bounded and connected in $\mathbb{R}^d$; $\partial\Omega$ is 'smooth enough' for the relevant Sobolev spaces and embeddings to make sense (e.g., satisfying a 'cone condition').
- All the functions $D_k(\cdots)$, $\mathcal{V}_k(\cdots)$, $S_k(\cdots)$ are measurable on $\mathcal{Q}$ and are continuous with respect to all other variables.

(H-1.ii) There is a constant $\delta > 0$ such that $\delta \le D_k \le \gamma$.

(H-1.iii) $\mathcal{V}_k$ is bounded: $|\mathcal{V}_k| \le \gamma$.

(H-1.iv) One has $|S_k(\cdot, r, \xi, \eta)| \le S^0(\cdot) + \gamma[|r| + |\xi|]$ for all η with some $S^0 \in L^2(\mathcal{Q})$.

(H-1.v) For fixed $\mathcal{E}(\cdot)$, S_k is (uniformly) Lipschitz continuous:

$$|S_k(\cdot, r, \xi, \eta) - S_k(\cdot, r', \xi', \eta)| \le \gamma[|r - r'| + |\xi - \xi'|].$$

(H-1.vi) The linear operator $\mathbf{L}$, defined by (2.2) with the homogeneous boundary conditions, is continuous from $L^2(\Omega)$ to $L^2(\Omega)$ (equivalently, from $L^2(\mathcal{Q})$ to $L^2(\mathcal{Q})$).

(H-1.vii) The data $\hat{u}_k$ are in $L^2(\to H^1) \cap H^1(\to H^{-1})$.

Under the stronger set of hypotheses **(H-2)** it will be possible to show, in Section 4, that a Picard iteration is contractive (with respect to a suitably chosen metric) giving uniqueness as well.

(H-2.i) = **(H-1.i)** with $d = 3$ so $H^1(\Omega) \hookrightarrow L^6(\Omega)$.

(H-2.ii) D_k is independent of $\mathcal{E}$ and $0 < \delta \le D_k \le \gamma$.

(H-2.iii) $\mathcal{V}_k$ is Lipschitzian: $|\mathcal{V}_k(\cdot, \eta) - \mathcal{V}_k(\cdot, \eta')| \le \gamma|\eta - \eta'|$ as well as bounded: $|\mathcal{V}_k| \le \gamma$.

(H-2.iv) = **(H-1.iv)**.

(H-2.v) S_k satisfies a Lipschitz condition of the form:

$$|S_k(\cdot, r, \xi, \eta) - S_k(\cdot, r', \xi', \eta')| \le \gamma(|r - r'| + |\xi - \xi'| + (1 + |r| + |\xi|)|\eta - \eta'|).$$

(H-2.vi) For some $p \le 6$, $\mathbf{L}$ is compact from $L^p(\Omega)$ to $C(\bar{\Omega})$.

(H-2.vii) With $2 \le p \le 6$ as in **(H-2.vi)**, one has $\hat{u}_k, \nabla\hat{u}_k \in L^p(\mathcal{Q})$ and $\dot{\hat{u}}_k \in L^p(\to L^{3p/(2p+1)})$.

We say that u is a (weak) solution of (2.3) if one has:

$$(2.5) \qquad (u - \hat{u}) \in \mathcal{Z}_0 := \{v \in L^2(\to \mathcal{H}^1_\Gamma) \cap H^1(\to H^{-1}) : v|_{t=0} = 0\},$$

$$(2.6) \qquad \int_\mathcal{Q} \dot{u}v + \int_\mathcal{Q} D\nabla u \cdot \nabla v = \int_\mathcal{Q} Sv - \int_\mathcal{Q} u\mathcal{V} \cdot \nabla v \qquad \forall v \in \mathcal{Z}_0.$$

Implicit in (2.5) are the initial conditions: $u = \hat{u}$ at $t = 0$ and Dirichlet conditions: $u = \hat{u}$ at $\Gamma = \Gamma_D$; implicit in (2.6) are the homogeneous flux conditions: $J \cdot \mathbf{n} = 0$ on $\partial\Omega \backslash \Gamma_D = \Gamma_N$.

3 Existence

In this section we prove, under **(H-1)**, the existence of a solution of the coupled system: (2.3) ($k = 1, \ldots, K$) with $\mathcal{E} = \mathbf{Lu}$ and (2.4). Our approach here is to define the map $\mathbf{T} : \mathcal{E} \mapsto \mathbf{u}$ by solving (2.3) with $\mathcal{E}$ taken as *given* and then to apply the Schauder Fixpoint Theorem to

$$(3.1) \qquad \mathbf{L} \circ \mathbf{T} : \quad \mathcal{E} \stackrel{\mathbf{T}}{\longmapsto} \mathbf{u} \stackrel{\mathbf{L}}{\longmapsto} \mathcal{E} : \; L^2(\mathcal{Q}) \longrightarrow \mathcal{X} \longrightarrow L^2(\mathcal{Q})$$

with $\mathcal{X} = \mathcal{X}_2 := L^\infty(\to L^2) \cap L^2(\to H^1)$.

Theorem 1: Under the hypotheses **(H-1)**, the system (2.3) for $k = 1, \ldots, K$ has a weak solution $[\mathcal{E}^*, \mathbf{u}^*]$ in $L^2(\mathcal{Q}) \times \mathcal{Z}$ with $\mathcal{E}^* = \mathbf{Lu}^*$ and the constitutive relations (2.4).

Proof: We divide the proof into the following five steps:

[**1**] $\mathbf{T}$ is well-defined; set $\mathcal{K}_1 := \{\mathbf{T}(\mathcal{E}) : \mathcal{E} \in L^2(\mathcal{Q})\} \subset \mathcal{X}$;
[**2**] $\mathcal{K}_1$ is bounded in $\mathcal{X}$;
[**3**] $\mathcal{K}_2 := \{\mathbf{Lu} : \mathbf{u} \in \mathcal{K}_1\}$ is precompact in $L^2(\mathcal{Q})$; set $\mathcal{K} := \overline{\text{co}}\mathcal{K}_2$;
[**4**] $\mathbf{T} : L^2(\mathcal{Q}) \to \mathcal{X}$ is continuous;
[**5**] $\mathbf{L} \circ \mathbf{T}$ has a fixpoint $\mathcal{E}^* \in \mathcal{K}$.

The desired solution pair is $[\mathcal{E}^*, \mathbf{u}^*]$ with $\mathbf{u}^* := \mathbf{T}(\mathcal{E}^*)$. Step [**1**] is fairly standard (detailed proof omitted); [**3**] and [**5**] are quick. The principal efforts are for [**2**] (the original estimate — exactly as in [10]) and the new argument for [**4**].

[**1**] We are defining $\mathbf{T}(\bar{\mathcal{E}})$ as the (unique) weak solution $\mathbf{u} \in [\bar{\mathbf{u}} + \mathcal{Z}_0]$ of

$$(3.2) \qquad \dot{u}_k - \nabla \cdot [\bar{D}_k \nabla u_k + u_k \bar{\mathcal{V}}_k] = \bar{S}_k(\cdot, \mathbf{u}, \nabla \mathbf{u}) \qquad (k = 1, \ldots, K)$$

where $\bar{D}$, $\bar{\mathcal{V}}$, $\bar{S}$ are defined by fixing $\mathcal{E} = \bar{\mathcal{E}}$ in (2.4). That such a solution exists is fairly standard in view of the assumed Lipschitz condition **(H-1.v)** with **(H-1.iv)** — one considers the Picard map: $\bar{\mathbf{u}} \mapsto \mathbf{u}$ defined by the *linear equation* obtained on replacing $\bar{S}(\cdot,\mathbf{u},\nabla\mathbf{u})$ in (3.2) by $\bar{S}(\cdot,\bar{\mathbf{u}},\nabla\bar{\mathbf{u}})$ (still with $\mathcal{E}$ fixed as $\bar{\mathcal{E}}$ so $\bar{D},\bar{\mathcal{V}}$ are now fixed) and shows contractivity with respect to an exponentially weighted norm (1.1), using the estimate: $\|\bar{S}(\cdot,\bar{\mathbf{u}},\nabla\bar{\mathbf{u}})-\bar{S}(\cdot,\bar{\mathbf{u}}',\nabla\bar{\mathbf{u}}')\|\le M\|\bar{\mathbf{u}}-\bar{\mathbf{u}}'\|_{(1)}$.

[2] Set $z=z_k=u_k-\hat{u}_k$ so, formally,

$$\dot{z}=\nabla\cdot[D\nabla z+z\mathcal{V}]+(\bar{S}-\nabla\cdot[D\hat{u}+\hat{u}\mathcal{V}]+\hat{u}^{\cdot}) \tag{3.3}$$

with a weak formulation corresponding to (2.6). We take $v=2e^{-rt}z|_{[0,\tau]}$ as test function in this weak formulation. Note that $v\dot{z}=([e^{-rt}z^2]^{\cdot}+re^{-rt}z^2)$ on $(0,\tau)$ and that **(H-1.iv)** gives:

$$\|S(\cdot,\mathbf{u},\nabla\mathbf{u})\|\le M[1+\|\mathbf{z}\|_{(1)}]. \tag{3.4}$$

We will also estimate: $\|\mathbf{z}\|_{(1)}\|v\|\le\frac{\delta}{MK}e^{-rt}\|\mathbf{z}\|_{(1)}^2+\frac{MK}{\delta}e^{-rt}\|z\|^2$, etc. Here and henceforth, M, etc., are generic 'constants' which may depend only on Ω, norm bounds on $\hat{\mathbf{u}}$ as in **(H-1.vii)**, the modulus of ellipticity δ, and the generic 'constant' γ of **(H-1)**; specifically, M will *not* depend on r — nor, of course, on $\mathcal{E}$ or $\mathbf{u}$. It follows that (replacing τ by t):

$$\begin{aligned}e^{-rt}\|z(t)\|^2+r\int_0^t e^{-rs}\|z\|^2+2\delta\int_0^t e^{-rs}\|z\|_{(1)}^2\\ \le\delta\int_0^t e^{-rs}\|z\|_{(1)}^2+\tfrac{\delta}{2K}\int_0^t e^{-rs}\|\mathbf{z}\|_{(1)}^2+\int_0^t e^{-rs}[M+M'\|z\|^2]\end{aligned}$$

for each k. Taking $r>M'$, this gives

$$e^{-rt}\|z_k(t)\|^2+\delta\int_0^t e^{-rs}\|z_k\|_{(1)}^2\le M+\tfrac{\delta}{2K}\int_0^t e^{-rs}\|\mathbf{z}\|_{(1)}^2$$

whence, summing over k, we obtain the 'final' estimate

$$e^{-rt}\|\mathbf{z}(t)\|^2+\tfrac{\delta}{2}\int_0^t e^{-rs}\|\mathbf{z}\|_{(1)}^2\le KM \tag{3.5}$$

which clearly bounds $\mathbf{z}$ (and so $\mathbf{u}=\mathbf{T}(\mathcal{E})=[\mathbf{z}+\hat{\mathbf{u}}]$) in $\mathcal{X}=\mathcal{X}_2$. Note that the bound we have obtained is independent of $\mathcal{E}\in L^2(\mathcal{Q})$ so $\mathcal{K}_1$ is a bounded set in $\mathcal{X}$. Note that, while we have only obtained such a bound when $r>M'$, these $\mathcal{X}_2$-norms are equivalent so the *fact* that $\mathcal{K}_1$ is a bounded set is independent of r; we will be free to choose r independently in any subsequent calculation.

[3] Given the bound above on $\mathbf{u}=\mathbf{T}(\mathcal{E})$ in $L^2(\to H^1)$, we have bounds on each S_k and J_k in the space $L^2(\mathcal{Q})$ and so a bound on $\dot{u}_k=S_k+\nabla\cdot J_k$ in $L^2(\to H^{-1})$. Since the embedding $H^1(\Omega)\hookrightarrow L^2(\Omega)$ is compact, application of the Aubin Compactness Theorem [1] gives precompactness of $\mathcal{K}_1$ in $L^2([0,T]\to L^2(\Omega))=L^2(\mathcal{Q})$. Since, by **(H-1.vi)**, $\mathbf{L}$ is continuous from $L^2(\mathcal{Q})$ to $L^2(\mathcal{Q})$, it follows that the image

$\mathcal{K}_2 = \mathbf{L}\mathcal{K}_1$ must be precompact so the set $\mathcal{K} = \overline{\mathrm{co}}\mathcal{K}_2$ is compact and convex in $L^2(\mathcal{Q})$. Note that $\mathbf{LT}(\mathcal{K}) \subset \mathcal{K}_2 \subset \mathcal{K}$ so $\mathcal{K}$ is invariant.

[4] Now suppose we have a sequence $\{\mathcal{E}^\nu\}$ in $L^2(\mathcal{Q})$ converging in $L^2(\mathcal{Q})$ to some $\mathcal{E}^0$; we write $\mathbf{u}^\nu := \mathbf{T}(\mathcal{E}^\nu)$, etc. We wish to show that $\mathbf{u}^\nu \to \mathbf{u}^0 = \mathbf{T}(\mathcal{E}^0)$ in the sense of $\mathcal{X}_2$.

The key to the argument is to assume, first, that $\mathcal{E}^\nu \to \mathcal{E}^0$ *pointwise* ae on $\mathcal{Q}$. This implies convergence ae on $\mathcal{Q}$ for $D, \mathcal{V}$ as well. It is not difficult to see that:

$$\|[D(\cdot,\mathcal{E}^\nu) - D(\cdot,\mathcal{E}^0)]\nabla u^0\|^2 = \textstyle\int_{\mathcal{Q}} |[D(\cdot,\mathcal{E}^\nu) - D(\cdot,\mathcal{E}^0)]\nabla u^0|^2 \to 0$$

since we have convergence to 0 ae on $\mathcal{Q}$ for the integrand and, with $\nabla \mathbf{u}^0$ fixed in $L^2(\mathcal{Q})$, **(H-1.ii)** gives the dominating bound $2\gamma^2|\nabla \mathbf{u}^0|^2$. Similarly, using **(H-1.iii)** and **(H-1.iv)** for domination, one obtains $L^2(\mathcal{Q})$ convergence:

$$u^0\mathcal{V}(\cdot,\mathcal{E}^\nu) \to u^0\mathcal{V}(\cdot,\mathcal{E}^0), \quad S(\cdot,\mathbf{u}^0,\nabla\mathbf{u}^0,\mathcal{E}^\nu) \to S(\cdot,\mathbf{u}^0,\nabla\mathbf{u}^0,\mathcal{E}^0).$$

We also recall that **(H-1.v)** gives:

$$\|S(\cdot,\mathbf{u}^\nu,\nabla\mathbf{u}^\nu,\mathcal{E}^\nu) - S(\cdot,\mathbf{u}^0,\nabla\mathbf{u}^0,\mathcal{E}^\nu)\| \le M\|\mathbf{z}^\nu\|_{(1)}.$$

Let M_ν be a bound for $\|[D(\mathcal{E}^\nu) - D(\mathcal{E}^0)]\nabla u^0\|^2$, etc.; note that we have just shown $M_\nu \to 0$.

Set $z = z^\nu = u^\nu - u^0$ much as in (3.3) and similarly take $v = 2e^{-rt}z|_{[0,\tau]}$ in the weak formulations (2.6) for u^ν, u^0; then subtract and estimate much as in **[2]** above. One obtains

$$\begin{aligned} &e^{-rt}\|z(t)\|^2 + r\textstyle\int_0^t e^{-rs}\|z\|^2 + 2\delta\int_0^t e^{-rs}\|z\|_{(1)}^2 \\ &\le \delta\textstyle\int_0^t e^{-rs}\|z\|_{(1)}^2 + \frac{\delta}{2K}\int_0^t e^{-rs}\|\mathbf{z}\|_{(1)}^2 + M_\nu + M'\int_0^t e^{-rs}\|z\|^2 \end{aligned} \tag{3.6}$$

from which, again choosing $r > M'$ and summing over k, one obtains a bound on $\|\mathbf{z}^\nu\|_{\mathcal{X}}$ which goes to 0 as $M_\nu \to 0$. Thus, $\mathcal{E}^\nu \to \mathcal{E}^0$ pointwise ae on $\mathcal{Q}$ implies $\mathbf{u}^\nu = \mathbf{T}(\mathcal{E}^\nu) \to \mathbf{u}^0 = \mathbf{T}(\mathcal{E}^0)$ in $\mathcal{X}$.

For arbitrary L^2 convergence: $\mathcal{E}^\nu \to \mathcal{E}^0$, we note that one can always find *subsequences* which converge ae on $\mathcal{Q}$. Since the limit is specified uniquely, one has the desired convergence for the full sequence: $\mathbf{T}(\mathcal{E}^\nu) \to \mathbf{T}(\mathcal{E}^0)$ in $\mathcal{X} = \mathcal{X}_2$, proving the continuity of $\mathbf{T} : L^2(\mathcal{Q}) \to \mathcal{X}_2$. *A fortiori*, noting **(H-1.vi)**, one has continuity of $\mathbf{L} \circ \mathbf{T}$ from $L^2(\mathcal{Q})$ to $\mathcal{K}_2 \subset L^2(\mathcal{Q})$.

Since $\mathcal{K}$ is compact and $\mathbf{T}$ is continuous, we note that $\mathcal{K}_3 := \{\mathbf{u} = \mathbf{T}(\mathcal{E}) : \mathcal{E} \in \mathcal{K}\}$ is compact in $\mathcal{X}_2$. For future reference, observe that $\{\nabla\mathbf{u} : \mathbf{u} \in \mathcal{K}_3\}$ is then compact in $L^2(\mathcal{Q})$ whence, setting $\sigma^2(t) = \sigma^2(t;\mathbf{u}) := 1 + \|\mathbf{u}(t,\cdot)\|_{(1)}^2$, we have σ in a compact subset of $L^2(0,T)$; thus, $\int_0^t e^{-\mu s}\sigma^2(s;\mathbf{u})\,ds \to 0$ uniformly in $[t,\mathbf{u}] \in [0,T] \times \mathcal{K}_3$ as $\mu \to \infty$. We have shown that, for any $\varepsilon > 0$,

$$0 \le \int_0^t e^{-\mu s}[1 + \sigma^2(s;\mathbf{u})]\,ds < \varepsilon \qquad 0 \le t \le T, \mathbf{u} \in \mathcal{K}_3 \tag{3.7}$$

for a suitably chosen $\mu = \mu(\varepsilon)$.

[5] Restricting, $\mathbf{L} \circ \mathbf{T}$ to the compact, convex, invariant set $\mathcal{K}$, the Schauder Theorem applies to give existence of a fixpoint $\mathcal{E}^* \in \mathcal{K}$. This completes the existence proof on setting $\mathbf{u}^* := \mathbf{T}(\mathcal{E}^*)$. □

Theorem 2: Let $\{\hat{\mathbf{u}}^\nu\}$ be as in (**H-1.vii**) with $\hat{\mathbf{u}}^\nu \to \hat{\mathbf{u}}^0$ and let $\{\mathbf{u}^\nu\}$ be corresponding solutions of the coupled system. Then (for a subsequence) we have $[\mathcal{E}^\nu, \mathbf{u}^\nu] \to [\mathcal{E}^0, \mathbf{u}^0]$ where $[\mathcal{E}^0, \mathbf{u}^0]$ is some solution corresponding to $\hat{\mathbf{u}}^0$.

Proof: From [**4**], above, we know $\mathcal{K} \times \mathcal{K}_3$ is compact so we can find a convergent subsequence $[\mathcal{E}^\nu, \mathbf{u}^\nu] \to [\mathcal{E}^0, \mathbf{u}^0]$ (for some $[\mathcal{E}^0, \mathbf{u}^0]$) with, in addition, pointwise convergence ae on $\mathcal{Q}$ for $\mathcal{E}^\nu, \nabla \mathbf{u}^\nu$. Essentially the same argument as in Step [**4**] of the proof for Theorem 1 shows that $\mathbf{u}^0$ is a solution of (2.3) corresponding to $\mathcal{E}^0, \hat{\mathbf{u}}^0$. Clearly we also have $\mathcal{E}^0 = \lim \mathbf{L}\mathbf{u}^\nu = \mathbf{L}\mathbf{u}^0$, completing the proof. □

4 Uniqueness

In this section we prove, under essentially the same set of hypotheses (**H-2**) used in [10], that the solution of the coupled system whose existence was shown in Section 3 is actually unique. Answering a question raised by Hans Weinberger at the original conference presentation of [10], our present approach, somewhat more constructive than there, applies the Contractive Mapping Theorem to the map

$$\mathbf{L} \circ \mathbf{T} : \quad \mathcal{E} \xmapsto{\mathbf{T}} \mathbf{u} \xmapsto{\mathbf{L}} \mathcal{E} : \ \mathcal{C}(\bar{\mathcal{Q}}) \longrightarrow \mathcal{X}_p \longrightarrow \mathcal{C}(\bar{\mathcal{Q}})$$

where $\mathcal{X}_p := L^\infty(\to L^p) \cap \mathcal{X}_2$ with p as in (**H-2.vi**). Note that the big difference in this setting from that of (3.1) is that we now wish to consider $\mathcal{E} \in \mathcal{C}(\bar{\mathcal{Q}})$ and so will need an estimate for $\mathbf{u}$ in $L^\infty(\to L^p(\Omega))$ to apply (**H-2.vi**).

Theorem 3: Under the hypotheses (**H-2**), the system (2.3) for $k = 1, \ldots, K$ has a unique weak solution $[\mathcal{E}^*, \mathbf{u}^*] \in L^2(\mathcal{Q}) \times \mathcal{Z}$ with $\mathcal{E}^* = \mathbf{L}\mathbf{u}^*$ and the constitutive relations (2.4). This solution is obtainable by iterating the map $\mathbf{L} \circ \mathbf{T}$, starting with any measurable $\mathcal{E}$.

Proof: We divide the proof into the following five steps, using notation from the proof of Theorem 1:

[**1**] $\mathcal{K}_1$ is bounded in $L^\infty(\to L^p)$, hence in $\mathcal{X}_p$;
[**2**] $\mathcal{K}$ is compact in $\mathcal{C}(\bar{\mathcal{Q}})$; after 2 iterations of $\mathbf{L} \circ \mathbf{T}$, starting with any measurable initial $\mathcal{E}$, the iterates are in $\mathcal{K}$;
[**3**] principal estimates for differences;
[**4**] $\mathbf{L} \circ \mathbf{T}$ is contractive on $\mathcal{K}$ with respect to a suitably chosen metric for $\mathcal{C}(\bar{\mathcal{Q}})$;
[**5**] There is a unique solution $[\mathcal{E}^*, \mathbf{u}^*]$, continuously dependent on the data $\hat{\mathbf{u}}$.

Here, [**2**] and [**5**] are quick; [**1**] is as in [10]; the interesting steps are [**3**] and [**4**]. For [**1**] and [**3**] we have somewhat similar estimation arguments, involving $\mathbf{z} = \mathbf{u} - \hat{\mathbf{u}}$ or $\mathbf{z} = \mathbf{u}' - \mathbf{u}''$, and will use a common notation. Setting

$$y = y_k := |z|^{p/2}, \qquad w := |z|^{p-1}\operatorname{sgn} z = z|z|^{p-2},$$

it follows that

$$
\begin{array}{llll}
(4.1) & zy & = |z|^p = w^2, & |z| = y^{\vartheta}, \\
(4.2) & \nabla w & = (2-\vartheta) y^{1-\vartheta}\nabla y, & w\nabla z = \vartheta\, y \nabla y, \\
(4.3) & p|\nabla w \cdot \nabla z| & = \delta' |\nabla w|^2 \text{ with } \delta' = 2(2-\vartheta)\delta, & \\
(4.4) & v\dot{z} & = [e^{-rt}w^2]^{\cdot} + re^{-rt}w^2 \text{ for } v := pe^{-rt}w|_{[0,\tau]} &
\end{array}
$$

where, for convenience, we have set $\vartheta := 2/p \leq 1$.

[1] Let $\mathbf{z} = \mathbf{u} - \hat{\mathbf{u}}$ as in (3.3) and take v as in (4.1) for the weak formulation. This gives

$$
\begin{aligned}
e^{-rt}\|y(t)\|^2 &+ r\int_0^t e^{-rs}\|y\|^2 + \delta'\int_0^t e^{-rs}\|y\|^2_{(1)} \\
&\leq p\int_0^t e^{-rs}\int_\Omega [Sw + \gamma|z||\nabla w| + |\hat{J}\,||\nabla w| + \hat{u}^{\cdot} w]
\end{aligned}
$$

and, as in obtaining (3.5), one gets the desired bound on suitably estimating the terms on the right. For example, since (3.4), (3.5) bound S in $L^2(\mathcal{Q})$, we estimate w in L^2 by $\|w\| \leq \|y\|_p \|y\|^{1-\vartheta}$ and then observe that $H^1(\Omega) \hookrightarrow L^p(\Omega)$ (since we have assumed $d = 3$ and $p \leq 6$) so $\|y\|_p \leq M\|y\|_{(1)}$. The same estimate would have treated the term $\hat{u}^{\cdot} w$ had we assumed $\hat{u} \in L^2(\mathcal{Q})$ in **(H-2.vii)**; for the hypothesis actually imposed we observe that $y^6 = |w|^{3p/(p-1)}$ so

$$
\int_0^t e^{-rs}\int_\Omega \hat{u}^{\cdot} w \leq \varepsilon\int_0^t e^{-rs}\|y\|^2_{(1)} + C_\varepsilon\int_0^t e^{-rs}\|\hat{u}^{\cdot}\|^p_{3p/(2p+1)}
$$

(note that still another mode of estimation would work if we were to assume $\hat{u}^{\cdot} \in L^1(\to L^p)$ instead and these could be combined for alternate possibilities). One similarly treats the term $|\hat{J}\,||\nabla w|$ by using **(H-2.vii)** to estimate $|\hat{J}\,|y^{1-\vartheta}$ in L^2. As in (3.5) one thus bounds y in $\mathcal{X}_2$; clearly, $\|y(t)\|^2 = \|z\|_p^p$ so this bounds $\mathbf{z}$ in $L^\infty(\to L^p)$ and therefore, recalling the bound on $\|\mathbf{u}\|_{(1)}$ from the proof of Theorem 1, bounds $\mathbf{u} = \mathbf{T}(\mathcal{E})$ in $\mathcal{X}_p$, independently of $\mathcal{E}$.

[2] From **(H-2.vi)** it follows that one can introduce a space Y such that the embedding $L^p(\Omega) \hookrightarrow Y$ is compact and $\mathbf{L} : Y \to \mathcal{C}(\bar{\Omega})$ continuous; we may take $Y \hookrightarrow H^{-1}(\Omega)$. Since, for $\mathbf{u} \in \mathcal{K}_1$, we have bounds on $\mathbf{u}$ in $L^\infty(\to L^p)$ and on $\dot{\mathbf{u}}$ in $L^2(\to H^{-1})$, it follows that $\mathcal{K}_1$ is contained in a compact subset of $\mathcal{C}([0,T] \to Y)$ [14][2] so $\mathcal{E} = \mathbf{L}\mathbf{u} \in \mathcal{K}_2$ is in a compact subset of $\mathcal{C}(\to \mathcal{C}(\bar{\Omega})) = \mathcal{C}(\bar{\mathcal{Q}})$ whence $\mathcal{K} := \overline{\text{co}}\mathcal{K}_2$ is also compact in $\mathcal{C}(\bar{\mathcal{Q}})$.

---

[2]Consider the embedding map $\phi : \mathcal{Y} \hookrightarrow \mathcal{Y}'$ where $\mathcal{Y}$ is the Y-closure of the image of the M-ball of $L^p(\Omega)$ and $\mathcal{Y}'$ the corresponding image in $H^{-1}(\Omega)$. $\mathcal{Y}$ is compact and ϕ is a continuous bijection, so ϕ^{-1} is continuous. The bounds give $\mathbf{u}$ in a compact subset of $\mathcal{C}([0,T] \to H^{-1})$ by Arzela-Ascoli and 'lifting' by ϕ^{-1} gives compactness where desired.

[3] Suppose we have $\bar{\mathcal{E}}', \bar{\mathcal{E}}'' \in K$ and, correspondingly, $\mathbf{u}' = \mathbf{T}(\bar{\mathcal{E}}')$, etc.; we set $\mathbf{z} := \mathbf{u}' - \mathbf{u}''$. It is also convenient to set

$$\begin{aligned}
\eta(t) = \eta_k(t) &= \|z_k\|_p = \|y_k\|^\vartheta \ \text{ so } \eta^p = \|y\|^2, \\
\varsigma(t) &= \|\mathbf{z}\|_p = (\textstyle\sum_k \eta_k^p)^{1/p}, \qquad \bar{\varsigma}(t) = \|\bar{\mathcal{E}}' - \bar{\mathcal{E}}''\|_\infty, \\
\hat{\varsigma}(\tau) &= \sup\{\varsigma(t) : 0 \le t \le \tau\}.
\end{aligned}$$

We also let $\tilde{\sigma} \geq 1$ bound pointwise *each* of the (finitely many functions $|u_k|, |\nabla u_k|$, etc.; then set $\sigma^2(t) = 1 + \|\tilde{\sigma}(t,\cdot)\|$. Note that the bound $\|\mathbf{Lu}\|_\infty \le \gamma_* \|\mathbf{u}\|_p$ given by **(H-2.vi)** will give

(4.5)
$$\|\mathcal{E}' - \mathcal{E}''\|_\infty \le \gamma_* \hat{\varsigma}(t)$$

for each t. One then has

$$\begin{aligned}
|u'\mathcal{V}' - u''\mathcal{V}''| &\le M[|z| + \sigma\bar{\varsigma}] \\
|S' - S''| &\le M[|\mathbf{z}, \nabla\mathbf{z}| + \sigma\bar{\varsigma}|.
\end{aligned}$$

from the Lipschitz conditions **(H-2.iii,v)**.

We proceed by taking $v = 2e^{-rt}z|_{[0,\tau]}$ — as in obtaining the continuity of $\mathbf{T}$ via (3.6) but here with the advantage of the Lipschitz conditions which are now imposed. Thus, we get

$$\begin{aligned}
e^{-rt}\|z(t)\|^2 &+ r\int_0^t e^{-rs}\|z\|^2 + 2\delta\int_0^t e^{-rs}\|z\|_{(1)}^2 \\
&\le 2\int_0^t e^{-rs}\int_\Omega (|u'\mathcal{V}' - u''\mathcal{V}''||\nabla z| + |S' - S''||z|) \\
&\le \delta\int_0^t e^{-rs}\|z\|_{(1)}^2 + \tfrac{\delta}{2K}\int_0^t e^{-rs}\|\mathbf{z}\|_{(1)}^2 \\
&\qquad + M'\int_0^t e^{-rs}\|z\|^2 + M\int_0^t e^{-rs}\sigma^2\bar{\varsigma}^2.
\end{aligned}$$

Taking r large enough and summing over k, this gives the first principal estimate:

(4.6)
$$\int_0^t e^{-rs}\|\mathbf{z}\|_{(1)}^2 \le M\int_0^t e^{-rs}\sigma^2\bar{\varsigma}^2.$$

Next take the test function $v = pe^{-rt}w|_{[0,\tau]}$ as in (4.1) and get

$$\begin{aligned}
e^{-rt}\eta^p(t) &+ r\int_0^t e^{-rs}\eta^p + \delta'\int_0^t e^{-rs}\|y\|_{(1)}^2 \\
&\le 2\int_0^t e^{-rs}\int_\Omega (|u'\mathcal{V}' - u''\mathcal{V}''|y^{1-\vartheta}|\nabla y| + |S' - S''|y^{2-\vartheta}) \\
&\le M\int_0^t e^{-rs}([\eta + \sigma\bar{\varsigma}]\eta^{p/2-1}\|y\|_{(1)} + [\|\mathbf{z}\|_{(1)} + \sigma\bar{\varsigma}]\|w\|); \\
e^{-rt}\varsigma^p(t) &\le M\int_0^t e^{-rs}(\sigma^2\bar{\varsigma}^2 + \|\mathbf{z}\|_{(1)}^2)\varsigma^{p-2}; \\
\varsigma^p(t) &\le Me^{rt}\Big(\int_0^t e^{-rs}\sigma^2\bar{\varsigma}^2\Big)\hat{\varsigma}^{p-2}(\tau) \quad (t \le \tau),
\end{aligned}$$

using (4.6) for the last, in which we now take the sup over $[0, \tau]$ — first on the right, then on the left. Dividing by $\hat{\varsigma}^{p-2}$, one then obtains the second principal estimate:

$$e^{-rt}\hat{\varsigma}^2(\tau) \leq M_* \int_0^t e^{-rs}\sigma^2\bar{\varsigma}^2. \tag{4.7}$$

[**4**] Now choose $\varepsilon > 0$ small enough: $\gamma_* M_* \varepsilon < 1$ with γ_* as in (4.5) and M_* as in (4.7); set $\mu = \mu(\varepsilon)$ as in (3.7), which still applies here. Then, multiplying (4.7) by $e^{-\mu t}$, we have

$$e^{-(r+\mu)t}\hat{\varsigma}^2(t) \leq M_* \int_0^t e^{-\mu(t-s)}\sigma^2(s)\, e^{-(r+\mu)s}\bar{\varsigma}^2(s)\, ds \tag{4.8}$$

$$\leq M_*\varepsilon \sup\{e^{-(r+\mu)s}\bar{\varsigma}^2(s) : 0 \leq s \leq \tau\} \tag{4.9}$$

for $t \leq \tau \leq T$. Fixing r large enough to justify (4.6), (4.7) and $\mu = \mu(\varepsilon)$ as above, we now select the rather unusual norm:

$$\|f\|_* := \max_{0 \leq \tau \leq T} \{e^{-(r+\mu)\tau} \max_{0 \leq t \leq \tau} \max_{x \in \bar{\Omega}} \{|f(t,x)|\}\}$$

for $C(\bar{\mathcal{Q}})$. Then, from (4.5) and (4.8) we get

$$\|\mathcal{E}' - \mathcal{E}''\|_* \leq \gamma_* M_* \varepsilon \|\bar{\mathcal{E}}' - \bar{\mathcal{E}}''\|_* \tag{4.10}$$

— which is just the desired contractivity estimate.

[**5**] From (4.10) and the Contractive Mapping Theorem, we obtain existence and *uniqueness* of a fixpoint $\mathcal{E}^*$ of the map $\mathbf{L} \circ \mathbf{T}$, giving a unique solution of the system. The continuous dependence on the data $\hat{\mathbf{u}}$ follows from Theorem 2. Theorem 2 really gives continuity to $\mathcal{X}_2$ but a compactness argument lifts that to $\mathcal{X}_p$. Indeed, estimation along the lines of Step [**3**] shows this is actually *Lipschitz* continuity. □

5 Discussion

We conclude with a few comments on the physical relevance and some possible extensions of these results. Some of these will be treated more fully in a forthcoming paper [11].

• It has been customary in the literature to write the drift velocity $\mathcal{V}_k$ in the form $\mu_k\mathcal{E}$ initially taking the *mobility* μ_k constant but then $\mathcal{E}$-dependent to allow for the observed *velocity saturation* (boundedness of $\mathcal{V}$ even for large fields, as in (**H-1.iii**)). The difficulty is that this is a phenomenological approach to finding an acceptable simplified version, fixing on the approximate average velocity, of a fuller treatment involving distributions in velocity/energy as well as position;

since the velocity average at t, x cannot really be simply a function of $\mathcal{E}(t,x)$, this leads to problems which have become important in a variety of problems of current technological interest. There are various models which tinker with the dependences of the μ_k in essentially *ad hoc* fashions, explicitly introduce averaged energies as dependent variables and take $D_k, \mathcal{V}_k, S_k$ to depend on these rather than directly on $\mathcal{E}$, etc. We fix on (2.4) as defining a class of problems of mathematical interest which covers most of the standard models. It seems likely that the methods here would handle the alternate models as well.

• Somewhat related to the consideration above is treatment of the *Einstein relation* connecting D_k to μ_k. For saturating velocities this would imply that one could not have uniform ellipticity as assumed in **(H-1.ii)** so the parabolic problem would become degenerate for large $|\mathcal{E}|$. (Indeed, there is some experimental indication of this for *Si* (but not for *GaAs*), although this is subject to interpretation.) It seems plausible that the present results might continue to be valid in a setting permitting certain such forms of degeneracy but, for the present, we are simply imposing **(H-1.ii)** as a general hypothesis.

• It is the condition **(H-2.vi)**, in particular, which is restrictive for applications since this amounts to a geometric condition on Ω, prohibiting consideration of regions with corners. Provided the boundary $\partial\Omega$ is smooth enough and one avoids 'change of type' for the boundary conditions (as here for Γ_D, Γ_N), this condition **(H-2.vi)** is a standard result — (2.2) gives $\psi \in W^{1,p}(\Omega)$ [7] for the Poisson equation and then $W^{1,p}(\Omega) \hookrightarrow \mathcal{C}(\bar{\Omega})$ when $p > d$. Unfortunately, this generally fails otherwise [4]. Some of the corners occurring in interesting geometries involve no such problems (due to symmetry considerations) but others do and this is a genuine restriction. In comparison with the argument in [10], we note that here one has at least the existence proof under the condition **(H-1.vi)**, almost trivially satisfied for $\mathbf{L}$ given by (2.2), without needing the restrictive condition **(H-2.vi)**. It would also be desirable to permit dependence of D_k on $\mathcal{E}$.

Such extensions of the uniqueness result should be available if one would have greater regularity (specifically, better gradient estimates) for the solution $\mathbf{u}$ than is available from the arguments presented here. It seems unlikely that the indicated regularity here can be significantly improved without strengthening the hypotheses somehow. It *does* seem plausible that useful results could be obtained by noting (for the situations of applied interest) that the material properties expressed in the dependences on t, x are actually *piecewise smooth* in $x \in \Omega$.

• The homogeneous 'no flux' condition used here is typical, arising from insulation or by symmetry. For expository simplicity (avoiding boundary terms) we have considered only this case here but these methods could also apply to nonlinear boundary conditions on Γ_N of the form: $J_k \cdot \mathbf{n} = \phi_k(\cdot, \mathbf{u})$ — or even $\phi_k(\cdot, \mathbf{u}, \mathcal{E})$ under the hypotheses **(H-2)**.

• The existence theorem could easily be 'improved' by permitting $D_k, \mathcal{V}_k$ to depend on **u** as well as on $\mathcal{E}$. This involves negligible modification of the arguments: one takes the argument of **T** now to be the *pair* $[\mathcal{E}, \mathbf{u}] \in L^2(\mathcal{Q})$, a modification which also permits imposing Lipschitz continuity only with respect to $\nabla \mathbf{u}$ in **(H-1.v)** (One could even permit a dependence on ∇u_k if this would satisfy a one-sided condition of monotonicity type). Including a dependence on **u** would almost 'automatically' include temperature coupling since the heat could be considered formally as just another 'species', satisfying much the same sort of equation. The only difference is that the source term for this, given by ohmic heating, is essentially $[\mathcal{E} \cdot J]$ which does not satisfy **(H-1.iv)** and so requires a certain amount of special treatment.

• Physically, of course, it would be meaningless for the densities $\{u_k\}$ to be negative. Imposing appropriate structural assumptions on S_k, easy maximum principle arguments can be used to show that the $\{u_k\}$ do remain non-negative. While simple maximum principle arguments do not seem to apply here to provide $L^\infty(\mathcal{Q})$ bounds for **u**, it seems plausible that (with some harmless modification of the hypotheses) these might be obtained for the present context by the techniques of [3]; one might then also expect $C(\bar{\mathcal{Q}})$ regularity for **u**. Unfortunately, neither those techniques, nor any others presently envisioned, seem likely to lead to the kind of bounds — *uniform as* $t \to \infty$ — which were obtained in [9] and which might lead to results for periodicity and/or steady states when one now includes the possibility of generation terms in S. It is not clear, physically, whether such results should reasonably be expected but we do note that the present analysis has made no use of such physically important properties as charge conservation and the relation of the direction of $\mathcal{V}_k$ to $q_k\mathcal{E}$ — perhaps there remains reason to hope that including these in the hypotheses could lead, as in [9] to such results.

• The contractivity argument in the proof of Theorem 3 worked with the map **T** which assumed an 'inner loop' as in Step [**1**] in the proof of Theorem 1. It would be interesting, instead, to iterate by starting with $\mathbf{u}, \nabla \mathbf{u}$, computing $\mathcal{E} = \mathbf{L}\mathbf{u}$, substituting in (2.4), and solving a linear equation to get the new $\mathbf{u}, \nabla \mathbf{u}$. It seems possible that contractivity for such an iteration might be demonstrated along the lines of the proof of Theorem 3.1 in [2]. Of much greater interest would be the adaptation of the arguments here to show convergence of appropriate discretizations. This is not quite a routine extension of the present results but the modifications do not seem to lead to any substantial new difficulties.

References

[1] J. P. Aubin, *Un théorème de compacité*, CRAS de Paris **265**, pp. 5042–5045 (1963).

[2] S. Belbas and T. I. Seidman, *Periodic solutions of a parabolic quasi-variational*

inequality from stochastic optimal control, Applicable Anal. (to appear).

[3] H. Gajewski and K. Gröger, *On the basic equations for carrier transport in semiconductors*, J. Math. Anal. Appl. **113**, pp. 12–35 (1986).

[4] P. Grisvard, *Elliptic Problems in Nonsmooth Domains*, Pitman APP, Boston (1985).

[5] J. Jerome *Consistency of semiconductor modeling: an existence/stability analysis for the stationary van Roosbroeck system*, SIAM J. Appl. Math. **45**, pp. 565-590 (1985).

[6] M. Mock *An initial value problem from semiconductor device theory*, SIAM J. Math. Anal. **5**, pp. 597-612 (1974).

[7] C. B. Morrey, jr. *Multiple Integrals in the Calculus of Variations*, Springer -Verlag, New York (1966).

[8] T. I. Seidman *Steady state solutions of a nonlinear diffusion-reaction system with electrostatic convection*, Nonlinear Anal.–TMA **4**, pp. 623-637 (1980).

[9] T. I. Seidman *Time-dependent solutions of a nonlinear system arising in semiconductor theory, II: boundedness and periodicity*, Nonlinear Anal.–TMA **10**, pp. 491–502 (1986).

[10] T. I. Seidman *The transient semiconductor problem with generation terms*, in *Computational Aspects of VLSI Design and Semiconductor Device Simulation*, Amer. Math. Soc., Providence (1988).

[11] T. I. Seidman *Time-dependent solutions of a diffusion-reaction system with with electrostatic convection*, in preparation

[12] T. I. Seidman and G. M. Troianniello *Time-dependent solutions of a nonlinear system arising in semiconductor theory*, Nonlinear Anal–TMA **9**, pp. 1137–1157 (1985).

[13] S. Selberherr *Analysis and Simulation of Semiconductor Devices*, Springer -Verlag, Wien (1984).

[14] J. Simon *Compact sets in the space $L^p(0,T;B)$*, Ann. Mat. pura appl. **CXLVI**, pp. 65-96 (1987).

[15] W. van Roosbroeck *Theory of flow of electrons and holes in germanium and other semiconductors*, Bell System Tech. J. **29**, pp. 560-607 (1950).

SWITCHING SYSTEMS AND PERIODICITY[1]

Thomas I. Seidman
Department of Mathematics and Statistics
University of Maryland Baltimore County
Baltimore, MD 21228

ABSTRACT

We consider a class of bimodal systems which alternate modes on hitting 'switching surfaces': the canonical model is a physical thermostat. A sufficient condition is obtained for the existence of *periodic* solutions. The significance for this theory of certain *anomalous points* on the switching surface is shown by an example with *no* periodic solutions. Further discussion is presented for the important case of *linear switching systems*, including a new existence theorem.

1 Introduction

The intention of this paper is to present the periodicity problem for switching systems and to indicate some of the results which have been obtained.

The abstract notion of *switching system* was introduced in [7] to generalize a model of a (physical) thermostat; for further detail, see [7], [8], [9], [10]. Here it will be sufficient to consider a much more intuitive restricted class of switching systems, including those which have been used for thermostat models. Thus, a switching system, here, will be an autonomous bimodal system in which one follows a *mode* given by a differential equation: $\dot{x} = f_1(x)$ until a *switching time* at which the trajectory hits the boundary $\mathcal{S}_1 = \partial\mathcal{R}_1$ of a *forbidden region* $\mathcal{R}_1$ and one then proceeds by similarly following the other mode $\dot{x} = f_2(x)$ until hitting the other *switching surface* $\mathcal{S}_2 := \partial\mathcal{R}_2$, etc. We will be somewhat more precise in the next section.

The initial stimulus to these investigations came from a conversation with K. Glashoff and J. Sprekels regarding the model [3]. Their computational experience suggested that, for any initial state, one rapidly settled down to a periodic regime, cycling between the two modes (ON/OFF). Direct experience with (physical) thermostats also suggests such behavior. An attempt to demonstrate analytically the existence of *exactly* periodic solutions for the rather different model considered in [3] both succeeded and failed: the model supports certain (physically spurious) constant solutions due to a convexification and it was not clear how to show existence of *nontrivial* periodic solutions. After developing in [7] the switching system model, the concern for periodicity remains. Also, compare [1].

---

[1]This research has been partially supported by the Air Force Office of Scientific Research under grants #AFOSR-87-0190 and #AFOSR-87-0350.

In the next section we discuss topologically general results for periodicity of switching systems. The principal thrust of this material is the importance to the theory of considering certain *anomalous points* on the switching surfaces. This is demonstrated, in particular, by the construction of an example of a switching system on $\mathbb{R}^2$ for which there exists no periodic solution at all although all the 'nice' topological hypotheses are verified (e.g., there is a compact, convex invariant set and the flows are smooth) except that there is a single anomalous point.

For a thermostat, the modes corresponding to [furnace ON] and [furnace OFF] are linear (given by the same pde with different inhomogeneities)and the switching surfaces $S_\jmath$ are defined by the temperature set-points (evaluated at the sensor). This will be our model example of the important subclass of *linear switching systems*, discussed in Section 3.

2 General Results

Our aim in this section is to consider the extent to which such standard techniques of analysis of periodicity for odes as the Schauder Fixpoint Theorem apply to 'general' (nonlinear) switching systems. Since we are considering autonomous systems, the 'natural' map to consider is that from an initial point to the position at a subsequent switching time.

For our present purposes we consider a somewhat less general definition of a switching system than is treated in [8], thus avoiding some technical details while still including the principal problems of interest for our discussion of the periodicity problem. Thus, we assume we have two *modes*, given by continuous semi-flows $\boldsymbol{\pi}_k$ indexed by $k = 1, 2$. (The formulation in terms of the solution maps $\boldsymbol{\pi}_k$ rather than differential equations simplifies consideration of hypotheses on the equations; the state space $\mathcal{X}$ will be a Banach space which we may think of as $\mathbb{R}^d$ but which may be infinite-dimensional for, e.g., an example such as the thermostat problem where the equation is actually a pde.) We introduce, also, a pair of *forbidden regions* $\mathcal{R}_k$ — open sets in $\mathcal{X}$ with disjoint closures whose boundaries will be the admitted *switching surfaces.* Thus, the quadruple $\sum = [\boldsymbol{\pi}_k, \mathcal{R}_k]_{k=1,2}$ 'is' a switching system.

DEFINITION: A **solution** of the switching system $\sum$ is then any function pair

$$[x, \jmath] : \mathbb{R}^+ \to \mathcal{X} \times \{1, 2\} \tag{2.1}$$

satisfying the following set of rules:

- On an interswitching interval $[t_1, t_2)$ for which $\jmath(\cdot)$ has the constant value k we have $x(\cdot)$ following the mode $\boldsymbol{\pi}_k$, i.e., $x(t) = \boldsymbol{\pi}_k(t - t_1)x(t_1)$ for $t_1 \leq t < t_2$.
- One is forbidden to have $\jmath(t) = 1$ if $x(t) \in \mathcal{R}_1$ and, similarly, $\jmath(t) = 2$ is forbidden if $x(t) \in \mathcal{R}_2$.

- $\jmath(\cdot)$ is piecewise constant (say, left-continuous) with a *switch* permitted *only* at $\partial\mathcal{R}_\jmath$ — i.e., one may have $\jmath(\tau-) = 1$ with $\jmath(\tau) = \jmath(\tau+) = 2$ only if $x(\tau) \in \partial\mathcal{R}_1$ and similarly for switching from $\jmath = 2$ to $\jmath = 1$.

We assume here that one is always assured of global solutions with separated switching times; see [8] or [10] for more detail as to sufficient hypotheses for this although this will be clear for the settings we now consider.

Implicit in this set of rules is a possible anomaly: if a trajectory for $\boldsymbol{\pi}_k$ (starting at a point $\xi_0 \in [\mathcal{X}\backslash\bar{\mathcal{R}}_k]$) could first intersect $\partial\mathcal{R}_k$ at a time τ and then continue without actually entering the open set $\mathcal{R}_k$, then these rules would, somewhat ambiguously, accept as solutions *both* those which switch at this τ and also those which do not! This indeterminacy is necessary to preserve as an underlying principle that: THE LIMIT OF SOLUTIONS IS AGAIN A SOLUTION.

DEFINITION: We call a point $\xi \in \partial\mathcal{R}_k$ **anomalous** (for $\boldsymbol{\pi}_k$) if there is some $\xi_0 \in [\mathcal{X}\backslash\bar{\mathcal{R}}_k]$ and $\tau > 0$ such that: $\boldsymbol{\pi}_k(\tau)\xi_0 = \xi$; $\boldsymbol{\pi}_k(t)\xi_0 \in [\mathcal{X}\backslash\bar{\mathcal{R}}_k]$ for $0 < t < \tau$; there are sequences $\xi_\nu \to \xi_0$ and $t_\nu \to \tau$ with $\boldsymbol{\pi}_k(t_\nu) \in \mathcal{R}_k$; there is also a sequence $\xi'_\nu \to \xi_0$ and $\varepsilon > 0$ for which $\boldsymbol{\pi}_k(t)\xi'_\nu \in [\mathcal{X}\backslash\bar{\mathcal{R}}_k]$ for $0 < t < \tau + \varepsilon$.

Thus, ξ is called anomalous if it is a (potential) switching point on a solution which is the limit both of a sequence of solutions for which (nearby) switching is mandatory *and also* of a sequence of solutions for which there cannot be any such switching— these are the points at which switching is optional. Even if the indeterminacy at such points were to be resolved by some (arbitrary) 'selection principle', the existence of anomalous points is of fundamental significance to our theory: see Example 2.1 below.

DEFINITION: Given $k = 1$ or 2 and some $\xi \in [\mathcal{X}\backslash\bar{\mathcal{R}}_k]$, suppose there is a $\tau > 0$ such that: $\boldsymbol{\pi}_k(\tau)\xi \in \partial\mathcal{R}_k$ with $\boldsymbol{\pi}_k(t)\xi \in [\mathcal{X}\backslash\bar{\mathcal{R}}_k]$ for $0 < t < \tau$. We set $\mathbf{t}(\xi) = \mathbf{t}_k(\xi) := \tau$ and $\mathbf{x}(\xi) = \mathbf{x}_k(\xi) := \boldsymbol{\pi}_k(\tau)\xi$. Thus, $\mathbf{x}$ is the first point at which the trajectory from ξ (in mode k) hits $\bar{\mathcal{R}}_k$ and $\mathbf{t}$ is the first hitting time.

Lemma 2.1 *Given $\xi \in [\mathcal{X}\backslash\bar{\mathcal{R}}_k]$, suppose $\mathbf{x}_k(\xi)$ is defined and is not an anomalous point. Then the functions $\mathbf{x}_k(\cdot)$ and $\mathbf{t}_k(\cdot)$ are defined and continuous in a neighborhood of ξ.*

The proof is a rather straightforward application of the definitions, recalling the continuity in t, ξ of $\boldsymbol{\pi}_k$. The somewhat messy detail will thus be omitted here. □

The principal *positive* result available for 'general' switching systems is essentially a corollary of this lemma.

DEFINITION: **A minimal periodic solution (mps)** has the following form: Suppose we were to have a point $\xi_0 \in \partial\mathcal{R}_2$ and imagine solving the switching system with initial data $[\xi_0, 1]$. If there is a solution segment which proceeds in mode $\jmath = 1$ until it switches to mode $\jmath = 2$ at a point $\xi_1 \in \partial\mathcal{R}_1$ and then continues in that mode until returning to ξ_0, then one can switch back to mode $\jmath = 1$ and continue periodically — repeating this 'two-phase' segment, alternating modes, 'forever'.

This is the simplest possible nontrivial periodic solution of the switching system.

Theorem 2.2 *Let $\mathcal{K}$ be a compact, topologically convex subset of $\mathcal{X}\backslash\bar{\mathcal{R}}_1$. Suppose $\mathbf{x}_1(\xi)$ is defined for each $\xi \in \mathcal{K}$ and set $\mathcal{K}' := \{\mathbf{x}_1(\xi) : \xi \in \mathcal{K}\} \subset \partial\mathcal{R}_2$. Next, suppose also that $\mathbf{x}_2(\xi')$ is defined for each $\xi' \in \mathcal{K}'$ and set $\mathcal{K}'' := \{\mathbf{x}_2(\xi') : \xi' \in \mathcal{K}'\}$. If there are no anomalous points in either $\mathcal{K}'$ or $\mathcal{K}''$ and if $\mathcal{K}'' \subset \mathcal{K}$, then there exists an mps of the switching system with 'initial value' $[\xi, 1]$ for some $\xi \in \mathcal{K}'' \subset \mathcal{K}$.*

PROOF: Define the map $F = (\mathbf{x}_1 \circ \mathbf{x}_2)$. Applying Lemma 2.1 twice, using the assumption that there are no anomalous points, we see that F is well-defined and continuous from $\mathcal{K}$ to $\mathcal{K}''$. By assumption F is then a continuous selfmap of $\mathcal{K}$ and applying the Schauder Theorem shows existence of a fixpoint $\xi \in \mathcal{K}$ (Necessarily, of course, $\xi \in \mathcal{K}''$.) so that, by our definition, $[\xi, 1]$ will be initial data for such an mps of the switching system. □

Example 2.1 *We conclude this section by constructing a class of examples to show that the hypothesis in Theorem 2.2 excluding anomalous points cannot be omitted.*

We take $\mathcal{X}$ to be the plane $\mathbb{R}^2$ and let $\mathcal{R}_k$ be a pair of separated (open) disks — say, with $\mathcal{R}_1$ on the 'left' as shown in Figure 1 — and take each semiflow $\boldsymbol{\pi}_k$ to have a global attractor in the corresponding $\mathcal{R}_k$ so all solutions of the switching system will necessarily alternate modes infinitely often.

It will be sufficient to describe the trajectories associated with certain critical semiflows; the rest will smoothly fill out the intervening space and 'speed' along the trajectories is irrelevant for our present considerations.

For the semiflow $\boldsymbol{\pi}_1$ there will be no anomalous points on $\partial\mathcal{R}_1$. Clearly, in discussing periodicity we are here only concerned with those orbits which intersect $\partial\mathcal{R}_2$. The critical orbits are the 'upper orbit' $[\alpha_1]$, which (moving from 'right infinity') touches $\partial\mathcal{R}_2$ tangentially at a point **a** before intersecting $\partial\mathcal{R}_1$ at the point **A**, and the 'lower orbit' $[\delta_1]$, which (again moving from 'right infinity') touches

$\partial\mathcal{R}_2$ tangentially at a point **d** before intersecting $\partial\mathcal{R}_1$ at the point **D**. We take $[\delta_1]$ to touch $\partial\mathcal{R}_2$ tangentially at a point **d′** and then loop away somewhat from $\partial\mathcal{R}_2$ before again touching at **d** and continuing to **D**. All orbits which intersect $\partial\mathcal{R}_2$ must then lie between $[\alpha_1]$ and $[\delta_1]$ and we identify, in particular, one such orbit $[\gamma_1]$ (slightly 'above' $[\delta_1]$) which enters $\mathcal{R}_2$, loops out and then re-enters at a point **c′** between **d′** and **d**, crosses $\mathcal{R}_2$ exiting at a point **c**, and finally enters $\mathcal{R}_1$ at a point **C** (slightly above **D**) in the arc $\overline{\mathbf{AD}}$.

Now consider the semiflow $\boldsymbol{\pi}_2$. There are now three critical orbits: the 'upper orbit' $[\alpha_2]$ passing through **A**, the 'lower orbit' $[\delta_2]$ passing through **D**, and an orbit $[\beta_2]$ containing the only anomalous point of the switching system. This orbit $[\beta_2]$ (moving from 'left infinity') passes through a point **B** of $\overline{\mathbf{AD}}$, then tangentially touches $\partial\mathcal{R}_2$ at the point **a** — which is thus anomalous — and loops somewhat away from $\partial\mathcal{R}_2$ before entering $\mathcal{R}_2$ at **d′**. The orbit $[\alpha_2]$ enters $\mathcal{R}_2$ at **c′** (swinging slightly above $[\beta_2]$ 'around' $\mathcal{R}_2$) while $[\delta_2]$ enters $\mathcal{R}_2$ at **c**.

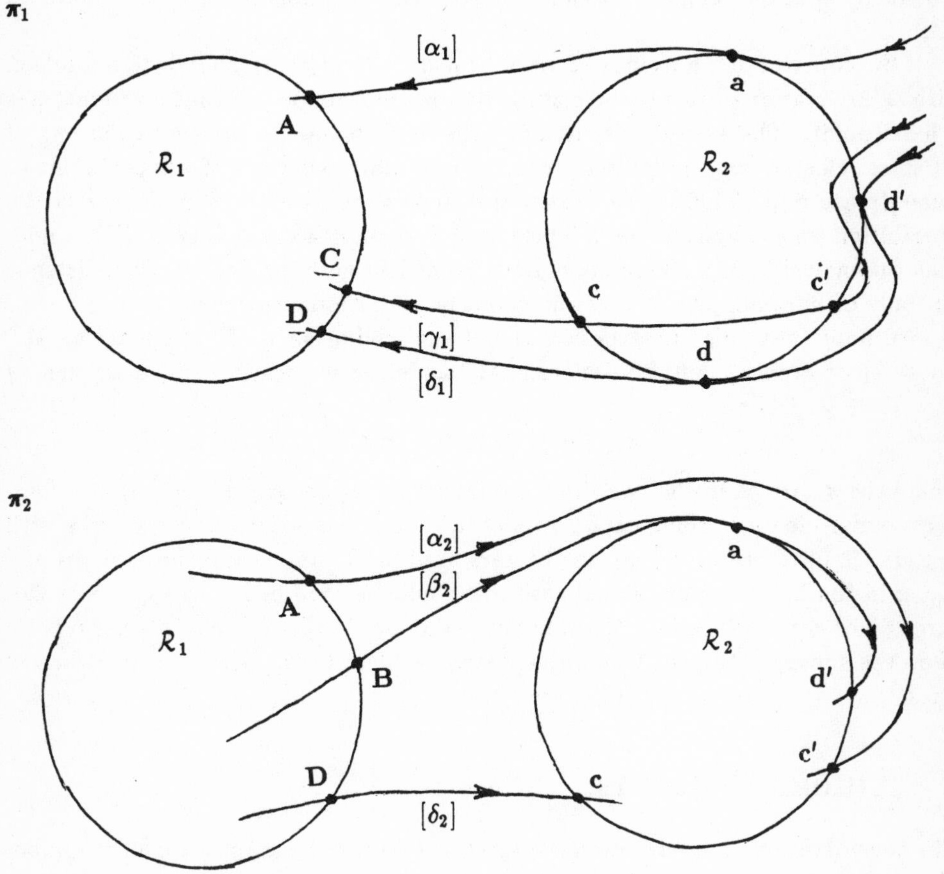

Figure 1

If one now *identifies* $\mathbf{A} \approx \mathbf{D}$, then $\overline{\mathbf{AD}}$ becomes, topologically, a circle. The 'round trip map' $(\mathbf{x}_2 \circ \mathbf{x}_1)$ gives $[\mathbf{A} \approx \mathbf{D}] \mapsto \mathbf{C}$ by construction and, with our identification, gives $\mathbf{B} \mapsto [\mathbf{A} \approx \mathbf{D}]$ (independently of the option selected at $\mathbf{a}$. Thus, this 'round trip' determines a well-defined map $\boldsymbol{\rho}$ of the 'circle' $\overline{\mathbf{AD}}$ to itself; one easily verifies, using the continuity of $\boldsymbol{\pi}_1, \boldsymbol{\pi}_2$, that this map $\boldsymbol{\rho}$ is continuous as well as orientation-preserving and injective, hence a homeomorphism as the circle is compact.

Within the framework above one can (homotopically) adjust the construction so that $(\mathbf{x}_2 \circ \mathbf{x}_1)$ is *any* desired (orientation-preserving) homeomorphism $\boldsymbol{\rho} : \overline{\mathbf{AD}} \to \overline{\mathbf{AD}}$. To see this, parametrize $\overline{\mathbf{AD}}$ as $[0, 2\pi]$ with $0 \approx 2\pi$ and modify $\boldsymbol{\pi}_2$ so the 'initial point' $\mathbf{B} \in \partial\mathcal{R}_2$ of the orbit $[\beta_2]$ is just $\boldsymbol{\rho}^{-1}(\mathbf{A} \approx \mathbf{D})$. One can then modify $\boldsymbol{\pi}_1$ (only in the 'rectangle' between $[\alpha_1]$ and $[\delta_1]$ and from $\overline{\mathbf{ad}}$ going into $\mathcal{R}_1$) so each 'initial point' in $\overline{\mathbf{AD}}$ is mapped by $(\mathbf{x}_2 \circ \mathbf{x}_1)$ as desired to produce $\boldsymbol{\rho}$. In particular, one can construct $\boldsymbol{\pi}_1, \boldsymbol{\pi}_2$ so $\boldsymbol{\rho}$ is a 'rotation' of the parametric circle through an arbitrarily specified angle ω. (This can even be accomplished with C^∞ flows on $\mathbb{R}^2$.)

If the construction has, indeed, been 'tuned' so that $\boldsymbol{\rho}$ corresponds to a rotation with $\omega/2\pi$ irrational, this is the classic example of a map such that no iterate has a fixed point. The resulting switching system then has *no periodic solutions at all* since, clearly, *any* periodic solution must entail existence of a fixed point of some iterate of $\boldsymbol{\rho}$. Thus, this construction shows the essential significance of the prohibition (in the hypotheses of Theorem 2.2) of anomalous points in $\mathcal{K}', \mathcal{K}''$; note that the inclusion of a single anomalous point has here 'prevented' the existence not only of *minimal* periodic solutions but of *all* periodic solutions.

We note that this construction is not really limited to $\mathbb{R}^2$. Let $\mathcal{X}$ be $\mathbb{R}^m$ $(m > 2)$ or even an infinite-dimensional Banach space which can be written as $\mathcal{X} = \mathbb{R}^2 \oplus \mathcal{Y}$. Define

$$\hat{\boldsymbol{\pi}}_k(t)[\xi, y] := [\boldsymbol{\pi}_k(t)\xi, \mathbf{S}(t)y]$$

where the $\boldsymbol{\pi}_k$ are as in the two-dimensional construction above and $\mathbf{S}(\cdot)$ is a compact semigroup on $\mathcal{Y}$ going exponentially to 0. The regions $\hat{\mathcal{R}}_k$ are, say, balls with centers $[\xi_k, 0]$ where the ξ_k are the centers used in $\mathbb{R}^2$ above and the radii are the same; in the infinite-dimensional case this could be modified to get compact closure. Since $\mathbf{S}(t) \to 0$ as $t \to \infty$, the only points of interest for possible periodicity lie in the subspace $(y = 0)$ where the system reduces to the previous construction for $\mathbb{R}^2$.

3 Linear Systems

The consideration of *linear switching systems* permits the application of a number of more specialized modes of analysis which provide the deeper results available in that case.

DEFINITION: For a **linear switching system** the dynamics are given by a pair of (abstract) linear odes:

$$\dot{x} = \mathbf{A}x + u_k \qquad (k = \pm 1) \tag{3.1}$$

where $\mathbf{A}$ is the infinitesimal generator of a C_0 linear semigroup $\mathbf{S}(\cdot)$ on the state space $\mathcal{X}$ (cf, e.g., [5]) and the *forbidden regions* are given by

$$\mathcal{R}_k = \begin{cases} \{\xi \in \mathcal{X} : \langle \Lambda, \xi \rangle > \vartheta_+\} & \text{if } k = 1 \\ \{\xi \in \mathcal{X} : \langle \Lambda, \xi \rangle < \vartheta_-\} & \text{if } k = -1 \end{cases} \tag{3.2}$$

for some $\Lambda \in \mathcal{X}^*$ and $\vartheta_- < \vartheta_+$. (Note that, for reasons which will shortly be apparent, we have chosen to index the modes by ± 1 when considering *linear* switching systems.)

We will restrict our attention to (exponentially) *stable* semigroups $\mathbf{S}(\cdot)$ so there are global attractors $z_k = -\mathbf{A}^{-1}u_k$ for the two modes and we assume $z_k \in \mathcal{R}_k$; then all solutions of the switching system will alternate modes infinitely often. Observe, also that without loss of generality we may make a (linear) change of variables to have $z_{\pm 1} = \pm z$ and $\langle \Lambda, z \rangle = 1$. From the dynamics (and noting that the assumed stability gives independence of the 'infinite past') we obtain the representation:

$$x(t) = \int_{-\infty}^{t} \mathbf{S}(t-s)[\jmath(s) z]\, ds. \tag{3.3}$$

If we set

$$\vartheta(t) := \langle \Lambda, x(t) \rangle, \qquad \varphi(t) := \langle \Lambda, \mathbf{S}(t) z \rangle, \tag{3.4}$$

then we see from (3.3) that the *sensor function* $\vartheta(\cdot)$ is the convolution of the *switching function* $\jmath(\cdot)$ and the *impulse response function* $\varphi(\cdot)$:

$$\vartheta(t) = \int_{-\infty}^{t} \varphi(t-s)\jmath(s)\, ds = \int_{0}^{\infty} \varphi(r)\jmath(r-t)\, dr. \tag{3.5}$$

One easily sees that the rules imply switching for such a linear system when $\vartheta(\cdot)$ crosses (touches?) the critical values $\vartheta_\pm$. Indeed, it is possible to construct an essentially equivalent new switching system in which the 'state' is the semi-infinite past history of ϑ, the dynamics is given by (3.5), and the functional Λ is now evaluation of ϑ at the current time. Thus the function $\varphi(\cdot)$ and the critical values $\vartheta_\pm$ completely characterize the behavior of such a stable linear switching system. See [8], [10].

We henceforth assume that φ is in $L^1(\mathbb{R})$ (defined as 0 for $t < 0$) and in $C_0(\mathbb{R}^+)$. This is automatically true for φ coming from an exponentially stable semigroup as above but also includes some cases in which the semigroup is merely asymptotically stable. We easily see from the above that our construction has imposed the normalizations:

(3.6) $$\int_0^\infty \varphi = 1, \qquad -1 < \vartheta_- < \vartheta_+ < 1.$$

It is interesting to consider the finite-dimensional situation $\mathcal{X} = \mathbb{R}^m$. The stability assumption makes it easy to obtain a bounded (and so a compact) invariant set but we must consider the possibility of anomalous points. An anomalous point ξ on the switching surface $\vartheta = \vartheta_+$ would typically be such that:

$$\langle \Lambda, \xi \rangle = \vartheta_+, \quad \langle \Lambda, \mathbf{A}\xi \rangle = 0, \quad \langle \Lambda, \mathbf{A}^2 \xi \rangle = \varsigma < 0.$$

If e.g. $\{\Lambda, \mathbf{A}^*\Lambda, [\mathbf{A}^*]^2\Lambda\}$ is linearly independent, then such points will always exist — indeed, they then form an affine half-space of dimension $(m-2)$. This need not make our general existence theorem above inapplicable but does make the determination of a suitable $\mathcal{K}$ much more difficult. For $m = 2$, however, one has linear dependence and an argument [7] using the quadratic characteristic polynomial of $\mathbf{A}$ shows that there cannot be any anomalous points and so one has an mps for 'arbitrary' $\vartheta_\pm$; recent work by Stoth [11] shows that this solution (for $\mathcal{X} = \mathbb{R}^2$) is always unique and stable.

It is worth noting a somewhat different analysis, used in [11] but treated more deeply in [2]. If one were to have an mps, say with period p, then (shifting so the switch from $\jmath = 1$ to $\jmath = -1$ occurs at $t = 0$) the switching function $\jmath(\cdot)$ will be -1 for $0 < t < \alpha$ and 1 for $\alpha < t < p$ (some switching time $0 < \alpha < p$) and, by (3.5), the sensor function is given by

(3.7) $$\vartheta(t) = \int_0^p \psi(\tau;p)\jmath(t-\tau)\,d\tau \text{ where } \psi(\tau;p) := \sum_{n=0}^{\infty} \varphi(\tau + np).$$

Now consider the function $F : [p, \alpha] \mapsto [u, v]$ given by

(3.8) $$\begin{aligned} u := \ \vartheta(0) \ &= \ 2\int_0^{p-\alpha} \psi(\cdot;p) - 1 = 1 - 2\int_{p-\alpha}^{p} \psi(\cdot;p) \\ v := \ \vartheta(\alpha) \ &= \ 1 - 2\int_0^{\alpha} \psi(\cdot;p) \end{aligned}$$

and note that this $\jmath(\cdot)$ gives a solution of the switching system through (3.3) if $u = \vartheta_+$, $v = \vartheta_-$ — provided that (3.7) gives

(3.9) $$\vartheta(\cdot) > v \text{ on } (0,\alpha), \qquad \vartheta(\cdot) < u \text{ on } (\alpha, p).$$

To seek an mps for such switching systems (for varying $\vartheta_\pm$) is thus an attempt to invert this function F, subject to (3.9).

In treating the (physical) thermostat problem as a switching system we would have a pde governing the dynamics: the heat equation with either of two source terms depending on whether the furnace is ON or OFF. Depending on the modeling, this source may appear either in the equation or in the boundary data. Especially in the latter case, the inhomogeneity may not be in the state space $\mathcal{X}$ but the appropriate semigroup is so strongly smoothing that we can expect $\mathbf{S}(t)z$ to be well-behaved for positive t. Similarly, the sensor measures temperature at a point; this is not a functional in $\mathcal{X}^*$ but nevertheless is well-behaved on the range of $\mathbf{S}(t)$ for $t > 0$. Assuming the sensor is not actually placed in the furnace, the impulse response function φ can be expected to be analytic for $t > 0$ (C^∞ vanishing at 0) and vanishing at ∞ with exponential decay. A physical thermostat has a pair of 'set points' whose effect in switching the furnace is, indeed, more-or-less[2] as presented here. In practical operation the separation of the switching values is very small compared to the potential range of variation. The desirable stability analysis of this situation is still lacking. The (one-dimensional) models for which results have been obtained regarding periodicity [6], [2] have been of this form (i.e., with dynamics given by (3.5) and switching according to the crossings of $\vartheta_\pm$ by $\vartheta(t)$). In addition, they have involved boundary conditions consistent with the use of the maximum principle: in particular, in each case one had

$$\varphi(t) > 0 \ (t > 0), \qquad \varphi'(t) < 0 \ (t > t_*). \tag{3.10}$$

Prüss [6] obtained existence when $\vartheta_\pm$ are far apart while, more recently, Friedman and Jiang [2] provided the best results currently available:

- Existence of an mps for *every* choice of switching values $\vartheta_\pm$.
- Uniqueness of this solution when $\vartheta_\pm$ are far enough apart.

We note that the paper [2] is not in the present framework and uses special properties of the pde setting, some of which do not correspond to hypotheses easily formulated in terms of φ.

The existence result in [6] is for $\vartheta_\pm$ far enough apart and another of the results in [11] is general existence when φ is strictly decreasing to 0 on $\mathbb{R}^+$. Note also a quite recent paper by Gripenberg [4] which proves existence, under quite general conditions, of periodic functions 'weakly controlled' by the thermostat; these need not be solutions in the sense considered here. We now conclude this paper by presenting a new 'general' existence result for periodic solutions.

Theorem 3.1 *Suppose $\varphi(\cdot) \in L^1$ is continuous on $[0, \infty)$; assume the normalization (3.6). Suppose, also, that there is some $t_* \geq 0$ such that: (i) $\varphi(\cdot)$ is (strictly)*

[2]Strictly speaking, this ignores unmodeled 'fast dynamics' within the thermostat and furnace. The interesting connection between switching systems and this sort of bifurcating singular perturbation problem is not under consideration here.

decreasing to 0 *on* (t_*, ∞) *and (ii)* Φ *is bounded away from* 0 *on* $[0, t_*]$ *where*

$$\Phi(t) := \int_t^\infty \varphi(s)\,ds = 1 - \int_0^t \varphi(s)\,ds; \tag{3.11}$$

without loss of generality we take t_* *so* $\Phi(t) > \Phi(t_*) =: \beta_*$ *for* $t < t_*$. *Then there always exists an mps of the switching system whenever* $\vartheta_\pm$ *are far enough apart.*

PROOF: Consider a sequence $\mathbf{s} := \{\sigma_1, \sigma_2, \ldots\}$ of positive numbers. With $s_0 = 0$, recursively set $s_k = s_{k-1} - \sigma_k$ and define $\jmath = \jmath_{\mathbf{s}}(\cdot)$ on $(-\infty, 0)$ as $(-1)^k$ on each interval (s_k, s_{k-1}). Assuming $\vartheta(0) < \vartheta_+$, extend $\jmath$ on $(0, \tau)$ as $+1$ until $\vartheta(\sigma) = \vartheta_+$ with $\vartheta = \vartheta_{\mathbf{s}}(\cdot)$ obtained from $\jmath_{\mathbf{s}}$ by (3.5); this defines $\sigma = \sigma_{\mathbf{s}}$. Similarly, we then continue to extend $\jmath_{\mathbf{s}}$ on $(\sigma, \sigma + \tau)$ as -1 until $\vartheta(\sigma + \tau) = \vartheta_-$, defining $\tau = \tau_{\mathbf{s}}$. Denote by $\mathbf{M}$ the map: $\mathbf{s} \to \mathbf{s}'$ where $\sigma_1' := \tau_{\mathbf{s}}$, $\sigma_2' := \sigma_{\mathbf{s}}$ and then $\sigma_k' := \sigma_{k-2}$ for $k = 2, 3, \ldots$; note that a fixpoint of $\mathbf{M}$ provides an mps of the switching system.

Our first task is to obtain a suitable lower bound for $\sigma = \sigma_{\mathbf{s}}$. Note that the construction of $\jmath_{\mathbf{s}}$ gives

$$\begin{aligned} \vartheta_+ = \vartheta(\sigma) &= \int_0^\sigma \varphi - \int_\sigma^{\sigma+\sigma_1} \varphi + \int_{\sigma+\sigma_1}^\infty \varphi(s)\jmath(\sigma - s) \\ &\le \int_0^\sigma \varphi - \int_\sigma^{\sigma+\sigma_1} \varphi + \int_{\sigma+\sigma_1}^\infty \varphi \\ &= 1 - 2\int_{\sigma+\sigma_1}^\infty \varphi \qquad\qquad \text{so} \end{aligned}$$

$$\Phi(\sigma) - \Phi(\sigma + \sigma_1) \le \frac{1 - \vartheta_+}{2} =: \alpha_+ \tag{3.12}$$

provided we can assume $\sigma_1 \ge t_*$ so φ will be positive on $(\sigma + \sigma_1, \infty)$. Now (temporarily) fix $t' > t_*$ so $\Phi(t') =: \beta < \beta_*$ and, for $\alpha \le \beta_* - \beta$, define

$$s(\alpha) = s(\alpha; t') := \min\{s \ge t_* : \Phi(s) - \Phi(s + t') \le \alpha\}. \tag{3.13}$$

Clearly, $s(\alpha)$ is well-defined and tends to 0 as $\alpha \to 0+$. Hence, for any $t' > t_*$ as here there will exist $\alpha(t')$ such that:

$$0 < \alpha(t') \le \beta_* - \Phi(t'), \qquad s(\alpha(t')) \ge t'; \tag{3.14}$$

without loss of generality we might fix $t' > t_*$ so $\alpha' = \alpha(t')$ is as large as possible. From this definition we easily see that if

$$\vartheta_+ \ge 1 - 2\alpha', \qquad \vartheta_- \le -1 + 2\alpha', \tag{3.15}$$

then $\sigma_1 \ge t' > t_*$ implies $\sigma_{\mathbf{s}} \ge t'$ and, repeating the analysis, further implies $\tau_{\mathbf{s}} \ge t'$. This shows, subject to (3.15), that $\{\mathbf{s} : \sigma_k \ge t'\}$ is invariant under the map $\mathbf{M}$. Now fixing $\vartheta_\pm$ subject to (3.15), a simple estimate like (3.12) gives

$$2\Phi(\sigma) \ge |1 - \vartheta_+|, \qquad 2\Phi(\tau) \ge |-1 - \vartheta_-|,$$

providing an upper bound: $\sigma, \tau \leq t''$; hence, $\mathbf{M}$ is a well-defined selfmap of $\mathcal{K}_* := [t', t'']^\infty$.

Note that $\mathcal{K}_*$ is convex in $\mathbb{R}^\infty$ and is compact by the Tychonov Theorem. The function: $\mathbf{s} \mapsto \sigma_\mathbf{s}$ will clearly be continuous — essentially as in Lemma 2.1 — if (3.5) gives $\dot{\vartheta}(\sigma-) > 0$ (with $\vartheta = \vartheta_\mathbf{s}$, $\sigma = \sigma_\mathbf{s}$). Assuming, for the moment, that φ is differentiable on $\mathbb{R}^+$ with $\varphi' \in L^1$, we have

$$\begin{aligned}
\dot{\vartheta}(t) &= \varphi(0) + \int_{-\infty}^{t} \varphi'(t-s)\jmath(s)\,ds \quad \text{for } 0 < t < \sigma, \\
\dot{\vartheta}(\sigma-) &= \varphi(0) + \int_0^\infty \varphi'(s)\jmath_\mathbf{s}(\sigma - s)\,ds \qquad (3.16)\\
&= \varphi(0) + \int_0^\sigma \varphi' - \int_\sigma^{\sigma+\sigma_1} \varphi' + - \ldots \\
&= 2\left([\varphi(t_1) - \varphi(t_2)] + [\varphi(t_3) - \varphi(t_4)] + \ldots\right)
\end{aligned}$$

where $t_1 = \sigma$ and then $t_{k+1} = t_k + \sigma_k$. (A density argument then shows that (3.16) holds without the differentiability assumption on φ; one can similarly avoid the assumption of *strict* decrease on (t_*, ∞).) Since our assumption (3.15) ensures that $t_1 = \sigma_\mathbf{s} \geq t' > t_*$, each term on the right of (3.16) is strictly positive by the hypotheses so $\dot{\vartheta}(\sigma-) > 0$ as desired. Essentially the same argument shows $\dot{\vartheta}(\sigma + \tau-) < 0$ so the function: $\mathbf{s} \mapsto \tau_\mathbf{s}$ is also continuous.

From the above, $\mathbf{M}$ is a *continuous* selfmap of $\mathcal{K}_*$ (subject to (3.15) — i.e., if the switching values $\vartheta_\pm$ are far enough apart as assumed) so, by the Schauder Theorem, there is necessarily a fixpoint giving the desired mps for the switching system. □

References

[1] H. W. Alt, *On the thermostat problem*, Control and Cybernetics **14** (1985), pp. 171–193.

[2] A. Friedman and L.-S. Jiang, *Periodic solutions for a thermostat control problem*, Comm. PDE, to appear.

[3] K. Glashoff and J. Sprekels, *An application of Glicksberg's theorem to set-valued integral equations arising in the theory of thermostats*, SIAM J. Math. Anal. **12** (1981), pp. 477–486; *The regulation of temperature by thermostats and set-valued integral equations*, J. Int. Eqns. **4**, pp. 95–112; (also, personal communication).

[4] G. Gripenberg, *On periodic solutions of a thermostat equation*, SIAM J. Math. Anal. **18** (1987), pp. 694–702.

[5] D. Henry, *Geometric Theory of Semilinear Parabolic Equations*, (Lect. Notes in Math. #840), Springer-Verlag, New York, 1981.

[6] J. Prüss, *Periodic solutions of the thermostat problem*, in *Differential Equations in Banach Spaces* (Lect. Notes in Math. #1223), Springer-Verlag, Berlin, 1986, pp. 216-226.

[7] T. I. Seidman, *Switching systems: thermostats and periodicity*, (Math. Res. Report **83-07**), UMBC, Baltimore, Nov., 1983.

[8] T. I. Seidman, *Switching systems, I*, to appear.

[9] T. I. Seidman, *Control of switching systems*, in *Proc. Conf. on Inf. Sci. and Systems*, Johns Hopkins Univ., Baltimore, 1987, pp. 485–489.

[10] T. I. Seidman, *Switching systems*, monograph in preparation.

[11] B. Stoth, diplomthesis: *Periodische Lösungen von linearen Thermostatproblemen*, (Report **SFB 256**), Univ. Bonn, 1987.

BREATHERS FOR THE SINE-GORDON EQUATION

Michael W. Smiley*
Department of Mathematics
Iowa State University
Ames, Iowa 50011

ABSTRACT - The problem of breathers, solutions of a nonlinear homogeneous wave equation that are nontrivial, time-dependent and T-periodic is considered. A manifold of such solutions is shown to exist in a distributional sense and some qualitative properties of these solutions are described.

In this article we treat the physically motivated problem of breathers for the sine-Gordon equation (with mass term)

$$u_{tt} - \Delta u + m^2 u + \alpha \sin u = 0, \qquad (t,x) \in \mathbf{R} \times \mathbf{R}^3, \tag{1.1}$$

in which u is to be a time-periodic function, with period T, having the property that $u(t,x) \to 0$ as $\|x\| \to \infty$. Specifically we say a solution $u(t,x)$ of (1.1), with these properties, is a breather if it is time-dependent. Thus the undulation of the spatial profile, if observed dynamically in time, will be suggestive of breathing. In our investigations we will consider only radially symmetric functions, thus $u(t,x) = U(t,r)$ where $r = \|x\|$, for some $U : \mathbf{R} \times \mathbf{R}^+ \to \mathbf{R}$. Here we use the notation $\mathbf{R}^+ = (0,+\infty)$. We will show that all solutions exhibit the property of localization in space in the sense that the rate of decay at infinity is of exponential order. In (1.1) we will restrict our attention to the more interesting case in which $\alpha > 0$, although our methods also apply when $\alpha < 0$. In order to adequately describe our results, and the nature of the solutions we find, we first introduce the change of variables $u(t,x) = rw(t,r)$ where $w : \mathbf{R} \times \mathbf{R}^+ \to \mathbf{R}$. We find that $w \in L^\infty(\mathbf{R} \times \mathbf{R}^+)$ and moreover $w(t,r) = O\big(\exp(-\delta r)\big)$ as $r \to \infty$, (a.e.) $(t,r) \in \mathbf{R} \times \mathbf{R}^+$, for some $\delta > 0$. The constant δ will depend on the proximity of $(m^2 + \alpha)$ to a point in the sequence of eigenvalues $\{(2\pi n/T)^2 : n = 0,1,2,\ldots\}$ for ∂_t^2 coupled with the T-periodic boundary conditions. Using the radial symmetry we determine that $w(t,r)$ is a solution of

* This research was sponsored by the Air Force Office of Scientific Research, Air Force Systems Command, U.S.A.F. under Grant 84-0252. The United States Government is authorized to reporduce and distribute reprints for Governmental purposes not withstanding any copyright notation therein.

(1.2) $w_{tt} - w_{rr} + m^2 w + \alpha r \sin(w/r) = 0, \qquad (t,r) \in \mathbf{R} \times \mathbf{R}^+,$

(1.3) $w(t+T,r) = w(t,r), \qquad (t,r) \in \mathbf{R} \times \mathbf{R}^+,$

(1.4) $w(t,\cdot) \in L^2(\mathbf{R}^+), \qquad t \in \mathbf{R},$

in which (1.4) represents a weak form of the decay condition at infinity. Our main result can be stated as follows.

THEOREM. Let $n \geq 1$ be an integer and suppose that m,α are such that $[2\pi n/T]^2 < m^2 + \alpha \leq [2\pi(n+1)/T]^2$. Then (1.2)-(1.4) has a (2n+1)-dimensional manifold of nontrivial solutions, and a 2n-dimensional submanifold of breathers.

The manifold of solutions alluded to above is of class C^0 and is local to the origin in a certain function space to be described subsequently. By a solution of (1.2)-(1.4) we mean a distributional solution with respect to the set of test functions $D_T = \{\phi \in C^\infty(\mathbf{R} \times \mathbf{R}^+) : \phi(t+T,r) = \phi(t,r)$ for $(t,r) \in \mathbf{R} \times \mathbf{R}^+$, and $\phi(t,\cdot) \in C_0^\infty(\mathbf{R}^+)$ for all $t \in \mathbf{R}\}$. We say that $w \in L^2((0,T) \times \mathbf{R}^+)$ is a solution if: i) $w(t+T),r) = w(t,r)$ (a.e.) $(t,r) \in \mathbf{R} \times \mathbf{R}^+$, ii) $w(t,\cdot) \in L^2(\mathbf{R}^+)$ (a.e.) $t \in \mathbf{R}$, and iii) for each $\phi \in D_T$ we have

$$\int_0^T \int_0^{+\infty} \{w[\phi_{tt} - \phi_{rr}] + m^2 w\phi + \alpha r \sin(w/r)\phi\} dr dt = 0.$$

This of course represents a weakening of the usual notion of distributional solution since all test functions vanish in a neighborhood of the origin. However, this definition allows for singular behavior at the origin. For example, if we interpret this in the 1-dimensional t-independent case, a function continuous on $\mathbf{R}$, satisfying the differential equation except at the origin, having decay at infinity, and satisfying the integrability condition at the origin is considered to be a solution of the problem even though it may have a discontinuity in its first derivative at the origin. In fact, for the general 1-dimensional problem, this point of view eliminates the one-sided nature of the only previous known existence result due to A. Weinstein [10], [11].

All other results known to the author regarding this problem are of a nonexistence character in the sense that they give conditions which insure that any solution of the problem must in fact be independent of t, so there can be no breathers. In some cases the solution's independence of t can be used to deduce that only the trivial solution exists. For example, this may follow by considering an ordinary differential equation. The first of these results concerned the 1-dimensional problem and was due to J. M. Coron [1]. He showed that if u is a T-periodic solution of class $C^2(\mathbf{R} \times \mathbf{R})$, having decay at infinity, and if $m^2 + \alpha < (2\pi/T)^2$ then in fact u must be independent of t. Actually, a general nonlinearity $g(u)$ was considered with the pertinent condition being $g'(0) < (2\pi/T)^2$. Further results of this type for the 1-dimensional problem were subsequently given by P. Vuillermot [9], who has explored various classes of nonlinearities.

The crucial comparison of $g'(0)$ to the constant $(2\pi/T)^2$ was further stressed by H. A. Levine in a paper [3] generalizing Coron's result to the n-dimensional radially symmetric case. Also considering the n-dimensional problem, but from a different perspective, W. Strauss [7] showed that only the trivial solution could exist under the key assumption that $ug(u) > 2G(u)$ for $u \neq 0$, where $G'(u) = g(u)$ and $G(0) = 0$.

A LINEARIZATION

The basic Hilbert space used in our approach will be denoted by $H_{T,\delta}$ and consists of those functions $w(t,r)$, $w : \mathbf{R} \times \mathbf{R}^+ \to \mathbf{R}$, which are T-periodic in t and of finite norm

$$\|w\| = \{\frac{1}{T}\int_0^T\int_0^{+\infty} |w(t,r)|^2\, e^{2\delta r}\, drdt\}^{1/2} < +\infty .$$

We also use $H^1_{T,\delta} = \{w \in H_{T,\delta} : w_t, w_r \in H_{T,\delta}\}$ where the derivatives are distributional derivatives with test funtions in D_T; the norm is taken to be $\|w\|_1 = \left(\|w\|^2 + \|w_t\|^2 + \|w_r\|^2\right)^{1/2}$. For the linearized problem

$$w_{tt} - w_{rr} + (m^2 + \alpha)w = h, \qquad (t,r) \in \mathbf{R} \times \mathbf{R}^+, \tag{2.1}$$

together with (1.3)-(1.4), where $h \in H_{T,\delta}$, we have the following result.

THEOREM 2.1. Let $n \geq 0$ be the integer such that $[2\pi n/T]^2 < m^2 + \alpha \leq [2\pi(n+1)/T]^2$ and let $0 < \delta < \sqrt{m^2 + \alpha - (2\pi n/T)^2}$. Then (2.1), (1.3)-(1.4) has a (2n+1)-dimensional affine space of solutions $w \in H^1_{T,\delta}$. All solutions have the form $w = w^* + w_0$, where $w_0 \in N = \{w \in H^1_{T,\delta} :\ w$ satisfies (2.1), (1.3)-(1.4) with h identically zero on $\mathbf{R} \times \mathbf{R}^+\}$ and $w^* \in N^\perp$ with the orthogonality taken in $H^1_{T,\delta}$. Moreover, there is a bounded linear map $K : H_{T,\delta} \to H^1_{T,\delta}$, defined by the correspondence $h \to w^*$, so that $w^* = Kh$; and a constant C which may depend on h such that

$$\left(\int_0^T \{|w(t,r)|^2 + |w_t(t,r)|^2 + |w_r(t,r)|^2\}dt\right)^{1/2} \leq C\exp(-\delta r), \quad \text{(a.e.)} \quad r > 0. \tag{2.2}$$

Remark: From estimate (2.2) and the Sobolev embedding theorem it follows that $|w(t,r)| \leq C\exp(-\delta r)$, (a.e.) $(t,r) \in \mathbf{R} \times \mathbf{R}^+$.

The above result is included in the recently developed linear theory presented in [5], and relies on the classical Paley-Wiener theorem [4]. We sketch the proof here. We begin by seeking approximate solutions in the form

$$(2.3)\qquad W_J(t,r) = \sum_{|j| \le J} w_j(r)\exp(i\theta_j t), \qquad \theta_j = 2\pi j/T, \quad J \geq 1.$$

In (2.3) we have used the complex notation for convenience. The functions $w_j(r)$ are required to satisfy the (weak) boundary value problem

$$(2.4)\qquad -w_j'' + (m^2 + \alpha - \theta_j^2)w = h_j = \frac{1}{T}\int_0^T h(t,r)\exp(-i\theta_j t)dt,$$

$$(2.5)\qquad w_j \in L^2(\mathbf{R}^+).$$

If $h \in H_{T,\delta}$ then $h_j \in L^2_\delta(\mathbf{R}^+)$, where the weight function $\exp(2\delta r)$ is again used in defining the norm on $L^2_\delta(\mathbf{R}^+)$. All solutions of (2.4) have Laplace transforms given by

$$(2.6)\qquad W_j(s) = \left(-H_j(s) + w_{0j}s + w_{1j}\right)/\left(s^2 + \theta_j^2 - m^2 - \alpha^2\right),$$

where $H_j(s) = \underset{\sim}{L}\{h_j(r)\}$ and $w_{0j} = w_j(0)$, $w_{1j} = w_j'(0)$. In order that $w \in L^2(\mathbf{R}^+)$ it must be true that (cf. [4]) $W_j(s)$ is analytic in $\mathrm{Re}(s) > 0$ and satisfies

$$(2.7)\qquad \sup_{0 < \sigma < +\infty} \frac{1}{2\pi}\int_{-\infty}^{+\infty} |W_j(\sigma + i\tau)|^2 d\tau < +\infty.$$

Hence any poles of $W_j(s)$ in $\mathrm{Re}(s) > 0$ must be removable if the integrability condition (2.7) is to be satisfied. This requirement uniquely determines w_{0j}, w_{1j} if $\theta_j^2 - \alpha - m^2 > 0$, and leaves one degree of freedom if $\theta_j^2 < \alpha + m^2$. If J is large this leads to approximate solutions of the form $W_J(t,r) = W_T^*(t,r) + w_0(t,r)$, where $w_0 \in N$ is independent of J and W_J^* is uniquely determined. A priori estimates then show that

$$(2.8)\qquad \|W_J\|_1 \le c(\delta)\|h\|,$$

where $c(\delta)$ is a constant independent of J and such that $c(\delta) \to +\infty$ as $\delta \to 0^+$ or $\delta \to \sqrt{m^2 + \alpha - \theta_j^2}$. It then follows that $W_J^* \to w^*$ in $H^1_{T,\delta}$ as $J \to \infty$, and that w^* is the uniquely determined (weak) solution of (2.1), (1.3)-(1.4) belonging to $N^\perp$. Further estimates lead to (2.2).

THE EXISTENCE OF BREATHERS

In treating the nonlinear problem (1.2)-(1.4) we shall linearize about the origin, although in the present situation the remaining terms $[w - r\sin(w/r)]$ are not truly of higher order. In fact the nonlinearity is not Fréchet differentiable at $w = 0$. Nonetheless we have the following.

LEMMA 3.1. The function $\psi(t,r) = \alpha[w - r\sin(w/r)]$ generates a continuous Nemytsky operator [8] on $H_{T,\delta}$. Moreover, for any $\lambda > 0$ there is an $\eta > 0$ and a ball $B(\eta) = \{w \in H^1_{T,\delta} : \|w\|_1 < \eta\}$ such that

$$\|\psi(\cdot,w) - \psi(\cdot,v)\| \leq \lambda\|w-v\|_1, \qquad \forall\, w,v \in B(\eta). \tag{3.1}$$

Proof: Considering ψ as a function of $(r,w) \in \mathbf{R}^+ \times \mathbf{R}$ we find that

$$\psi(r,w) - \psi(r,v) = \alpha\left[w - v - 2r\cos\left(\frac{w+v}{2r}\right)\sin\left(\frac{w-v}{2r}\right)\right]. \tag{3.2}$$

In particular it follows that $|\psi(r,w) - \psi(r,v)| \leq 2\alpha|w - v|$, for all $w,v \in \mathbf{R}$ and $r > 0$. If $r \geq r_0 > 0$ then we may add $\pm(w - v)\cos(w+v/2r)$ and use the inequality $|x - b\sin(x/b)| \leq x^2/2b$, in which $b > 0$, to obtain

$$\begin{aligned} |\psi(r,w) - \psi(r,v)| &\leq \alpha(2|w-v||w+v| + |w-v|^2)/4r_0 \\ &\leq 3\alpha(|w| + |v|)|w-v|/4r_0 \end{aligned} \tag{3.3}$$

Hence for any $w,v \in H^1_{T,\delta}$ we have

$$\|\psi(\cdot,w) - \psi(\cdot,v)\|^2 \leq \frac{1}{T}\int_0^T\int_0^{r_0} 4\alpha^2|w-v|^2\, e^{2\delta r}\, drdt$$
$$+ \frac{1}{T}\int_0^T\int_{r_0}^{+\infty} \left(\frac{3\alpha}{4r_0}\right)^2 (|w| + |v|)^2|w-v|^2\, e^{2\delta r}\, drdt$$

The first integral is bounded by $4\alpha^2\sqrt{r_0}\, C^2\|w - v\|_1$, where C is an embedding constant. The second integral is bounded by $(3\alpha C/2r_0)^2(\|w\|_1^2 + \|v\|_1^2)\|w - v\|_1$, where again C denotes an embedding constant. We now choose $r_0 > 0$ sufficiently small and then $\eta > 0$ so that (3.1) is satisfied whenever $w,v \in B(\eta)$.

THEOREM 3.2. Let $n \geq 0$ be the integer such that $[2\pi n/T]^2 < m^2 + \alpha \leq [2\pi(n+1)/T]^2$. Then there are norm balls $B_0(\eta_0) = \{w \in N : \|w\|_1 < \eta_0\}$ and $B_1(\eta_1) = \{w \in N^\perp : \|w\|_1 < \eta_1\}$ such that for each $w_0 \in B_0(\eta_0)$ there is a uniquely determined $w^* \in B_1(\eta_1)$ with the property that $w = w^* + w_0$ is a solution of (1.2)-(1.4).

Proof: We write (1.2) in the form $w_{tt} - w_{rr} + (m^2 + \alpha)w = \psi(r,w)$, and then use Theorem 2.1 to see that $w \in H^1_{T,\delta}$ is a solution of (1.2)-(1.4) if and only if $w = w^* + w_0$ where $w^* = K\psi(r,w^* + w_0)$ and $w_0 \in N$. Since K is bounded and ψ is a Lipshitz map from $H^1_{T,\delta}$ to $H_{T,\delta}$ with constant as small as we please, as determined by our choice of η_0, η_1 according to the previous lemma, it follows that with w_0 held fixed we obtain a contraction mapping $F(w^*) = K\psi(\cdot, w^* + w_0)$ on $B_1(\eta_1)$. By the contraction mapping principle it follows that F has a unique fixed point $w^* \in B_1(\eta_1)$, and this is true for each $w_0 \in B_0(\eta_0)$.

Remark: It is not difficult to extend this argument [6] to show that the correspondence $w_0 \to w^*$ is continuous in the $H^1_{T,\delta}$ topology and defines a continuous map $w^* = \tau(w_0)$. Hence there is a C^0 manifold structure on the set of solutions we have found. In addition w^* satisfies (2.2) by the nature of K and w_0 automatically satisfies (2.2). Hence all solutions are bounded on $\mathbf{R} \times \mathbf{R}^+$ and have exponential decay as $r \to +\infty$.

FURTHER REMARKS

We reflect on some questions that remain to be resolved. First we point out that our analysis applies to the 1 and 3 dimensional cases. Our conjecture is that the same result is true in n-dimensions when $n < 5$. However, even in the 2-dimensional case the appearance of singular terms (i.e., Bessel functions of the 2nd kind) in the problems analogous to (2.4) prevents a direct application of the method of Laplace transforms.

The nature of the solutions in 3 dimensions is such that $u(t,r) = w(t,r)/r$ where $w \in L^\infty(\mathbf{R} \times \mathbf{R}^+)$. Hence $u(t,r) = O(1/r)$ as $r \to 0^+$. Clearly $u \in L^\infty(\mathbf{R} \times \mathbf{R}^+)$ if and only if $w(t,r) = O(r)$ as $r \to 0^+$. Since $w = w^* + w_0$ we may say without rigor that this implies that $w_0 = \Sigma w_{0j} \exp(-\beta_j r + i\theta_j t) = O(r)$ as $r \to 0^+$, where $\beta_j = \sqrt{m^2 + \alpha - \theta_j^2}$ and the sum is over $|j| < n$, and hence that $w_{0j} = 0$ for $|j| < n$. It would then follow that $w_1 = \tau(w_0) = 0$ and that none of the nontrivial solutions are bounded at the origin. However this argument is at best heuristic. In the 1-dimensional case there is a similar concern at the origin. All solutions in this case are bounded but may have discontinuities in their first spacial derivatives at the origin. Are there some which have more smoothness? The answer is an emphatic yes. There is a known [2] breather in this case with $m = 0$, $\alpha > 1$, and $T = 2\pi$. It is the smooth function $u(t,x) = 4 \arctan(b \sin t/\cosh bx)$, where $b = \sqrt{\alpha - 1}$. That this solution is on the solution manifold given by our existence results has been verified numerically, and corresponds to a certain path in the manifold parameterized through the coefficient of $\exp(-br)\sin(t)$, which is a basis function in N.

Let $H^1_r(\mathbf{R}^3)$ be the set of functions $u \in H^1(\mathbf{R}^3)$ which are radially symmetric. If $u_0 \in H^1_r(\mathbf{R}^3)$ and $u_1 \in L^2_r(\mathbf{R}^3)$ we may consider the Cauchy problem consisting of (1.1) together with the initial conditions $u(0,x) = u_0(x)$, $u_t(0,x) = u_1(x)$. It is natural to ask whether any of the breathers we have shown to exist are stable. More precisely, are there attracting breather states, and neighborhoods of initial values

$u_0(x)$, $u_1(x)$ which are drawn into these states? These questions and many others remain to be resolved.

ACKNOWLEDGEMENT: The author would like to thank Prof. H. A. Levine for numerous discussions on this problem and for the many penetrating questions he posed which helped refine our results. The author would also like to thank Prof. R. K. Miller for pointing out the classical Paley-Wiener theorem.

REFERENCES

[1] J.M. Coron, Periode minimale pour une corde vibrante de longueur infinite, C.R. Acad. Sci. Paris Ser. A, 294 (1982), 127-129.

[2] G.L. Lamb, Elements of Soliton Theory, John Wiley and Sons, New York, 1980.

[3] H.A. Levine, Minimal periods or solutions of semilinear wave equations in exterior domains and for solutions of the equations of nonlinear elasticity, to appear in J. Math. Anal. and Appl.

[4] R. Paley and N. Wiener, Fourier Transforms in the Complex Domain, A.M.S. Colloquium Publications, Vol. 19, Providence, R.I., 1934.

[5] M.W. Smiley, Time-periodic solutions of wave equations on $\mathbb{R}^1$ and $\mathbb{R}^3$, to appear in Math. Meth. Appl. Sci.

[6] M.W. Smiley, Breathers and Forced Oscillations of Nonlinear Wave Equations on $\mathbb{R}^3$, submitted to Comm. in Math. Phys.

[7] W.A. Strauss, Stable and unstable states of nonlinear wave equations, in Nonlinear Partial Differential Equations, Contemporary Mathematics, Vol. 17, A.M.S., Providence, R.I., 1983.

[8] M. Vainberg, Variational Methods for the Study of Nonlinear Operators, (translated by A. Feinstein), Holden-Day, San Francisco, 1964.

[9] P. Vuillermot, Nonexistence of spatially localized free vibrations for certain semilinear wave equations on R^2, C.R. Acad. Sci. Paris, Ser. I, 9 (1986), 395-398.

[10] A. Weinstein, Periodic nonlinear waves on a half-line, Comm. in Math. Phys. 99 (1985), 385-388.

[11] A Weinstein, Erratum, Periodic nonlinear waves on a half-line, Comm. in Math. Phys. 107 (1986), 177.

THE RICCATI EQUATION REVISITED

Andrew Vogt
Department of Mathematics
Georgetown University
Washington, D. C. 20057

At the First Howard University Symposium the author proposed a generalization of the Riccati equation: a (generalized) Riccati equation is a differential equation in x that arises from a linear system in u and v by the transformation $x = v^{-1}(u)$. In this note we review some consequences of this definition and elaborate some special cases.

§1. THE GENERAL FORM

The equation $x = v^{-1}(u)$ has to be interpreted. We take x and u to be members of a Banach space X, and v to be an invertible element of a subalgebra A of the algebra L(X) of bounded linear transformations from X into X. The algebra A is assumed to be closed and to contain the unit e of L(X). The field of scalars is taken to be the reals, and I denotes an open interval of the real line.

A Riccati equation associated with X, A, and I, it is shown in [4] where precise definitions, statements, and proofs are given, is any quadratic differential equation of the form:

$$dx/dt = a(t)(x)(x) + b(t)(x) + c(t) \tag{1.1}$$

where $t \longmapsto a(t), b(t), c(t)$ are continuous functions from I into M(X,A), $\mathcal{L}(A)$, and X respectively. The set M(X,A) consists of all continuous linear transformations from X into A satisfying

$$v \circ a(x)(x) = a(v(x))(x) \tag{1.2}$$

for all x in X and v in A, and $\mathcal{L}(A)$ is the set of all members b of L(X) such that

$$(b \circ v - v \circ b) \text{ is in A for all v in A.} \tag{1.3}$$

The linear system in X⊕A associated with the Riccati equation (1.1) is:

$$\begin{cases} du/dt = b_2(t)(u) + v(c(t)) \\ dv/dt = -a(t)(u) - v \circ b_1(t) + b_2(t) \circ v - v \circ b_2(t). \end{cases} \tag{1.4}$$

Here $b(t) = b_1(t) + b_2(t)$ is any decomposition of b into two continuous functions with values in A and $\mathcal{L}(A)$ respectively. For example, $b_1 = 0$, $b_2 = b$.

Any solution (u(t),v(t)) of (1.4) with v(t) invertible in A yields a solution of (1.1) by the substitution $x(t) = v(t)^{-1}(u(t))$. Since the initial value of (u,v) can be set equal to (x_0,e) and (1.1) has a unique solution with initial value x_0, all solutions of (1.1) come from solutions of (1.4).

The above construction can be used to generate Riccati equations. We take as inputs a Banach space X, a subalgebra A of L(X), and an interval I; and the outputs are all functions a, b, and c of the variable t in I having the properties noted above, and hence all Riccati equations. Included among the outputs are the familiar scalar and matrix Riccati equations (see [4]).

Potential benefits from this point of view are application of linear methods to a larger class of equations and greater insight into the nature of Riccati equations. One may also hope ultimately to arrive at a simpler description of Riccati equations - e.g., one in which the role of the algebra A is deemphasized, or most desirable of all a criterion that would enable one to tell by inspection whether a given equation is a Riccati equation or not. Progress on these issues is reported below.

§2. THE AUTONOMOUS CASE

To avoid unnecessary notational complexity, let us assume that the coefficients are independent of t. Then the Riccati equation has the form:

$$dx/dt = a(x)(x) + b(x) + c \tag{2.1}$$

where a belongs to M(X,A), b belongs to $\mathcal{L}$(A), and c belongs to X.

It is evident from (1.2) and (1.3) that the function in (2.1) is not the most general continuous quadratic function from X into X. As the algebra A varies in L(X), the requirements for membership in M(X,A) and $\mathcal{L}$(A) vary. As A gets larger, it is easier to find linear transformations a from X into A but harder to ensure that (1.2) holds for all v in A. Likewise, there is a trade-off associated with membership in $\mathcal{L}$(A): as A gets larger, it is easier to find members b of L(X) for which $(b\circ v - v\circ b)$ in A when v is a given element of A, but there are more elements v of A for which this condition must be true.

Proposition 2.1: If A is either $\{\lambda e : \lambda \text{ is in } R\}$ or L(X), then $M(X,A) \simeq X^*$ and $\mathcal{L}(A) = L(X)$.

Proof: For either algebra, equality of $\mathcal{L}$(A) and L(X) follows trivially from (1.3).

Suppose that A consists of the scalar multiples of the identity e in L(X). For f in the dual space X^*, define $\hat{f} : X \longrightarrow A$ by $\hat{f}(x) = f(x)e$. It is easily seen that $\hat{f}$ is continuous and linear and satisfies (1.2). Moreover, the mapping $f \longmapsto \hat{f}$ is a continuous linear monomorphism from X^* into M(X,A). The mapping is onto since any member a of M(X,A) obviously satisfies $a(x) = f(x)e$ for some f in X^*.

Alternatively, suppose A = L(X). For f in X^* define $\tilde{f} : X \longrightarrow A$ by $\tilde{f}(x) = f(\ )x = f\otimes x$. Continuity and linearity are clear, while (1.2) reduces to the observation that for v in L(X), $v\circ\tilde{f}(x)(x) = f(x)v(x) = \tilde{f}(v(x))(x)$. One-to-oneness of the mapping $f \longmapsto \tilde{f}$ is clear, and it remains only to establish surjectivity.

Let a be any member of M(X,A). Assume that for some nonzero x in X a(x)(x) is independent of x. Given y and z in X, let v be a member of L(X) taking x to y and a(x)(x) to z (e.g., let $v = f\otimes y + g\otimes z$ where f and g are members of X^* with f zero on a(x)(x) and one on x, and g vice versa). By (1.2) z = a(y)(x). Yet this is impossible since the same equation may be obtained with a different z but the same x and y. So our assumption must be false: a(x)x = f(x)x where f is some fixed scalar-valued function of x such that f(0) = 0.

By (1.2) $a(v(x))(x) = v \circ a(x)(x) = f(x)v(x)$. For vectors x and y in X, with x nonzero, choose v in L(X) with v(x) = y. Then a(y)(x) = f(x)y. Since a(y) is in L(X), f must be a member of X^*. Then a(y) = $f \otimes y = \tilde{f}(y)$, and surjectivity is established. ■

The Riccati equations associated with the algebras in Proposition 2.1 are of the form

$$dx/dt = f(x)x + b(x) + c \tag{2.2}$$

where f is in X^*, b is in L(X), and c is in X. The linear term is completely general but the quadratic one is very special.

Among subalgebras A in L(X), the algebras $\{\lambda e : \lambda \text{ is in } R\}$ or L(X) are extreme cases. There is reason to believe that M(X,A) gets larger, and $\mathcal{L}(A)$ smaller, as the algebra A moves in size away from these two extremes.

The commutant A' of a set of operators A in L(X) is defined by

$$A' = \{b : b \text{ is in } L(X), \text{ and } b \circ v = v \circ b \text{ for all } v \text{ in } A\}.$$

The commutant is a closed Banach subalgebra of L(X) containing the unit e. It has other properties of note: $A_1 \subseteq A_2 \Rightarrow A_2' \subseteq A_1'$, $A \subseteq A''$, $A \cup A' \subseteq \mathcal{L}(A)$ when A is an algebra, and the following.

<u>Proposition 2.2</u>: Let A be a subalgebra of L(X). Then $\mathcal{L}(A) \subseteq \mathcal{L}(A')$.

<u>Proof</u>: Let b be in $\mathcal{L}(A)$, u in A', and v in A. Then $(b \circ u - u \circ b) \circ v = (b \circ v - v \circ b) \circ u + v \circ b \circ u - u \circ b \circ v = u \circ (b \circ v - v \circ b) + v \circ b \circ u - u \circ b \circ v = v \circ (b \circ u - u \circ b)$. Hence $(b \circ u - u \circ b)$ is in A' and b is in $\mathcal{L}(A')$. ■

Whenever an algebra A satisfies A = A" (e.g., our two extreme cases or any von Neumann algebra), then $\mathcal{L}(A) \subseteq \mathcal{L}(A') \subseteq \mathcal{L}(A'') = \mathcal{L}(A)$. Hence $\mathcal{L}(A) = \mathcal{L}(A')$.

Let B be a real Banach algebra with unit. Then l and r respectively denote the canonical algebra homomorphism and antihomomorphism from B into L(B) defined by $l(b)(c) = bc$ and $r(b)(c) = cb$ for b,c in B. The sets l(B) and r(B) are Banach subalgebras of L(B) sharing its unit, and l and r are isometries between B and these subalgebras. By Der B we denote the set of derivations on B, i.e., members δ of L(B) satisfying $\delta(uv) = u\delta(v) + \delta(u)v$ for all u and v in B.

Proposition 2.3: Let X be a Banach algebra B with unit e, and let $A = l(B)$. Then $M(X,A) \simeq B$, $A' = r(B)$, and $\mathcal{L}(A) = \text{Der } B \oplus r(B)$.

Proof: If b is in B, let $\hat{b} : X \longrightarrow A$ be defined by $\hat{b}(x) = l(xb)$. For x and y in B, (1.2) holds: $l(y)\circ\hat{b}(x)(x) = yxbx = \hat{b}(l(y)(x))(x)$. It is also easy to see that the map $b \longmapsto \hat{b}$ is a continuous linear monomorphism from B into M(X,A). Let a be an element of M(X,A). For each x in B there is an element $a'(x)$ in B such that $a(x) = l(a'(x))$. Set $b = a(e)(e)$. Then $0 = a(l(x)(e))(e) - l(x)\circ a(e)(e) = a(x)(e) - xb = a'(x) - xb$. So $a(x) = l(xb) = \hat{b}(x)$, and the monomorphism is an isomorphism.

Now $A' = \{w : w \text{ is in } L(B),\ w\circ (x) = l(x)\circ w \text{ for all } x \text{ in } B\}$. If y and z are in B, $r(z)\circ l(x)(y) = xyz = l(x)\circ r(z)(y)$. Thus $r(z)\circ l(x) = l(x)\circ r(z)$, r(z) is in A', and $r(B) \subseteq A'$. Conversely, if w is in A', then $0 = w\circ l(x)(e) - l(x)\circ w(e) = w(xe) - xw(e) = w(x) - r(w(e))(x)$ for all x in B. Hence $w = r(w(e))$, and $A' = r(B)$.

If δ is in $\text{Der } B \subseteq L(B)$, for u and v in B, $\delta\circ l(u)(v) - l(u)\circ\delta(v) = \delta(uv) - u\delta(v) = \delta(u)v = l(\delta(u))(v)$. So $\delta\circ l(u) - l(u)\circ\delta$ is in A, δ is in $\mathcal{L}(A)$, and $\text{Der } B \subseteq \mathcal{L}(A)$. If δ is in $\text{Der } B \cap r(B)$, then $0 = \delta\circ l(x) - l(x)\circ\delta = l(\delta(x))$ for x in B. Hence $\delta(x) = 0$ for all x, and $\delta = 0$. Thus $\text{Der } B \oplus r(B) = \text{Der } B \oplus A' \subseteq \mathcal{L}(A)$.

For w in $\mathcal{L}(A)$ let $\delta_w : B \longrightarrow B$ be defined by $\delta_w(v) = w(v) - vw(e)$. Then δ_w is linear and continuous. For u and v in B,

$$
\begin{aligned}
&\delta_w(uv) - u\delta_w(v) - \delta_w(u)v = \\
&w(uv) - uvw(e) - uw(v) + uvw(e) - w(u)v + uw(e)v = \\
&\{w(uv) - uw(v)\} - \{w(u) - uw(e)\}v = \\
&(w\circ l(u) - l(u)\circ w)(v) - \{(w\circ l(u) - l(u)\circ w)(e)\}v = \\
&l(x)(v) - \{l(x)(e)\}v = 0
\end{aligned}
$$

where x is an element of B such that $w\circ l(u) - l(u)\circ w = l(x)$. Thus δ_w is in Der B. For u and v in B, $\{(w - \delta_w)\circ l(u) - l(u)\circ(w - \delta_w)\}(v) = (w - \delta_w)(uv) - u(w - \delta_w)(v) = uvw(e) - u(vw(e)) = 0$. Thus $w - \delta_w$ is in

A', $w = \delta_W + (w - \delta_W)$ is in Der B ⊕ A' = Der B ⊕ r(B), and the latter set is identical with $\mathcal{L}$(A). ■

The Riccati equation in a Banach algebra B with unit accordingly has a quadratic term of the form l(xa)(x) where a is some element of B, and a linear term $(r(b) + \delta)(x)$ where b belongs to B and δ is a derivation. Thus (2.1) becomes:

$$dx/dt = xax + xb + \delta(x) + c$$

Der B includes inner derivations, i.e., maps of the form $x \longmapsto b_1x - xb_1$. So the Riccati equation can be rewritten in the form

$$dx/dt = xax + b_1x + xb_2 + \delta(x) + c \qquad (2.3)$$

where a, b_1, b_2, and c are arbitrary members of B and δ is either the zero map or a derivation on B that is not inner. Many Banach algebras have derivations that are not inner. Equation (2.3) without δ is the standard form of Riccati equation in a Banach algebra.

§3. THE HOMOGENEOUS CASE

Now let us examine the homogeneous autonomous equation

$$dx/dt = a(x)(x) \qquad (3.1)$$

with a in M(X,A).

The condition for membership in M(X,A) implies a more general relation.

<u>Proposition 3.1</u>: Let a be a member of M(X,A). Then

$$v \circ a(u(x)) \circ w(x) = a((v \circ u)(x)) \circ w(x) \qquad (3.2)$$

for all u, v, w in A and x in X.

<u>Proof</u>: Let λ be a nonzero scalar such that $q = e + \lambda w$ is invertible in

A. Then by linearity

$$v \circ a(u(x)) \circ w(x) = (1/\lambda)\{v \circ a(u(x)) \circ q(x) - v \circ a(u(x))(x)\} .$$

Omitting $(1/\lambda)$, we may rewrite the first term on the right side as:

$$\begin{aligned} v \circ a(u(x)) \circ q(x) &= v \circ a((u \circ q^{-1})(q(x)))(q(x)) \\ &= v \circ u \circ q^{-1} \circ a(q(x))(q(x)) \\ &= a((v \circ u \circ q^{-1})(q(x)))(q(x)) \\ &= a((v \circ u)(x))(q(x)) \\ &= a((v \circ u)(x))(x) + \lambda a((v \circ u)(x))(w(x)) \\ &= v \circ u \circ a(x)(x) + \lambda a((v \circ u)(x)) \circ w(x) \\ &= v \circ a(u(x))(x) + \lambda a((v \circ u)(x)) \circ w(x) . \end{aligned}$$

Combining this calculation with the previous equation, we get (3.2). ∎

Let x_0 be a fixed nonzero element of X. Then $Ax_0 = \{v(x_0) : v \text{ is in } A\}$ is an invariant subspace of X under A. If $\hat{a}(x)$ denotes the restriction of $a(x)$ to this subspace, by Proposition 3.1

$$v \circ \hat{a}(y) = \hat{a}(v(y)) \tag{3.3}$$

for y in Ax_0 and v in A. This suggests that instead of requiring that the continuous linear transformation $a: X \longrightarrow L(X)$ satisfy

$$v \circ a(x)(x) = a(v(x))(x) , \tag{3.4}$$

we might ask that

$$v \circ a(x) = a(v(x)) \tag{3.5}$$

for all x in X and v in A.

Example: (3.5) implies (3.4), but the converse implication does not hold. Let $X = R^3$, and let A be the algebra of all real matrices of the form

$$\begin{bmatrix} v_1 & v_2 & v_3 \\ 0 & v_1 & 0 \\ 0 & 0 & v_1 \end{bmatrix} . \tag{3.6}$$

Define $a: R^3 \longrightarrow A$ as follows: for a column vector with entries x, y, z, let a(x,y,z) equal

$$\begin{bmatrix} sy+tz & sx+b_1y+b_2z & tx+c_1y+c_2z \\ 0 & sy+tz & 0 \\ 0 & 0 & sy+tz \end{bmatrix}$$

where s, t, b_1, b_2, c_1, c_2 are given scalars.

The Riccati equation corresponding to this transformation is:

$$\begin{cases} dx/dt = 2(sy+tz)x + (b_1y+b_2z)y + (c_1y+c_2z)z \\ dy/dt = (sy+tz)y \\ dz/dt = (sy+tz)z\ . \end{cases} \tag{3.7}$$

This system is easily solved by first passing to a scalar Riccati equation in the variable sy + tz.

The transformation a satisfies (3.4) but not (3.5) unless s = t = 0. Indeed, if v is a member of A of the form (3.6), a short computation shows that va(x,y,z) - a(v(x,y,z)) equals

$$(v_2t - v_3s) \cdot \begin{bmatrix} 0 & z & -y \\ 0 & 0 & 0 \\ 0 & 0 & 0 \end{bmatrix} .$$

This is the zero transformation for all choices of v if and only if s = t = 0. On the other hand, applied to the vector (x,y,z) it yields the zero vector regardless of the values of s, t or v.

It should be noted that equation (3.7) can be derived from a different mapping a that does satisfy (3.5).

As long as we contemplate adjustments in (3.4), we propose yet another. Suppose that $a : X \longrightarrow L(X)$ is a continuous linear transformation satisfying:

$$a(x) \circ a(y) = a(a(x)y) \tag{3.8}$$

for all x and y in X. Then the set $A = \{w : w \text{ is in } L(X) \text{ and } w \circ a(x) = a(w(x)) \text{ for all } x \text{ in } X\}$ is easily seen to be a closed subalgebra of L(X) containing the unit e and containing range a. In particular, (3.5) is true if a satisfies (3.8) and A is the algebra just introduced. This enables us to free ourselves from a preassigned algebra by requiring that $a : X \longrightarrow L(X)$ satisfy (3.8).

Indeed, in [4] we showed that when a satisfies (3.8), we can write down an explicit solution of the initial value problem

$$dx/dt = a(x)(x),\ x(0) = x_0$$

namely, $x(t) = (e - ta(x_0))^{-1}(x_0)$.

On the Banach space X we now define a binary operation * by:

$$x * y = a(x)y \tag{3.9}$$

for x and y in X. Then (3.8) is merely the requirement that this multiplication be associative: $(x * y) * z = a(a(x)y)z = a(x)\circ a(y)z = x * (y * z)$. Bilinearity of the multiplication is a consequence of (3.9).

Proposition 3.2: Let $a : X \longrightarrow L(X)$ be a continuous linear transformation satisfying (3.8). Then X is a Banach algebra under the multiplication * given by (3.9). Furthermore, every Banach algebra may be obtained in this manner.

Proof: X is a Banach space, and (3.9) equips X with a bilinear associative multiplication that is continuous in each variable separately. By [3], p. 5, X is a Banach algebra (its norm may need to be replaced by an equivalent norm to guarantee that $\|x * y\| \leq \|x\|\|y\|$ for all x and y). Conversely, given a Banach algebra X with multiplication *, (3.9) can be used to define a(x) and a, and these transformations are easily seen to be continuous and linear, with a satisfying (3.8).∎

When X is equipped with the multiplication *, the mapping $a : X \longrightarrow L(X)$ can be regarded as an algebra homomorphism: $a(x * y) = a(a(x)y) = a(x)\circ a(y)$. Moreover, range a is a subalgebra of L(X), although it may not be closed and may not contain the unit of L(X). The homogeneous Riccati equation (3.1) can be rewritten as:

$$dx/dt = x * x. \tag{3.10}$$

Thus Riccati equations based on (3.5) or (3.8) - and these include the matrix Riccati equations as well as (2.3) - all reduce to the elementary form (3.10), reminiscent of the scalar case.

To find the multiplication that accomplishes this reduction, we identify X with any known Banach algebra A to which it is linearly homeomorphic. (If X is finite-dimensional, we can use any Banach algebra A of the same dimension.) If $L : X \longrightarrow A$ is the linear homeomorphism, X is a Banach algebra under the multiplication $x * y = L^{-1}(L(x)L(y))$. Given one multiplication $*$ on X making X into a Banach algebra, others can be obtained by applying a continuous linear automorphism $L : X \longrightarrow X$ as above, or by defining a new multiplication $\odot$ by $x \odot y = x * a * y$ where a is any fixed element of X. In function spaces two types of Riccati equations obviously follow this pattern:

$$\partial u/\partial t = a(x)u(x)^2 \qquad \text{and} \qquad \partial u/\partial t = a(x) * u(x) * u(x)$$

where $*$ denotes convolution.

The Banach algebra structure induced on X by Proposition (3.2) need not be of any special form: the multiplication $*$ need not be commutative and X may have no unit element. Nor are there restrictions on the dimension of X in order for it to possess nontrivial Banach algebra structures, as will be seen below.

§4. FINITE-DIMENSIONAL CYCLIC CASES

From now on we restrict our attention to equations of type (3.1) with a satisfying (3.8). The next proposition yields a method for producing a large class of examples of such equations.

Proposition 4.1: Let $a : X \longrightarrow L(X)$ be a continuous linear transformation satisfying (3.8). Let v be a member of L(X) such that $a(v(x)) = v \circ a(x)$ for any x in X. Then if p is any polynomial with real coefficients and x is any element of X,

$$a(p(v)x) = p(v) \circ a(x) \; . \tag{4.1}$$

<u>Proof</u>: By linearity it suffices to establish that $a(v^n(x)) = v^n \circ a(x)$ for nonnegative integers n. Obviously this is true for $n = 0$. The inductive step is: $a(v^{n+1}(x)) = a(v\circ(v)^n(x)) = v\circ a(v^n(x)) = v\circ v^n \circ a(x) = v^{n+1}\circ a(x)$. ■

(4.1) suggests the following device. Suppose that X has finite dimension n. Choose if possible an element T in L(X) such that $a(T(x)) = T\circ a(x)$ for all x and such that T has a cyclic vector x_0, i.e., one with span $\{x_0, Tx_0, \ldots, T^{n-1}x_0\} = X$. By (4.1) $a(p(T)x_0) = p(T)a(x_0)$ for any polynomial p. Since any vector x in X can be expressed as $p(T)x_0$ for some polynomial p, a is completely determined by the value of $a(x_0)$. For simplicity we assume $a(x_0) = e$. Let m be the minimal (monic) polynomial of T. Since T has a cyclic vector, m has degree n and coincides with the characteristic polynomial of T.

Define $a: X \longrightarrow L(X)$ by

$$a(p(T)x_0) = p(T).$$

Then a is well-defined: if $p(T)x_0 = q(T)x_0$, $p(T) = q(T)$. Furthermore, $a(a(x)y) - a(x)\circ a(y) = a(a(p(T)x_0)q(T)x_0) - a(p(T)x_0)\circ a(q(T)x_0) = a(p(T)q(T)x_0) - p(T)q(T) = 0$. So a yields a Riccati equation. In fact, if $c_0, \ldots, c_{n-1}$ are the linear functionals dual to the basis vectors $x_0, Tx_0, \ldots, T^{n-1}x_0$, so that

$$x = \sum_{i=0}^{n-1} c_i(x)T^i x_0,$$

then the corresponding Riccati equation of type (3.1) is:

$$dx/dt = \sum_{i=0}^{n-1} c_i(x)T^i x. \qquad (4.2)$$

A necessary and sufficient condition for a linear operator T on a finite-dimensional space X to have a cyclic vector is that in its Jordan form all Jordan blocks corresponding to a given eigenvalue have different sizes. (If we require that the decomposition be real, a similar statement applies to the real canonical form. See [2], p. 130 and thereabouts.)

<u>Example</u> It is instructive to consider the Riccati equation corresponding to a single Jordan block. Equation (4.2) will uncouple into a sys-

tem of such equations in the appropriate coordinate system. Let the matrix of T be

$$\begin{bmatrix} \lambda_0 & 1 & 0 & \dots & 0 \\ 0 & \lambda_0 & 1 & \dots & 0 \\ \dots & \dots & \dots & \dots & \dots \\ 0 & & \dots\dots & & \lambda_0 \end{bmatrix}$$

relative to the standard basis $e_1, e_2, \dots, e_n$ for $X = R^n$.

Then the minimal polynomial of T is $m(\lambda) = (\lambda - \lambda_0)^n$, and e_n is a cyclic vector for R^n. Each element of the standard basis can be obtained from the equation $e_j = (T - \lambda_0)^{n-j} e_n$ for $1 \le j \le n$. Let $x_1, x_2, \dots, x_n$ denote the components of a vector x in R^n with respect to the standard basis. Then the Riccati equation (4.2) takes the form

$$dx/dt = p(T)x$$

where $p(T)e_n = x = \sum_j x_j e_j = \sum_j x_j (T - \lambda_0)^{n-j} e_n$. Thus $p(T) = \sum_j x_j (T - \lambda_0)^{n-j}$, and

$$\begin{aligned} dx/dt &= \sum_j x_j (T - \lambda_0)^{n-j} \left(\sum_i x_i (T - \lambda_0)^{n-i} e_n\right) \\ &= \sum_{j,i} x_j x_i (T - \lambda_0)^{2n-j-i} e_n \\ &= \sum_{k=1}^{n} \left(\sum_{j=k}^{n} x_j x_{n+k-j}\right)(T - \lambda_0)^{n-k} e_n \\ &= \sum_{k=1}^{n} \left(\sum_{j=k}^{n} x_j x_{n+k-j}\right) e_k \, . \end{aligned}$$

Writing this vector equation as a system of scalar equations, we get:

$$dx_k/dt = x_k x_n + x_{k+1} x_{n-1} + \dots + x_n x_k \quad \text{for } k = 1, \dots, n \, . \tag{4.3}$$

If n = 1, this gives $dx/dt = x^2$; if n = 2 or 3, the systems

$$\begin{cases} dx_1/dt = x_1 x_2 + x_2 x_1 \\ dx_2/dt = (x_2)^2 \end{cases} \quad \text{and} \quad \begin{cases} dx_1/dt = x_1 x_3 + (x_2)^2 + x_3 x_1 \\ dx_2/dt = x_2 x_3 + x_3 x_2 \\ dx_3/dt = (x_3)^2 \; ; \end{cases}$$

and so forth.

Evidently these equations unpeel. If one solves the bottom equation (a scalar Riccati equation), one can then solve the preceding one and continue upward, solving them all. Successive equations after the bottom one, with substitutions from equations already solved, are linear but have time-dependent coefficients and nonhomogeneous terms.

The algebraic structure induced by the map $a(p(T)x_0) = p(T)$ on the Banach space X is straight-forward. For polynomials p and q, the product of $x = p(T)x_0$ and $y = q(T)x_0$ is $x * y = p(T)q(T)x_0$. Hence X is isomorphic to the quotient ring $R[\lambda]/(m(\lambda))$, and the isomorphism is the map taking $p(T)x_0$ to $p(\lambda)$ mod $m(\lambda)$. Since any nth degree polynomial m is the minimal polynomial of a linear operator T in an n-dimensional space (for example, take T to be given with respect to some basis by the companion matrix of $m(\lambda)$ - see [1], pp. 316-318), the number of algebraic structures possible is large.

Indeed, the methods used to arrive at (4.2) can be generalized somewhat and explicitly described in terms of polynomials.

<u>THEOREM 4.2</u>: Let X be of finite dimension n, let $T : X \longrightarrow X$ be a linear transformation possessing a cyclic vector x_0, let m be the minimal polynomial of T.

If $a : X \longrightarrow L(X)$ is a linear transformation satisfying $a(y) \circ a(x) = a(a(y)x)$ for all x and y in X, and T is a member of range a, then either

i) there exists a polynomial k in $R[\lambda]$ of degree less than n such that g.c.d $\{k,m\} = 1$ or λ, and for any polynomial p

$$a(p(T)x_0) = p(T)k(T); \tag{4.4}$$

or alternatively

ii) there exist polynomials j and k in $R[\lambda]$ and a nonzero real number c such that $m(\lambda) = c(j(\lambda) - \lambda)\lambda$, degree $k < n$, g.c.d $\{k,m\} = 1$,

$$\text{m is a factor of } j \circ j - j \text{ and of } k \circ j - k, \tag{4.5}$$

and for any polynomials p and q in $R[\lambda]$

$$a(p(T)x_0) = p(T)S \tag{4.6}$$

where $S : X \longrightarrow X$ is defined by

$$S(q(T)x_0) = q \circ j(T)k(T)x_0 \ . \tag{4.7}$$

Conversely, if polynomials are given satisfying the conditions of i) or ii) and if a is defined by (4.4) or by (4.6) and (4.7), then a satisfies all the conditions stated at the outset, and range a is a commutative or noncommutative algebra according as i) or ii) holds.

Remarks: Before embarking upon a proof, let us note that polynomials j, k, and m satisfying the requirements of ii), particularly (4.5), can be found readily. One example is $j(\lambda) = \lambda^2$, $k(\lambda) = \lambda^2 + 1$, $m(\lambda) = \lambda(\lambda^2 - \lambda)$. More generally, one can begin by constructing j: choose j so that the polynomial

$$\frac{j(j(\lambda)) - j(\lambda)}{j(\lambda) - \lambda}$$

is divisible by λ. Then choose k so that

$$\frac{k(j(\lambda)) - k(\lambda)}{j(\lambda) - \lambda}$$

is divisible by λ and g.c.d. $\{k(\lambda), \lambda(j(\lambda) - \lambda)\} = 1$. Take $m(\lambda)$ to be a monic scalar multiple of $\lambda(j(\lambda) - \lambda)$. A trivial choice of k is the constant function $k(\lambda) \equiv 1$.

Proof of the Theorem: Given a, choose y_0 such that $a(y_0) = T$. Let $S = a(x_0)$, and let j and k be the (unique) polynomials of degree < n such that $S(x_0) = k(T)x_0$ and $S(y_0) = j(T)x_0$.

By Proposition 4.1 $a(p(T)x_0) = p(T)a(x_0) = p(T)S$ for every polynomial p. The equation $a(x_0) \circ a(y_0) = a(a(x_0)y_0)$ translates into $ST = a(Sy_0) = a(j(T)x_0) = j(T)S$. Hence $ST^k = (j(T))^k S$ for every nonnegative integer n, and by linearity $Sp(T) = p(j(T))S$ for any polynomial p. It follows that

$$S(q(T)x_0) = q \circ j(T)Sx_0 = q \circ j(T)k(T)x_0 \tag{4.8}$$

for every polynomial q. This equation defines S completely in terms of j and k.

The element $y_0 = u(T)x_0$ for some polynomial u. Hence $T = a(y_0) = a(u(T)x_0) = u(T)S$. It follows that $Tx_0 = u(T)S(x_0) = u(T)k(T)x_0$ and that $\lambda = u(\lambda)k(\lambda) \bmod m(\lambda)$. So there is a polynomial v such that $\lambda = u(\lambda)k(\lambda) + v(\lambda)m(\lambda)$. Thus g.c.d $\{k(\lambda), m(\lambda)\} = 1$ or λ. Likewise, $T^2 = u(T)ST = u(T)j(T)S = j(T)u(T)S = j(T)T$, so that $(j(\lambda) - \lambda)\lambda = 0 \bmod m(\lambda)$.

Suppose that $k(0) = 0$. Then for some polynomial k_0 $k(\lambda) = \lambda k_0(\lambda)$. In (4.8) one obtains $q\circ j(T)k(T) = q\circ j(T)Tk_0(T) = q(T)Tk_0(T) = q(T)k(T)$ since $q\circ j(\lambda)\lambda - q(\lambda)\lambda$ is divisible by $(j(\lambda) - \lambda)\lambda$ and hence by $m(\lambda)$. Thus in (4.8) $S(q(T)x_0) = q(T)k(T)x_0 = k(T)(q(T)x_0)$ for all polynomials q. So $S = k(T)$, and a is given by (4.4), in part i) of the Theorem. The same result follows trivially if $j(\lambda) = \lambda$.

Assume now that $k(0) \neq 0$ and that $j(\lambda)$ is not the first degree polynomial λ. The equation $(j(\lambda) - \lambda)\lambda = 0 \bmod m(\lambda)$ then implies since degree $j < n$ that degree $j = n - 1$ and that $m(\lambda) = c(j(\lambda) - \lambda)\lambda$ for some nonzero real number c.

The equation $a(x_0)\circ a(x_0) = a(a(x_0)x_0)$ translates into $S^2 = a(Sx_0) = a(k(T)x_0) = k(T)S$. It follows from (4.8) that $(k(T))^2x_0 = k(T)S(x_0) = S^2(x_0) = S(k(T)x_0) = k\circ j(T)k(T)x_0$. Hence, $(k\circ j - k)k = 0 \bmod m$. Similarly, $k(T)j(T)k(T)x_0 = k(T)STx_0 = S^2Tx_0 = S(j(T)k(T)x_0) = j\circ j(T)k\circ j(T)k(T)x_0$. So $k^2j = (j\circ j)(k\circ j)k \bmod m$. Combining this with the previous mod m result, we find that $(j\circ j - j)k^2 = 0 \bmod m$. Now $(k\circ j - k)$ and $(j\circ j - j)$ are both divisible by the polynomial $j(\lambda) - \lambda$. Since $k(0) \neq 0$ and $m(\lambda) = c(j(\lambda) - \lambda)\lambda$, (4.5) is true. Since equation (4.8) and (4.7) coincide, we have established case ii).

Now consider case i) of the converse. If a is defined by (4.4), a is obviously a linear map from X into L(X). Then $a(a(p(T)x_0)q(T)x_0) = a(p(T)k(T)q(T)x_0) = p(T)k(T)q(T)k(T) = a(p(T)x_0)\circ a(q(T)x_0)$ for arbitrary p and q, and (3.8) holds. Since g.c.d $\{k,m\}$ equals 1 or λ, there are polynomials u and v such that $\lambda = u(\lambda)k(\lambda) + v(\lambda)m(\lambda)$. Hence, $T = u(T)k(T) = a(u(T)x_0)$ is in range a. Commutativity of range a follows trivially from (4.4).

For case ii) of the converse, suppose j and k satisfy the conditions of case ii). S as in (4.7) is a well-defined member of L(X): since $m\circ j = c(j\circ j - j)j$ is divisible by m, different representations of the domain variable for S do not affect the value of the image. And so a as in (4.6) is a well-defined linear transformation from X into L(X).

To obtain (3.8), let p, q, and r be polynomials. Then

$$a(p(T)x_0)\circ a(q(T)x_0)(r(T)x_0) = p(T)Sq(T)S(r(T)x_0) =$$
$$p(T)Sq(T)r\circ j(T)k(T)x_0 = p(T)q\circ j(T)r\circ j\circ j(T)k\circ j(T)k(T)x_0,$$

and this can be compared with

$$a(a(p(T)x_0)(q(T)x_0))(r(T)x_0) = a(p(T)Sq(T)x_0)(r(T)x_0) =$$

$$a(p(T)q\circ j(T)k(T)x_0)(r(T)x_0) = p(T)q\circ j(T)k(T)r\circ j(T)k(T)x_0.$$

The only factors that appear to differ in the final versions of each expression are $r\circ j\circ j(T)k\circ j(T)$ and $r\circ j(T)k(T)$. However,

$$\begin{aligned}(r\circ j\circ j)(k\circ j) - (r\circ j)k &= (r\circ j\circ j)k - (r\circ j)k \bmod m \\ &= (r\circ j\circ j - r\circ j)k \bmod m .\end{aligned}$$

Since $r\circ j\circ j - r\circ j$ is divisible by $j\circ j - j$ and the latter is divisible by m, the factors are equal. That T is in range a follows as in case i).

Finally we show that in case ii) range a is noncommutative. Since S = $a(x_0)$ is in range a, it suffices to show that $ST(x_0)$ and $TS(x_0)$ are distinct. Suppose to the contrary that they are equal. Then 0 = $ST(x_0) - TS(x_0) = j(T)k(T)x_0 - Tk(T)x_0$. So $(j(\lambda) - \lambda)k(\lambda)$ is divisible by $m(\lambda) = c(j(\lambda) - \lambda)\lambda$, but then $k(\lambda)$ is divisible by λ and the requirement that g.c.d. $\{k,m\} = 1$ fails. ∎

A simple calculation shows that the homogeneous Riccati equations resulting from (4.4) and (4.6) have the forms:

$$dx/dt = k(T)p_x(T)x$$

and

$$dx/dt = k(T)p_x\circ j(T)x$$

where the subscript x is a reminder that the polynomial p depends on x through the relation $p_x(T)x_0 = x$.

References

[1] G. Birkhoff and S. Mac Lane, A Survey of Modern Algebra, Revised Edition, Macmillan, New York, 1953.

[2] M. W. Hirsch and S. Smale, Differential Equations, Dynamical Systems, and Linear Algebra, Academic Press, New York, 1974.

[3] C. E. Rickart, General Theory of Banach Algebras, Van Nostrand, Princeton, N. J., 1960.

[4] A. Vogt, The Riccati equation: when nonlinearity reduces to linearity, in Nonlinear Semigroups, Partial Differential Equations and Attractors, T. L. Gill and W. W. Zachary, eds., Lecture Notes in Mathematics 1248, Springer-Verlag, Berlin, 1987, pp. 169-185.

LECTURE NOTES IN MATHEMATICS
Edited by A. Dold and B. Eckmann

Some general remarks on the publication of proceedings of congresses and symposia

Lecture Notes aim to report new developments - quickly, informally and at a high level. The following describes criteria and procedures which apply to proceedings volumes. The editors of a volume are strongly advised to inform contributors about these points at an early stage.

§1. One (or more) expert participant(s) of the meeting should act as the responsible editor(s) of the proceedings. They select the papers which are suitable (cf. §§ 2, 3) for inclusion in the proceedings, and have them individually refereed (as for a journal). It should not be assumed that the published proceedings must reflect conference events faithfully and in their entirety. Contributions to the meeting which are not included in the proceedings can be listed by title. The series editors will normally not interfere with the editing of a particular proceedings volume - except in fairly obvious cases, or on technical matters, such as described in §§ 2, 3. The names of the responsible editors appear on the title page of the volume.

§2. The proceedings should be reasonably homogeneous (concerned with a limited area). For instance, the proceedings of a congress on "Analysis" or "Mathematics in Wonderland" would normally not be sufficiently homogeneous.

One or two longer survey articles on recent developments in the field are often very useful additions to such proceedings - even if they do not correspond to actual lectures at the congress. An extensive introduction on the subject of the congress would be desirable.

§3. The contributions should be of a high mathematical standard and of current interest. Research articles should present new material and not duplicate other papers already published or due to be published. They should contain sufficient information and motivation and they should present proofs, or at least outlines of such, in sufficient detail to enable an expert to complete them. Thus resumes and mere announcements of papers appearing elsewhere cannot be included, although more detailed versions of a contribution may well be published in other places later.

Surveys, if included, should cover a sufficiently broad topic, and should in general not simply review the author's own recent research. In the case of surveys, exceptionally, proofs of results may not be necessary.

"Mathematical Reviews" and "Zentralblatt für Mathematik" require that papers in proceedings volumes carry an explicit statement that they are in final form and that no similar paper has been or is being submitted elsewhere, if these papers are to be considered for a review. Normally, papers that satisfy the criteria of the Lecture Notes in Mathematics series also satisfy this

.../...

requirement, but we would strongly recommend that the contributing authors be asked to give this guarantee explicitly at the beginning or end of their paper. There will occasionally be cases where this does not apply but where, for special reasons, the paper is still acceptable for LNM.

§4. Proceedings should appear soon after the meeeting. The publisher should, therefore, receive the complete manuscript within nine months of the date of the meeting at the latest.

§5. Plans or proposals for proceedings volumes should be sent to one of the editors of the series or to Springer-Verlag Heidelberg. They should give sufficient information on the conference or symposium, and on the proposed proceedings. In particular, they should contain a list of the expected contributions with their prospective length. Abstracts or early versions (drafts) of some of the contributions are very helpful.

§6. Lecture Notes are printed by photo-offset from camera-ready typed copy provided by the editors. For this purpose Springer-Verlag provides editors with technical instructions for the preparation of manuscripts and these should be distributed to all contributing authors. Springer-Verlag can also, on request, supply stationery on which the prescribed typing area is outlined. Some homogeneity in the presentation of the contributions is desirable.

Careful preparation of manuscripts will help keep production time short and ensure a satisfactory appearance of the finished book. The actual production of a Lecture Notes volume normally takes 6 -8 weeks.

Manuscripts should be at least 100 pages long. The final version should include a table of contents and as far as applicable a subject index.

§7. Editors receive a total of 50 free copies of their volume for distribution to the contributing authors, but no royalties. (Unfortunately, no reprints of individual contributions can be supplied.) They are entitled to purchase further copies of their book for their personal use at a discount of 33.3 %, other Springer mathematics books at a discount of 20 % directly from Springer-Verlag. Contributing authors may purchase the volume in which their article appears at a discount of 33.3 %.

Commitment to publish is made by letter of intent rather than by signing a formal contract. Springer-Verlag secures the copyright for each volume.

Vol. 1232: P.C. Schuur, Asymptotic Analysis of Soliton Problems. VIII, 180 pages. 1986.

Vol. 1233: Stability Problems for Stochastic Models. Proceedings, 1985. Edited by V.V. Kalashnikov, B. Penkov and V.M. Zolotarev. VI, 223 pages. 1986.

Vol. 1234: Combinatoire énumérative. Proceedings, 1985. Edité par G. Labelle et P. Leroux. XIV, 387 pages. 1986.

Vol. 1235: Séminaire de Théorie du Potentiel, Paris, No. 8. Directeurs: M. Brelot, G. Choquet et J. Deny. Rédacteurs: F. Hirsch et G. Mokobodzki. III, 209 pages. 1987.

Vol. 1236: Stochastic Partial Differential Equations and Applications. Proceedings, 1985. Edited by G. Da Prato and L. Tubaro. V, 257 pages. 1987.

Vol. 1237: Rational Approximation and its Applications in Mathematics and Physics. Proceedings, 1985. Edited by J. Gilewicz, M. Pindor and W. Siemaszko. XII, 350 pages. 1987.

Vol. 1238: M. Holz, K.-P. Podewski and K. Steffens, Injective Choice Functions. VI, 183 pages. 1987.

Vol. 1239: P. Vojta, Diophantine Approximations and Value Distribution Theory. X, 132 pages. 1987.

Vol. 1240: Number Theory, New York 1984–85. Seminar. Edited by D.V. Chudnovsky, G.V. Chudnovsky, H. Cohn and M.B. Nathanson. V, 324 pages. 1987.

Vol. 1241: L. Gårding, Singularities in Linear Wave Propagation. III, 125 pages. 1987.

Vol. 1242: Functional Analysis II, with Contributions by J. Hoffmann-Jørgensen et al. Edited by S. Kurepa, H. Kraljević and D. Butković. VII, 432 pages. 1987.

Vol. 1243: Non Commutative Harmonic Analysis and Lie Groups. Proceedings, 1985. Edited by J. Carmona, P. Delorme and M. Vergne. V, 309 pages. 1987.

Vol. 1244: W. Müller, Manifolds with Cusps of Rank One. XI, 158 pages. 1987.

Vol. 1245: S. Rallis, L-Functions and the Oscillator Representation. XVI, 239 pages. 1987.

Vol. 1246: Hodge Theory. Proceedings, 1985. Edited by E. Cattani, F. Guillén, A. Kaplan and F. Puerta. VII, 175 pages. 1987.

Vol. 1247: Séminaire de Probabilités XXI. Proceedings. Edité par J. Azéma, P.A. Meyer et M. Yor. IV, 579 pages. 1987.

Vol. 1248: Nonlinear Semigroups, Partial Differential Equations and Attractors. Proceedings, 1985. Edited by T.L. Gill and W.W. Zachary. IX, 185 pages. 1987.

Vol. 1249: I. van den Berg, Nonstandard Asymptotic Analysis. IX, 187 pages. 1987.

Vol. 1250: Stochastic Processes – Mathematics and Physics II. Proceedings 1985. Edited by S. Albeverio, Ph. Blanchard and L. Streit. VI, 359 pages. 1987.

Vol. 1251: Differential Geometric Methods in Mathematical Physics. Proceedings, 1985. Edited by P.L. García and A. Pérez-Rendón. VII, 300 pages. 1987.

Vol. 1252: T. Kaise, Représentations de Weil et GL_2 Algèbres de division et GL_n. VII, 203 pages. 1987.

Vol. 1253: J. Fischer, An Approach to the Selberg Trace Formula via the Selberg Zeta-Function. III, 184 pages. 1987.

Vol. 1254: S. Gelbart, I. Piatetski-Shapiro, S. Rallis. Explicit Constructions of Automorphic L-Functions. VI, 152 pages. 1987.

Vol. 1255: Differential Geometry and Differential Equations. Proceedings, 1985. Edited by C. Gu, M. Berger and R.L. Bryant. XII, 243 pages. 1987.

Vol. 1256: Pseudo-Differential Operators. Proceedings, 1986. Edited by H.O. Cordes, B. Gramsch and H. Widom. X, 479 pages. 1987.

Vol. 1257: X. Wang, On the C*-Algebras of Foliations in the Plane. V, 165 pages. 1987.

Vol. 1258: J. Weidmann, Spectral Theory of Ordinary Differential Operators. VI, 303 pages. 1987.

Vol. 1259: F. Cano Torres, Desingularization Strategies for Three-Dimensional Vector Fields. IX, 189 pages. 1987.

Vol. 1260: N.H. Pavel, Nonlinear Evolution Operators and Semigroups. VI, 285 pages. 1987.

Vol. 1261: H. Abels, Finite Presentability of S-Arithmetic Groups. Compact Presentability of Solvable Groups. VI, 178 pages. 1987.

Vol. 1262: E. Hlawka (Hrsg.), Zahlentheoretische Analysis II. Seminar, 1984–86. V, 158 Seiten. 1987.

Vol. 1263: V.L. Hansen (Ed.), Differential Geometry. Proceedings, 1985. XI, 288 pages. 1987.

Vol. 1264: Wu Wen-tsün, Rational Homotopy Type. VIII, 219 pages. 1987.

Vol. 1265: W. Van Assche, Asymptotics for Orthogonal Polynomials. VI, 201 pages. 1987.

Vol. 1266: F. Ghione, C. Peskine, E. Sernesi (Eds.), Space Curves. Proceedings, 1985. VI, 272 pages. 1987.

Vol. 1267: J. Lindenstrauss, V.D. Milman (Eds.), Geometrical Aspects of Functional Analysis. Seminar. VII, 212 pages. 1987.

Vol. 1268: S.G. Krantz (Ed.), Complex Analysis. Seminar, 1986. VII, 195 pages. 1987.

Vol. 1269: M. Shiota, Nash Manifolds. VI, 223 pages. 1987.

Vol. 1270: C. Carasso, P.-A. Raviart, D. Serre (Eds.), Nonlinear Hyperbolic Problems. Proceedings, 1986. XV, 341 pages. 1987.

Vol. 1271: A.M. Cohen, W.H. Hesselink, W.L.J. van der Kallen, J.R. Strooker (Eds.), Algebraic Groups Utrecht 1986. Proceedings. XII, 284 pages. 1987.

Vol. 1272: M.S. Livšic, L.L. Waksman, Commuting Nonselfadjoint Operators in Hilbert Space. III, 115 pages. 1987.

Vol. 1273: G.-M. Greuel, G. Trautmann (Eds.), Singularities, Representation of Algebras, and Vector Bundles. Proceedings, 1985. XIV, 383 pages. 1987.

Vol. 1274: N. C. Phillips, Equivariant K-Theory and Freeness of Group Actions on C*-Algebras. VIII, 371 pages. 1987.

Vol. 1275: C.A. Berenstein (Ed.), Complex Analysis I. Proceedings, 1985–86. XV, 331 pages. 1987.

Vol. 1276: C.A. Berenstein (Ed.), Complex Analysis II. Proceedings, 1985–86. IX, 320 pages. 1987.

Vol. 1277: C.A. Berenstein (Ed.), Complex Analysis III. Proceedings, 1985–86. X, 350 pages. 1987.

Vol. 1278: S.S. Koh (Ed.), Invariant Theory. Proceedings, 1985. V, 102 pages. 1987.

Vol. 1279: D. Ieşan, Saint-Venant's Problem. VIII, 162 Seiten. 1987.

Vol. 1280: E. Neher, Jordan Triple Systems by the Grid Approach. XII, 193 pages. 1987.

Vol. 1281: O.H. Kegel, F. Menegazzo, G. Zacher (Eds.), Group Theory. Proceedings, 1986. VII, 179 pages. 1987.

Vol. 1282: D.E. Handelman, Positive Polynomials, Convex Integral Polytopes, and a Random Walk Problem. XI, 136 pages. 1987.

Vol. 1283: S. Mardešić, J. Segal (Eds.), Geometric Topology and Shape Theory. Proceedings, 1986. V, 261 pages. 1987.

Vol. 1284: B.H. Matzat, Konstruktive Galoistheorie. X, 286 pages. 1987.

Vol. 1285: I.W. Knowles, Y. Saitō (Eds.), Differential Equations and Mathematical Physics. Proceedings, 1986. XVI, 499 pages. 1987.

Vol. 1286: H.R. Miller, D.C. Ravenel (Eds.), Algebraic Topology. Proceedings, 1986. VII, 341 pages. 1987.

Vol. 1287: E.B. Saff (Ed.), Approximation Theory, Tampa. Proceedings, 1985–1986. V, 228 pages. 1987.

Vol. 1288: Yu. L. Rodin, Generalized Analytic Functions on Riemann Surfaces. V, 128 pages, 1987.

Vol. 1289: Yu. I. Manin (Ed.), K-Theory, Arithmetic and Geometry. Seminar, 1984–1986. V, 399 pages. 1987.

Vol. 1290: G. Wüstholz (Ed.), Diophantine Approximation and Transcendence Theory. Seminar, 1985. V, 243 pages. 1987.

Vol. 1291: C. Mœglin, M.-F. Vignéras, J.-L. Waldspurger, Correspondances de Howe sur un Corps p-adique. VII, 163 pages. 1987

Vol. 1292: J.T. Baldwin (Ed.), Classification Theory. Proceedings, 1985. VI, 500 pages. 1987.

Vol. 1293: W. Ebeling, The Monodromy Groups of Isolated Singularities of Complete Intersections. XIV, 153 pages. 1987.

Vol. 1294: M. Queffélec, Substitution Dynamical Systems – Spectral Analysis. XIII, 240 pages. 1987.

Vol. 1295: P. Lelong, P. Dolbeault, H. Skoda (Réd.), Séminaire d'Analyse P. Lelong – P. Dolbeault – H. Skoda. Seminar, 1985/1986. VII, 283 pages. 1987.

Vol. 1296: M.-P. Malliavin (Ed.), Séminaire d'Algèbre Paul Dubreil et Marie-Paule Malliavin. Proceedings, 1986. IV, 324 pages. 1987.

Vol. 1297: Zhu Y.-l., Guo B.-y. (Eds.), Numerical Methods for Partial Differential Equations. Proceedings. XI, 244 pages. 1987.

Vol. 1298: J. Aguadé, R. Kane (Eds.), Algebraic Topology, Barcelona 1986. Proceedings. X, 255 pages. 1987.

Vol. 1299: S. Watanabe, Yu.V. Prokhorov (Eds.), Probability Theory and Mathematical Statistics. Proceedings, 1986. VIII, 589 pages. 1988.

Vol. 1300: G.B. Seligman, Constructions of Lie Algebras and their Modules. VI, 190 pages. 1988.

Vol. 1301: N. Schappacher, Periods of Hecke Characters. XV, 160 pages. 1988.

Vol. 1302: M. Cwikel, J. Peetre, Y. Sagher, H. Wallin (Eds.), Function Spaces and Applications. Proceedings, 1986. VI, 445 pages. 1988.

Vol. 1303: L. Accardi, W. von Waldenfels (Eds.), Quantum Probability and Applications III. Proceedings, 1987. VI, 373 pages. 1988.

Vol. 1304: F.Q. Gouvêa, Arithmetic of *p*-adic Modular Forms. VIII, 121 pages. 1988.

Vol. 1305: D.S. Lubinsky, E.B. Saff, Strong Asymptotics for Extremal Polynomials Associated with Weights on $\mathbb{R}$. VII, 153 pages. 1988.

Vol. 1306: S.S. Chern (Ed.), Partial Differential Equations. Proceedings, 1986. VI, 294 pages. 1988.

Vol. 1307: T. Murai, A Real Variable Method for the Cauchy Transform, and Analytic Capacity. VIII, 133 pages. 1988.

Vol. 1308: P. Imkeller, Two-Parameter Martingales and Their Quadratic Variation. IV, 177 pages. 1988.

Vol. 1309: B. Fiedler, Global Bifurcation of Periodic Solutions with Symmetry. VIII, 144 pages. 1988.

Vol. 1310: O.A. Laudal, G. Pfister, Local Moduli and Singularities. V, 117 pages. 1988.

Vol. 1311: A. Holme, R. Speiser (Eds.), Algebraic Geometry, Sundance 1986. Proceedings. VI, 320 pages. 1988.

Vol. 1312: N.A. Shirokov, Analytic Functions Smooth up to the Boundary. III, 213 pages. 1988.

Vol. 1313: F. Colonius, Optimal Periodic Control. VI, 177 pages. 1988.

Vol. 1314: A. Futaki, Kähler-Einstein Metrics and Integral Invariants. IV, 140 pages. 1988.

Vol. 1315: R.A. McCoy, I. Ntantu, Topological Properties of Spaces of Continuous Functions. IV, 124 pages. 1988.

Vol. 1316: H. Korezlioglu, A.S. Ustunel (Eds.), Stochastic Analysis and Related Topics. Proceedings, 1986. V, 371 pages. 1988.

Vol. 1317: J. Lindenstrauss, V.D. Milman (Eds.), Geometric Aspects of Functional Analysis. Seminar, 1986–87. VII, 289 pages. 1988.

Vol. 1318: Y. Felix (Ed.), Algebraic Topology – Rational Homotopy. Proceedings, 1986. VIII, 245 pages. 1988

Vol. 1319: M. Vuorinen, Conformal Geometry and Quasiregular Mappings. XIX, 209 pages. 1988.

Vol. 1320: H. Jürgensen, G. Lallement, H.J. Weinert (Eds.), Se groups, Theory and Applications. Proceedings, 1986. X, 416 pa 1988.

Vol. 1321: J. Azéma, P.A. Meyer, M. Yor (Eds.), Séminaire Probabilités XXII. Proceedings. IV, 600 pages. 1988.

Vol. 1322: M. Métivier, S. Watanabe (Eds.), Stochastic Analy Proceedings, 1987. VII, 197 pages. 1988.

Vol. 1323: D.R. Anderson, H.J. Munkholm, Boundedly Contro Topology. XII, 309 pages. 1988.

Vol. 1324: F. Cardoso, D.G. de Figueiredo, R. Iório, O. Lopes (Ec Partial Differential Equations. Proceedings, 1986. VIII, 433 pa 1988.

Vol. 1325: A. Truman, I.M. Davies (Eds.), Stochastic Mechanics Stochastic Processes. Proceedings, 1986. V, 220 pages. 1988.

Vol. 1326: P.S. Landweber (Ed.), Elliptic Curves and Modular Form Algebraic Topology. Proceedings, 1986. V, 224 pages. 1988.

Vol. 1327: W. Bruns, U. Vetter, Determinantal Rings. VII,236 pag 1988.

Vol. 1328: J.L. Bueso, P. Jara, B. Torrecillas (Eds.), Ring Thee Proceedings, 1986. IX, 331 pages. 1988.

Vol. 1329: M. Alfaro, J.S. Dehesa, F.J. Marcellan, J.L. Rubio Francia, J. Vinuesa (Eds.): Orthogonal Polynomials and their Appli tions. Proceedings, 1986. XV, 334 pages. 1988.

Vol. 1330: A. Ambrosetti, F. Gori, R. Lucchetti (Eds.), Mathemati Economics. Montecatini Terme 1986. Seminar. VII, 137 pages. 198

Vol. 1331: R. Bamón, R. Labarca, J. Palis Jr. (Eds.), Dynami Systems, Valparaiso 1986. Proceedings. VI, 250 pages. 1988.

Vol. 1332: E. Odell, H. Rosenthal (Eds.), Functional Analysis. P ceedings, 1986–87. V, 202 pages. 1988.

Vol. 1333: A.S. Kechris, D.A. Martin, J.R. Steel (Eds.), Cabal Semi 81–85. Proceedings, 1981–85. V, 224 pages. 1988.

Vol. 1334: Yu.G. Borisovich, Yu. E. Gliklikh (Eds.), Global Analy – Studies and Applications III. V, 331 pages. 1988.

Vol. 1335: F. Guillén, V. Navarro Aznar, P. Pascual-Gainza, F. Puer Hyperrésolutions cubiques et descente cohomologique. XII, 1 pages. 1988.

Vol. 1336: B. Helffer, Semi-Classical Analysis for the Schröding Operator and Applications. V, 107 pages. 1988.

Vol. 1337: E. Sernesi (Ed.), Theory of Moduli. Seminar, 1985. VIII, 2 pages. 1988.

Vol. 1338: A.B. Mingarelli, S.G. Halvorsen, Non-Oscillation Domai of Differential Equations with Two Parameters. XI, 109 pages. 1988.

Vol. 1339: T. Sunada (Ed.), Geometry and Analysis of Manifold Procedings, 1987. IX, 277 pages. 1988.

Vol. 1340: S. Hildebrandt, D.S. Kinderlehrer, M. Miranda (Eds. Calculus of Variations and Partial Differential Equations. Proceeding 1986. IX, 301 pages. 1988.

Vol. 1341: M. Dauge, Elliptic Boundary Value Problems on Corne Domains. VIII, 259 pages. 1988.

Vol. 1342: J.C. Alexander (Ed.), Dynamical Systems. Proceeding 1986–87. VIII, 726 pages. 1988.

Vol. 1343: H. Ulrich, Fixed Point Theory of Parametrized Equivaria Maps. VII, 147 pages. 1988.

Vol. 1344: J. Král, J. Lukeš, J. Netuka, J. Veselý (Eds.), Potenti Theory – Surveys and Problems. Proceedings, 1987. VIII, 271 page 1988.

Vol. 1345: X. Gomez-Mont, J. Seade, A. Verjovski (Eds.), Holomorph Dynamics. Proceedings, 1986. VII, 321 pages. 1988.

Vol. 1346: O. Ya. Viro (Ed.), Topology and Geometry – Rohl Seminar. XI, 581 pages. 1988.

Vol. 1347: C. Preston, Iterates of Piecewise Monotone Mappings o an Interval. V, 166 pages. 1988.

Vol. 1348: F. Borceux (Ed.), Categorical Algebra and its Applications Proceedings, 1987. VIII, 375 pages. 1988.

Vol. 1349: E. Novak, Deterministic and Stochastic Error Bounds i Numerical Analysis. V, 113 pages. 1988.